Lecture Notes in Bioinformatics 6674

Edited by S. Istrail, P. Pevzner, and M. Waterman

Subseries of Lecture Notes in Computer Science

W0038566

Jianer Chen   Jianxin Wang
Alexander Zelikovsky (Eds.)

Bioinformatics Research and Applications

7th International Symposium, ISBRA 2011
Changsha, China, May 27-29, 2011
Proceedings

 Springer

Series Editors

Sorin Istrail, Brown University, Providence, RI, USA
Pavel Pevzner, University of California, San Diego, CA, USA
Michael Waterman, University of Southern California, Los Angeles, CA, USA

Volume Editors

Jianer Chen
Texas A&M University, Department of Computer Science
College Station, TX 77843-3112, USA
E-mail: chen@cs.tamu.edu

Jianxin Wang
Central South University, School of Information Science and Engineering
Changsha, 410083, China
E-mail: jxwang@mail.csu.edu.cn

Alexander Zelikovsky
Georgia State University, Department of Computer Science
Atlanta, GA 30303, USA
E-mail: alexz@cs.gsu.edu

ISSN 0302-9743   e-ISSN 1611-3349
ISBN 978-3-642-21259-8   e-ISBN 978-3-642-21260-4
DOI 10.1007/978-3-642-21260-4
Springer Heidelberg Dordrecht London New York

Library of Congress Control Number: Applied for

CR Subject Classification (1998): J.3, H.2.8, H.3-4, F.1, I.5

LNCS Sublibrary: SL 8 – Bioinformatics

Typesetting: Camera-ready by author, data conversion by Scientific Publishing Services, Chennai, India

Printed on acid-free paper

Springer is part of Springer Science+Business Media (www.springer.com)

Preface

The seventh edition of the International Symposium on Bioinformatics Research and Applications (ISBRA 2011) was held during May 27-29, 2011 at Central South University in Changsha, China. The symposium provides a forum for the exchange of ideas and results among researchers, developers, and practitioners working on all aspects of bioinformatics and computational biology and their applications.

The technical program of the symposium included 36 contributed papers, selected by the Program Committee from 92 full submissions received in response to the call for papers. Additionally, the symposium included poster sessions and featured invited keynote talks by four distinguished speakers: Bernard Moret from the Swiss Federal Institute of Technology in Lausanne, Switzerland, spoke on phylogenetic analysis of whole genomes, David Sankoff from the University of Ottawa spoke on competing formulations of orthologs for multiple genomes, Russell Schwartz from Carnegie Mellon University spoke on phylogenetics of heterogeneous samples, Liping Wei from Peking University spoke on the critical role of bioinformatics in genetic and pharmacogenetic studies of neuropsychiatric disorders, and Eric Xing from Carnegie Mellon University spoke on the a structured sparse regression approach to diseases.

We would like to thank the Program Committee members and external reviewers for volunteering their time to review and discuss symposium papers. We would like to extend special thanks to the Steering and General Chairs of the symposium for their leadership, and to the Finance, Publication, Publicity, and Local Organization Chairs for their hard work in making ISBRA 2011 a successful event. Last but not least we would like to thank all authors for presenting their work at the symposium.

May 2011

Jianer Chen
Jianxin Wang
Alex Zelikovsky

Symposium Organization

Steering Chairs

Dan Gusfield	University of California, Davis, USA
Yi Pan	Georgia State University, USA
Marie-France Sagot	INRIA, France

General Chairs

Mona Singh	Princeton University, USA
Lijian Tao	Central South University, China
Albert Y. Zomaya	The University of Sydney, Australia

Program Chairs

Jianer Chen	Texas A&M University, USA and Central South University, China
Jianxin Wang	Central South University, China
Alex Zelikovsky	Georgia State University, USA

Publicity Chairs

Ion Mandoiu	University of Connecticut, USA
Yanqing Zhang	Georgia State University, USA

Publication Chair

Raj Sunderraman	Georgia State University, USA

Finance Chair

Yu Sheng	Central South University, China

Local Organization Chairs

Min Wu	Central South University, China
Zhiming Yu	Central South University, China

Local Organizing Committee

Jianxin Wang	Central South University, China
Guojun Wang	Central South University, China
Dongjun Huang	Central South University, China
Fei Li	Central South University, China
Min Li	Central South University, China
Gang Chen	Central South University, China

Program Committee

Srinivas Aluru	Iowa State University, USA
Danny Barash	Ben-Gurion University, Israel
Robert Beiko	Dalhousie University, Canada
Anne Bergeron	Université du Québec à Montréal, Canada
Daniel Berrar	University of Ulster, UK
Paola Bonizzoni	Universitá de Studi di Milano-Bicocca, Italy
Daniel Brown	University of Waterloo, Canada
Tien-Hao Chang	National Cheng Kung University, Taiwan
Chien-Yu Chen	National Taiwan University, Taiwan
Jianer Chen	Texas A&M University, USA
Bhaskar Dasgupta	University of Illinois at Chicago, USA
Amitava Datta	University of Western Australia, Australia
Guillaume Fertin	Université de Nantes, France
Vladimir Filkov	University of California Davis, USA
Jean Gao	University of Texas at Arlington, USA
Katia Guimarães	Universidade Federal de Pernambuco, Brazil
Jieyue He	Southeast University, China
Jinling Huang	East Carolina University, USA
Lars Kaderali	University of Heidelberg, Germany
Iyad Kanj	DePaul University, USA
Ming-Yang Kao	Northwestern University, USA
Yury Khudyakov	Centers for Disease Control and Prevention, USA
Danny Krizanc	Wesleyan University, USA
Jing Li	Case Western Reserve University, USA
Zhiyong Liu	Chinese Academy of Science, China
Ion Mandoiu	University of Connecticut, USA
Fenglou Mao	University of Georgia, USA
Osamu Maruyama	Kyushu University, Japan
Ion Moraru	University of Connecticut Health Center, USA
Craig Nelson	University of Connecticut, USA
Andrei Paun	Louisiana Tech University, USA
Maria Poptsova	University of Connecticut, USA
Sven Rahmann	Technical University of Dortmund, Germany

Shoba Ranganathan	Macquarie University, Australia
Isidore Rigoutsos	IBM Research, USA
Cenk Sahinalp	Simon Fraser, Canada
Russell Schwartz	Carnegie Mellon University, USA
João Carlos Setubal	Virginia Polytechnic Institute and State University, USA
Jens Stoye	Universität Bielefeld, Germany
Raj Sunderraman	Georgia State University, USA
Wing-Kin Sung	National University of Singapore, Singapore
Sing-Hoi Sze	Texas A&M University, USA
Haixu Tang	Indiana University, USA
Gabriel Valiente	Technical University of Catalonia, Spain
Stéphane Vialette	Université Paris-Est Marne-la-Vallée, France
Jianxin Wang	Central South University, China
Li-San Wang	University of Pennsylvania, USA
Lusheng Wang	City University of Hong Kong, China
Carsten Wiuf	University of Aarhus, Denmark
Richard Wong	Kanazawa University, Japan
Yufeng Wu	University of Connecticut, USA
Alex Zelikovsky	Georgia State University, USA
Fa Zhang	Chinese Academy of Science, China
Yanqing Zhang	Georgia State University, USA
Leming Zhou	University of Pittsburgh, USA

External Reviewers

Aggarwala, Varun
Al Seesi, Sahar
Amberkar, Sandeep
Andres, Stephan Dominique
Astrovskaya, Irina
Berman, Piotr
Bernauer, Julie
Blin, Guillaume
Cao, Kajia
Cliquet, Freddy
Cohen-Boulakia, Sarah
Della Vedova, Gianluca
Dondi, Riccardo
Dörr, Daniel
Evans, Patricia
Fang, Ming
Gherman, Marius
Grudinin, Sergei
Hwang, Yih-Chii

Jahn, Katharina
Jiang, Minghui
Jiao, Dazhi
Kang, Min Gon
Kiani, Narsis Aftab
Kim, Dong-Chul
Kim, Wooyoung
Kopczynski, Dominik
Köster, Johannes
Lara, James
Lee, Byoungkoo
Leung, Fanny
Li, Fan
Li, Shuo
Li, Weiling
Lin, Chiao-Feng
Macdonald, Norman
Malhotra, Samta
Mao, Xizeng

Marschall, Tobias
Martin, Marcel
Mayampurath, Anoop
Mi, Qi
Missirian, Victor
Nakhleh, Luay
Nguyen, Minh Q.
Olman, Victor
Rizzi, Raffaella
Rizzi, Romeo
Skums, Pavel
Srichandan, Bismita
Tan, Cheemeng

Tang, Xiaojia
Tannier, Eric
Wang, Wenhui
Willing, Eyla
Wittler, Roland
Wu, Yin
Yang, Xiao
Yu, Chuan-Yih
Zhang, Chi
Zhang, Xinjun
Zhang, Yanqing
Zhou, Fengfeng
Zola, Jaroslaw

Table of Contents

Phylogenetics of Heterogeneous Samples
(Keynote Talk)

Russell Schwartz

Carnegie Mellon University
Pittsburgh, PA 15213 USA

Phylogenetics, or the inference of evolutionary trees, is one of the oldest and most intensively studied topics in computational biology. Yet it remains a vibrant area of research, in part because advances in our ability to gather data for phylogenetic inference continue to create novel and more challenging variants of the phylogeny problem. In this talk, I will discuss a particular challenge underlying some important phylogenetic problems in the genomic era: reconstructing evolutionary histories from samples of heterogeneous populations, each of which may contain contributions from multiple evolutionary stages or pathways. This problem combines the challenges of identifying common population subgroups from large variation data sets and reconstructing a history of those subgroups. Methods for solving the problem thus end up drawing from a variety of computational techniques, including classic discrete algorithmic approaches to phylogenetics, and machine learning and statistical inference methods for finding robust structure within large, noisy data sets. In this talk, I will present two examples of the problem on very different scales. I will describe the use of phylogenetic methods for inferring evolutionary histories of cell lineages within tumors from genome-wide assays, where cell-to-cell heterogeneity within individual tumors complicates analysis. I will further describe the application of similar concepts for inferring histories of human population groups from genetic variation data, where variability within populations and admixture between them present similar difficulties. Collectively, these problems illustrate some of the special challenges of phylogenetic inference on heterogeneous samples and some of the breadth of techniques needed to address these challenges.

References

[1]   Schwartz, R., Schackney, S.: Applying unmixing to gene expression data for tumor phylogeny inference. BMC Bioinf. 11, 42 (2010)
[2]   Tsai, M.-C., Blelloch, G., Ravi, R., Schwartz, R.: A consensus tree approach for reconstructing human evolutionary history and detecting population substructure. In: Borodovsky, M., Gogarten, J.P., Przytycka, T.M., Rajasekaran, S. (eds.) ISBRA 2010. LNCS (LNBI), vol. 6053, pp. 167–178. Springer, Heidelberg (2010)
[3]   Tolliver, D., Tsourakakis, C., Subramanian, A., Shackney, S., Schwartz, R.: Robust unmixing of tumor states in array comparative genomic hybridization data. Bioinformatics Proc. 26(12), i106–i114 (2010)
[4]   Tsai, M.-C., Blelloch, G., Ravi, R., Schwartz, R.: A consensus-tree approach for reconstructing human evolutionary history and identifying population substructure. IEEE/ACM Trans. Comp. Biol. Bioinf. (2011) (in press)

J. Chen, J. Wang, and A. Zelikovsky (Eds.): ISBRA 2011, LNBI 6674, p. 1, 2011.
© Springer-Verlag Berlin Heidelberg 2011

OMG! Orthologs for Multiple Genomes - Competing Formulations
(Keynote Talk)

David Sankoff

Department of Mathematics and Statistics, University of Ottawa

Multiple alignment of the gene orders in sequenced genomes is an important problem in comparative genomics [1]. A key aspect is the construction of disjoint orthology sets of genes, in which each element is orthologous to all other genes (on different genomes) in the same set. Approaches differ as to the nature and timing and relative importance of sequence alignment, synteny block construction, and paralogy resolution in constructing these sets. We argue that these considerations are best integrated in the construction of *pairwise* synteny blocks as a first step, followed by the conflation of the pairwise orthologies into larger sets. The two advantages of this are: first, the availability of finely tuned pairwise synteny block software (e.g., SYNMAP in the CoGE platform [2,3]) and second, the opportunity to dispense with parameters, thresholds or other arbitrary settings during the construction of the orthology sets themselves. The orthology sets problem becomes a pure graph algorithm problem.

The pairwise homologies SYNMAP provides for all pairs of genomes constitute the set of edges E of the *homology graph* $H = (V, E)$, where V is the set of genes in any of the genomes participating in at least one homology relation. Ideally all the genes in a connected component of H should be orthologous. We should allow *at most* one gene from each genome in such an orthology set, or at most two duplicate genes for genomes that descend from a WGD event. In practice, however, there will be many conflicts in H leading to many apparent paralogies that we must consider as erroneous. The problem we are addressing, in collaboration with Chunfang Zheng and Krister Swenson, is how to convert H into a new graph, respecting the data as far as possible, but with the property desired of all connected components, namely containing no paralogs other than relics of WGD.

Our first approach is simply to delete a minimum number of edges, such that each component in the resulting graph contains at most one gene from each genome (at most two in WGD descendants). This is a NP-hard graph problem, namely minimum orthogonal partition, for which there are good approximation algorithms [4]. The idea is to conserve as many of the homologies in the data as possible.

We define an empty pair (g, G) to consist of a gene g and a genome G not containing g and containing no homolog to G. Our second approach is to delete edges such that each component in the resulting graph contains at most one gene from each genome (at most two in WGD descendants), but which minimizes the

J. Chen, J. Wang, and A. Zelikovsky (Eds.): ISBRA 2011, LNBI 6674, pp. 2–3, 2011.

number of empty gene-genome pairs. This focus on conserving genes (vertices) rather than homologies (edges) in the graph.

The difference between these two objectives, though subtle, may have important consequences in practice. To investigate these differences, we implemented the algorithm in [4] for the minimum orthogonal partition problem and designed a greedy algorithm [5] for the gene-conserving approach.

We obtained genomic data on 10 annotated dicots in CoGE and and used SynMap to produce sets of synteny blocks for all 45 pairs of genomes. We extracted all the homology relations from all the synteny blocks in each pair, and put them all together to form the graph H. We then tested the two approaches.

We show how the gene-conserving solution requires deleting far more edges than the minimum orthogonal partitioning, and how the latter discards far more genes than the gene-conserving approach. This has important consequences for gene-order phylogeny based on the orthology sets. This is illustrated in terms of the branch lengths of the phylogenetic tree constructed on the basis of the gene orders.

References

1. Fostier, J., et al.: A greedy, graph-based algorithm for the alignment of multiple homologous gene lists. Bioinformatics 27, 749–756 (2011)
2. Lyons, E., Freeling, M.: How to usefully compare homologous plant genes and chromosomes as DNA sequences. Plant J. 53, 661–673 (2008)
3. Lyons, E., et al.: Finding and comparing syntenic regions among Arabidopsis and the outgroups papaya, poplar and grape: CoGe with rosids. Plant Phys. 148, 1772–1781 (2008)
4. He, G., Liu, J., Zhao, C.: Approximation algorithms for some graph partitioning problems. Journal of Graph Algorithms and Applications 4, 1–11 (2000)
5. Zheng, C., Sankoff, D.: Gene order in Rosid phylogeny, inferred from pairwise syntenies among extant genomes. In: Chen, J., Wang, J., Zelikovsky, A. (eds.) ISBRA 2011. LNCS (LNBI), vol. 6674, pp. 99–110. Springer, Heidelberg (2011)

Phylogenetic Analysis of Whole Genomes
(Keynote Talk)

Bernard M.E. Moret

Laboratory for Computational Biology and Bioinformatics,
Swiss Federal Institute of Technology (EPFL),
EPFL-IC-LCBB, INJ 230, Station 14, CH-1015 Lausanne, Switzerland
bernard.moret@epfl.ch

1 Introduction

The rapidly increasing number of sequenced genomes offers the chance to resolve long-standing questions about the evolutionary history of certain groups of organisms, to develop a better understanding of evolution, to make substantial advances in functional genomics, and to start bridging genomics and genetics. Comparative genomics is the term used today for much of the work carried out in whole-genome analysis, correctly emphasizing that the "guilt-by-association" approach used in the analysis of gene and regulatory sequences remains the fundamental tool in the analysis of whole genomes. However, the limitations of pairwise comparisons are even more severe in whole-genome analysis than in sequence analysis and, of course, pairwise comparisons have little to tell us about evolution. Thus we are witnessing a significant increase in phylogenetic research based, not on sequence data, but on larger-scale features of the genome, such as genomic rearrangements, duplications and losses of genomic regions, regulatory modules and networks, chromatin structure, etc.

However, phylogenetic analysis for whole genomes remains very primitive when compared to the analysis of, e.g., coding sequences. In part, this is due to immaturity: while such an analysis was in fact conducted as far back as the 1930s (by Sturtevant and Dobzhansky, using chromosomal banding data and hand computations), the first serious computational attempts are less than 20 years old. The major reason, however, is simply the very complex nature of the data, which makes it very difficult to design good simple models and which causes most questions framed in even the simplest of models to be computationally intractable. A few examples will suffice to illustrate this point. Sequence analysis typically uses character positions as its basic units and assumes some form of independence among the positions, but we lack even a good definition of the basic unit (the syntenic block) most commonly used in comparative genomics. While the phylogenetic community frequently deplores the lack of good tools for the multiple alignment of sequence data, we simply have no tool capable of aligning multiple whole genomes unless they are all very closely related. Whereas computing a parsimony score on any given phylogenetic tree is solvable in linear time, computing a parsimony score on a tree of three leaves under most extant models of genomic rearrangements or duplications and losses is intractable (NP-hard). And while no systematics journal would publish an inferred phylogeny without some form of statistical

J. Chen, J. Wang, and A. Zelikovsky (Eds.): ISBRA 2011, LNBI 6674, pp. 4–7, 2011.

support (bootstrapping values or log-likelihood scores), we have as yet no way of boot-strapping phylogenies built from whole-genome data nor sufficiently good stochastic models to derive likelihood scores.

Fortunately, more and more research groups are working on phylogenetic analysis of whole genomes, so that rapid progress is being made. In this presentation, I will briefly survey the main computational problems, summarize the state of the art for each, and present some recent results from my group that take us closer to a solution to some of these problems.

2   Some Extant Problems

A comparative analysis of complete genomes starts by the identification of syntenic blocks, that is, contiguous regions that are shared, to within some tolerance factor, across the genomes. Ideally, syntenic blocks should be defined in an evolutionary set-ting, but, as in the case of gene orthology, practical implementations so far have used a variety of heuristics—based on the identification of shared anchors such as genes or other markers and on guidelines about the desired size of such blocks and the amount of dissimilarity tolerable within the blocks. The most recent and ambitious package for the identification of synteny blocks is DRIMM-Synteny which takes into account both duplications and rearrangements. Still missing from the literature is a formal evolution-ary definition of synteny, in the spirit of definitions of homology and orthology, and accompanying criteria for selection of appropriate amounts of internal dissimilarity.

Genomic alignment needs much more work. Miller *et al.* developed a pipeline for the alignment of vertebrate genomes in the UCSC Genome Browser. The approach used (an initial star alignment against the human genome, followed by a progressive alignment to place all genomes on the same reference indexing) precludes its extension to more distantly related organisms. The package progressiveMauve is, like DRIMM-Synteny, an improved version of an earlier package, designed to take into account duplications in addition to rearrangements; it computes a multiple alignment, at the sequence level, of several genomes, not relying on any particular reference genome. Its target is clearly the smaller genomes of, e.g., bacteria. Aligning multiple genomes that are only distantly related may require a tree alignment rather than the conventional common indexing of character positions. Events such as chromosomal fusion, fission, or linearization remain to be taken into account.

Constructing a phylogenetic tree based on whole-genome data has seen significant progress. The first published packages, BPAnalysis and GRAPPA, worked only with unichromosomal genomes and were limited in the number of taxa as well as the size of the genomes (the number of syntenic blocks); MGR, which could handle multichro-mosomal data, scaled poorly, as did the Bayesian package Badger. With the DCJ model of rearrangements, new work was started on pairwise distance estimation and phyloge-netic reconstruction, the latter using both parsimony-based methods and distance-based methods. In addition, the use of distance methods led to the first reliable method for confidence assessment. Still missing are robust and scalable methods for phylogenetic reconstruction in the presence of duplicate syntenic blocks, maximum-likelihood meth-ods, and better bootstrapping. All of these methods rely on the prior identification of syntenic blocks; yet, in the case of distantly related taxa, these blocks may have to be defined in a phylogenetic setting. Simultaneous tree inference and sequence alignment is still in its infancy, so it is no surprise that there has been very little work so far on simultaneous tree inference and syntenic block identification.

Extending the analysis of whole genomes from genomic structure to function starts with regulatory networks and chromatin structure. The former have mostly been studied in single species, but recent work at the Broad (Arboretum) and in my group (ProphyC) have shown that an evolutionary approach can significantly improve the quality of inference, both for entire networks and for modules. Projects for phylogenetic analyses of epigenomic data (such as histone modifications) are starting up everywhere. But models for connecting chromatin structure and regulation remain to be devised and current models for the evolution of regulatory networks leave much to be desired.

3   Some Encouraging Results from My Group

The DCJ (double-cut-and-join) model has considerably simplified the handling of rearrangements and been used in attempts to reconstruct parts of ancestral genomes for yeasts, among other organisms. My group developed a very accurate statistical estimator that takes as input the edit (shortest) distance between two arrangements and returns a maximum-likelihood estimate of the true distance. Later, we gave an ML estimator based on a slight variant of the DCJ model that takes into account duplication of blocks; on simulated data under deliberately mismatched models, the estimator stays within 10% of the true distance in almost all cases and under 5% in most cases. Using this estimator with the FastME program for distance-based reconstruction produces very accurate reconstructions on instances with up to 500 taxa and genomes of up to 20'000 syntenic blocks.

We have also used the DCJ model and work on so-called adequate subgraphs (substructures of the graph representation of rearrangements) to improve the computation of rearrangement medians, the basic step in computing parsimony scores for phylogenetic trees based on rearrangement data. Here the assumption of unique syntenic blocks remains necessary, but with this assumption we demonstrated fast and accurate computations on high-resolution genomic data (10'000 to 20'000 syntenic blocks) as well as very accurate scoring under simulations.

Our incursion into distance methods for rearrangement data suggested introducing perturbations into the distance matrices themselves, yielding the first usable method for evaluating the robustness of a phylogenetic reconstruction. In recent work, we have resampled the adjacencies and obtain discrimination comparable to that demonstrated by conventional phylogenetic bootstrapping for sequence data. While we designed this bootstrapping approach for distance-based methods, extending them to parsimony-based methods is straightforward, although computationally intensive. Thus a serious and longstanding impediment to the use of rearrangement data in phylogenetic inference is nearly overcome.

Functional inferences from large-scale genomic data often involves the inference of regulatory networks for various genes. The difficulty in obtaining comparable data across various species has long restricted such studies to single species, although comparisons were made across various tissues from the same host. Whole-genome RNA-Seq inventories are now providing a richer and more easily comparable source of expression data across many organisms, thus motivating the development of inference methods based on phylogenetic approaches. My group developed the ProPhyC software to refine inferred networks through the use of phylogenetic relationships and of a simple evolutionary model for regulatory networks. Using this software on inferred networks for closely related species (or tissues) produces significantly improved networks in terms of topological accuracy and thus demonstrates the power of a phylogenetic approach to the analysis of these systems.

4 Conclusions

Phylogenetic inference and, more generally, phylogenetic methods are assuming a greater role in the analysis of whole-genome data. A logical extension of the pairwise comparative approach, phylogenetic methods, while often complex, provide important advantages:

Genetic and Pharmacogenetic Studies of Neuropsychiatric Disorders: Increasingly Critical Roles of Bioinformatics Research and Applications

(Keynote Talk)

Liping Wei[1,2,3]

[1] Center for Bioinformatics,
State Key Laboratory of Protein and Plant Gene Research
College of Life Sciences, Peking University, Beijing, 100871, China
[2] Laboratory of Personalized Medicine,
School of Chemical Biology and Biotechnology
Shenzhen Graduate School of Peking University, Shenzhen, 518055, China
[3] National Institute of Biological Sciences, Zhongguancun Life Science Park,
Beijing 102206, China
weilp@mail.cbi.pku.edu.cn

For most neuropsychiatric disorders there is a lack of good cellular model and animal model. Thus genetic and pharmacogenetic studies of these complex disorders are powerful approaches to identify the underlying genes and pathways. Recent advances in high-throughput sequencing technologies enable such studies at genome scale. Due to the huge amount of data involved, bioinformatic research and applications play increasingly critical roles. For instance, one may uncover novel functional candidates by screening the genomic and genetic data using bioinformatic criteria; one may discover novel global patterns through genome-wide integration of multiple sources of data and across multiple species; and last but not least, one needs to develop new bioinformatic software and databases in order to handle the data effectively.

To illustrate bioinformatic research and applications in the genetic and pharmacogenetic studies of neuropsychiatric disorders I will review several examples from our own lab. First, we have studied the genetic susceptibility factors underlying addiction using data from Genome-Wide Association Studies in collaboration with NIH/DIDA. We identified a surprising case of human-specific de novo protein-coding gene involved in nicotine addiction, FLJ33706 (alternative gene symbol C20orf203) [1]. Cross-species analysis revealed interesting evolutionary paths of how this gene had originated from noncoding DNA sequences: insertion of repeat elements especially Alu contributed to the formation of the first coding exon and six standard splice junctions on the branch leading to humans and chimpanzees, and two subsequent changes in the human lineage escaped two stop codons and created an open reading frame of 194 amino acids. We experimentally verified FLJ33706s mRNA and protein expression in the brain. Real-Time PCR in multiple tissues demonstrated that FLJ33706 was most abundantly expressed in brain. Human polymorphism data suggested that FLJ33706

J. Chen, J. Wang, and A. Zelikovsky (Eds.): ISBRA 2011, LNBI 6674, pp. 8–10, 2011.

encodes a protein under purifying selection. A specifically designed antibody detected its protein expression across human cortex, cerebellum and midbrain. Immunohistochemistry study in normal human brain cortex revealed the localization of FLJ33706 protein in neurons. FLJ33706 is one of the first discovered cases of motherless or de novo protein-coding genes that originated from noncoding DNA sequences. We have since identified 31 other such genes in the human genome. Our results suggest that de novo protein-coding genes may be an under-investigated source of species-specific new phenotypes.

When we compared genes associated with addiction by genetic technologies with those by other molecular biology technologies, we found that different technologies tend to discover different types of genes. Towards a complete picture of the molecular network underlying addiction, we re-analyzed and integrated data that linked genes to addiction by multiple experimental technologies platforms published in the past 30 years. We compiled a list of 396 genes that were supported by at least two independent pieces of evidence [2]. Next, we developed a bioinformatic software, named KOBAS, that mapped the genes to pathways and calculated the statistically significantly enriched pathways [3]. We found that five pathways were common to addiction to four different substances, cocaine, opioids, alcohol, and nicotine. Two of the common pathways, GnRH signaling pathway and Gap Junction pathway, had been linked to addiction for the first time. Finally, using the pathways as scaffold, we constructed a molecular network underlying addiction. These common pathways and network may be potential attractive drug targets to treat addiction. Our results demonstrate that an integrative bioinformatic analysis can discover novel patterns that elude single experimental technologies.

One of the unprecedented opportunities brought by the next-generation sequencing technologies is to study how individuals genetic variations affect their dosage, response, and adverse effect of drugs, and to use this knowledge towards personalized medicine. In our own lab, we had investigated the mysterious neuropsychiatric and skin adverse effect reported in Japan after administration of Tamiflu (oseltamivir phosphate). We identified a nonsynonymous SNP (Single Nucleotide Polymorphism) in dbSNP database, R41Q, near the enzymatic active site of human cytosolic sialidase, a homologue of virus neuraminidase that is the target of oseltamivir [4]. This SNP occurred in 9.29% of Asian population and none of European and African American population. Our structural analyses and Ki measurements using in vitro sialidase assays indicated that this SNP could increase the unintended binding affinity of human sialidase to oseltamivir carboxylate, the active form of oseltamivir, thus reducing sialidase activity. In addition, this SNP itself resulted in an enzyme with an intrinsically lower sialidase activity, as shown by its increased Km and decreased Vmax values. Theoretically administration of oseltamivir to people with this SNP might further reduced their sialidase activity. We noted the similarity between the reported neuropsychiatric side effects of oseltamivir and the known symptoms of human sialidase-related disorders. We proposed that this Asian-enriched sialidase variation caused by the SNP, likely in homozygous form, may be associated with certain severe adverse reactions to oseltamivir. Preliminary results from initial samples collected in a case-control study appear to support the hypothesis, although continued sample collection is required and ongoing.

References

1. Li, C.Y., Zhang, Y., Wang, Z., Zhang, Y., Cao, C., Zhang, P.W., Lu, S.J., Li, X.M., Yu, Q., Zheng, X., Du, Q., Uhl, G.R., Liu, Q.R., Wei, L.: A human-specific de novo protein-coding gene associated with human brain functions. PLoS Computational Biology 6(3), e1000734 (2010)
2. Li, C.Y., Mao, X., Wei, L.: Genes and (Common) Pathways Underlying Drug Addiction. PLoS Computational Biology 4(1), e2 (2008)
3. Mao, X., Cai, T., Olyarchuk, J.G., Wei, L.: Automated Genome Annotation and Pathway Identification Using the KEGG Orthology (KO) As a Controlled Vocabulary. Bioinformatics 21(19), 378793 (2005)
4. Li, C.Y., Yu, Q., Ye, Z.Q., Sun, Y., He, Q., Li, X.M., Zhang, W., Luo, J., Gu, X., Zheng, X., Wei, L.: A nonsynonymous SNP in human cytosolic sialidase in a small Asian population results in reduced enzyme activity: potential link with severe adverse reactions to oseltamivir. Cell Research 17(4), 357–362 (2007)

Genome-Phenome Association Analysis of Complex Diseases a Structured Sparse Regression Approach
(Keynote Talk)

Eric Xing

School of Computer Science
Carnegie Mellon University
Pittsburgh, PA 15213, USA
epxing@cs.cmu.edu

Genome-wide association (GWA) studies have recently become popular as a tool for identifying genetic variables that are responsible for increased disease susceptibility. A modern statistical method for approaching this problem is through model selection (or structure estimation) of Structured Input-Output Regression Models (SIORM) fitted on genetic and phenotypic variation data across a large number of individuals.

The inputs of such models bear rich structure, because the cause of many complex disease syndromes involves the complex interplay of a large number of genomic variations that perturb disease-related genes in the context of a regulatory network. Likewise, the outputs of such model are also structured, as patient cohorts are routinely surveyed for a large number of traits such as hundreds of clinical phenotypes and genome-wide profiling for thousands of gene expressions that are interrelated. A Structured Input-Output Regression Model nicely captures all these properties, but raises severe computational and theoretical challenge on consistent model identification.

In this talk, I will present models, algorithms, and theories that learn Sparse SIORMs of various kinds in very high dimensional input/output space, with fast and highly scalable optimization procedures, and strong statistical guarantees. I will demonstrate application of our approaches to a number of complex GWA scenarios, including associations to trait networks, to trait clusters, to dynamic traits, under admixed populations, and with epistatic effects.

This is joint work with Seyoung Kim, Mladen Kolar, Seunghak Lee, Xi Chen, and Kriti Puniyani and Judie Howrylak.

J. Chen, J. Wang, and A. Zelikovsky (Eds.): ISBRA 2011, LNBI 6674, p. 11, 2011.
© Springer-Verlag Berlin Heidelberg 2011

Prediction of Essential Proteins by Integration of PPI Network Topology and Protein Complexes Information

Jun Ren[1,2], Jianxin Wang[1,*], Min Li[1], Huan Wang[1], and Binbin Liu[1]

[1] School of Information Science and Engineering, Central South University,
Changsha, 410083, China
[2] College of Information Science and Technology, Hunan Agricultural University,
Changsha, 410128, China
jxwang@mail.csu.edu.cn, renjun19@163.com, mli@cs.gsu.edu

Abstract. Identifying essential proteins is important for understanding the minimal requirements for cellular survival and development. Numerous computational methods have been proposed to identify essential proteins from protein-protein interaction (PPI) network. However most of methods only use the PPI network topology information. HartGT indicated that essentiality is a product of the protein complex rather than the individual protein. Based on these, we propose a new method ECC to identify essential proteins by integration of subgraph centrality (SC) of PPI network and protein complexes information. We apply ECC method and six centrality methods on the yeast PPI network. The experimental results show that the performance of ECC is much better than that of six centrality methods, which means that the prediction of essential proteins based on both network topology and protein complexes set is much better than that only based on network topology. Moreover, ECC has a significant improvement in prediction of low-connectivity essential proteins.

Keywords: essential proteins, protein complexes, subgraph centrality.

1 Introduction

A protein is said to be essential for an organism if the organism cannot survive without it [1]. Identifying essential proteins is important for understanding the minimal requirements for cellular survival and development. Research experiments [2-3] detected that essential proteins evolve much slower than other proteins, which suggested they play key roles in the basic functioning of living organisms. Based on it, some biologists suggested that essential proteins of lower organisms are associated with human disease genes. *Kondrashov FA et al.* detected that the essential proteins of Drosophila melanogaster are fairly similar to human morbid genes.[4] *Furney SJ et al.* detected that essential proteins tend to have higher correlation with dominant and recessive mutants of disease genes [5]. So people can identify human disease genes based on identifying essential proteins of lower organisms.

---

* Corresponding author.

J. Chen, J. Wang, and A. Zelikovsky (Eds.): ISBRA 2011, LNBI 6674, pp. 12–24, 2011.

However, experimental methods for identifying putative essential proteins, such as creating conditional knockouts, cannot find many essential proteins in one experiment. And it costs a lot to do an experiment. As a result, many essential proteins are still unknown, especially in human. Meanwhile, other biological data, such as protein-protein interactions (PPIs), are increasing fast and available conveniently with high-throughput identification. In 2001, *Jeong H et al.* had already shown that proteins with high degree in a PPI network have more possibility to be essential than those selected by chance and provided the centrality-lethality rule [6]. The centrality-lethality rule demonstrates a high correlation between a node's topological centrality in a PPI network and its essentiality. Since then, much attention has been given to the study of proteins with high centrality, such as high-degree nodes and hubs in PPI networks [7-9]. *Ernesto E.* summarized six centralities, degree centrality (*DC*), betweenness centrality (*BC*), closeness centrality (*CC*), subgraph centrality(*SC*), eigenvector centrality(*EC*) and information centrality(*IC*), and used them in identification of essential proteins in yeast PPI networks [10]. *Ernesto E* pointed out that compared to the other centrality measures, *SC* has superior performance in the selection of essential proteins and explained why.

The use of centrality measures based on network topology has become an important method in identification of essential proteins. However recent research works pointed out that many essential proteins have low connectivity and are difficult to be identified by centrality measures [7-9,11]. Many research works focus on identification essential proteins by integration PPI networks and other biological information, such as cellular localization, gene annotation, genome sequence, and so on [7,11,12]. *Acencio ML et al.* demonstrated that network topological features, cellular localization and biological process information are all reliable predictors of essential genes [12]. *Hart GT et al.* pointed out that essentiality is a product of the protein complex rather than the individual protein [13].

Our research on yeast also shows that protein complexes have high correlation with essential proteins. Based on it, we propose a new method, *ECC*, to identify essential proteins by integration of PPI network topology and protein complexes information. *ECC* proposes two kinds of centralities, subgraph centrality and complex centrality. Subgraph centrality of a protein is used to characterize its importance in the total network and complex centrality is used to characterize its importance in the protein complex set. *ECC* defines a harmonic centrality to integrate the two centralities and identifies essential proteins by ranking proteins according to it.

We apply our method *ECC* and six centrality methods on the yeast PPI network and evaluate their performance by comparing their identified proteins with the gold standard essential proteins. As protein complexes of most species are not all identified by experimental methods, here *ECC* uses two kinds of protein complex sets. One is identified by experimental methods. Another is identified by *CMC* algorithm in the PPI network [14]. No matter which protein complex set *ECC* uses, experimental results show that comparison the results of *ECC* method with the optimum result of six centrality methods, 3.6% to 17.3% improvements are obtained. Experimental results also show that *ECC* method can identify much more low-connectivity essential proteins than six centrality methods. The percentage of low-connectivity essential proteins identified by *SC* method is highest in six centrality methods and that identified by *ECC* method is as 1.9 to 7.6 times as it.

2 Method

2.1 Six Centrality Measures

A PPI network is represented as an undirected graph $G(V, E)$ with proteins as nodes and interactions as edges. The protein centrality in a PPI network is used to characterize the importance or contribution of an individual protein to the global structure or configuration of the PPI network. Many research works indicated that PPI networks have characters of "small-world behavior" and "centrality-lethality" [6,15]. So the removal of nodes with high centrality makes the PPI network collapse into isolated clusters, which possibly means the biological system collapse. This may be why a protein's essentiality has high correlation with its centrality.

Six centralities are commonly used for predicting a protein's essentiality[7, 10, 16, 17], degree centrality (DC), betweenness centrality (BC), closeness centrality (CC), eigenvector centrality(EC), subgraph centrality(SC), and information centrality(IC). They are defined as follows.

Definition 1. The degree centrality of a protein i in a PPI network G, $DC(i)$, is the number of proteins interacting with i.

$$DC(i) = | N(i) | \tag{1}$$

where $N(i)$ is the set of neighbors of protein i.

Definition 2. The betweenness centrality of a protein k in a PPI network G, $BC(k)$, is equal to the fraction of shortest paths going through the protein k.

$$BC(k) = \sum_i \sum_j \rho(i, k, j) / \rho(i, j) \qquad i \neq j \neq k \tag{2}$$

where $\rho(i, j)$ is the number of shortest paths from protein i to protein j, and $\rho(i, k, j)$ is the number of these shortest paths that pass through protein k.

Definition 3. The closeness centrality of a protein i in a PPI network G, $CC(i)$, is the sum of graph-theoretic distances from all other proteins in the G.

$$CC(i) = (N - 1) \Big/ \sum_j d(i, j) \tag{3}$$

where the distance $d(i, j)$ from a protein i to another j is defined as the number of interactions in the shortest path from i to j, N is the number of proteins in G.

Definition 4. The eigenvector centrality of a protein i in a PPI network G, $EC(i)$, is defined as the ith component of the principal eigenvector of A, where A is the adjacency matrix of G. The defining equation of an eigenvector is $\lambda e = Ae$, where λ is an eigen value and e is the eigenvector of A. $EC(i) = e_1(i)$, where e_1 corresponds to the largest eigen value of A.

Definition 5. The subgraph centrality of a protein i in a PPI network G, $SC(i)$ counts the total number of closed walks in which i takes part and gives more weight to closed walks of short lengths.

$$SC(i) = \sum_{l=0}^{\infty} \mu_l(i)/l! = \sum_{j=1}^{N} [v_j(i)]^2 e^{\lambda_j} \tag{4}$$

where $\mu_l(i)$ is the number of closed walks of length l starting and ending at protein i, $(v_1, v_2,..., v_N)$ is an orthonormal basis of R^N composed by eigenvectors of A associated to the eigenvalues $(\lambda_1, \lambda_2,..., \lambda_N)$, and $v_j(i)$ is the ith component of v_j.

Definition 6. The information centrality IC is based on the information transmitted between any two points in a connected network. The matrix B is defined as $B = D$ $A + J$, where A is the adjacency matrix of a PPI network G, D is a diagonal matrix of the degree of each protein in G, and J is a matrix with all its elements equal to one. The element of information matrix I of G is defined as $I_{ij} = (c_{ii} + c_{jj} - c_{ij})^{-1}$, where c_{ij} is the element of matrix C and $C=B^{-1}$. The information centrality of a protein i in G, $IC(i)$, is then defined as follows:

$$IC(i) = [\frac{1}{N}\sum_j \frac{1}{I_{ij}}]^{-1} \tag{5}$$

Ernesto E compared the correlation of the six centralities and the proteins' essentiality in yeast PPI network. All centralities have positive correlation with essentiality and *SC* has the highest correlation than other centralities [10]. As closed walks are related to the network subgraph, *SC* accounts for the number of subgraphs in which a protein participates and gives more weights to smaller subgraphs [18]. These subgraphs, particularly triangles, have been previously identified as important structural motifs in biological networks [19]. So the knock-out of a protein with high *SC* value, makes more structural motifs collapse, which possibly result in the collapse of the biological system. This is why the *SC* has highest correlation.

2.2   The Correlation of Protein Complexes and Essential Proteins

The protein complex is a muti-protein organism that appears in same time and same location with certain structure and certain function. *Hart GT* pointed out that protein complexes have high correlation with essential proteins [13]. We also verify it in the yeast. We download yeast PPI network from DIP database [20], its essential proteins and standard protein complexes from MIPS database [21]. In 4746 proteins of yeast PPI network, 1130 proteins of them are inessential protein set, 1042 proteins of them are in standard protein complex set, and 487 of them are in both two sets. The probability of a protein to be an essential protein is only 23.8% when it is selected randomly and is 46.7% when it is selected from protein complex set. The probability of a protein in a protein complex is only 22.0% when it is selected randomly and is 43.1% when it is an essential protein. Both these indicates that protein complexes have high correlation with essential proteins.

As we know, protein complex is a muti-protein organism with certain structure and function. A knock-out of a protein in a protein complex may result in the destruct of the protein complex and the loss of its function. Obviously, a protein complex has more possibility to be destroyed if the ill protein connects more other proteins in it. So we use *in-degree* proposed by Radicchi to characterize the importance of a protein to the protein complex which include it [22]. *In-degree* ($k^{in}(i, C)$) of a protein i in a protein complex C is defined as the number of interactions which connect i to other proteins in C.

$$k^{in}(i,C) = |\{e(i, j) | i, j \in V(C)\}| \qquad (6)$$

where $e(i, j)$ is the interaction which connect i to j, and $V(C)$ is the vertex set of C. We also know that a protein can be included in several protein complexes. So a knock-out of this kind of protein would result in the loss of several functions. So if a protein is included in more protein complexes and has higher *in-degree* value in these protein complexes, the knock-out of it would make more protein complexes destroyed, which makes more functions loss and results in lethality or infertility of the whole organism. Based on these, we define the complex centrality of a protein i, *Complex_C(i)*, as the sum of *in-degree* value of i in all protein complexes which include it.

$$Complex\_C(i) = \sum_{C_i \in CS \; \&\& \; i \in C_i} k^{in}(i, C_i) \qquad (7)$$

where CS is the protein complexes set, C_i is a protein complex which include i. If a protein has a higher *Complex_C* value, it either has a high *in-degree* value in a protein complex or is included in several protein complexes or both. So a knock-out of it would make more functions loss. Thus it has more possibility to be an essential protein.

2.3 Algorithm ECC

Though the complex set has high correlation with essential proteins, it cannot include all essential proteins. For example, in all 1130 essential proteins, the gold standard yeast complex set includes487 essential proteins and there are 643 essential proteins are not included in it. The reason is that protein's essentiality also has high correlation with the global structure of the PPI network, for example "centrality- lethality" rule [6]. However protein complexes only reveal the local character. Moreover, there also have some protein complexes not been discovered now. So we use both the complex centrality and the global centrality to discovery essential proteins. The global centrality adopts subgraph centrality (SC) as it has the best performance in identifying essential proteins in all six centrality methods [10]. To integrate $SC(i)$ and *Complex_C(i)*, we define a harmonic centrality (HC) of protein i as follows:

$$HC(i) = a * SC(i)/SC_{max} + (1-a) * Complex\_C(i)/Complex\_C_{max} \qquad (8)$$

where a is a proportionality coefficient and takes value in range of 0 to 1, SC_{max} is the maximum SC value, $Complex\_C_{max}$ is the maximum $Complex\_C$ value. Based on ranking proteins by their HC values and outputting a certain top number of proteins, a new method *ECC* (find **E**ssential proteins based on protein **C**omplexes and **C**entrality) is proposed as follows. Here a PPI network is described as an undirected graph G and a protein complex set is described as a subgraph set of graph G.

Algorithm *ECC*:
Input: Undirected graph $G=(V(G), E(G))$,
       Subgraph set $Cset=\{C_i=(V(C_i), E(C_i))|C_i \subset G\}$, parameter a and k;
Output: Identified proteins
1, **For** each vertex $v_i \in V(G)$ **do** $Complex\_C(v_i) = 0$
    // initialize all $Complex\_C(v_i)$ as zero
2, **For** each subgraph $C_i \in Cset$ **do**
For vertex $v_i \in V(C_i)$ **do** $Complex\_C(v_i)= Complex\_C(v_i)+ k^{in}(v_i, C_i)$
    // calculate complex centrality of each vertex in $Cset$
3, $Complex\_C_{max}$=the maximum value in all $Complex\_C(v_i)$
4, **For** each vertex $v_i \in V(G)$ **do** calculate its SC value: $SC(v_i)$
5, SC_{max}=the maximum value in all $SC(v_i)$
6, **For** each vertex $v_i \in V(G)$ **do**
       $HC(v_i) = a*SC(v_i)/SC_{max}+ (1-a)*Complex\_C(v_i)/Complex\_C_{max}$
7, Sort HC in decreasing order and output the first k proteins with highest HC value

Fig. 1. The description of algorithm ECC

3   Results

To evaluate the performance of our method in identifying essential proteins, we implement it in yeast for its well characterized by knockout experiments and widely used in previous works. The PPI network of yeast is downloaded from DIP database and named as *YDIP* in the paper [20]. *YDIP* includes 4746 proteins and 15166 interactions in total without self-interactions and repeated interactions. As protein complexes identified by experimental methods are only parts of all protein complexes, here we use two kinds of protein complex sets. One is identified by experimental methods and downloaded from MIPS database [21]. Its map in *YDIP* is named as *YGS_PC* (**Y**east **G**old **S**tandard **P**rotein **C**omplex set). *YGS_PC* includes 209 protein complexes and 1042 proteins. Another is identified by CMC algorithm in *YDIP* and named as *YCMC_PC*. *YCMC_PC* includes 623 protein complexes and 1538 proteins. Here we choose CMC algorithm to identify protein complexes for its good performance and short running time[14]. The gold standard essential protein set of yeast is also downloaded from MIPS database [21]. Its map in *YDIP* is named as *YGS_EP* in the paper. *YGS_EP* includes 1130 essential proteins.

In this section, we first discuss the performance of our method *ECC* and the effect of parameter *a* on the result. Then, we compare *ECC* method with six centrality methods to verify that the accuracy in identifying essential proteins can be improved by adding protein complexes information. At last, we compare *ECC* method with six centrality methods in identifying low-connectivity essential proteins.

3.1   Identification of Essential Proteins by Integration of PPI Network Topology and Protein Complexes Information

We apply *ECC* method to identify essential proteins by using the yeast PPI network and yeast protein complex set. Tab.1 and Tab.2 show the effect of the variation of parameter *a* and *k* on *ECC* when the protein complex set is *YGS_PC* and *YCMC_PC* respectively. Parameter k is the number of identified proteins. In most paper of identifying essential proteins, the number of identified proteins is usually no more than 25% number of total proteins [7,10]. We choose k value of 200, 400, 600, 800, 1000, 1200 according to it. Parameter a is the proportionality coefficient to harmonize SC value and $Complex\_C$ value. ECC identifies essential proteins only by the SC value when $a=0$, and only by the $Complex\_C$ value when $a=1$. Accuracy of the result is defined as the ratio of the number of essential proteins in identified proteins to the number of identified proteins.

Both Tab.1 and Tab.2 show, no matter which value parameter a takes, the number of essential proteins is increasing with k value increase. Meanwhile the accuracy is decreasing, which means that a protein with high centrality has more probability to be an essential protein.

Both Tab.1 and Tab.2 show, no matter which value parameter k takes, the accuracy is lowest when $a=0$. It means that no matter which protein complex set is used, the accuracy is improved by adding the protein complex set information. Tab.1 and Tab.2 show that when a in the range of 0.2 to 0.8, the improvement is from 6.2% to 9.8% by using YGS_PC and from 3.5% to 9.5% by using $YCMC\_PC$. The improvement when using YGS_PC is a little more than that when using $YCMC\_PC$ because YGS_PC has higher confidence.

Table 1. The effect of the variation of parameter a and k on ECC by using YGS_PC

Number of essential proteins

a k	0	0.1	0.2	0.3	0.4	0.5	0.6	0.7	0.8	0.9	1
200	125	135	138	139	140	139	139	138	138	140	140
400	219	254	255	256	254	253	249	248	247	246	242
600	294	348	353	342	343	345	340	341	338	333	327
800	372	423	438	443	441	439	431	422	422	415	415
1000	436	502	518	519	519	519	519	519	519	519	444
1200	475	578	584	584	584	584	584	584	584	584	478

Accuracy(%)

a k	0	0.1	0.2	0.3	0.4	0.5	0.6	0.7	0.8	0.9	1
200	62.5	67.5	69.0	69.5	70.0	69.5	69.5	69.0	69.0	70.0	70.0
400	54.8	63.5	63.8	64.0	63.5	63.3	62.3	62.0	61.8	61.5	60.5
600	49.0	58.0	58.8	57.0	57.2	57.5	56.7	56.8	56.3	55.5	54.5
800	46.5	52.9	54.8	55.4	55.1	54.9	53.9	52.8	52.8	51.9	51.9
1000	43.6	50.2	51.8	51.9	51.9	51.9	51.9	51.9	51.9	51.9	44.4
1200	39.6	48.2	48.7	48.7	48.7	48.7	48.7	48.7	48.7	48.7	39.8

Table 2. The effect of the variation of parameter a and k on ECC by using $YCMC\_PC$

Number of essential proteins

a \\ k	0	0.1	0.2	0.3	0.4	0.5	0.6	0.7	0.8	0.9	1
200	125	126	132	136	140	141	142	143	142	142	142
400	219	244	253	256	256	257	256	256	256	256	258
600	294	344	343	343	344	343	343	343	343	343	340
800	372	418	412	413	411	411	411	412	412	412	413
1000	436	474	486	490	488	490	488	488	489	489	485
1200	475	550	550	553	552	552	552	552	552	552	541

Accuracy(%)

a \\ k	0	0.1	0.2	0.3	0.4	0.5	0.6	0.7	0.8	0.9	1
200	62.5	63.0	66.0	68.0	70.0	70.5	71.0	71.5	71.0	71.0	71.0
400	54.8	61.0	63.3	64.0	64.0	64.3	64.0	64.0	64.0	64.0	64.5
600	49.0	57.3	57.2	57.2	57.3	57.2	57.2	57.2	57.2	57.2	56.7
800	46.5	52.3	51.5	51.6	51.4	51.4	51.4	51.5	51.5	51.5	51.6
1000	43.6	47.4	48.6	49.0	48.8	49.0	48.8	48.8	48.9	48.9	48.5
1200	39.6	45.8	45.8	46.1	46.0	46.0	46.0	46.0	46.0	46.0	45.1

Tab.1 shows, no matter which value parameter k takes, the difference of accuracy is less than 2% when a in the range of 0.2 to 0.8 and the optimum a value is also in the range. Tab.2 shows, when $k \geq 400$, the difference of accuracy is less than 0.3% when a in the range of 0.3 to 0.9. The optimum a value is also in the range when $k=600$ and $k \geq 1000$. The optimum accuracy is 64.5% with $a=1$ when $k=400$ and 52.3% with $a=0.1$ when $k=800$. Comparison the optimum accuracy with the accuracy when a in the range of 0.3 to 0.9, the difference is no more than 0.5% when $k=400$ and no more than 0.9% when $k=800$. When $k=200$, the difference of accuracy is less than 1% when a in the range of 0.5 to 1 and the optimum a value is also in the range. Conclusion above, we can see that, no matter which protein complex set is used, the result of ECC method is not much difference with the variation of a in the range of 0.3 to 0.8. So parameter a has a good robustness in the range of 0.3 to 0.8. The possible reason is that many proteins with high $Complex\_C$ value also have high SC value. As accuracy values in this range are closed to the optimum accuracy value, the optimum a value is usually selected as its medium value of 0.5.

3.2 Comparison with Six Centrality Methods in Identifying Essential Proteins

We apply our method ECC and six centrality methods on the yeast PPI network $YDIP$ and compare their performance in Tab.3 and Tab.4. Here, two results of ECC are showed for two protein complex sets respectively. One is named as $ECC1$ with using YGS_PC. Another is named as $ECC2$ with using $YCMC\_PC$. The optimum result of six centrality methods is named as *OC*. As discussed in the section 3.1, parameter *a* in both *ECC1* and *ECC2* are set as 0.5. Tab.3 shows the number of essential proteins selected by *ECC* and six centrality methods. Tab.4 compares the accuracy of *ECC* method with that of six centrality methods.

Tab.3 shows that, no matter which value *k* takes, the number of essential proteins of *ECC1* and *ECC2* are both more than that of all six centrality methods. It means that, no matter which protein complex set is used, the performance of *ECC* method is better than that of six centrality methods. This is because *ECC* integrate both PPI network topology information and protein complexes information but six centrality methods only consider the topology of PPI network.

Tab.4 shows that, no matter which value *k* takes, accuracy of ECC1 and ECC2 are both higher than that of six centrality methods. The improvement of *ECC1* to *OC* is from 9.7% to 15.5%. The improvement of *ECC2* to *OC* is from 3.6% to 17.3%. The accuracy of both *ECC1* and *ECC2* are improved most when *k* in the range of 400 to 600, which means *ECC* method has best performance when *k* takes medium value. Tab.4 shows that when *k*<800, the improvement of *ECC1* and *ECC2* are not much difference. The possible reason is that proteins with high *Complex_C* value of *YGS_PC* overlap a lot with those of *YCMC_PC*. Tab.4 also shows that when *k*>=800, the improvement of *ECC1* is much higher than that of *ECC2*. The possible reason is that the confidence of protein complexes of *YGS_PC* is higher than that of *YCMC_PC* as *YGS_PC* is obtained from experimental method and *YCMC_PC* is obtained from computational method.

Table 3. Number of essential proteins selected by *ECC*(a=0.5) and six centrality methods

k	DC	IC	EC	SC	BC	CC	OC	ECC1	ECC2
200	103	109	123	125	83	90	125 (OC =SC)	139	141
400	203	208	216	219	164	178	219 (OC =SC)	253	257
600	298	302	293	294	247	263	302 (OC =IC)	345	343
800	381	390	369	372	323	334	390 (OC =IC)	439	411
1000	458	453	441	436	394	392	458 (OC =DC)	519	490
1200	533	528	490	475	460	453	533 (OC =DC)	584	552

Table 4. Accuracy of *ECC*(a=0.5) and six centrality methods

k	200	400	600	800	1000	1200
Accuracy (OC)	62.5%	54.8%	50.3%	48.8%	45.8%	44.4%
Accuracy (ECC1)	69.5%	63.3%	57.5%	54.9%	51.9%	48.7%
Accuracy (ECC2)	70.5%	64.3%	57.2%	51.4%	49.0%	46.0%
Accuracy (ECC1)- Accuracy (OC)	11.2%	15.5%	14.3%	12.5%	13.3%	9.7%
Accuracy (ECC2)- Accuracy (OC)	12.8%	17.3%	13.7%	5.3%	7.0%	3.6%

Note: Accuracy(ECC1) is the accuracy of ECC1.

3.3   Comparison with Six Centrality Methods in Identifying Low-Connectivity Essential Proteins

Many essential proteins are low-connectivity proteins. For example, in all 1130 essential proteins of *YDIP*, 577 of them (51%) are low-connectivity proteins. Here a protein whose degree is less than the average degree of *YDIP* is considered as a low-connectivity protein. However six centrality methods are connectivity-based detection methods, which results low-connectivity essential proteins are neglected by them.

Tab.5 shows the number and the percentage of low-connectivity essential proteins that are selected by our method *ECC* and six centrality methods. Tab.5 shows that in six centrality methods, *SC* has the best performance in selection low-connectivity essential proteins. The number of low-connectivity essential proteins identified by *SC* is from 0 to 65 and that of *ECC1* is from 3 to 149. It is as 2.3 to 9 times as that identify by *SC* methods. The percentage of low-connectivity essential proteins identified by *SC* is from 0% to 13.7% and that of *ECC1* is from 2.2% to 26.0%. It is as 1.9 to 7.6 times as that identify by *SC* method. All these show that the number and percentage of low-connectivity essential proteins of *ECC1* are both much more than those of six centrality methods. However Tab.5 also shows that both them of *ECC2* have not much difference to those identified by SC method. Moreover, both them of *ECC2* are less than those identified by *SC* method when $k \leq 800$.

Table 5. Number and percentage of low-connectivity essential proteins selected by *ECC* and six centrality methods

	k	DC	IC	EC	SC	BC	CC	ECC1	ECC2
Number of low-connectivity essential protein	200	0	0	0	0	0	0	3	0
	400	0	0	3	3	0	4	27	0
	600	0	0	14	13	4	8	76	4
	800	0	0	27	30	15	19	112	14
	1000	0	0	49	48	27	35	135	53
	1200	0	4	65	65	47	51	149	68
Percentage of low-connectivity essential protein	200	0.0%	0.0%	0.0%	0.0%	0.0%	0.0%	2.2%	0.0%
	400	0.0%	0.0%	1.4%	1.4%	0.0%	2.2%	10.7%	0.0%
	600	0.0%	0.0%	4.8%	4.4%	1.6%	3.0%	22.0%	1.2%
	800	0.0%	0.0%	7.3%	8.1%	4.6%	5.7%	25.5%	3.4%
	1000	0.0%	0.0%	11.1%	11.0%	6.9%	8.9%	26.0%	10.8%
	1200	0.0%	0.8%	13.3%	13.7%	10.2%	11.3%	25.5%	12.3%

As we know, *Complex_C* is a local centrality calculated only by protein complex set. A protein with a high *Complex_C* value probably has a low degree. These proteins may be low-connectivity essential proteins and can be identified by *ECC* method as they have high *Complex_C* value. As many proteins in *YGS_PC* have low degree, *ECC* can identify many proteins with high *Complex_C* value and low degree by using *YGS_PC*.

CMC algorithm identifies protein complexes as dense subgraphs, which results proteins with high *Complex_C* values most have high degree when the complex set of *ECC* is *YCMC_PC*. This is why *ECC1* can identify more low-connectivity essential proteins than six centrality methods but *ECC2* cannot.

Tab.6 analyses low-connectivity essential proteins identified by *ECC1* and *SC*. Here, we name the set of essential proteins in *ECC1* but not in *SC* as *ECC-SC*. As shown in Tab.6, most low-connectivity essential proteins in *ECC1* are in *ECC1-SC*. For example, when $k<=600$, all low-connectivity essential proteins in *ECC1* are in *ECC1-SC*. It means most low-connectivity essential proteins of *ECC1* are not identified by centrality methods. Tab.6 also shows that the percentage of low-connectivity essential proteins in *ECC1-SC* is increasing with k value increase and when $k \geq 800$ more than half of proteins in *ECC1-SC* are low-connectivity essential proteins.

ECC identify essential proteins by ranking proteins according to their *HC* value and outputting the top k proteins. *HC* is a harmonic centrality of *SC* and *Complex_C*. So proteins with high *Complex_C* value and low *SC* value have medium *HC* value and are in the middle queue. These proteins may be low-connectivity essential proteins and can be identified by *ECC* method but not by *SC* method. This is why the percentage and number of low-connectivity essential proteins in *ECC1-SC* are increasing with k value increase.

Table 6. Analysis low-connectivity essential proteins identified by *ECC1* and *SC*

k	200	400	600	800	1000	1200
Number of low-connectivity essential proteins in ECC1	3	27	76	112	135	149
Number of low-connectivity essential proteins in ECC1-SC	3	27	76	109	130	127
Number of essential proteins in ECC1-SC	91	151	194	227	227	208
Percentage of low-connectivity essential proteins in ECC1-SC	3.3%	17.9%	39.2%	48.0%	57.3%	61.1%

4   Conclusions and Future Work

By research the correlation of protein complexes and essential proteins, we find that the proteins in complexes have more possibility to be essential proteins than proteins selected by random. Thus, we define a local centrality, *Complex_C*, to evaluate the importance of a protein in a complexes set. As protein's essentiality also has high correlation with the global structure of the PPI network, we integrate *Complex_C* value and *SC* value into *HC* value to consider both protein complex information and PPI network topology. Based on these, we propose a new method, *ECC*, to identify essential protein by ranking proteins according to *HC* value. We apply *ECC* method and six centrality methods on *YDIP* and compare their performance. The experimental results show that:

1. Proportionality coefficient a of ECC has a good robustness. The performance of ECC is not much difference when a in the range of 0.3 to 0.8 and the typical value of a is selected as 0.5.

2. The prediction of essential proteins based on both network topology and protein complex set is significantly better than that only based on network topology. Comparison the result of ECC method with the optimum result of the six centrality methods, 3.6% to 17.3% improvements are obtained.

3. Our method ECC has a significant improvement in the prediction of low-connectivity essential proteins when using YGS_PC. The percentage of low-connectivity essential proteins identified by SC method is highest in six centrality methods and that identified by ECC method is as 1.9 to 7.6 times as it. Moreover, most low-connectivity essential proteins identified by ECC method can not be identified by SC method.

Acknowledgments. The authors would like to thank Liu GM and his colleagues for sharing the tool of CMC. This work is supported in part by the National Natural Science Foundation of China under Grant No.61003124 and No.61073036, the Ph.D. Programs Foundation of Ministry of Education of China No.20090162120073, the Freedom Explore Program of Central South University No.201012200124, the U.S. National Science Foundation under Grants CCF-0514750, CCF-0646102, and CNS-0831634.

References

1. Kamath, R.S., Fraser, A.G., et al.: Systematic functional analysis of the Caenorhabditis elegans genome using RNAi. Nature 421, 231–237 (2003)

2. Pal, C., Papp, B., Hurst, L.: Genomic function:Rate of evolution and gene dispensability. Nature 411(6841), 1046–1049 (2003)

3. Zhang, J.Z., He, X.L.: Significant impact of protein dispens ability on the instantaneous rate of protein evolution. Mol. Biol. Evol. 22(4), 1147–1155 (2005)

4. Kondrashov, F.A., Ogurtsov, A.Y., Kondrashov, A.S.: Bioinformatical assay of human gene morbidity. Nucl. Acids Res. 32(5), 1731–1737 (2004)

5. Furney, S.J., Alba, M.M., Lopez-Bigas, N.: Differences in the evolutionary history of disease genes affected by dominantor recessive mutations. BMC Genomics 7, 165 (2006)

6. Jeong, H., Mason, S.P., Barabasi, A.L., Oltvai, Z.N.: Lethality and centrality in protein-networks. Nature 411, 41–42 (2001)

7. Li, M., Wang, J.X., Wang, H., Pan, Y.: Essential Proteins Discovery from Weighted Protein Interaction Networks. In: Borodovsky, M., Gogarten, J.P., Przytycka, T.M., Rajasekaran, S. (eds.) ISBRA 2010. LNCS, vol. 6053, pp. 89–100. Springer, Heidelberg (2010)

8. He, X.L., Zhang, J.Z.: Why Do Hubs Tend to Be Essential in Protein Networks? PloS Genetics 2(6), 826–834 (2006)

9. Zotenko, E., Mestre, J., O'Leary, D.P., Przytycka, T.M.: Why Do Hubs in the Yeast Protein Interaction Network Tend To Be Essential: Reexamining the Connection between the Network Topology and Essentiality. PLoS Comput. Biol. 4(8), 1–16 (2008)

10. Ernesto, E.: Virtual identification of essential proteins within the protein interaction network of yeast. Proteomics 6(1), 35–40 (2006)

11. Chua, H.N., Tew, K.L., Li, X.L., Ng, S.-K.: A Unified Scoring Scheme for Detecting Essential Proteins in Protein Interaction Networks. In: 20th ICTAI, vol. 2, pp. 66–73 (2008)

12. Acencio, M.L., Lemke, N.: Towards the prediction of essential genes by integration of nework topology, cellular localization and biological process information. BMC Bioinformatics 10, 290 (2009)
13. Hart, G.T., Lee, I., Marcotte, E.: A high-accuracy consensus map of yeast protein complexes reveals modular nature of gene essentiality. BMC Bioinformatics 8, 236 (2007)
14. Liu, G.M., Wong, L., Chua, N.: Complex Discovery from Weighted PPI Networks. Bioinformatics 25(15), 1891–1897 (2009)
15. Maslov, S., Sneppen, K.: Specificity and stability in topology of protein networks. Science 296(5569), 910–913 (2002)
16. Jacob, R., Koschtzki, D., Lehmann, K.A., et al.: Algorithms for Centrality Indices. In: Brandes, U., Erlebach, T. (eds.) Network Analysis. LNCS, vol. 3418, pp. 62–82. Springer, Heidelberg (2005)
17. Mason, O., Verwoerd, M.: Graph theory and networks in biology. IET Systems Biology 1(2), 89–119 (2006)
18. Estrada, E., Rodríguez-Velázquez, J.: Subgraph centrality in complex networks. Phys. Rev. E. 71(5), 056103 (2005)
19. Milo, R., Itzkovitz, S., Kashtan, N., et al.: Super families of designed and evolved networks. Science 303(5663), 1538–1542 (2004)
20. Xenarios, I., Salwínski, L., Duan, X.J., et al.: DIP, the Database of Interacting Proteins: a research toolfor studying cellular networks of protein interactions. Nucleic Acids Res. 30, 303–305 (2002)
21. Mewes, H.W., Frishman, D., Gruber, C., et al.: MIPS: a database for genomes and protein sequences. Nucleic Acids Res. 28, 37–40 (2000)
22. Radicchi, F., Castellano, C., Cecconi, F., et al.: Defining and identifying communities in networks. Proc. Natl. Acad. Sci. USA 101(9), 2658–2663 (2004)

Computing the Protein Binding Sites

Fei Guo[1] and Lusheng Wang[2],*

[1] School of Computer Science and Technology, Shandong University,
Jinan 250101, Shandong, China
[2] Department of Computer Science, City University of Hong Kong,
83 Tat Chee Avenue, Kowloon, Hong Kong
cswangl@cityu.edu.hk

Abstract. Identifying the location of binding sites on proteins is of
fundamental importance for a wide range of applications including molecular docking, de novo drug design, structure identification and comparison of functional sites. Structural genomic projects are beginning to
produce protein structures with unknown functions. Therefore, efficient
methods are required if all these structures are to be properly annotated.
When comparing a complete protein with all complete protein structures
in the PDB database, experiments show that all the existing approaches
have recall values less than 50%. This implies that more than 50% of
real binding sites cannot be reported by those existing approaches. We
develop an efficient approach for finding binding sites between two proteins. Our approach consists of three steps, local sequence alignment,
protein surface detection, and 3D structures comparison. Experiments
show that the average recall value of our approach is 82% and the precision of our approach is also significantly better than the existing approaches. The software package is available at http://sites.google.
com/site/guofeics/bsfinder.

Keywords: 3D protein structure, binding site prediction, surface detection, rigid transformation.

1 Introduction

Many methods have been proposed for identifying the location of binding sites
on proteins. Laurie[1] gave an energy-based method for the prediction of protein-
ligand binding sites. Bradford[2] have combined a support vector machine (SVM)
approach with surface patch analysis to predict protein-protein binding sites.
Chen[3] developed a tool 3D-partner for inferring interacting partners and binding models. 3D-partner first utilizes IMPALA to identify homologous structures
(templates) of a query protein sequence from heterodimer profile library. The sequence profiles of those templates are then used to search interacting candidates
of the query from protein sequence databases by PSI-BLAST. Lo[4] developed
a method for predicting helix-helix interaction from residue contacts in membrane proteins. They first predict contact residues from sequences. Their paring

---

* Corresponding author.

J. Chen, J. Wang, and A. Zelikovsky (Eds.): ISBRA 2011, LNBI 6674, pp. 25–36, 2011.

relationships are further predicted in the second step via statistical analysis on contact propensities and sequence and structural information. Li[8] proposed an approach for finding binding sites for groups of proteins. It contains the following steps: finding protein groups as bicliques of protein-protein interaction networks (PPI), identifying conserved motifs, and searching domain-domain interaction databases. Liu[13] extended the method of Li[8] and considered comparing 3D local structures.

SuMo is a system for finding similarities in arbitrary 3D structures or substructures of proteins. It is based on a unique representation of macromolecules using selected triples of chemical groups [16]. The web server pdbFun analyzes the structure and the function of proteins at the residue level. It has very flexible and strong query functions that a query can involve all solvent-exposed, hydrophilic residues that are not in alpha-helices and are involved in nucleotide binding [18]. SiteEngine is a method that recognize the regions on the surface of one protein that are similar to the binding sites of another. SiteEngine uses geometric hashing triangles for transferring the input sites into the recognized region [19]. When comparing a complete protein with all complete protein structures in the PDB database, experiments show that all the existing approaches [16,18,19] have recall values less than 50% implying that more than 50% of real binding sites cannot be reported by those existing approaches.

In this paper, we develop an efficient approach for finding binding sites between two proteins. Our approach consists of three steps, local sequence alignment, protein surface detection, and 3D structures comparison. Experiments show that the average recall value of our approach is 82% and the precision of our approach is also significantly better than the existing approaches.

2   Methods

Given two complete protein structures, our task is to find the binding sites between the two proteins. Our method contains three steps. Firstly, we do local sequence alignment at the atom level to get the alignments of conserved regions. Those alignments of conserved regions may contain some gaps. Secondly, among the conserved regions obtained in Step 1, we use the 3D structure information to identify the surface segments. Finally, for any pair of the surface segments identified in Step 2, we compute a rigid transformation to compare the similarity of the two substructures in 3D space and output the qualified pairs as binding sites.

When computing the rigid transformations, we treat each protein as a molecule with some volume and introduce a method to ensure that the two whole protein 3D structures have no overlap under such a rigid transformation in 3D space.

2.1   Step 1: Local Sequence Alignment

In PDB format files, each residue (amino acids) is represented in the traditional order of atom records N, CA, C, O, followed by the side chain atoms (CB, CG1, CG2 . . .) in order first of increasing remoteness, and then branch. The whole protein sequence of residues can be translated into a sequence of atoms based

```
1tu4D  C    O    CA   C    O    CB   N    CA   C    O    CB   N    CA   C    O
       E28  E28  S29  S29  S29  S29  A30  A30  A30  A30  A30  V31  V31  V31  V31
5p21A  C    O    CA   C              N    CA   C    O         N    CA   C    O
       A11  A11  G12  G12            G13  G13  G13  G13       V14  V14  V14  V14

1tu4D  CB   CG2  N    CA   C    O    N    CA   C    CB   CG   CD   CE   N    CA
       V31  V31  G32  G32  G32  G32  K33  K33  K33  K33  K33  K33  K33  S34  S34
5p21A  CB   CG2  N    CA   C    O    N    CA   C    CB   CG   CD   CE   N    CA
       V14  V14  G15  G15  G15  G15  K16  K16  K16  K16  K16  K16  K16  S17  S17
                                              0    0    0    0    0

1tu4D  C    CB   OG   N    CA   C    CB   OG   N    CA   C    CB   CG   OD1  ND2
       S34  S34  S34  S35  S35  S35  S35  S35  L36  G78  N133 N133 N133 N133 N133
5p21A  C    CB   OG   N    CA   C    CB                  C         CG   OD1  ND2
       S17  S17  S17  A18  A18  A18  A18                 N116      N116 N116 N116

1tu4D  N    CA   C    O    CB   CG   CD   CE   NZ   N    CB   CG   OD1  OD2  CB
       K134 K134 K134 K134 K134 K134 K134 K134 K134 A135 D136 D136 D136 D136 L137
5p21A  N    CA   C              CB   CG   CD   CE   NZ   N         CG   OD1  OD2
       K117 K117 K117           K117 K117 K117 K117 K117 C118      D119 D119 D119

1tu4D  CG   CD1  CD2  O    N    CA   C    O    CB   OG   N    CA   C    CB   N
       L137 L137 L137 T162 S163 S163 S163 S163 S163 S163 A164 A164 A164 A164 K165
5p21A  CD2            N    CA   C    O    CB   OG   N    CA   C    CB   N
       L120           T144 S145 S145 S145 S145 S145 S145 A146 A146 A146 A146 K147

1tu4D  CA   CB   CG   CD   CE   NZ
       K165 K165 K165 K165 K165 K165
5p21A  CA   CB   CG   CE
       K147 K147 K147 K147
```

Fig. 1. The pairs of binding sites between 1tu4D and 5p21A

on this representation. The sequences of binding sites between two proteins are usually conserved at the atom level. When looking at the SitesBase, we know that the pair of binding sites forms a conserved region that are well aligned at the atom level, where atoms of the same types are matched and all the unmatched atoms correspond to gaps. Fig. 1 is the result of SitesBase for proteins 1tu4D and 5p21A, where the sequences of atoms of the pairs of binding sites are aligned.

Based on this observation, we use the standard Smith-Waterman's local alignment algorithm [21] to find the conserved regions, where a matched pair of atoms with the same type has a score 1, a mismatched two atoms with different types has a score $-\infty$, a mismatch between an atom and a space has a score -2. The local alignment algorithm will return a set of conserved segments in the alignment of the two protein sequences of atoms.

We have done many experiments and find that the set of conserved segments output by the local sequence alignment algorithm always contains all the pairs of binding sites in the SitesBase. The only problem is that the local sequence alignment algorithm outputs too many matched pairs of atoms. For example, for the two proteins 1tu4D and 5p21A, the local sequence alignment algorithm outputs four segments with 279 atoms. The first segment consists of residues 20-37 from 1tu4D and residues 3-20 from 5p21A, where 116 pairs of atoms are matched. The second segment consists of residues 75-81 from 1tu4D and residues 57-63 form 5p21A, where 49 pairs of atoms are matched. The third segment consists of residues 130-139 from 1tu4D and 113-122 from 5p21A, where 66 pairs of atoms of them are matched. The fourth segment consists of residues 161-167 from 1tu4D and residues 143-149 from 5p21A, where 48 pairs of atoms are matched. The first pair of confirmed binding site in SitesBase, residues 28-35 from 1tu4D and residues 11-18 from 5p21A is included in the first segment output by the local sequence alignment algorithm, the second pair of binding site in SitesBase, residues 133-137 from 1tu4D and residues 116-120 from 5p21A, is included in the third segment output by the local sequence alignment algorithm, and the third pair of confirmed binding site in SitesBase, residues 162-165 from 1tu4D and residues 144-147 from 5p21A, is included in the fourth segment.

After obtained the set of conserved segments from the local sequence alignment, we will focus on the columns with identical pairs of atoms and ignore the rest of columns in the remaining steps. Next we will further reduce the matched pairs of atoms by using the following steps.

2.2   Step 2: Identifying Surface Segments

Firstly, we discretize the 3D Euclidean space by setting a grid of size 1Å. A grid point is *protein* point if the point is within 2Å distance of an atom in the protein. A grid point is an *empty* point if it is not *protein* point. A grid point is an *interior* point if all its six neighbor grid points are the *protein* points. A grid point is a *surface* point if at least one of the six neighbor grid points is not the *protein* point. An atom in the protein is a *surface* atom if it is within distance 1.5Å of a *surface* point. Fig. 2 gives an example, where the dark grid points are surface points.

For a conserved segment output by the local sequence alignment, we consider all its subsegments containing at least 15 matched pairs of atoms. For such a subsegment, if both sequences on this subsegments have at least 2/3 atoms as the surface atoms, we then treat such a subsegment as a candidate binding site for further process in the next step.

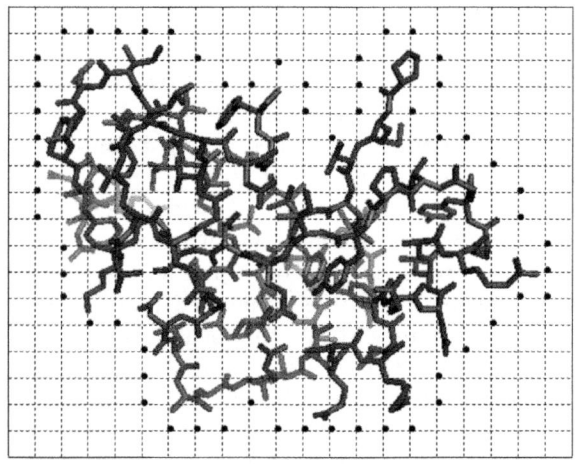

Fig. 2. The surface grid points are indicated by the dark points

2.3   Step 3: Computing Rigid Transformations to Match Candidate Binding Sites

For any candidate binding sites obtained after Step 1 and Step 2, we will further test if the pair of 3D substructures can match well on such a site. Precisely, we will find the set of subsegments of a given segment of alignment $\mathcal{A}$ that there exists a rigid transformation such that the distance between the two atoms in the same column of the subsegment is at most d, where d is a parameter given by the user. This requires us to solve the following 3D protein structure matching problem:

Input: A segment of sequence alignment $\mathcal{A}$ of two proteins, where each position in the alignment has two identical atoms, the 3D coordinate of each atom in the alignment, and a threshold d,

Goal: find the set of subsegment of $\mathcal{A}$ such that for each output subsegment the Euclidean distance between the two atoms in the same column of the alignment is at most d.

The 3D protein structure matching problem can be solved in several ways. Here we use the method in [22] which is a faster version of the method in [9] to solve the problem. The method in [22] can compute a rigid transformation such that the distance between each matched pair of atoms is at most $(1 + \epsilon)d$, where $\epsilon = 0.1$ is a parameter to control the precision of the transformation. This is just an approximate rigid transformation, and it is good enough in practice.

Testing the overlap of the two proteins in 3D space. When computing the rigid transformation, we also require that the two proteins do not overlap under the rigid transformation. For each rigid transformation that can match the two substructures of the candidate subsegment, we test if the two proteins have overlap in 3D space under such a transformation as follows:

1. construct the grid in 3D space and name each grid point as *interior* point, *surface* point and *empty* point as in Step 2 with respect to each of the two given proteins.

2. Let X be the number of grid points that are *interior* points for both proteins and X_1 and X_2 be the number of *interior* point of the first protein and the second protein, respectively. If $X \leq 0.05 \times min\{X_1, X_2\}$, then we say that there is no overlap between the two proteins under the current rigid transformation and we output the matched pair of substructures as the predicted binding sites. Otherwise, we have to give up the rigid transformation.

3   Implementation

We have implemented the algorithms in Java. The software package can run on both Windows and Linux. Bsfinder can complete four different operations: (1) find the binding sites between two different proteins; (2) find the binding sites between a given protein and all the proteins in a database; (3) search the sites in a complete protein structure; and (4) find the binding sites between different chains of the same protein. Our program outputs the matched 3D structures in the PDB format, and users can view them by Jmol.

4   Results

4.1   Comparison with Existing Methods

In this section, we compare our program BsFinder with three existing programs
SiteEngine, SuMo, and pdbFun. They used different methods to predict the
binding sites of given proteins. SiteEngine (`http://bioinfo3d.cs.tau.ac.il/`
`SiteEngine/`) is a method that recognize the regions on the surface of one
protein that are similar to the binding sites of another, and geometric hashing
triangles are used for transferring the input sites into the recognized region. [19].
SuMo (`http://sumo-pbil.ibcp.fr/cgi-bin/sumo-welcome`) is a system for
finding binding sites onto query structures, by comparing the structure of triplets
of chemical groups against the binding sites found in PDB database [16]. The web
server pdbFun (`http://pdbfun.uniroma2.it`) locates binding sites in proteins
at the residue level, and it analyzes structural similarity between any pair of
residue selections [18].

To compare our program BsFinder with the three existing systems, we use
all the proteins in PDB database and select 55 proteins to search the whole
database. Note that the Structural Classification of Proteins (SCOP) database
(`http://scop.mrc-lmb.cam.ac.uk/scop/index.html`) in [15] aims to provide
a detailed and comprehensive description of the structural and evolutionary relationships between all proteins whose structures are known. It provides 11 classes
to separate all known protein folds. Each class contains several different families. We choose 5 proteins (containing binding sites) from each class in different
families such that there is only one entry from each family, and the 55 proteins
are shown in Table. 1. We use these 55 proteins to search the whole database.
Since BsFinder allows user to give the value of d, we set the threshold $d = 1.5\text{Å}$
and output the matched sites with at least 15 atoms.

Table 1. 55 proteins from PDB database

1c52	8gss	256b	8ick	4vhb	2bpv	2rto	2trm	2xat	1jju
4fx2	5p21	2dub	3man	6dfr	1j6w	3pyp	1elv	1oiy	3bu4
1t9g	7cat	1jx4	1cy6	1sk6	1h2s	1ddt	1u19	1ppj	1ntm
7ins	1ki0	1ptr	1gmn	1f4l	1g9b	1jsh	1mg1	1s1c	1kwx
1izl	1dwl	1ffx	3ldh	2yhx	1go9	1hth	1lxf	2prg	1h2k
1g8x	1jy4	1k09	1abz	1l6x					

4.2   Evaluation of Prediction

To calculate the precision and recall for each approach, we need to know which
pair of binding sites output by the programs is real. Here we look at SitesBase
(`http://www.modelling.leeds.ac.uk/sb/`) in [17], which holds the set of the
known binding sites found in PDB. The *precision* is defined as the number of
sites output by the program that are confirmed in SitesBase divided by the total

number of sites output by the program, where a output pair of sites is *confirmed* in SitesBase if at least two complete residues of the output pair of sites are the same as the pair of binding sites in SitesBase. As the pairs of sites output by SuMo are very short, a pair of sites output by SuMo is *confirmed* if it has at least one residue which is identical to that in SitesBase. Ideally, all the pairs of sites output by the program are confirmed in SitesBase, in the case, the precision is 100%. Apparently, the bigger the precision is, the better the program is. The *recall* is defined as the number of sites output by the program that are confirmed in SitesBase divided by the total number of binding sites for the input proteins in SitesBase. If all the binding sites for the pair of input proteins in the SitesBase can be output by the program, then the recall is 100%. Again, the bigger the recall is, the better the program is.

We use the 55 selected proteins described in Section 4.1 to search the whole PDB database. The average numbers of the sites output by BsFinder, SiteEngine, SuMo, and pdbFun are 6425, 6003, 6329, and 1936, respectively. On average, pdbFun outputs the smallest number of sites and the other three systems output approximately the same number of sites. The average numbers of the confirmed sites output by BsFinder, SiteEngine, SuMo, and pdbFun are 2218, 1265, 674, and 281, respectively. See Fig. 3(a).

We calculate the precision and recall for 55 proteins output by four programs. Apparently, BsFinder has the biggest precision and recall for most of the cases. On average, the precision of BsFinder is 34% while the precision for SiteEngine, SuMo, and pdbFun are 21%, 11%, and 15%, respectively. The average recall of BsFinder is 82% while the average recall for SiteEngine, SuMo, and pdbFun are 47%, 25%, and 11%, respectively. See Fig. 3(b). The value of recall is very important in practice. The value 11% of recall means that 89% of real binding sites in SitesBase are not output by the program. From the experiments results, we know that for the existing programs, the biggest problem is their lower values of recall. Though BsFinder has a recall value 82%, still there are 18% of real binding sites in SitesBase are missing. Most of the 18% missing binding sites of BsFinder are "blur" sites, e.g., most of the missing matched pairs contain lots (more than 5 atoms) of gaps. It is challenging to increase the value of recall.

4.3   Comparison of Running Time

To compare the running times of different programs, we use a Pentium(R) 4 (CPU of 2.40GHz) to run all the four programs. Based on the 55 selected proteins, the average running times of BsFinder, SiteEngine, SuMo, and pdbFun for searching the whole PDB database are roughly 50 minutes, 70 minutes, 30 minutes, and 5 minutes, respectively. See Table. 2. Thus, BsFinder is the second slowest program. However, it is still faster than SiteEngine which has the highest average precision and recall among the three existing programs.

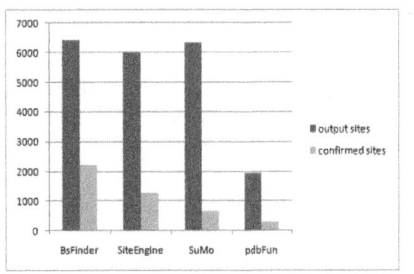 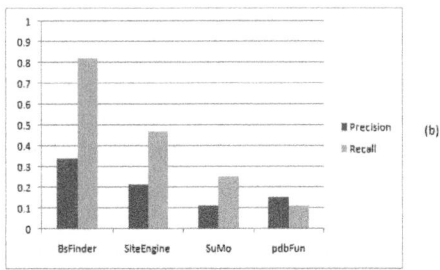

Fig. 3. (a)The average numbers of the output sites (black bar) and the confirmed sites (gray bar) for BsFinder, SiteEngine, SuMo, and pdbFun; (b)The average precision (black bar) and recall (gray bar) for BsFinder, SiteEngine, SuMo, and pdbFun

Table 2. Comparison of four programs

	RunningTime	Precision	Recall
BsFinder	50 minutes	34%	82%
SiteEngine	70 minutes	21%	47%
SuMo	30 minutes	11%	25%
pdbFun	5 minutes	15%	11%

4.4 Performance of Programs for Different Families

To see the performance of programs for different protein families, we select five proteins from three different families (G proteins family in P-loop folds, PYP-like family in Profilin-like folds, and FAD-linked reductases family in FAD/NAD(P)-binding folds). The average numbers of pairs of matched sites output by BsFinder for three families are 7680, 5289, and 7892, respectively. The average confirmed output pairs of sites in SitesBase for three families are 3487, 1132, and 4138, respectively. The average values of the precision for the three families are 45%, 21% and 53%, respectively. The average values of the recall for the three families are 94%, 60% and 96%, respectively. The results are shown in Fig. 4.

G proteins family in P-loop folds

We select 5 proteins (1a2b, 1cxz, 1dpf, 1ftn, 1s1c) from G proteins family in P-loop folds. The precision values of BsFinder (48%, 46%, 43%, 42% and 47%) are bigger than that of other three programs. The recall values of BsFinder (95%, 93%, 92%, 91% and 99%) are more than 90%, while the recall values of the other three programs are almost less than 40%.

PYP-like family in Profilin-like folds

We select 5 proteins (1d7e, 1f9i, 1kou, 1nwz, 2phy) from PYP-like family in Profilin-like folds. The precision values of BsFinder (17%, 18%, 24%, 25% and 21%) are similar to that of the other three programs. The recall values of BsFinder (58%, 64%, 59%, 63% and 57%) are bigger than that of other three programs.

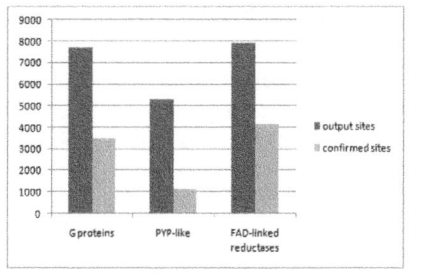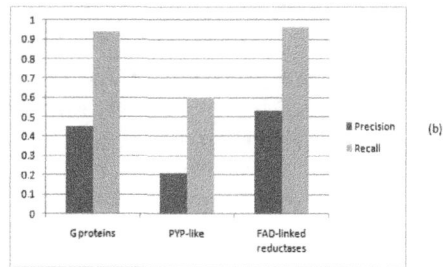

Fig. 4. (a)The average numbers of the sites output by BsFinder (black bar) and the confirmed sites (gray bar) for three different families; (b)The average precision (black bar) and recall (gray bar) for three different families

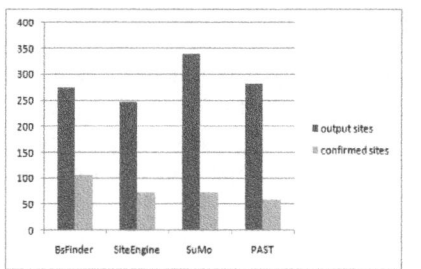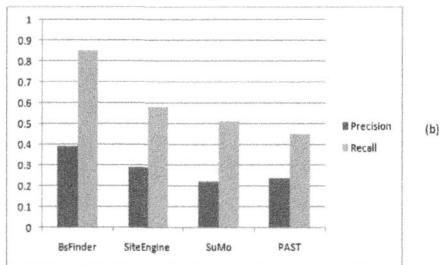

Fig. 5. (a)The average numbers of the output sites (black bar) and the confirmed sites (gray bar) for BsFinder, SiteEngine, SuMo, and PAST; (b)The average precision (black bar) and recall (gray bar) for BsFinder, SiteEngine, SuMo, and PAST

FAD-linked reductases family in FAD/NAD(P)-binding folds

We select 5 proteins (1b4v, 1b8s, 1coy, 1ijh, 3cox) from FAD-linked reductases family in FAD/NAD(P)-binding folds. The precision values of BsFinder (54%, 52%, 53%, 53% and 54%) are all more than 50%. The recall values of BsFinder (97%, 96%, 96%, 96% and 98%) are very close to 100%.

4.5   Search a Binding Site in PDB

The four programs, BsFinder, SiteEngine, SuMo, and PAST [20] can search a binding site on a set of complete proteins. PAST (http://past.in.tum.de/) is a program for finding the binding sites from the protein structures similar to the given binding site. It is based on an adaptation of the generalized suffix tree and relies on a linear representation of the protein backbone [20].

We randomly select the 100 binding sites from the SitesBase and search the whole PDB database. The average numbers of the sites output by BsFinder, SiteEngine, SuMo, and PAST are 274, 266, 399, and 281, respectively. The average numbers of the confirmed sites output by BsFinder, SiteEngine, SuMo, and PAST are 106, 73, 72, and 58, respectively. BsFinder finds a relatively small number of output sites, and the number of confirmed sites output by BsFinder is the biggest. The average number of output sites and confirmed sites are shown in Fig. 5(a).

The average precision of BsFinder is 39% while the average precision for SiteEngine, SuMo, and PAST are 27%, 22%, and 24%, respectively. The average recall of BsFinder is 86% while the average recall for SiteEngine, SuMo, and PAST are 58%, 51%, and 45%, respectively. The average precision and recall of the 100 binding sites are shown in Fig. 5(b).

5   Discussion

In the first step of our algorithm, we do sequence alignment where each letter is an atom. This allows the matched pairs of sites to have some missed atoms. The following statistic data show that Step 1 is important.

The gaps in binding sites
The gap distribution of the matched pairs of sites reported by BsFinder is shown in Fig. 6. Among the output matched pairs of sites, 67127 of them do not contain any gap, 63593 of them contain one gap, 77725 contain two gaps, 81259 contain three gaps, 38863 contain four gaps, 21198 contain five gaps and 3533 contain more than five gaps. The gap distribution of the confirmed sites are 18285 (no gap), 19504 (one gap), 26809 (two gaps), 26809 (three gaps), 15847 (four gaps), 12197 (five gaps) and 2452 (more than five gaps).

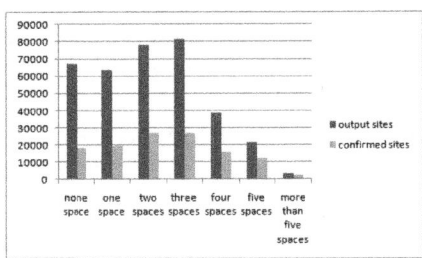

Fig. 6. The gap distributions of the output sites (black bar) and the confirmed sites (gray bar)

The power of surface detection
In Step 2 of our algorithm, we identify the surface atoms in the input proteins and rule out the substructures in which less than 2/3 of atoms are the surface atoms for further calculation of the rigid transformation. To demonstrate the effect of Step 2, we compare the final version of BsFinder with the version without Step 2. By adjusting the parameters, the final version of BsFinder has improved precision while the recall value remains essentially unchanged. The precision values for BsFinder without Step 2 and the final version of BsFinder are 29% and 34%, respectively. The recall values for BsFinder without Step 2 and the final version of BsFinder are 83% and 82%, respectively. See Fig. 7. Therefore, by doing Step 2 the precision value can be improved by about 5%. This is a significant improvement.

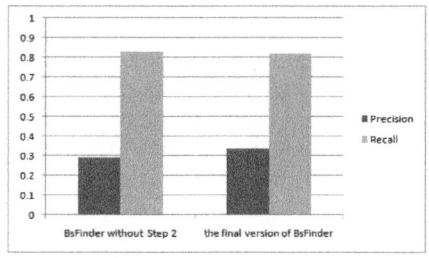

Fig. 7. The average precision (black bar) and recall (gray bar) for BsFinder without Step 2 and the final version of BsFinder

6 Conclusion

We have developed a program for finding pairs of binding sites for input proteins. Our method uses the 3D structure information to detect the similar regions. Experiments show that our program outperforms all existing programs.

Acknowledgement

FG and LW are fully supported by a grant from the Research Grants Council of the Hong Kong Special Administrative Region, China [Project No. CityU 121608].

References

1. Laurie, A.T., Jackson, R.M.: Q-SiteFinder: an energy-based method for the prediction of protein-ligand binding sites. Bioinformatics 21, 1908–1916 (2005)
2. Bradford, J.R., Westhead, D.R.: Improved prediction of protein-protein binding sites using a support vector machines approach. Bioinformatics 21, 1487–1494 (2004)
3. Chen, Y.-C., Lo, Y.-S., Hsu, W.-C., Yang, J.-M.: 3D-partner: a web server to infer interacting partners and binding models. Nucleic Acids Research 35, W561–W567 (2007)
4. Lo, A., Chiu, Y.-Y., Rodland, E.A., Lyu, P.-C., Sung, T.-Y., Hsu, W.-L.: Predicting helix-helix interactions from residue contacts in membrane proteins. Bioinformatics 25, 996–1003 (2008)
5. Siew, N., Elofsson, A., Rychlewski, L., Fischer, D.: Maxsub: an automated measure for the assessment of protein structure prediction quality. Bioinformatics 16(9), 776–785 (2000)
6. Zemla, A.: LGA: a method for fnding 3D similarities in protein structures. Nucl. Acids Res. 31(13), 3370–3374 (2003)
7. Zhang, Y., Skolnick, J.: Scoring function for automated assessment of protein structure template quality. Proteins: Structure, Function, and Bioinformatics 57(4), 702–710 (2004)

8. Li, H., Li, J., Wong, L.: Discovering motif pairs at interaction sites from protein sequences on a proteome-wide scale. Bioinformatics 22(8), 989–996 (2006)
9. Li, S.C., Bu, D., Xu, J., Li, M.: Finding largest well-predicted subset of protein structure models. In: Ferragina, P., Landau, G.M. (eds.) CPM 2008. LNCS, vol. 5029, pp. 44–55. Springer, Heidelberg (2008)
10. Pennec, X., Ayache, N.: A geometric algorithm to find small but highly similar 3D substructures in proteins. Bioinformatics 14, 516–522 (1998)
11. Fischer, D., Bachar, O., Nussinov, R., Wolfson, H.: An efficient automated computer vision based technique for detection of three dimensional structural motifs in proteins. J. Biomol. Struct. Dynam. 9, 769–789 (1992)
12. Bachar, O., Fischer, D., Nussinov, R., Wolfson, H.: A computer vision based technique for sequence independent structural comparison of proteins. Protein Eng. 6, 279–288 (1993)
13. Liu, X., Li, J., Wang, L.: Modeling protein interacting groups by quasi-bicliques: complexity, algorithm and application. IEEE/ACM Transactions on Computational Biology and Bioinformatics 7(2), 354–364 (2010)
14. Jambon, M., Imberty, A., Deleage, G., Geourion, C.: A new bioinformatic approach to detect common 3D sites in protein structures. Proteins 52, 137–145 (2003)
15. Murzin, G., Brenner, E., Hubbard, T., Chothia, C.: SCOP: a structural classification of proteins database for the investigation of sequences and structures. J. Mol. Biol. 247, 536–540 (1995)
16. Jambon, M.: The SuMo server: 3D search for protein functional sites. Bioinformatics 21, 3929–3930 (2005)
17. Gold, N.D., Jackson, R.M.: SitesBase: a database for structure-based protein-ligand binding site comparisons. Nucleic Acids Res. 34(suppl. 1), D231–D234 (2006)
18. Ausiello, G.: pdbFun: mass selection and fast comparison of annotated PDB residues. Nucleic Acids Res. 137, W133–W137 (2005)
19. Shulman-Peleg, A.: Recognition of functional sites in protein structures. J. Mol. Biol. 339, 607–633 (2004)
20. Täubig, H., Buchner, A., Griebsch, J.: PAST: fast structure-based searching in the PDB. Nucleic Acids Res. 34, W20–W23 (2006)
21. Smith, T.F., Waterman, M.S.: Identification of Common Molecular Subsequences. Journal of Molecular Biology 147, 195–197 (1981)
22. Guo, F., Wang, L., Yang, Y., Lin, G.: Efficient Algorithms for 3D Protein Substructure Identification. In: The International Conference on Bioinformatics and Biomedical Engineering (2010)
23. Huang, B., Schröder, M.: LIGSITE^{csc}: predicting ligand binding sites using the Connolly surface and degree of conservation. BMC Struct. Biol. 6, 19–29 (2006)
24. Henrich, S.: Computational approaches to identifying and characterizing protein binding sites for ligand design. Journal of Molecular Recognition 23(2), 209–219 (2010)

SETTER - RNA SEcondary sTructure-based TERtiary Structure Similarity Algorithm

David Hoksza[1] and Daniel Svozil[2]

[1] Charles University in Prague, FMP, Department of Software Engineering,
Malostranské nám. 25, 118 00, Prague, Czech Republic
hoksza@ksi.mff.cuni.cz
http://siret.ms.mff.cuni.cz/hoksza
[2] Institute of Chemical Technology Prague, Laboratory of Informatics and
Chemistry, Technická 5, 166 28 Prague, Czech Republic
daniel.svozil@vscht.cz
http://ich.vscht.cz/~svozil

Abstract. The recent interest in function of various RNA structures, reflected in the growth of solved RNA structures in PDB, calls for methods for effective and efficient similarity search in RNA structural databases. Here, we propose a method called **SETTER** (RNA **SE**condary sTructure-based **TER**tiary structure similarity) based on partitioning of RNA structures into so-called generalized secondary structure units (GSSU). We introduce a fast similarity method exploiting RMSD-based algorithm allowing to assess distance to a pair of GSSU, and a method for aggregating these partial distances into a final distance corresponding to structural similarity of the examined RNA structures. Our algorithm yields not only the distance but also a superposition allowing to visualize the structural similarity. Comparative experiments show that our proposed method is competitive with the best existing solutions, both in terms of effectiveness and efficiency.

Keywords: RNA, RNA secondary structure, RNA tertiary structure, RNA structural similarity.

1 Introduction

The primary components of living organisms - nucleic acids and proteins - are biopolymers, long linear molecules composed from the sequence of building blocks called monomers. While proteins are the active elements of cells, the instruction for their synthesis is stored in deoxyribonucleic acid (DNA). DNA is a biopolymer consisting of four types of units called nucleotides. Each nucleotide is composed from three parts: one of four possible bases (adenine (A), guanine (G), cytosine (C), thymine (T)), a sugar deoxyribose, and a phosphate group. Gene - the DNA sequence serving as a prescription for protein synthesis (expression) - determines which protein will be expressed in the organism in the given time at the given place. The bases within base-pairs are stabilized at their positions by the chemical interaction called hydrogen bond. In DNA the bases

J. Chen, J. Wang, and A. Zelikovsky (Eds.): ISBRA 2011, LNBI 6674, pp. 37–48, 2011.
© Springer-Verlag Berlin Heidelberg 2011

are complementary meaning that A pairs always with T, and C with G forming the so-called canonical (or Watson-Crick) base-pairs.

DNA is too valuable material to be used directly in the protein expression. Instead, the genetic information is first transcribed into another type of nucleic acid - ribonucleic acid RNA. The basic building blocks of RNA are similar to that of DNA with two important exceptions: thymine is substituted by uracil (U), and deoxyribose by ribose. Unlike DNA, most RNA molecules are single-stranded. However, the RNA chain is not stretched in biological conditions, instead it maintains a distinct 3D arrangement called conformation (or fold). The biological function of RNA is directly related to its conformation, and the study of 3D structure of biopolymers generally is very important for better understanding of the inner workings of living organisms. Resolved structures (i.e. xyz coordinates of all atoms in the molecule) are deposited into the PDB database [4] that is available free of charge to broad scientific community.

Single-stranded RNA molecules adopt very complex 3D structures, as the presence of ribose introduces additional hydrogen bonding site allowing for formation of various non-canonical base pairs [22]. RNA structure is hierarchical [5], and can be divided into primary (RNA sequence), secondary, tertiary and quaternary levels. RNA secondary structure motifs [17], that are stable independently of their 3D folds, can be defined as double helices combined with various types of loop structures, and they can be categorized based on the mutual positions of these simple elements. A single loop connecting the end of helix is a hairpin loop, two single strands linking a pair of double-helical segments comprise an internal loop (if one of these links is of zero length a bulge loop is formed), and three or more double-helical segments linked by a single-strand sequences form a junction loop. RNA motifs have been classified according to function, 3D structure or tertiary interaction in the SCOR (Structural classification of RNA) database [20,27]. The SCOR classification system is based on the Directed Acyclic Graph (DAG) to reflect the fact, that RNA structural elements can have several distinct features and may belong to multiple classes. Characterization of secondary RNA motifs is important and it finds application in such areas as RNA design [18,9], RNA structure prediction [25], RNA modeling [10] or RNA gene finding [7]. RNA plays a variety of essential roles in many cellular processes, including enzymatic activity [26], protein synthesis regulation [12], gene transcriptional regulation [2,11] and chromosome replication [12,16]. The knowledge of RNA 3D structure is indispensable for characterizing of such functions, and thus the ultimate goal remains the prediction of the tertiary structure.

Currently (January 2011), the PDB database stores 1980 RNA structures. Such a wealth of data allows the analysis and characterization of the RNA structural space, which may help to characterize RNA function. Since 3D structure is typically more evolutionary conserved than sequence, detecting structural similarities between RNA molecules can bring insights into their function that would not be detected by sequence information alone. The development of automatic tools capable of efficient and accurate RNA structural alignment and comparison has become an important part of structural bioinformatics of RNA. Detecting structural similarities between two RNA (or protein) molecules at the tertiary level is a difficult task that has been shown to be NP-hard [21]. Therefore currently available software tools for comparing two RNA 3D structures, such as ARTS [13,14], DIAL [15], iPARTS [28], SARA [6] , SARSA [8] or LaJolla [3], are all based on some heuristic approaches.

The best existing approaches SARA and iPARTS to which SETTER is compared will now be briefly described. The SARA program represents distances among selected atoms as unit vectors existing in the unit spheres. All-to-all unit-vector RMS distances of consecutive unit spheres are computed and used as scoring matrix for subsequent dynamic programming based global alignment. Dynamic programming is also employed in iPARTS algorithm in which 3D RNA structures are represented as 1D sequences of 23 possible symbols, each of which corresponds to the distinct backbone conformational family.

In this paper, we propose a new pairwise RNA comparison method based on 3D similarity of the so-called general secondary structure units (GSSU) resembling secondary structure motifs. Each of the compared RNA structures is divided into non-intersecting set of GSSUs. For a pair of GSSUs, similarity measure is introduced based on executing multiple RMSD transformations on particular subsets from the GSSUs. The measure is then normalized to obtain the resulting distance/similarity (we will use the terms distance and similarity interchangeably throughout the text) of a pair of GSSUs. If the compared RNA structures contain more GSSUs, all-to-all distances are computed and aggregation takes place resulting in the pairwise RNA structure comparison. We show in the experimental section that our method outperforms SARA and iPARTS both in accuracy and runtime. Moreover, in SETTER there is essentially no limit on the size of aligned structures. This is in contrast with SARA and iPARTS which are (due to the use of dynamic programming) limited to structures having at most 1,000 and 1,900 nucleotides, respectively.

2   Method Principles

For the purpose of our method, each nucleotide in an RNA structure is represented by its $C4'$ atom although any other backbone atom could be utilized. RNA structure is represented as a set of GSSUs that can be regarded as fundamental units of RNA structure. In contrast to the basic secondary structure motifs, GSSUs contain more information by comprising larger subsets of RNA. These subsets represent meaningful RNA partitioning being easy to work with.

Definition 1. *Let $\mathcal{R}$ be an RNA structure with nucleotide sequence $\{n_i\}_{i=1}^{n}$ and let $\mathcal{WC} \subset \mathcal{R}$ denote set of n_i participating in a Watson-Crick base pair. By **generalized secondary structure unit (GSSU)** $\mathcal{G}$, we understand a pair of substrings of $\mathcal{R}$, $\{n_i\}_{i=i_1}^{i_2}$ and $\{n_i\}_{i=j_1}^{j_2}$ ($i_1 \leq i_2 < j_1 \leq j_2$) of maximum lengths such that each nucleotide n_x:*

- $i_1 \leq x \leq i_2$: $n_x \notin \mathcal{WC}$ or n_x is paired with n_y where $j_1 \leq y \leq j_2$
- $j_1 \leq x \leq j_2$: $n_x \notin \mathcal{WC}$ or n_x is paired with n_y where $i_1 \leq y \leq i_2$

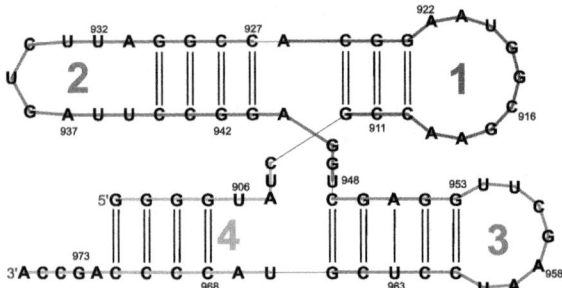

Fig. 1. 4 extracted GSSUs for RNA structure with PDB code 1EXD. The sequence starts at the 5' end and the colored numbers denote order of GSSU generation (number color corresponds with the respective GSSU's color). Note that, as GSSU 4 indicates, GSSU does not have to be comprised of a continuous chain of nucleotides but it has to correspond to the conditions of the Def. 1.

Let i_{max} and j_{min} be highest indices of the Watson-Crick paired bases. We define **loop** *as* $\mathcal{L} = \{n_i\}_{i=i_{max}+1}^{j_{min}-1} \subset \mathcal{R}$ *and* **stem** *as* $\mathcal{R} \setminus \mathcal{L}$ *and* **neck** *as the pair* $\{n_{i_{max}}, n_{j_{min}}\}$.

Note that even a structure without a single Watson-Crick pair has a GSSU which is identical with the structure itself. Usually, a GSSU looks like a hairpin motif but compared to hairpin, GSSU can contain bulges and internal loops within its stem part (see e.g. Fig. 1).

Due to the limited space, we will only briefly describe the GSSU extraction algorithm instead of showing its exact version. In general, extraction processes an RNA structure in the order of its sequence generating GSSUs based on the presence/absence of Watson-Crick hydrogen bonding pattern of each nucleotide[1]. We differentiate two states — GSSU generation is proceeding and GSSU generation does not take place. If GSSU is not being generated, the nucleotides are pushed on the stack to be processed later. If a nucleotide hydrogen-bonded to the nucleotide in the stack is identified during the process of GSSU generation, all non hydrogen-bonded nucleotides lying between them and the boundary nucleotides are added to the GSSU. The process of GSSU $\mathcal{G}$ is finished when a pair $\{nt_1, nt_2\}$ is found where $nt_2 \notin \mathcal{G}$. An example of GSSUs found in the structure of glutamine tRNA (PDB code 1EXD) is shown in Fig. 1.

2.1   Single GSSU Pairwise Comparison

When SETTER compares structures consisting of multiple GSSUs, pairwise GSSU comparison is employed. Therefore, single GSSU comparison can be viewed as the principle component of SETTER.

Each GSSU is represented by the ordered set of 3D coordinates enhanced with bonding and nucleotide/atom type information. The common way how to assess

---

[1] To obtain hydrogen bonding information from PDB files, we used the 3DNA utility [24,23].

similarity of two sets X and Y of points is to define pairing between them. The sets are then superposed by finding such translation and rotation that the mutual distances of individual paired points are minimized. Usually, the root mean square deviation (RMSD) is chosen as the distance measure, because there exists a polynomial time algorithm able to optimally superpose two structures given a pairing/alignment [19]. However, finding the optimal alignment is a hard problem. To evaluate the quality of alignments that can potentially be a part of global alignment SETTER uses Kabsch [19] RMSD algorithm. The search for the optimal superposition (including search for the optimal alignment) is NP-hard [21]. Because trying each possible alignment is not computationally feasible, suitable alignments with potential to participate in optimal alignment should be identified. That is the principle idea behind SETTER's structure comparison process.

The nucleotides participating in necks of two GSSUs should not be missed in the optimal alignment. Otherwise stated, to superpose two GSSUs means to match their loops which implies also matching their necks. By matching necks, one can unambiguously superpose the structures in two dimensions but since in reality GSSUs exist in three dimensional space, at least three points are needed to define the superposition. We call these points *triplet*, and an *alignment* is formed by matching these points between two processed structures. Two matched points are further referred to as a "pair". Therefore, SETTER aligns necks and then tries to align each pair of loops' nucleotides one by one. The loop pair defines final pair in the triplet necessary to superpose the GSSUs. For example, if two GSSUs having loops consisting of n and m nucleotides to be aligned, $n \times m$ alignments are generated (see Figure 2)

For each of the proposed alignments, a rotation matrix and a translation vector defining optimal superposition of the triplets is generated and subsequently used to superpose the whole GSSU. After that, nearest neighbor from the second GSSU in 3D space is identified for each nucleotide, and their distance is added to the overall distance of the two GSSUs. Finally the distance is normalized. The whole process can be formalized by equation 1.

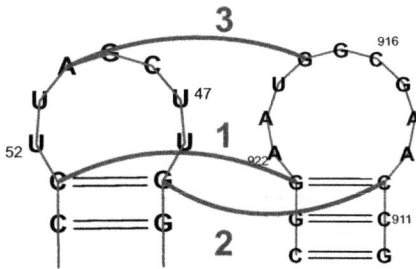

Fig. 2. Alignment of GSSU from tRNA domain of transfer-messenger RNA (PDB code 1P6V) with GSSU from glutamine tRNA (PDB code 1EXD). The final structural alignment is defined by three nucleotide pairs forming a triplet (the red lines 1, 2, and 3). To find an optimal superposition for the given neck pairs (lines 1 and 3), the position of the middle pair is varied (line 2).

$$NN_\zeta(x, \mathcal{G}) = \begin{cases} \min_{1 \le i \le |\mathcal{G}|} \{d_{nt}(x, \mathcal{G}_i)\} \times \zeta & \text{if } x = y \\ \min_{1 \le i \le |\mathcal{G}|} \{d_{nt}(x, \mathcal{G}_i)\} & \text{otherwise} \end{cases}$$

$$\gamma(\mathcal{G}^A, \mathcal{G}^B) = \sum_{i=1}^{|\mathcal{G}^A|} \begin{cases} 1 & \text{if } NN_1(\mathcal{G}^A{}_i, \mathcal{G}^B) \le \epsilon \\ 0 & \text{otherwise} \end{cases}$$

$$(1)$$

$$\delta(\mathcal{G}^A, \mathcal{G}^B) = \min_{t \in T} \left\{ \sum_{i=1}^{|\mathcal{G}^A|} NN_\alpha(\mathcal{G}^A{}_i, \tau(\mathcal{G}^B, t)) \right\}$$

$$\Delta(\mathcal{R}^A, \mathcal{R}^B) = \Delta(\mathcal{G}^A, \mathcal{G}^B) = \frac{\frac{\delta(\mathcal{G}^A, \mathcal{G}^B)}{\min\{|\mathcal{G}^A|, |\mathcal{G}^B|\}} \times (1 + \frac{||\mathcal{G}^A| - |\mathcal{G}^A||}{\min\{|\mathcal{G}^A|, |\mathcal{G}^B|\}})}{\gamma(\mathcal{G}^A, \tau(\mathcal{G}^B, t_{opt}))}$$

In the formula, $\mathcal{G}^A$ (identified with an RNA structure $\mathcal{R}^A$) and $\mathcal{G}^B$ (identified with an RNA structure $\mathcal{R}^B$) represent the GSSUs to be compared, $\mathcal{G}_i$ stands for i-th nucleotide in the nucleotide sequence of $\mathcal{G}$ and $|\mathcal{G}|$ for its length. $NN(x, \mathcal{G})$ is the Euclidean distance from a nucleotide x to its nearest neighbor in $\mathcal{G}$. If x and its nearest neighbor have identical type, the distance is modified by factor ζ. δ computes the raw distance - T is a set of transpositions resulting from the candidate alignments and $\tau(\mathcal{G}, t)$ transposes GSSU $\mathcal{G}$ using the transposition t. The normalized distance Δ then employs function γ counting number of nearest neighbors within the distance ϵ of the optimal transposition t_{opt}.

The whole process can be summarized in the following four steps:

1. Identify candidate set of alignments of triplet pairs (two nucleotides from neck, one from loop).
2. Compute superpositions (i.e. set of rotation matrices and translations vectors) for each of the alignments.
3. For each rotation matrix and translation vector superpose the structures.
4. For each superposition identify nearest neighbors, sum the distances to get δ and normalize it to obtain the final distance Δ.

Sometimes identification of hydrogen bonds may not be correct and the real neck position within the GSSU is shifted. Therefore, SETTER also tries to simulate the neck shift by aligning the residues next to (under) the necks. Finally, when aligning the neck $\{n_1^A, n_2^A\}$ of a GSSU $\mathcal{A}$ with then neck $\{n_1^B, n_2^B\}$ of a GSSU $\mathcal{B}$ it is not clear in which direction the loops are oriented in 3D space (whether the correct alignment is $\{\{n_1^A, n_1^B\}, \{n_2^A, n_2^B\}\}$ or $\{\{n_1^A, n_2^B\}, \{n_2^A, n_1^B\}\}$) and therefore both possibilities are investigated. These tweakings are necessary for accurate GSSU comparison, however they slightly increase the running time of SETTER.

Though in most cases the GSSU comprises of a stem and a loop, it is not a strict rule, as demonstrated in Fig. 1, GSSU number 4. Two particular situations can occur — GSSU has a zero-sized loop or the RNA does not have a single hydrogen bond (i.e., it does not have a secondary structure at all). In case of GSSU without the loop we select the third nucleotide for triplet alignment from the stem and we vary its position within the stem. When dealing with a GSSU having no secondary structure, several triplets covering whole structure are formed and used for the alignment.

2.2   Multiple GSSU Structure Comparison

For the comparison of RNA structures containing multiple GSSUs we utilized a straightforward solution. Consider the comparison of RNA structures $\mathcal{R}^{\mathcal{A}}$ and $\mathcal{R}^{\mathcal{B}}$ consisting of $n_{\mathcal{A}}$ and $n_{\mathcal{B}}$ GSSUs. We modify the Δ definition in the following way:

$$\Delta(\mathcal{R}^{\mathcal{A}}, \mathcal{R}^{\mathcal{B}}) = \min_{\substack{1 \leq i \leq n_A \\ 1 \leq j \leq n_B}} \{\mathcal{G}_i^{\mathcal{A}}, \mathcal{G}_j^{\mathcal{B}}\} \times (1 + |n_{\mathcal{A}} - n_{\mathcal{B}}|) \times \beta \qquad (2)$$

We compute all-to-all distances between the GSSUs and we choose the pair with minimal distance. Moreover, we multiply the distance by the difference in GSSUs count to favor structures with similar number of GSSU. The parameter β allows more distinct separation. Increasing value of β more noticeably favors similar-sized structures (in our experiment, we use $\beta = 2$).

SETTER uses only pairwise GSSU comparisons for matching RNA structures of any size including the largest ones such as ribosomal subunits. Since the mutual GSSU positions are rigid, the optimal superposition for a pair of GSSUs defines superposition for the whole structures which can be easily visualized (Fig. 3).

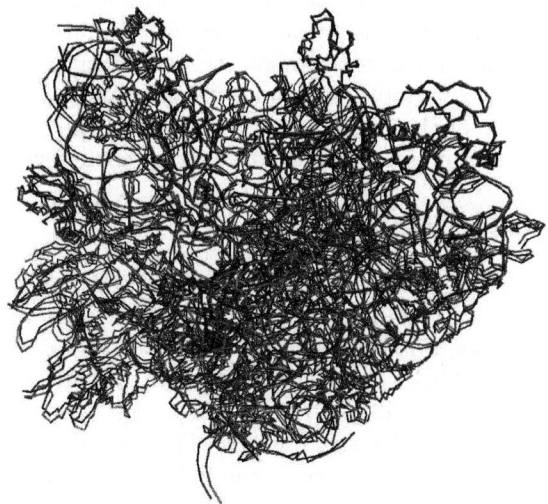

Fig. 3. The superposition of structures from 23S ribosomal RNA having PDB codes 1NWY:0 (2880 nucleotides, 84 GSSUs, blue) with 1SM1:0 (2880 nucleotides, 83 GSSUs, red) - RMSD = 2.43.

Though our solution follows the KISS principle (Keep It Simple and Stupid), it has several advantages over more elaborate approaches based for example on finding maximal common subgraphs in the network of interactions between individual GSSUs. Not only it is much faster, but it also allows to use effective early termination mechanism leading to additional speed improvements. This mechanism is introduced in the following section.

3   Speed Up

The nearest neighbor search process needed for Δ computation is highly expensive since it has $O(n^2)$ time complexity with respect to the GSSU's length. Moreover, the process has to be done for each of the candidate alignment, noticeably decreasing efficiency of SETTER. Therefore we implemented simple early termination condition into the SETTER's GSSU comparison process. We identify alignments that are not likely part of the optimal superposition and for these alignments the nearest neighbor search is skipped. Because the superposition was optimized for the aligned triplets, their distance will be low compared to other nucleotides in the structure and they will very likely stay nearest neighbors also after the superposition of the whole GSSUs. Thus, Δ of the triplet-based GSSUs will be probably lower then Δ of the GSSU from which they come. If we align a triplet $\mathcal{T}^A \subset \mathcal{G}^A$ with a triplet $\mathcal{T}^B \subset \mathcal{G}^B$ with $\Delta(\mathcal{G}^A, \mathcal{G}^B) = \chi$ being the best result so far, the comparison computation can be terminated (i.e., we do not identify all the nearest neighbors) if $\Delta(\mathcal{T}^A, \mathcal{T}^B) \times \lambda > \chi$. Since the early termination is a heuristic ($\Delta(\mathcal{T}^A, \mathcal{T}^B) < \Delta(\mathcal{G}^A, \mathcal{G}^B)$ does not have to be valid), we strengthen the early termination condition by introducing the parameter $\lambda \geq 1$. By varying the λ parameter, the trade-off between accuracy and speed can be set. In case of multiple GSSU comparison, the speed-up can be even more noticeable. The scope of χ variable can span multiple GSSU pairwise comparisons since we are searching for the minimum distance among all pairs of GSSUs. Such an approach can further emphasize the effect of the early termination condition.

4   Experimental Results

In order to evaluate SETTER and to compare it with other solutions we run test on datasets introduced in [6]. It contains three datasets — FSCOR, T-FSCOR, R-FSCOR based on functional classification obtained from the SCOR database [20], version 2.0.3. The FSCOR contains all RNA chains with more than three nucleotides with unique functional classification. The R-FSCOR is a structurally dissimilar subset of the FSCOR. The T-FSCOR set contains structures from the FSCOR set not present in the R-FSCOR set. Using these datasets we can evaluate quality of RNA similarity method in terms of functional assignment/classification ability. The task is to assign the functional (i.e. SCOR) classification to the query RNA structure by comparison with a database of classified RNA structure. Specifically, we performed two experiments — a leave-one-out test on the FSCOR dataset and a test assigning functions to structures from the T-FSCOR with the R-FSCOR serving as the database set.

When comparing functions of two RNA structures, we differentiate between possessing identical and possessing similar function. Two structures have identical function if they share the deepest SCOR classification. If they do not agree at the deepest level but share classification at the parent level, they are said to have similar function.

In our experiments, we compute ROC curves and their AUC (area under the ROC curve) that is considered to be a robust indicator of quality of a classifier [1]. ROC is computed such that for each query we identify the most similar database structure (nearest neighbor) and the distance to it. The nearest neighbors for all queries are sorted according to their distances, and a distance threshold is varied from the most similar to the most dissimilar pair to generate points of the ROC curve. For a given threshold we identify number of structure pairs above the threshold with identical/similar function and denote them as true positives (TP). Rest of the above-threshold structures are denoted as false positives (FP). If P (positives) is the number of pairs with identical/similar function in the whole result set and N (negatives) the number of pairs with different functions, then $\frac{FP}{N}$ is called *false positive ratio* and $\frac{TP}{P}$ *true positive ratio*. The ROC curve consists of false positive ratio (x-axis) vs true positive ratio (y-axis) points.

Throughout the experimental section SETTER is used with following settings — $\alpha = 0.2$, $\beta = 2$, $\epsilon = 4\,\text{Å}$ and $\lambda = 1$ (see sections 2.1, 2.2 and 3 for details).

In Fig. 4, ROC curves of SETTER and iPARTS on FSCOR (Fig. 4a) and T-FSCOR (Fig. 4a) datasets are compared. We can see that on the FSCOR dataset, SETTER outperforms iPARTS with AUC equal to 0.74 (identical function) and 0.93 (similar function) in case of SETTER. iPARTS achieves AUC of 0.72 for identical function and of 0.92 for similar function. For SARA, only AUC values were presented in [6] being 0.61 and 0.83, respectively. When testing the T-FSCOR set against the R-FSOR set, SETTER is outperformed by iPARTS as is demonstrated by ROC curves in Fig. 4a. Specifically, AUC values are equal to 0.70 and 0.88 in case of SETTER, and to 0.77 and 0.90 in case of iPARTS. The results of SARA on the T-FSCOR set were again worse then the results of both SETTER and iPARTS — 0.58 for identical function and 0.85 for similar function.

We also carried out experiments measuring running time of SARA, SETTER and iPARTS. The runtime of SETTER was measured on Linux machine with 4 Intel(R) Xeon(R) CPUs E7540, 2GHz (the algorithm is not parallelized) and 132 GB of RAM (although the average memory size needed for an RNA structure from the FSCOR set is less then 3.3 MB) running Red Hat Linux. Runtime of SARA and iPARTS were taken from the output of their web interfaces. Thus, the comparison is only approximate. However, the variations between SETTER and SARA/iPARTS are substantial (Tab. 1) and can not be attributed to the hardware differences only.

Table 1 shows times of all-to-all comparisons on four data sets. Note the difference between SETTER and iPARTS for growing structure size. It can be seen that SETTER's runtime grows more or less linearly with the size of the structure, in contrast to iPARTS where the growth is quadratical. That stems

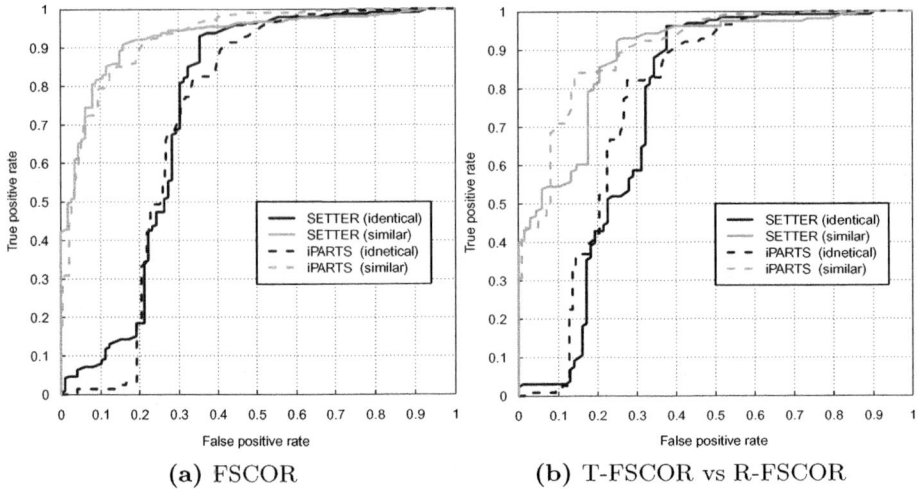

(a) FSCOR   (b) T-FSCOR vs R-FSCOR

Fig. 4. ROC curves of SETTER and iPARTS

Table 1. Runtime comparison of iPARTS, SARA and SETTER. The *tRNA* set contains structures 1EHZ:A, 1H3E:B, 1I9V:A, 2TRA:A and 1YFG:A structures (average length 76 nucleotides), *Ribozyme P4-P6 domain* contains 1GID:A, 1HR2:A and 1L8V:A (average length 157 nucleotides), *Domain V of 23S rRNA* contains 1FFZ:A and 1FG0:A (average length 496 nucleotides) and *16S rRNA* contains 1J5E:A and 2AVY:A (average length 1522 nucleotides).

data set	iPARTS	SARA	SETTER
tRNA	1.1 s	1.7 s	0.1 s
Ribozyme P4-P6 domain	2.6 s	9.2 s	1.8 s
Domain V of 23S rRNA	17.0 s	?	2.1 s
16S rRNA	2.8 min	?	8.1 s

In time of writing this paper, the SARA program was not able to handle sets *Domain V of 23S rRNA* and *16S rRNA*.

from iPARTS use of dynamic programming when searching for the alignment of 1D representations of the compared structures. In its original version, SETTER also uses $O(n^2)$ nearest neighbor identification procedure, but since it employs the speed optimization ($\lambda = 1$), the runtime of the algorithm, especially for large structures, is noticeably downsized.

5   Conclusion

In this paper, we have proposed a fast method for effective comparison of two RNA structures. The comparison is based on reasonably selected subsets of the nucleotide sequence resembling common secondary structure motifs. These subsets are then compared in three-dimensional space. Our method outperforms best existing solutions while maintaining high search speed.

In future, we would like to improve efficiency of our method by designing more sophisticated pruning method. We would also like to improve the effectivity by implementing multiple GSSU alignment and by introducing statistical methods, such as expectancy, into the classification process.

Acknowledgments. This work was supported by Czech Science Foundation project Nr. 201/09/0683 and by the Ministry of Education of the Czech Republic MSM6046137302.

References

1. Baldi, P., Brunak, S.A., Chauvin, Y., Andersen, C.A.F., Nielsen, H.: Assessing the accuracy of prediction algorithms for classification: an overview. Bioinformatics 16(5), 412–424 (2000)
2. Bartel, D.: MicroRNAs: Genomics, Biogenesis, Mechanism, and Function. Cell 116(2), 281–297 (2004)
3. Bauer, R.A., Rother, K., Moor, P., Reinert, K., Steinke, T., Bujnicki, J.M., Preissner, R.: Fast structural alignment of biomolecules using a hash table, n-grams and string descriptors. Algorithms 2(2), 692–709 (2009)
4. Berman, H.M., Westbrook, J.D., Feng, Z., Gilliland, G., Bhat, T.N., Weissig, H., Shindyalov, I.N., Bourne, P.E.: The protein data bank. Nucleic Acids Res. 28(1), 235–242 (2000)
5. Brion, P., Westhof, E.: Hierarchy and dynamics of rna folding. Annual Review of Biophysics and Biomolecular Structure 26(1), 113–137 (1997)
6. Capriotti, E., Marti-Renom, M.A.: Rna structure alignment by a unit-vector approach. Bioinformatics 118, i112–i118 (2008)
7. Carter, R.J., Dubchak, I., Holbrook, S.R.: A computational approach to identify genes for functional RNAs in genomic sequences. Nucl. Acids Res. 29(19), 3928–3938 (2001)
8. Chang, Y.-F., Huang, Y.-L., Lu, C.L.: Sarsa: a web tool for structural alignment of rna using a structural alphabet. Nucleic Acids Res. 36(Web-Server-Issue), 19–24 (2008)
9. Chworos, A., Severcan, I., Koyfman, A.Y., Weinkam, P., Oroudjev, E., Hansma, H.G., Jaeger, L.: Building programmable jigsaw puzzles with RNA. Science 306(5704), 2068–2072 (2004)
10. Ditzler, M.A., Otyepka, M., Šponer, J., Walter, N.G.: Molecular Dynamics and Quantum Mechanics of RNA: Conformational and Chemical Change We Can Believe In. Accounts of Chemical Research 43(1), 40–47 (2010)
11. Dorsett, Y., Tuschl, T.: siRNAs: applications in functional genomics and potential as therapeutics. Nature Rev. Drug Discovery 3, 318–329 (2004)
12. Doudna, J.A.: Structural genomics of RNA. Nat. Struct. Biol. 7 suppl., 954–956 (2000)
13. Dror, O., Nussinov, R., Wolfson, H.: ARTS: alignment of RNA tertiary structures. Bioinformatics 21(suppl. 2) (September 2005)
14. Dror, O., Nussinov, R., Wolfson, H.J.: The ARTS web server for aligning RNA tertiary structures. Nucleic Acids Res. 34(Web Server issue) (July 2006)
15. Ferrè, F., Ponty, Y., Lorenz, W.A., Clote, P.: Dial: a web server for the pairwise alignment of two rna three-dimensional structures using nucleotide, dihedral angle and base-pairing similarities. Nucleic Acids Res. 35(Web-Server-Issue), 659–668 (2007)

16. Hannon, G.J., Rivas, F.V., Murchison, E.P., Steitz, J.A.: The expanding universe of noncoding RNAs. Cold Spring Harb. Symp. Quant Biol. 71, 551–564 (2006)
17. Hendrix, D.K., Brenner, S.E., Holbrook, S.R.: RNA structural motifs: building blocks of a modular biomolecule. Q. Rev. Biophys. 38(3), 221–243 (2005)
18. Jaeger, L., Westhof, E., Leontis, N.B.: TectoRNA: modular assembly units for the construction of RNA nano-objects. Nucleic Acids Res. 29(2), 455–463 (2001)
19. Kabsch, W.: A solution for the best rotation to relate two sets of vectors. Acta Crystallographica Section A 32(5), 922–923 (1976)
20. Klosterman, P.S., Tamura, M., Holbrook, S.R., Brenner, S.E.: SCOR: a Structural Classification of RNA database. Nucleic Acids Res. 30(1), 392–394 (2002)
21. Kolodny, R., Linial, N.: Approximate protein structural alignment in polynomial time. Proc. Natl. Acad. Sci. USA 101(33), 12201–12206 (2004)
22. Leontis, N.B., Westhof, E.: Geometric nomenclature and classification of RNA base pairs. RNA 7(4), 499–512 (2001)
23. Lu, X.-J., Olson, W.K.: 3DNA: a versatile, integrated software system for the analysis, rebuilding and visualization of three-dimensional nucleic-acid structures. Nature Protocols 3(7), 1213–1227 (2008)
24. Lu, X.-J.J., Olson, W.K.: 3DNA: a software package for the analysis, rebuilding and visualization of three-dimensional nucleic acid structures. Nucleic Acids Res. 31(17), 5108–5121 (2003)
25. Shapiro, B., Yingling, Y., Kasprzak, W., Bindewald, E.: Bridging the gap in RNA structure prediction. Current Opinion in Structural Biology 17(2), 157–165 (2007)
26. Staple, D.W., Butcher, S.E.: Pseudoknots: RNA structures with diverse functions. PLoS Biology 3(6) (June 2005)
27. Tamura, M., Hendrix, D.K., Klosterman, P.S., Schimmelman, N.R., Brenner, S.E., Holbrook, S.R.: SCOR: Structural Classification of RNA, version 2.0. Nucleic Acids Res. 32(Database issue) (January 2004)
28. Wang, C.-W., Chen, K.-T., Lu, C.L.: iPARTS: an improved tool of pairwise alignment of rna tertiary structures. Nucleic Acids Res. 38(suppl.), W340–W347 (2010)

Prediction of Essential Genes by Mining Gene Ontology Semantics

Yu-Cheng Liu[1], Po-I Chiu[1], Hsuan-Cheng Huang[2], and Vincent S. Tseng[1,3]

[1] Department of Computer Science and Information Engineering, National Cheng Kung University, No.1, University Road, Tainan City 701, Taiwan, R.O.C.
[2] Institute of Biomedical Informatics, Center for Systems and Synthetic Biology, National Yang-Ming University, No.155, Sec.2, Linong Street, Taipei, 112 Taiwan, R.O.C.
[3] Institute of Medical Informatics, National Cheng Kung University, No.1, University Road, Tainan City 701, Taiwan, R.O.C.
uchenliu@gmail.com, rtgo@idb.csie.ncku.edu.tw,
hsuancheng@ym.edu.tw, tsengsm@mail.ncku.edu.tw

Abstract. Essential genes are indispensable for an organism's living. These genes are widely discussed, and many researchers proposed prediction methods that not only find essential genes but also assist pathogens discovery and drug development. However, few studies utilized the relationship between gene functions and essential genes for essential gene prediction. In this paper, we explore the topic of essential gene prediction by adopting the association rule mining technique with Gene Ontology semantic analysis. First, we proposed two features named GOARC *(Gene Ontology Association Rule Confidence)* and GOC-BA *(Gene Ontology Classification Based on Association)*, which are used to enhance the classifier constructed with the features commonly used in previous studies. Secondly, we use an association-based classification algorithm without rule pruning for predicting essential genes. Through experimental evaluations and semantic analysis, our methods can not only enhance the accuracy of essential gene prediction but also facilitate the understanding of the essential genes' semantics in gene functions.

Keywords: Data Mining, Gene Ontology, Essential Gene, Association Rule Mining.

1 Introduction

Essential genes are required for sustaining cellular life. A cell will decease if we delete any one of these genes. Owing to this significant reason, they will be the key to understanding the levels of organization of living systems. Furthermore, it will help direct drug development if we can discriminate these genes in pathogens.

Most of previous studies for essential gene prediction were focused on sequence features (such as phyletic retention, gene size, and others) and topological characteristics (of various biological networks including protein-protein interaction network, gene regulatory network, and metabolic network). However, few studies utilized the relationship between gene functions and essentiality for essential gene prediction. As

J. Chen, J. Wang, and A. Zelikovsky (Eds.): ISBRA 2011, LNBI 6674, pp. 49–60, 2011.

we know, it could be difficult since the description of gene functions is complicated and their relationship to gene essentiality is still unclear.

In this paper, we explore the topic of essential gene prediction by adopting association rule mining techniques with Gene Ontology [9] semantic analysis. We proposed two novel features named *GOARC (Gene Ontology Association Rule Confidence)* and *GOCBA (Gene Ontology Classification Based on Association)* and used them to enhance the classifier that utilizes the features reported previously. Besides, we applied the CBA (Classification Based on Associations) [13] algorithm without rule pruning to increase the capability of essential gene prediction. Through experimental evaluations and semantic analysis, it is shown that our methods can not only increase the accuracy of essential gene prediction but also facilitate the understanding of the essential genes' semantics in gene functions. Furthermore, our results could help biologists elucidate the gene functions closely related to gene essentiality.

The remainder of this paper is organized as follows. In Section 2, we give a brief review of the related work. We demonstrate our proposed method for essential gene prediction in Section 3. Section 4 shows the results of performance evaluation and discussion of the discovered association rules. Concluding remarks are made in Section 5.

2   Related Work

In this section we will introduce four studies on essential gene prediction for *Saccharomyces cerevisiae* and some researches related to our work.

2.1   Essential Gene Prediction

Gustafson *et al.* (2006) [7] used feature selection with conditional mutual information [5] to integrate comparative genomics, sequence information, protein-protein interaction network and so on. Afterwards, they used Naïve Bayes classifier based on the Orange machine learning package [4] to predict the essential genes after the feature selection process. Phyletic retention is the most discriminative feature that they utilized in this study. In the analysis result of a later study [12], the gene duplicability which describes the likelihood of a gene having more paralogs has obviously relationship on essential gene.

Seringhaus *et al.* (2006) [15] used the *Saccharomyces cerevisiae* protein sequence information to build an essential gene prediction model named Caveats. This model includes 7 classifier based on WEKA software package [8]. Afterwards, they applied it to the closely related yeast *Saccharomyces mikatae* which is relatively unstudied on gene essentiality. In this study, they proposed CLOSE_STOP_RATIO and RARE_AA_RATIO. Unfortunately, these did not have an obvious effect on the prediction.

Acencio *et al.* (2009) [1] proposed an integrated network of gene interactions that combined the protein-protein interaction network, gene regulation network and metabolism network. Afterwards, they use the high level GO term from GO-Slim and the topological feature on the integrated network of gene interactions to build the classifier. They used a meta-classifier "Vote", a WEKA's implementation of the voting algorithm that combines the output of 8 decision tree classifiers. Base on the experimental result, they found that the number of protein physical interactions, the nuclear localization of proteins and the number of regulating transcription factors are the most critical factors for the essential gene determination. But, this study has some disadvantages. First, not all of biologic network are complete. Besides, the metabolic network did not have an especially good ability for prediction. Secondly, these biologic networks have much regulation information can be discuss on it more than topology. Thirdly, only using some GO terms of GO-Slim is insufficient. Taking in to consider the relation between GO terms and other features may have a strong probability to improve the prediction ability.

Hwang *et al.* (2009) [10] used a lot of topological features based on protein-protein interaction network. Afterwards, they used SVM (support vector machine) to predict the essential gene base on these features. The proposed clique level, essentiality index, neighbors' intra degree and GO common function degree have good prediction ability. Besides, clique level has a similar concept with protein complex. Therefore, there is a strong probability to have a relationship between essential genes and gene functions.

2.2 Association Rule Mining

Association rule mining was first proposed by Agrawal *et al.* [2] [3]. It was used to analyze large databases to discover meaningful hidden patterns and relationships. An association rule is an explication in the form X => Y, where X and Y are different sets of items. X is the LHS (left hand side) and Y is the RHS (right hand side) of the rule. This means that Y can possibly occur where X occurs. There are two crucial thresholds to examine the importance of a rule, minimum support and minimum confidence. The support of a rule is the frequency that X and Y occur together in a transaction. If the frequency is larger than the minimum support, it is a frequent itemset. The confidence of a rule is the frequency that Y occurs when X occurs. If the frequency is larger than the minimum confidence, it becomes an association rule.

2.3 Frequent Closed Itemset Mining

Too many redundant rules are the main disadvantage of association rule mining. Therefore, frequent closed itemset mining is a usual way to solve this problem. An itemset Z is a frequent closed itemset if it is frequent and there exists no proper itemset containing Z with the same support. Therefore, closed itemset mining can obtain simple rules with the same information meaning. Besides, CLOSET [14] and CLOSET+ [16] are popular and efficient algorithms for this purpose.

2.4 Classification Based on Associations

Utilizing association rules to solve the prediction problem was first proposed by Liu *et al.* [13] with CBA (Classification Based on Associations) algorithm. It has two stages: the CAR (classification association rule) discovering and classification model building of CBA algorithm.

In the first stage, CBA utilizes the Apriori-based strategy to discover the frequent itemset from frequent 1 to frequent k which fit with the minimum support threshold. Afterwards CBA utilize the frequent pattern to discover the CARs. Besides, it utilizes the rule pruning strategy of C4.5 to delete the unsuitable rules for the prediction.

In the second stage, we utilize the CBAM1 strategy of CBA to given precedence for each CAR from last stage. The precedence is the kernel in this stage. The first precedence of GOCBA is the rule with higher confidence is more significant. If the rule have the same confidence value, then the rule with higher support is more significant. If the rules still have the same order, then the rule with fewer items in the LHS is more significant. Owing to the rule has less items in the LHS is more easily to be matched. Afterwards, when process the prediction of each unknown gene. It will try to match the rule in the order of precedence.

3   Materials and Methods

In this study, we utilized association rule mining to discover the gene functions that are required for sustaining cellular life. We can understand the essential genes' semantics in gene functions. Furthermore, base on these rules to extract two features named GOARC and GOCBA and appling CBA algorithm for the essential gene prediction. Detailed illustrations of the method are given below.

3.1   GO Association Rule Classifier

We use Gene Ontology for the essential gene prediction, which can be divided in to two stages as follow: "transaction transformation" and "association rule mining".

Transaction transformation: In this stage, we transform the annotated GO term of each gene in to association rule transaction data format. The annotated GO term of each gene is assigned as a transaction. Each annotated GO term in the transaction is assigned as a transaction item. Besides, essentiality for each gene is also assigned as a transaction item in its transaction. For example, suppose we have an annotated information table for each gene as the Table 1 indicates. Besides, we encode annotated GO terms and essential or non-essential as a unique id as the Table 2 indicates. Therefore, we can transform the annotate information table as transaction data format as the Table 3 indicates. For example, YIL100W is an essential gene and annotated in GO term GO:0000910 and GO:0005515 in Table 1. Suppose, Essential, GO:0000910 and GO:0005515 had been encoded as 1, 3 and 7 reference from the encoded table in Table 2. Therefore, YIL100W will be a transaction which has encoded items 1, 3 and 7 as shown in Table 3. And so on, we can transform all the genes in Table 1 into Table 3 as transaction data for the association rule mining process in the next stages.

Table 1. An example of annotate information table

Gene	Annotated GO terms	Class
YIL100W	GO:0000910,GO:0005515	Essential
YNR067C	GO:0005576,GO:0007109,GO:0009277,GO:0030428, GO:0042973	Non-Essential
YBL083C	GO:0003674,GO:0005575,GO:0008150	Non-Essential
YJL152W	GO:0003674,GO:0005575,GO:0008150	Non-Essential
YNR044W	GO:0000752,GO:0009277,GO:0050839	Essential

Table 2. An example of encode table

Item	Item Code
Non- Essential	0
Essential	1
GO:0000910	2
...	...
GO:0050839	n

Table 3. Example of transaction data format

Gene	Transaction Data
YIL100W	1,3,7
YNR067C	0,2,156,333,468,2991
...	...
YNR044W	1,8,156

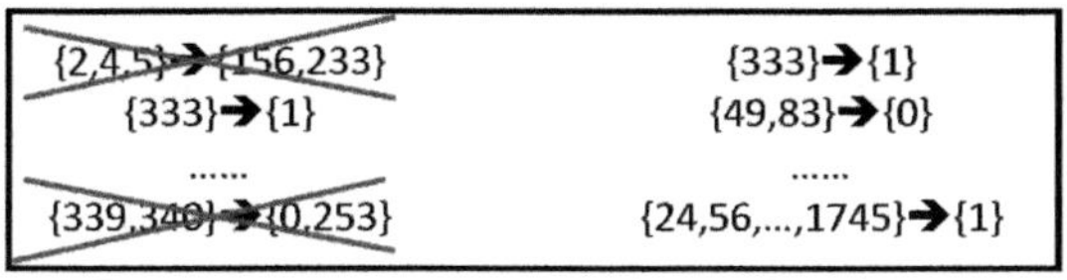

{2,4,5}→{156,233}	{333}→{1}
{333}→{1}	{49,83}→{0}
......	......
{339,340}→{0,253}	{24,56,...,1745}→{1}

Fig. 1. GO association rule extraction example

Association rule mining. In this stage, we utilize the association rule algorithm such as CLOSET+ to discover the close itemsets and association rules. Owing to the main purpose of this stage is utilizing association rules for essential gene prediction. Therefore, we extract the rules whose RHS are only contain essential or non-essential. As the Figure 1 indicates, item code "0" represent the non-essential and item code "1" represent the essential. In the prediction stage, we can use these rules to estimate whether the unknown gene is essential or non-essential. For example, the rule {2,4,5}→{156,233} without contain item code "0" or "1". Besides, the rule {339,340}→{0,253} contains item code "253" beyond "0" and "1". After the association rule mining stage, we can prediction the unknown gene based on these rules.

3.2 Features GOARC and GOCBA

In this paragraph we introduce two features named GOARC and GOCBA extracted from the GO association rules. These can be utilized for many classifiers, for example like SVM. The procedure of GOARC and GOCBA can be divided as "rule filtering", "sort rule set" and "rule matching" three stages as fallow.

Rule filtering. Owing to the main purpose of GOARC and GOCBA feature extraction is utilizing association rue for essential gene prediction. Therefore, for the next stage, we only extract the positive rules (RHS is essential) from the rules set after GO association rule mining (GO Association Rule Classifier) process.

Sort rule set. Because of each gene could possibly be matched with several positive rules in the rule matching stage. Therefore, we must sort all the positive rules depending on the precedence of each rule. In the rule matching stage, we can extract the GOARC or GOCBA feature value from the matched rule which has most precedence. The precedence of sorting is the main difference between GOARC and GOCBA. The first precedence of GOCBA is the rule has higher confidence is more significant. If the rule has the same confidence value, then the rule has higher support is more significant. If the rules still have the same order, then the rule has less items in the LHS is more significant. Owing to the rule has less items in the LHS is more easily to be matched. Besides, the first precedence of GOARC is the rule has more items in the LHS is more significant. If the rule have the same quantity in the LHS, then the rule with higher confidence is more significant. GOARC is preferred to find the specific rules.

Rule matching. In this stage, the predicted gene will compare its encoded annotated GO term with the LHS of each positive rule in order until matched. Furthermore, the GOCBA feature value is obtain from the confidence of first matched rule, as the example in the Figure 2 indicates. Besides, the GOARC feature value is obtain from the matched rule which has the maximum value of formula 1. The matched percentage increase, the predict gene is more fit with the matched rule, as the example in the Figure 3 indicates.

$$GOARC(i) = \max_{j}\{\frac{n_{GO}(Rule_j)}{n_{GO}(Gene_i)} \times Confidence \ of \ Rule_j\} \tag{1}$$

Gene A	<GO:0000003, GO:0000005, GO:0000007, GO:0000143, GO:0000247>	
LHS		**Conf.**
<GO:0000003, GO:0000007>		1
<GO:0000005, GO:0000143, GO:0000247>		1
...		...
<GO:0000003>		0.8
...		...
GOCBA(YAR015W) = 1		

Fig. 2. The example of GOCBA extraction

Gene A	<GO:0000003, GO:0000005, GO:0000007, GO:0000143, GO:0000247>		
LHS		**Conf.**	**GOARC**
<GO:0000005, GO:0000143, GO:0000247>		1	0.6
<GO:0000003, GO:0000007>		1	0.4
<GO:0000003>		0.8	0.16
GOARC(YAR015W) = 0.6			

Fig. 3. The example of GOARC extraction

3.3   Data for Evaluation

The information of *Saccharomyces cerevisiae* like sequence information, GO, GO-Slim and so on are downloaded from *Saccharomyces* Genome Database [2]. The essential gene information of *Saccharomyces cerevisiae* is referenced in previous study [6] [17]. The dataset has 3606 genes, composed of 957 essential and 3606 non-essential ones. On average, each gene was annotated in 4.9 GO terms.

Table 4. Encode table of previous studies

Symbol	Previous works
S	Seringhaus *et al.*'06
G	Gustafson *et al.*'06
H	Hwang et *al.*'09
A	Acencio *et al.*'09

We adopt the LibSVM [6] with RBF kernel as the SVM algorithm. Furthermore, utilize CLOSET+ algorithm that from Illinois Data Mining System [1] as association rule mining method. The support count threshold is defined as 3 (approximate to 0.1%). Besides, minimum confidence is defined as 70%.

We divided 70% of the original data of both target class to train the prediction method, the remainder data is used as testing data to measure the prediction ability. Owing to the number of essential genes is obviously less than non-essential genes in the data. We adopt the down sampling method for the training data in the sampling condition. The process is repeated (sampling, training and testing) 100 times to get the average evaluation result for each prediction method.

Besides, to have a clear evaluation result in the table, we encode each previous study as shown in Table 4. In the evaluation result table, SVM(X) represents the evaluation result of SVM classifier that used the features in that column.

3.4 Evaluation Method

Precision and F-measure (F_1-measure) are the evaluation methods of this research. That is base on confusion matrix as Table 5 shows. The precision value is obtained as the formula 2 indicate. Furthermore, the F-measure is obtained from precision and recall as the formula 4 display. Besides, the recall value is as the formula 3 demonstrate.

$$Precision = \frac{True\ Postive}{True\ Postive + False\ Postive} \tag{2}$$

$$Recall = \frac{True\ Postive}{True\ Postive + False\ Nagtive} \tag{3}$$

$$F_1 = \frac{2 \times Precision \times Recall}{Precision + Recall} \tag{4}$$

Table 5. Confusion matrix

		Condition	
		Positive	**Negative**
Prediction	**Positive**	True Positive	False Positive
	Negative	False Negative	True Negative

Precision is the usual evaluation method on prediction problems to determine the hit ratio of the prediction results. However, as quantity of prediction classes is less than the other class like essential gene data, the precision value is easy to raise. Moreover, recall is the hit ratio of the target. However, as quantity of prediction classes is more than the other class, the precision value is easy to rise. Therefore, only measuring the prediction ability with only one of these has a strong probability to have a biased result. For that reason, F-measure is the method that famous on evaluation precision and recall in the same time. It is a well know evaluation method on prediction problem. Therefore, F-measure will be the main evaluation method to discover large and accurate essential gene in this study. The precision that previous studies used will be another evaluation method.

4   Results and Discussion

4.1   Evaluation of Individual Features

Phyletic retention (PR) is the most discriminative feature in the past research on essential gene prediction [7] [10]. In this paragraph, we will compare the prediction ability of GOCBA, GOARC with PR. As the evaluation results in Table 6 indicate, GOCBA and GOARC both had better prediction power than PR, no matter what kind of estimate method and data sampling condition were used. Besides, GOCBA performed better than GOARC in most cases.

Table 6. Prediction capability of individual features

	F-measure (%)		Precision (%)	
	Original	**Down Sampling**	**Original**	**Down Sampling**
SVM(PR)	37.46	51.33	58.00	38.86
SVM(GOARC)	60.63	57.01	70.41	45.52
SVM(GOCBA)	59.28	61.14	72.99	54.45

4.2   Performance Enhancement by GOARC and GPCBA

In this paragraph, we will demonstrate the result of the SVM classifier that is based on the features that previous study proposed and our added features respectively. F-measure evaluation results are as Table 7 indicates, GOARC enhances the classifier based on any previous features obviously has the best result in both data conditions. Besides, GOCBA has the better result in most situations. The precision evaluation has the similar result. GOARC enhances the classifier has the best precision in any situation as the Table 8 indicates. GOARC is out standing on classifier enhancement.

Table 7. Enhancement of GOCBA and GOARC features as evaluated by F-measure

F-measure (%)	Original				Down Sampling			
	S	G	H	A	S	G	H	A
SVM(X)	6.63	42.19	50.93	20.18	43.71	56.77	58.25	47.02
SVM(X+GOARC)	56.45	57.43	57.73	56.41	57.98	63.94	62.63	58.70
SVM(X+GOCBA)	62.79	62.23	63.07	62.37	56.30	56.18	60.53	56.27

Table 8. Enhancement of GOCBA and GOARC features as evaluated by precision

Precision (%)	Original				Down Sampling			
	S	G	H	A	S	G	H	A
SVM(X)	57.26	66.28	71.25	58.67	35.53	45.27	46.75	36.10
SVM(X+GOARC)	75.78	76.30	75.90	76.18	51.16	55.86	54.88	50.89
SVM(X+GOCBA)	67.45	66.21	68.68	66.17	42.83	42.63	48.72	42.72

4.3 Classifier Evaluation

In this study, we applied CBA algorithm without rule pruning for predicting essential genes. In this paragraph, we will compare prediction ability with SVM base on previous features. As shown in Table 9 and Table 10, our CBA classifier obviously has better ability in any situation.

Table 9. Performance of CBA classifier as evaluated by F-measure

F-measure (%)	Original				Down Sampling			
	S	G	H	A	S	G	H	A
SVM(X)	6.63	42.19	50.93	20.18	43.71	56.77	58.25	47.02
CBA	56.70	56.70	56.70	56.70	62.31	62.31	62.31	62.31

Table 10. Performance of CBA classifier as evaluated by precision

Precision (%)	Original				Down Sampling			
	S	G	H	A	S	G	H	A
SVM(X)	57.26	66.28	71.25	58.67	35.53	45.27	46.75	36.10
CBA	81.85	81.85	81.85	81.85	54.71	54.71	54.71	54.71

4.4 Discussion of GO Association Rules

In this paragraph, we will focus on the rules discovered by CBA and discuss the GO terms associated with essential genes as revealed by these rules. We list part of the positive rules in Table 11. Each row is a GO association rule. The first column is the annotated GO term for essential genes. Support count represents the number of essential genes that match a rule. Confidence represents the percentage of essential genes within all the genes that match the LHS of a rule.

Table 11. Positive GO association rules discovered by CBA

Annotated GO Terms	Support Count	Conf. (%)
GO:0000472, GO:0032040, GO:0005730, GO:0000480, GO:0000447	19	100
GO:0000472, GO:0005730, GO:0000480, GO:0000447, GO:0030686	18	100
GO:0006364, GO:0005730, GO:0030686	17	100
GO:0032040, GO:0005730	34	97.14
GO:0000472, GO:0005730, GO:0000480, GO:0000447	24	92.31
GO:0005730, GO:0030686	41	91.11

The location of gene or the location of process had been used for essential gene prediction in previous studies [1] [7] [15]. Acencio *et al.* [1] had suggested that a gene annotated with the "nucleus" GO term tend to be essential. Since "nucleolus" (GO: 000573) is a part of "nucleus" (GO: 0005634) in the ontology, we listed all the positive rules with high confidence that contain "nucleolus" (GO: 000573) in Table 11. Following these rules, we listed all the associated GO terms with "nucleolus" (GO: 000573) in Table 12. We found that genes annotated to "endonucleolytic cleavage" or "preribosome" on the "nucleolus" are very likely to be essential. Most of the discovered GO association rules by CBA are either consistent with previous studies or make biological sense. Thus, the proposed method could not only predict essential genes, but also reveal the GO semantics closely related to essential genes.

Table 12. Associated GO terms with "nucleolus" (GO: 000573) following the rules listed in Table 11

GO Term	Description	Domain
GO:0000447	endonucleolytic cleavage in ITS1 to separate SSU-rRNA from 5.8S rRNA and LSU-rRNA from tricistronic rRNA transcript (SSU-rRNA, 5.8S rRNA, LSU-rRNA)	BP
GO:0000462	maturation of SSU-rRNA from tricistronic rRNA transcript (SSU-rRNA, 5.8S rRNA, LSU-rRNA)	BP
GO:0000472	endonucleolytic cleavage to generate mature 5'-end of SSU-rRNA from (SSU-rRNA, 5.8S rRNA, LSU-rRNA)	BP
GO:0000480	endonucleolytic cleavage in 5'-ETS of tricistronic rRNA transcript (SSU-rRNA, 5.8S rRNA, LSU-rRNA)	BP
GO:0006364	rRNA processing	BP
GO:0030686	90S preribosome	CC
GO:0030687	preribosome, large subunit precursor	CC
GO:0032040	small-subunit processome	CC
GO:0042273	ribosomal large subunit biogenesis	BP

5 Conclusions and Future Work

We have proposed a novel method for connecting gene functions (GO terms) to essential gene prediction by utilizing association rules. The proposed new features, GOARC and GOCBA, could enhance the prediction performance significantly. Using GO association rules, CBA classifier showed great performance. Besides, the discovered GO association rules not only are consistent with known biological knowledge, but also could reveal the GO semantics related to essential genes. Our results may provide deeper insights for biologists on the essential gene researches. In the future, we will utilize the hierarchical structure between GO terms to improve the prediction accuracy and integrate the rules with similar functions in GO.

References

1. Acencio, M.L., Lemke, N.: Towards the prediction of essential genes by integration of network topology, cellular localization and biological process information. BMC Bioinformatics 10, 290 (2009)
2. Agrawal, R., Imieliński, T., Swami, A.: Mining association rules between sets of items in large databases. In: Proceedings of the 1993 ACM SIGMOD International Conference on Management of Data, Washington, D.C., United States, pp. 207–216 (1993)
3. Agrawal, R., Srikant, R.: Fast Algorithms for Mining Association Rules in Large Databases. In: Proceedings of the 20th International Conference on Very Large Data Bases, pp. 487–499 (1994)
4. Demšar, J., Zupan, B., Leban, G., et al.: Orange: From experimental machine learning to interactive data mining. In: Boulicaut, J.-F., Esposito, F., Giannotti, F., Pedreschi, D. (eds.) PKDD 2004. LNCS (LNAI), vol. 3202, pp. 537–539. Springer, Heidelberg (2004)
5. Fleuret, F.: Fast Binary Feature Selection with Conditional Mutual Information. J. Mach. Learn. Res. 5, 1531–1555 (2004)
6. Giaever, G., Chu, A.M., Ni, L., et al.: Functional profiling of the *Saccharomyces cerevisiae* genome. Nature 418, 387–391 (2002)
7. Gustafson, A.M., Snitkin, E.S., Parker, S.C., et al.: Towards the identification of essential genes using targeted genome sequencing and comparative analysis. BMC Genomics 7, 265 (2006)
8. Hall, M., Frank, E., Holmes, G., et al.: The WEKA data mining software: an update. SIGKDD Explor. Newsl. 11, 10–18 (2009)
9. Harris, M.A., Clark, J., Ireland, A., et al.: The Gene Ontology (GO) database and informatics resource. Nucleic Acids Res. 32, D258-D261 (2004)
10. Hwang, Y.C., Lin, C.C., Chang, J.Y., et al.: Predicting essential genes based on network and sequence analysis. Mol. Biosyst. 5, 1672–1678 (2009)
11. Kittler, J., Hatef, M., Duin, R.P.W., et al.: On combining classifiers. IEEE Transactions on Pattern Analysis and Machine Intelligence 20, 226–239 (1998)
12. Liang, H., Li, W.H.: Gene essentiality, gene duplicability and protein connectivity in human and mouse. Trends Genet. 23, 375–378 (2007)
13. Liu, B., Hsu, W., Ma, Y.: Integrating Classification and Association Rule Mining. In: Proceedings of the Fourth International Conference on Knowledge Discovery and Data Mining, New York City, New York, USA, pp. 80–86 (1998)

14. Pei, J., Han, J., Mao, R.: CLOSET: An Efficient Algorithm for Mining Frequent Closed Itemsets. In: ACM SIGMOD Workshop on Research Issues in Data Mining and Knowledge Discovery, pp. 21–30 (2000)
15. Seringhaus, M., Paccanaro, A., Borneman, A., et al.: Predicting essential genes in fungal genomes. Genome Res. 16, 1126–1135 (2006)
16. Wang, J., Han, J., Pei, J.: CLOSET+: searching for the best strategies for mining frequent closed itemsets. In: Proceedings of the ninth ACM SIGKDD International Conference on Knowledge Discovery and Data Mining, Washington, D.C., pp. 236–245 (2003)
17. Winzeler, E.A., Shoemaker, D.D., Astromoff, A., et al.: Functional characterization of the S. cerevisiae genome by gene deletion and parallel analysis. Science 285, 901–906 (1999)
18. The IlliMine Project, http://illimine.cs.uiuc.edu
19. Saccharomyces Genome Database, http://downloads.yeastgenome.org/
20. LIBSVM: a library for support vector machines,
    http://www.csie.ntu.edu.tw/~cjlin/libsvm

High-Performance Blob-Based Iterative Reconstruction of Electron Tomography on Multi-GPUs

Xiaohua Wan[1,2], Fa Zhang[1], Qi Chu[1,2], and Zhiyong Liu[1]

[1] Institute of Computing Technology
[2] Graduate University, Chinese Academy of Sciences
Beijing, China
{wanxiaohua,chuqi,zyliu}@ict.ac.cn, zf@ncic.ac.cn

Abstract. Three-dimensional (3D) reconstruction of electron tomography (ET) has emerged as a leading technique to elucidate the molecular structures of complex biological specimens. Blob-based iterative methods are advantageous reconstruction methods for 3D reconstruction of ET, but demand huge computational costs. Multiple Graphic processing units (multi-GPUs) offer an affordable platform to meet these demands, nevertheless, are not efficiently used owing to a synchronous communication scheme and the limited available memory of GPUs. We propose a multilevel parallel scheme combined with an asynchronous communication scheme and a blob-ELLR data structure. The asynchronous communication scheme is used to minimize the idle GPU time. The blob-ELLR data structure only needs nearly 1/16 of the storage space in comparison with ELLPACK-R (ELLR) data structure and yields significant acceleration. Experimental results indicate that the multilevel parallel scheme allows efficient implementations of 3D reconstruction of ET on multi-GPUs, without loss any resolution.

Keywords: electron tomography (ET), three-dimensional (3D) reconstruction, iterative methods, blob, multi-GPUs.

1 Introduction

In biosciences, electron tomography (ET) uniquely enables the study of complex cellular structures, such as cytoskeletons, organelles, viruses and chromosomes [1]. From a set of projection images taken from a single individual specimen, 3D structure can be obtained by means of tomographic reconstruction algorithms [2]. Blob-based iterative methods (e.g., Simultaneous Algebraic Reconstruction Technique (SART) [3] and Simultaneous Iterative Reconstruction Technique (SIRT) [4]) are attractive reconstruction methods for ET in terms of robustness against noise [5], but have not been extensively used due to their high computational cost [6]. The need for high resolution makes ET of complex biological specimens use large projection images, which also yields large reconstructed

J. Chen, J. Wang, and A. Zelikovsky (Eds.): ISBRA 2011, LNBI 6674, pp. 61–72, 2011.
© Springer-Verlag Berlin Heidelberg 2011

volumes after an extensive use of computational resources and considerable processing time [7]. Graphics processing units (GPUs) offer an attractive alternative platform to address such computational requirements in terms of the high peak performance, cost effectiveness, and the availability of user-friendly programming environments, e.g. NVIDIA CUDA [8]. Recently, several advanced GPU acceleration frameworks have been proposed to allow 3D ET reconstruction to be performed on the order of minutes [9]. These parallel reconstructions on GPUs only adopt traditional voxel basis functions which are less robust than blob basis functions under noisy situations. Our previous work focuses on the blob-based iterative reconstruction on single GPU. However, the blob-based iterative reconstruction on single GPU is still time-consuming. Single GPU cannot meet the requirements of the computational resources and the memory storage of 3D-ET reconstruction since the size of the projection images increases constantly. The architectural notion of a CPU serviced by multi-GPUs is an attractive way of increasing the power of computations and the storage of memory.

Achieving the blob-based iterative reconstruction on multi-GPUs can be challenging: first, because of the overlapping nature of blobs, the use of blobs as basis functions needs the communication between multi-GPUs during the process of iterative reconstructions. CUDA provides a synchronous communication scheme to handle the communication between GPUs efficiently [8]. But the downside of the synchronous communication is that each GPU must stop and sit idle while data is exchanged. The GPU sits idle is a waste of resources which has a negative impact on performance. Besides, as data collection strategies and electron detectors improve, the memory demand of a sparse weighted matrix involved with blob-based iterative methods in ET reconstruction rapidly increases. Due to the limited available memory of GPUs, storing such a large sparse matrix is extremely difficult for most GPUs. Thus, computing the weighted matrix on the fly is an efficient alternative to storing the matrix in the previous GPU-based ET implementations [10]. But it could bring the redundant computations since the weighted matrix has to be computed for two times in each iteration.

To address the problems of blob-based iterative ET reconstruction on multi-GPUs, in this paper, we make the following contributions: first, we present a multilevel parallel strategy for blob-based iterative reconstructions of ET on multi-GPUs, which can achieve higher speedups than the parallel reconstruction on single GPU. Second, we develop an asynchronous communication scheme on multi-GPUs to minimize idle GPU time by asynchronously overlapping communications with computations. Finally, a data structure named blob-ELLR using several symmetry optimization techniques is developed to significantly reduce the storage space of the weighted matrix. It only needs nearly 1/16 of the storage space in comparison with ELLPACK-R (ELLR). Also, the blob-ELLR format can achieve optimal coalesced global memory access, and is suitable for 3D-ET reconstruction algorithms on multi-GPUs. Furthermore, we implement all the above techniques on a NVIDIA GeForce GTX 295, and experimental results show that the parallel strategy greatly reduces memory requirements and exhibits a significant acceleration, without loss any resolution.

The rest of the paper is organized as follows: Section 2 reviews relevant backgrounds both on reconstruction algorithms and on GPU hardware. Section 3 presents the multilevel parallel strategy of 3D-ET reconstruction on multi-GPUs. Section 4 introduces the asynchronous communication scheme between multi-GPUs. Section 5 describes the blob-ELLR data structure. Section 6 shows and analyzes the experimental results in detail. Then we summarize the paper and present our future work in Section 7.

2   Related Work

In ET, the projection images are acquired from a specimen through the so-called single-axis tilt geometry. The specimen is tilted over a range, typically from $-60°$ (or $-70°$) to $+60°$ (or $+70°$) due to physical limitations of microscopes, at small tilt increments ($1°$ or $2°$). An image of the same object area is recorded at each tilt angle and then the 3D reconstruction of the specimen is obtained from a set of projection images with blob-based iterative methods. In this section, we give a brief overview of blob-based iterative reconstruction algorithms, describe a kind of iterative method called SIRT, and present a GPU computational model.

2.1   Blob-Based Iterative Reconstruction Methods

Iterative methods represent 3D volume f as a linear combination of a limited set of known and fixed basis functions b_j, with appropriate coefficients x_j, i.e.

$$f(r, \phi) \approx \sum_{j=1}^{N} x_j b_j (r, \phi).$$ (1)

The use of blob basis functions provides iterative methods with better performance than traditional voxel basis functions due to their overlapping nature [11]. The basis functions that developed in [11] are used for the choice of the parameters of the blob in our algorithm (e.g., the radius of the blob is 2). In 3D-ET, the model of the image formation process is expressed by the following linear system:

$$p_i \approx \sum_{j=1}^{N} w_{ij} x_j,$$ (2)

where p_i denotes the ith measured image of f and w_{ij} the value of the ith projection of the jth basis function. Under such a model, the element w_{ij} can be calculated according to the projected procedure as follows:

$$w_{ij} = 1 - (rf_{ij} - int(rf_{ij})), \quad rf_{ij} = projected(x_j, \theta).$$ (3)

where rf_{ij} is the projected value of the pixel x_j with an angle θ_i. Here, W is defined as a sparse matrix with M rows and N columns and w_{ij} is the element of W. In general, the storage demand of the weighted matrix W rapidly increases as the size and the number of projection images increase. It is hard to store such a large matrix in GPUs due to the limited memory of GPUs.

Simultaneous iterative reconstruction technique (SIRT) begins with an initial $X^{(0)}$ obtained by the back projection technique (BPT) and repeats the iterative processes [4]. In iterations, the residuals, i.e. the differences between the actual projections P and the computed projections P' of the current approximation $X^{(k)}$ (k is the iterative number), are computed and then $X^{(k)}$ is updated by the backprojection of these discrepancies. The SIRT algorithm is typically written by the following expression:

$$x_j^{(k+1)} = x_j^{(k)} + \frac{1}{\sum_{i=1}^{M} w_{ij}} \sum_{i=1}^{M} \frac{w_{ij}(p_i - \sum_{h=1}^{N} w_{ih} x_h^{(k)})}{\sum_{h=1}^{N} w_{ih}}. \tag{4}$$

2.2   GPU Computation Model

Our algorithm is based on NIVIDIA GPU architecture and compute unified device architecture (CUDA) programming model. GPU is a massively multi-threaded data-parallel architecture, which contains hundreds of scalar processors (SPs). Eight SPs are grouped into a Streaming Multiprocessor (SM), and SPs in the same SM execute instructions in a Single Instruction Multiple Thread (SIMT) fashion [8]. During execution, 32 threads from a continuous section are grouped into a warp, which is the scheduling unit on each SM. NVIDIA provides the programming model and software environment of CUDA. CUDA is an extension to the C programming language. A CUDA program consists of a host program that runs on CPU and a kernel program that executes on GPU itself. The host program typically sets up data and transfers it to and from the GPU, while the kernel program processes that data. Kernel, as a program on GPUs, consists of thousands of threads. Threads have a three-level hierarchy: grid, block, thread. A grid is a set of blocks that execute a kernel, and each block consists of hundreds of threads. Each block can only be assigned to and executed on one SM. CUDA provides a synchronous communication scheme (i.e. cudaThreadSynchronize()) to handle the communication between GPUs. With the synchronous scheme, all of threads on GPUs must be blocked until the data communication has been completed. CUDA devices use several memory spaces including global, local, shared, texture, and registers. Of these different memory spaces, global and texture memory are the most plentiful. Global memory loads and stores by a half warp (16 threads) are coalesced in as few as one transaction (or two transactions in the case of 128-bit words) when certain access requirements are met. Coalesced memory accesses deliver a much higher efficient bandwidth than non-coalesced accesses, thus greatly affecting the performance of bandwidth-bound programs.

3   Multilevel Parallel Strategy for Blob-Based Iterative Reconstruction

The processing time of 3D reconstruction with blob-based iterative methods remains a major challenge due to large reconstructed data volume in ET. So parallel computing on multi-GPUs is becoming paramount to cope with the computational requirement. We present a multilevel parallel strategy for blob-based iterative reconstruction and implement it on the OpenMP-CUDA architecture.

3.1 Coarse-Grained Parallel Scheme Using OpenMP

In the first level of the multilevel parallel scheme, a coarse-grained parallelization is straightforward in line with the properties of ET reconstruction. The single-tilt axis geometry allows data decomposition into slabs of slices orthogonal to the tilt axis. For this decomposition, the number of slabs equals the number of GPUs, and each GPU reconstructs its own slab. Consequently, the 3D reconstruction problem can be decomposed into a set of 3D slabs reconstruction sub-problems. However, the slabs are interdependent due to the overlapping nature of blobs. Therefore, each GPU has to receive a slab composed of its corresponding unique slices together with additional redundant slices reconstructed in neighbor slabs. The number of redundant slices depends on the blob extension. In a slab, the unique slices are reconstructed by the corresponding GPU and require information provided by the redundant slices. During the process of 3D-ET reconstruction, each GPU has to communicate with other GPUs for the additional redundant slices.

We have implemented the 3D-ET reconstruction based on a GeForce GTX 295 which consists of two GeForce GTX 200 GPUs. Thus the first level parallel strategy makes use of two GPUs to perform the coarse-grained parallelization of the reconstruction. As shown in Fig. 1, the 3D volume data is halved into two slabs, and each slab contains its unique slices and a redundant slice. According to the shape of the blob adopted (the blob radius is 2), only one redundant slice is included in each slab. Each slab is assigned to and reconstructed on each individual GTX 200 on GTX 295 in parallel. Certainly, the parallel strategy can be applied on GPU clusters (e.g. Tesla-based cluster). In a GPU cluster, the number of slabs equals the number of GPUs for the decomposition described above. A shared-memory parallel programming scheme (OpenMP) is employed to fork two threads to control the separated GPU. Each slab is reconstructed by each parallel thread on each individual GPU. Consequently, the partial results attained by GPUs are combined to complete the final result of the 3D reconstruction.

3.2 Fine-Grained Parallel Scheme Using CUDA

In the second level of the multilevel parallel scheme, 3D reconstruction of one slab, as a fine-grained parallelization, is implemented with CUDA on each GPU. In the 3D reconstruction of a slab, the generic iterative process is described as follows:

- *Initialization:* compute the matrix W and make a initial value for $X^{(0)}$ by BPT;
- *Reprojection:* estimate the computed projection data P' based on the current approximation X;
- *Backprojection:* backproject the discrepancy ΔP between the experimental and calculated projections, and refine the current approximation X by incorporating the weighted backprojection ΔX.

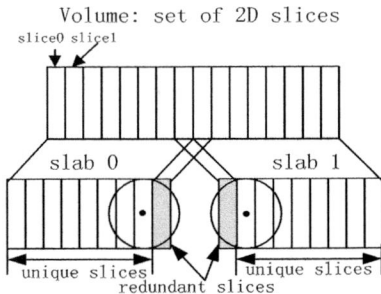

Fig. 1. Coarse-grained parallel scheme using blob. 3D volume is decomposed into slabs of slices. The number of slabs equals the number of GPUs. Each GPU receives a slab. Each slab includes a set of unique slices (in white) and an additional redundant slice (in gray) according to the shape of the blob.

4   Asynchronous Communication Scheme

As described above in the multilevel parallel scheme, there must be two communications between neighbor GPUs during one iterative reconstruction. One is to exchange computed projections of redundant slices after reprojection. The other is to exchange reconstructed pixels of redundant slices after backprojection. CUDA provides a synchronous communication scheme (i.e. cudaThreadSynchronize()) to handle the communication between GPUs. With the synchronous communication scheme, GPUs must sit idle while data is exchange. The GPU sits idle is a waste of resources which has a negative impact on performance.

To minimize the idle-time, we develop an asynchronous communication scheme in which different streams are used to perform asynchronous GPU execution and CPU-GPU memory access. The communication scheme splits GPU execution and memory copies into separate streams. Execution in one stream can be performed at the same time as a memory copy from another. As shown in Fig. 2, in one slab reconstruction, Reprojection of the redundant slices, memory copies and Backprojection of the redundant slices are performed in one stream. The executions (i.e. Reprojection and Backprojection) of the unique slices are performed in the other stream. This can be extremely useful for reducing GPU idle time. In section 6, we will give experiments to analyze in detail.

```
Decidemap<<< >>> ;
BPT<<<>>> for every initial pixel;
    for k in K iterations
        Reprojection<<<>>> for computed projections of redundant slices (Stream 1);
        Reprojection<<<>>> for computed projections of unique slices (Stream 2);
        copy computed projections of redundant slices from own GPU to host (Stream 1);
        copy computed projections of redundant slices from host to neighbor GPU (Stream 1);
        Backprojection<<<>>> for reconstructed pixel of redundant slices (Stream 1);
        Backprojection<<<>>> for reconstructed pixel of unique slices (Stream 2);
        copy reconstructed pixel of redundant slices from own GPU to host (Stream 1);
        copy reconstructed pixel of redundant slices from host to neighbor GPU (Stream 1);
```

Fig. 2. Pseduo code for a slab reconstruction with the asynchronous communication scheme

5   Blob-ELLR Format with Symmetric Optimization Techniques

In the parallel blob-based iterative reconstruction, another problem is the lack of memory on GPUs for the sparse weighted matrix. Recently, several data structures have been developed to store sparse matrices. Compressed row storage (CRS) is the most extended format to store the sparse matrix on CPUs [12]. ELLPACK can be considered as an approach to outperform CRS [13]. Vazquez et al. has proved that a variant of the ELLPACK format called ELLPACK-R (ELLR) is more suited for the sparse matrix data structure on GPUs [14]. ELLR consists of two arrays, $A\,[]$ and $I\,[]$ of dimension $N \times MaxEntriesbyRows$, and an additional N-dimensional integer array called $rl\,[]$ is included in order to store the actual number of nonzeroes in each row [14][15]. With the size and number of projection images increasing, the memory demand of the sparse weighted matrix rapidly increases. The weighted matrix involved is too large to load into most of GPUs due to the limited available memory, even with the ELLR data structure.

In our work, we present a data structure named blob-ELLR with several geometric related symmetry relationships. The blob-ELLR data structure decreases the memory requirement and then accelerates the speed of ET reconstruction on GPUs. As shown in Fig. 3, the maximum number of the rays related to each pixel is four on account of the radius of the blob (i.e., $a = 2$). To store the weighted matrix W, the blob-ELLR includes two 2D arrays: one float $A\,[]$ to save the entries, and the other integer $I\,[]$ to save the columns of every entry (see Fig. 3 middle). Both arrays are of dimension $(4B) \times N$, where N is the number of columns of W and $4B$ is the maximum number of nonzeroes in the columns (B is the number of the projection angles). Because the percentage of zeros is low in the blob-ELLR, it is not necessary to store the actual number of nonzeroes in each column. Thus the additional integer array $rl\,[]$ is not included in the blob-ELLR. Although the blob-ELLR without symmetric techniques can reduce the storage of the sparse matrix W, the number of $(4B) \times N$ is rather large especially when the number of N increases rapidly. The optimization takes advantage of the symmetry relationships as follows:

- *Symmetry 1:* Assume that the jth column elements of the matrix W in each view are w_{0j}, w_{1j}, w_{2j} and w_{3j}. The relationship among the adjacent column elements is:

$$w_{0j} = 1 + w_{1j}; w_{2j} = 1 - w_{1j}; w_{3j} = 2 - w_{1j}. \tag{5}$$

So, only w_{1j} is stored in the blob-ELLR, whereas the other elements are easily computed based on w_{1j} . This scheme can reduce the storage spaces of A and I to 25%.

- *Symmetry 2:* Assume that a point (x, y) of a slice is projected to a point r ($r = project(x, y, \theta)$) in the projection image corresponding to the tilt angle θ and project (x, y, θ) is shown as follows:

$$projection(x, y, \theta) = x \cos \theta + y \sin \theta \tag{6}$$

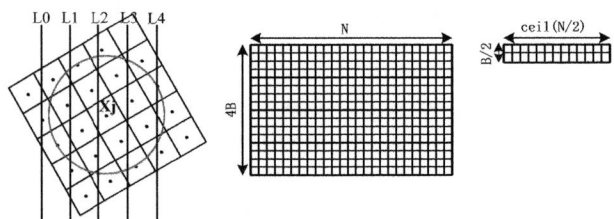

Fig. 3. Blob-ELLR format. In the blob (the radius $a = 2$), a projected pixel x_j contributes to four neighbor projection rays ($L1$, $L2$, $L3$ and $L4$) using only one view (left). The nonzeroes of W are stored in a 2D array A of dimension $(4B) \times N$ in blob-ELLR without symmetric techniques (middle). The symmetric optimization techniques are exploited to reduce the storage space of A to almost 1/16 of original size (right).

It is easy to see that the point $(-x, -y)$ of a slice is then projected to a point $r1$ ($r1 = -r$) in the same tilt angle θ. The weighted value of the point $(-x, -y)$ can be computed according to that of the point (x, y). Therefore, it is not necessary to store the weighted value of almost a half of the points in the matrix W so that the space requirements for A and I are further reduced by nearly 50%.

– *Symmetry 3:* In general, the tilt angles used in ET are halved by 0°. Under the condition, a point $(-x, y)$ with a tilt angle $-\theta$ is projected to a point $r2$ ($r2 = -r$). Therefore, the projection coefficients are shared with the projection of the point (x, y) with the tilt angle θ. This further reduces the storage spaces of A and I by nearly 50% again.

With the three symmetric optimizations mentioned above, the size of the storage for two arrays in the blob-ELLR is almost $(B/2) \times (N/2)$ reducing to nearly 1/16 of original size.

6   Result

In order to evaluate the performance of the multilevel parallel strategy, the blob-based iterative reconstructions of the caveolae from the porcine aorta endothelial (PAE) cell have been performed. Three different experimental datasets are used (denoted by small-sized, medium-sized, large-sized) with 56 images of 512×512 pixels, 112 images of 1024×1024 pixels, and 119 images of 2048×2048 pixels, to reconstruct tomograms of $512 \times 512 \times 190$, $1024 \times 1024 \times 310$ and $2048 \times 2048 \times 430$ respectively. All the experiments are carried out a machine running Ubuntu 9.10 32-bit with a CPU based on Intel Core 2 Q8200 at 2.33GHz, 4GB RAM and 4MB L2 cache, and a NVIDIA GeForce GTX 295 card including two GPUs, each GPU with 30 SMs of 8 SPs (i.e. 240 SPs) at 1.2GHz, 896MB of memory and compute capability 1.3, respectively. To clearly evaluate the performance of the asynchronous communication scheme and the blob-ELLR data structure respectively, we have performed two sets of experiments. The details of the experiments are introduced below.

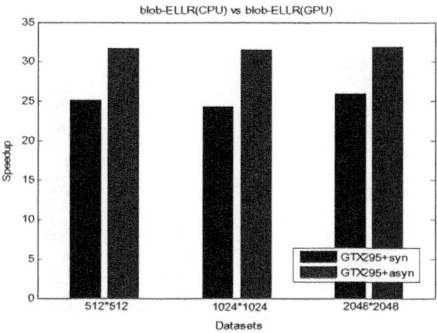

Fig. 4. Speedup factors showed by both implementations on GTX295 (with synchronous and asynchronous communication scheme respectively) over the reconstructions on the CPU

In the first set of experiments, we have implemented and compared the blob-based iterative reconstruction of the three experimental datasets using two methods: multi-GPUs with the synchronous communication scheme (named $GTX295 + syn$), and multi-GPUs with the asynchronous communication scheme respectively (named $GTX295 + asyn$). In the experiments, the blob-ELLR developed in our work is adopted to storage the weighted matrix in the reconstruction. Fig. 4 shows the speedups using the two methods for all the three datasets. As shown in Fig. 4, the speedups of $GTX295 + asyn$ are larger than those of $GTX295 + syn$ for the three experimental datasets. The asynchronous communication scheme provides the better performance than the synchronous scheme for the reason of asynchronous overlapping of communications and computations.

In the second set of experiments, to compare the blob-ELLR data structure with other methods used for the weighted matrix, we have implemented the blob-based iterative reconstructions where the weighted matrices are computed on the fly (named standard matrix), pre-computed and stored with ELLR (named ELLR matrix), pre-computed and stored with blob-ELLR (named blob-ELLR matrix) respectively. Fig. 5 shows the memory demanded by the sparse data structure (i.e. ELLR and blob-ELLR on the GPU respectively). In general, the requirements rapidly increase with the dataset size, approaching 3.5 G in the large dataset. This amount turns out to be a problem owing to the limited memory of GPUs. The upper boundary imposed by the memory available in GTX295 precludes addressing problem sizes requiring more than 896 MB of memory. However, in the blob-ELLR matrix structure, three symmetry relationships can greatly decrease the memory demands and make them affordable on GPUs.

In order to estimate the performance of the matrix methods (i.e. ELLR matrix and blob-ELLR matrix), the speedup factors against the standard method are showed in Fig. 6. From Fig. 6(a), we can see that the ELLR matrix method

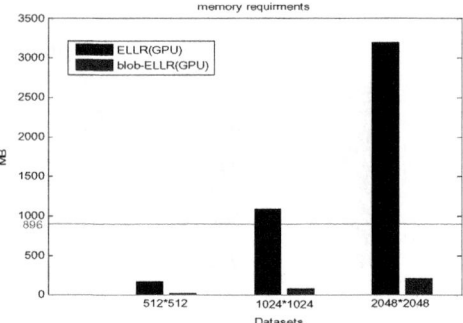

Fig. 5. Memory requirements of the different implementations for the datasets. The limit of 896 MB is imposed by the memory available in the GPU used in the work. The use of the blob-ELLR data structure reduces the memory demands, making most of the problems affordable.

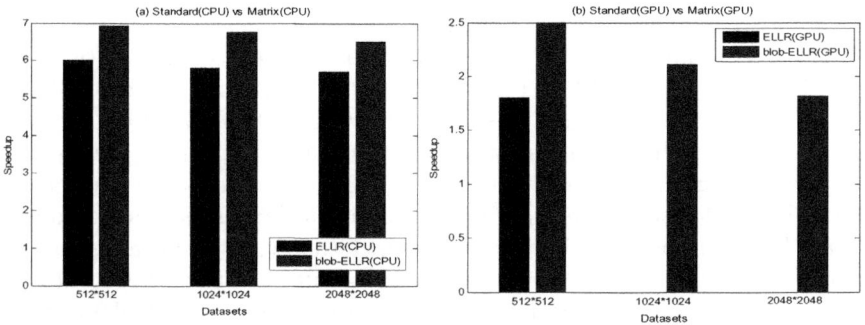

Fig. 6. Speedup factors showed by the different matrix methods (ELLR and blob-ELLR) over the standard method on the CPU (*a*) and the GPU (*b*)

exhibits acceleration factors approaching to 6×, and the blob-ELLR matrix method obtains higher speedup factors almost 7× on the CPU. In order to compare the blob-ELLR matrix method with the ELLR matrix and standard methods on the GPU, the acceleration factors of the matrix methods over the standard method on the GPU are presented in Fig. 6(b). It is clearly seen that the blob-ELLR matrix method yields better speedups than the ELLR matrix method on the GPU. Fig. 7 compares the speedup factors of different methods on the GPU vs the standard method on the CPU. The blob-ELLR matrix method exhibits excellent acceleration factors compared with the other methods. In the standard method, the speedup is almost 100× for three datasets. In the ELLR matrix method, the speedup is almost 160× for the first dataset. And in the case of the blob-ELLR matrix method, the acceleration factors increase and reach up to 200× for three datasets.

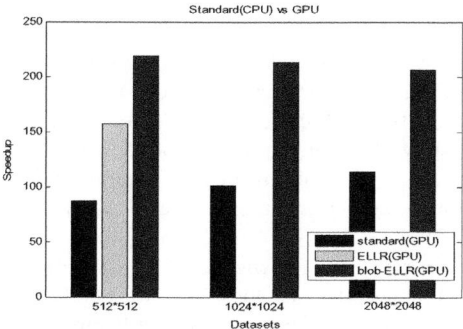

Fig. 7. Speedup factors derived from the different methods (the standard, the ELLR matrix and the blob-ELLR matrix methods) on the GPU compared to the standard method on the CPU

7 Conclusion

ET allows elucidation of the molecular architecture of complex biological specimens. Blob-based iterative methods yield better results than other methods, but are not used extensively in ET because of their huge computational demands. Multi-GPUs have emerged as powerful platforms to cope with the computational requirements, but have the difficulties due to the synchronous communication and limited memory of GPUs. In this work, we present a multilevel parallel scheme combined with an asynchronous communication scheme and a blob-ELLR data structure to perform high-performance blob-based iterative reconstruction of ET on multi-GPUs. The asynchronous communication scheme is used to minimize the idle GPU time, and the blob-ELLR structure only needs nearly 1/16 of the storage space in comparison with the ELLR storage structure and yields significant acceleration compared to the standard and ELLR matrix methods. Also, the multilevel parallel strategy can produce results of 3D-ET reconstruction without loss any resolution. In the future work, we will further investigate and implement the multilevel parallel scheme on a many-GPU cluster. The asynchronous communication scheme described in this work performs communication between GPUs using CPU memory, thus the communication latency is relatively low. In a many-GPU cluster, the proper use of the asynchronous communication scheme is vital to fully utilize computing resources.

Acknowledgments. We would like to thank Prof. Fei Sun and Dr. Ka Zhang in Institute of biophysics for providing the experimental datasets. Work supported by grants National Natural Science Foundation for China (60736012 and 61070129); Chinese Academy of Sciences knowledge innovation key project (KGCX1-YW-13).

References

1. Frank, J.: Electron Tomography: Methods for Three-dimensional Visualization of Structures in the Cell, 2nd edn. Springer, New York (2006)
2. Herman, G.T.: Image Reconstruction from Projections: the Fundamentals of Computerized Tomography, 2nd edn. Springer, London (2009)
3. Andersen, A.H., Kak, A.C.: Simultaneous Algebraic Reconstruction Technique (SART): a Superior Implementation of the ART Algorithm. Ultrasonics Imaging 6, 81–94 (1984)
4. Gilbert, P.: Iterative Methods for the 3D Reconstruction of an Object from Projections. Journal of Theoretical Biology 76, 105–117 (1972)
5. Lewitt, R.M.: Alternatives to Voxels for Image Representation in Iterative Reconstruction Algorhms. Physics in Medicine and Biology 37, 705–716 (1992)
6. Andreyev, A., Sitek, A., Celler, A.: Acceleration of Blob-based Iterative Reconstruction Algorithm using Tesla GPU. IEEE NSS/MIC (2009)
7. Fernandez, J.J., Garcia, I., Garazo, J.M.: Three-dimensional Reconstruction of Cellular Structures by Electron Microscope Tomography and Parallel Computing. Journal of Parallel and Distributed Computing 64, 285–300 (2004)
8. NVIDIA, CUDA Programming Guide (2008), http://www.nvidia.com/cuda
9. Castano-Diez, D., Mueller, H., Frangakis, A.S.: Implementation and Performance Evaluation of Reconstruction Algorithms on Graphics Processors. Journal of Structural Biology 157, 288–295 (2007)
10. Bilbao-Castro, J.R., Carazo, J.M., Garcia, I., Fernandze, J.J.: Parallelization of Reconstruction Algorithms in Three-dimensional Electron Microscopy. Applied Mathmatical Modelling 30, 688–701 (2006)
11. Matej, S., Lewitt, R.M.: Efficient 3D Grids for Image-reconstruction using Spherically-symmetrical Volume Elements. IEEE Trans. Nucl. Sci. 42, 1361–1370 (1995)
12. Bisseling, R.H.: Parallel Scientific Computation. Oxford University Press, Oxford (2004)
13. John, R.R., Ronald, F.B.: Solving Elliptic Problems using ELLPACK. Springer, New York (1985)
14. Vazquez, F., Garzon, E.M., Fernandez, J.J.: Accelerating Sparse Matrix-vector Product with GPUs. In: Proceedings of CMMSE 2009, pp. 1081–1092 (2009)
15. Vazquez, F., Garzon, E.M., Fernandez, J.J.: A Matrix Approach to Tomographic Reconstruction and its Implementation on GPUs. Journal of Structural Biology 170, 146–151 (2010)

Component-Based Matching for Multiple Interacting RNA Sequences

Ghada Badr[1,2] and Marcel Turcotte[1]

[1] School of Information Technology and Engineering,
University of Ottawa,
Ontario, K1N 6N5, Canada
[2] IRI - Mubarak city for Science and Technology,
University and Research District,
P.O. 21934 New Borg Alarab, Alex, Egypt

Abstract. RNA interactions are fundamental to a multitude of cellular processes including post-transcriptional gene regulation. Although much progress has been made recently at developing fast algorithms for predicting RNA interactions, much less attention has been devoted to the development of efficient algorithms and data structures for locating RNA interaction patterns.

We present two algorithms for locating all the occurrences of a given interaction pattern in a set of RNA sequences. The baseline algorithm implements an exhaustive backtracking search. The second algorithm also finds all the matches, but uses additional data structures in order to considerably decrease the execution time, sometimes by one order of magnitude. The worst case memory requirement for the later algorithm increases exponentially with the input pattern length and does not depend on the database size, making it practical for large databases. The performance of the algorithms is illustrated with an application for locating RNA elements in a Diplonemid genome.

1 Introduction

RNA interactions are one of the fundamental mechanisms of the genome's regulatory code. Interacting RNA molecules are important players in translation, editing, gene silencing, but also to synthetic RNA molecules designed to self-assemble. Progress has been made recently at predicting the base pair patterns formed by interacting RNA molecules [1,9,10,16]. These approaches predict the intra- and inter- molecular base pairs with high accuracy. However, less attention has been paid to the development of matching algorithms. Once an RNA interaction pattern has been characterized, one would like to find occurrences within the same genome, or related ones. However, efficient methods for localizing known RNA interaction patterns on a genomic scale are lacking.

The work described here is part of an ongoing collaborative research project to discover the mechanism by which gene fragments in *Diplonema* are joined together to form an mRNA. In *Diplonema papillatum*, the mitochondrial genome

J. Chen, J. Wang, and A. Zelikovsky (Eds.): ISBRA 2011, LNBI 6674, pp. 73–86, 2011.
© Springer-Verlag Berlin Heidelberg 2011

```
P= (p1,p2,p3)
p1="[[[",  p2 = "]]].(((..))).[[",  p3 = "]]"

S = (s1, s2, s3)
s1 = ''CTATATATATG'',   s2 = ''TTTATAAGAGATCTCTCGC'',   s3 = ''TCGCGCGGAAC''
```

Fig. 1. Illustration of the pattern and sequences

consists of approximately one hundred circular chromosomes [11], each encoding a single gene fragment. For instance, the *cox1* gene consists of nine fragments encoded on nine different chromosomes. One of the research questions investigated is: do conserved match-maker RNA elements exist, encoded in either the mitochondrial or nuclear genome of *Diplonema*, that guide the assembly of the gene fragments?". Such expressed RNA elements should consists of two segments that are the reverse complement of two neighboring fragments. In order for biologists to investigate such a research question, efficient matching algorithms are needed.

Problem definition: We define the RNA interaction pattern localization problem and propose two algorithms for finding all the occurrences of a pattern in a set of RNA sequences. The pattern P and the sequences S from Figure 1 will be used throughout the text as a running example. An RNA interaction pattern for m sequences consists of m sub-patterns, each one comprising intra- and inter-molecular base pairs. In Figure 1, $m = 3$, matching square brackets represent inter-molecular base pairs, while matching round brackets represent intra-molecular base pairs. Crossings between intra- and inter-molecular base pairs are allowed, but not the crossings within intra- or inter-molecular base pairs. The dots represent unpaired nucleotides.

Background: This paper presents a new algorithm for the localization of RNA interaction motifs in multiple sequences. The algorithm uses a Trie-based structure called linked list of prefixes (LLP) [3]. Tries, or the related suffix trees and suffix arrays, have found many applications in bioinformatics. Suffix trees, with help of additional data structures, have been used for extracting conserved structured motifs from a set of DNA sequences [6,12,15]. Suffix arrays have been used to localize RNA secondary structure motifs consisting of intra-molecular base pairs [2]. Strothmann proposed an efficient implementation of a data structure, affix arrays, that stores the suffixes of both the text and its reverse [17]. The affix arrays were used to search RNA structure motifs consisting of intra-molecular base pairs. Finally, suffix arrays were used to identify inter-molecular base pairs in the software program GUUGle [8].

Contributions: We present the first application of linked list of prefixes (LLP) to the problem of locating interaction patterns in RNA sequences. As far as we know, there are no tools available to locate interaction patterns. We compare the performance of the proposed algorithm to that of an exhaustive backtracking search algorithm. The proposed algorithm decreases considerably the execution time whilst being practical regarding to memory usage.

(a) Simple structure:
I5(3) BR I3(3) SS(1) H5(3) SS(2) H3 SS(1) I5(2) BR I3

(b) Component-based structure:
P= (p1,p2,p3)
p1 = (3, {INTERM1}, {}),   INTERM1 = (1,1,2,3),
P2 = (15, {INTERM2}, {INTRAM2}),   INTERM2 = (14,1,3,2), INTRAM2 = (5,10,3)
P3 = (2, {}, {})

Fig. 2. (a) The simple structure used in the backtracking approach and (b) the component-based structure used in the proposed LLP approach for the pattern shown in Figure 1.

2   Backtracking Approach

This section outlines the baseline algorithm that we used for comparison.

2.1   Simple Structure for Patterns

Here, a pattern is an expression consisting of 6 types of terms. I5 and I3 terms represent the left and the right hand side of a stem formed by inter-molecular base pairs. H5 and H3 terms represent the left and the right hand side of a stem formed by intra-molecular base pairs. SS represents an unpaired region (single-stranded). Finally, BR delimits the boundary between two sub-patterns. Figure 2 (a) shows the simple structure representation of the pattern introduced in Figure 1.

2.2   Backtracking Algorithm

The algorithm uses two stacks, iStack and hStack, in order to store the location of matched left hand side of stems, for inter-molecular and intra-molecular base pairs respectively. An expression is a list of terms. The top level of this algorithm will attempt matching the first term of the expression list at every location of the first input sequence. If the term can be successfully matched at that location, the algorithm proceeds matching the next term. Otherwise, this is a failure and the algorithm will proceed with the next starting position. When matching the left hand side of stem, the starting position of the match and its length are pushed onto the appropriate stack depending on the type of the term, I5 or H5. SS matches a fixed length segment. Upon encountering a BR term, the algorithm will attempt matching the next term of the expression list at every location of the next input string. Finally, matching I3 and H3 terms require removing from the appropriate stack, iStack or hStack, the starting location and length of the corresponding match. I3 and H3 succeed only if the segment starting at the current location is the reversed complement of the corresponding match. GU base pairs are allowed. This simple algorithm exhaustively enumerates all the locations of the pattern P in the input set of sequences S.

3   Component-Based Structure for Patterns

In this section we will update the pattern structure so we can handle the search more efficiently. The components structure is very simple that it allows to define any pattern. A pattern $P = \{p_1, p_2, \ldots, p_m\}$, can be uniquely defined by its sub-patterns p_j, $0 < i \leq m$. Each p_j consists of inter-molecular (INTERM) and intra-molecular (INTRAM) components as described in the previous section.

Each sub-pattern, p_j, can be uniquely defined by its length and by its two sets of $INTERM$ and $INTRAM$ components. Each component is defined by its Opening Brackets (OB) and Closing Brackets (CB) lengths and relative locations within the sub-patterns. Components can be defined as follows:

- $INTERM$ component: In this components OB and CB resides in different sub-patterns. OB resides in p_j and CB resides in another pattern p_k, where $k > j$ and $1 \leq k \leq m$. This implies that we should have at least two sub-patterns in P to be able to have at least one INTERM component. Both OB and CB are described by their relative position to the beginning of p_j, $OBoffset$, and p_k, $CBoffset$, respectively, and by their length, len. So, an $INTERM$ component in p_j can be defined as: $INTERM_j = (OBoffset, CBoffset, k, len)$.
- $INTRAM$ component: In this components OB and CB resides in the same sub-pattern, p_j where $1 < j \leq m$. Both OB and CB are described by their relative position to the beginning of p_j, $OBoffset$ and $CBoffset$ respectively, and by their length, len. So, an $INTRAM$ component can be defined as: $INTRAM_i = (OBoffset, CBoffset, len)$.

Hence, for a given pattern $P = \{p_1, p_2, \ldots, p_m\}$, each $p_j = (len_j, \{INTERM_1, INTERM_2, \ldots, INTERM_r\}, \{ITRAM_1, INTRAM_2, \ldots, INTRAM_q\})$, where any p_j, $1 \leq j \leq m$, should participle in at least one component. For example, for a pattern P, with three sub-patterns $p_1 = "[[["$, $p_2 = "]]].(((..))).[["$, and $p_3 = "]]"$, as shown in Figure 1, the corresponding component based structure for $P = (p_1, p_2, p_3)$ is shown in Figure 2 (b).

4   Trie-Based LLP Structure for Sequences

Tries have been used extensively in literatures for string searching (matching) algorithms [3,4,5,14]. Storing the sequence in the Trie structure allows to simultaneously search all possible starting points in the sequence for any occurrence of the given pattern. In [3], many exact and approximate matching algorithms were proposed in which the Tries were used to store the databases for strings. Tries are based on a simple splitting scheme, which is based on letters encountered in the strings. This rule can be described as follows:

If S is a set of strings, and $A = \{a_j\}_{j=1}^r$ is the alphabet, then the Trie associated with S is defined recursively by the rule [7]:

$$Trie(S) = (Trie(S \backslash a_1), \ldots, Trie(S \backslash a_r)).$$

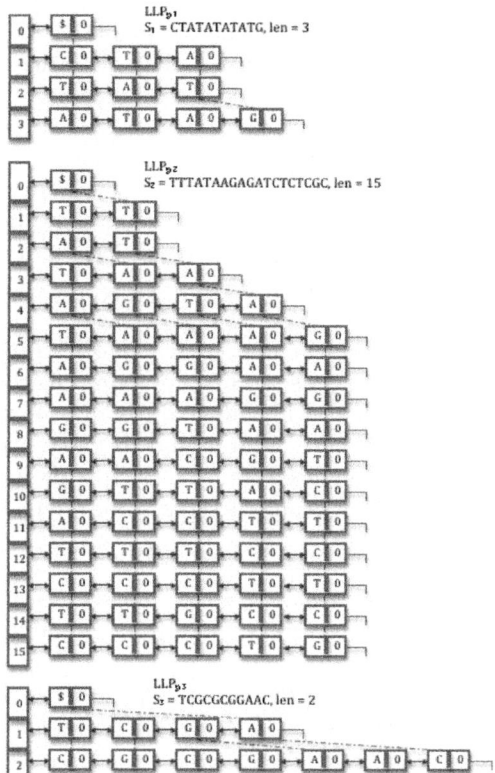

Fig. 3. An example for three LLP structures for the example provided in Figure 1, where $S = (s_1, s_2, s_3)$ and built for pattern $P = (p_1, p_2, p_3)$ with lengths (3,15,2) respectively. The zero values corresponds to the initial values of the reference counter. Red links are last child links. For simplicity, parent links are not shown.

where $S \backslash a_i$ means the subset of S consisting of strings that start with a_i, stripped of their initial letter a_i. The above recursion is halted as soon as S contains less than two elements. The advantage of the Trie is that it only maintains the minimal prefix set of characters that is necessary to distinguish all the elements of S.

In Section 2, we saw that in order to localize a structured pattern p in a given sequence s, we need to check p against all possible starting points of s for a maximum length equals the pattern length. Hence, to build a Trie structure for s, we need to extract all possible subsequences where a match can be found and store them in a Trie structure. Using the Trie structure allows matches to be done simultaneously. These subsequences can be obtained for all possible starting points in s and with length equals the pattern length. In this case, the maximum number of subsequences that can be obtained from s equals $|s| - |p|$. The resulting Trie is built in time proportional to the sum of all the subsequence lengths, $|p||s|$.

In order to search for the components of the given pattern, we need to apply a Breadth-First-Search (BFS) on the corresponding Trie. An updated Trie Structure was used in [3,14], namely the Linked Lists of Prefixes (LLP), to enhance searching, when the nodes in the Trie need to be traversed in level-by-level. The LLP can be built by implementing the Trie as linked lists and connecting all lists in the same level together with the LLP structure.

The LLP consists of a double-linked lists of levels, *where each level is a level in the corresponding Trie*. Each level, in turn, consists of a double-linked list of all nodes that correspond to prefixes with the same length. The levels are ordered in an increasing order of the length of the prefixes, exactly as in the case of the Trie levels.

The LLP structure is used during our search algorithm where it dramatically prunes the search space in sequences, when compared to the backtracking algorithm. Hence, the LLP structure will change during the search process. In order to facilitate accessing parents and children and to make pruning more efficient, we store additional information at each node. Each node in LLP will be used to store the following information, some of them were in the original structure as in [3]:

- *Char: Represents the corresponding letter represented by this node.*
- *Next node and previous node links: Maintain the double linked list at each level.*
- *Last child link (LastChild) and the number of children (NumChild): Maintain the list of children for each node. Last node links will help in pruning out the sub-Trie rooted at any given node quickly.*
- *Reference counter: It is used during the search algorithm to prune the search space when processing INTERM components. The value of zero at some point of the search process indicates that this node is not matched (referenced) even at least once during a previous match process.*
- *Parent links: Maintain an array of skip pointers to all parents in previous levels, starting from the direct parent to this node. This will be created once during the building process and will not change during the search process. It will help in efficiently match brackets for a given INTRAM component.*

Figure 3 shows the corresponding LLP structures for the example provided in Figure 1. The zero values represent the initial values of the reference counter. For simplicity, the parent links are not shown in the figures. For complete details on how to build the LLP structure from a given sequence (or dictionary of words), please refer to [3]. The space required for the Trie is proportional to $|A|^m$, where $|A|$ is the cardinality of the symbol alphabet and m is the length of pattern (subsequence), regardless of the sequence database size. In this paper, $A = \{A, C, G, T\}$ and so $|A| = 4$. Thus the worst case memory requirement of the algorithm will only depend on the pattern length, m, making it very practical for large databases. The execution time depends on the pattern size and structure.

5   The LLP Approach

In this approach, the search process will pass through two phases: preprocessing phase and matching phase. In the preprocessing phase, pattern components will be extracted, and the corresponding sequences will be processed to construct the corresponding LLPs according to the corresponding pattern sizes, as described in the previous section. In the matching phase, different pattern components will be matched against the corresponding LLPs. Pruning the search space will be done progressively through manipulating both INTRAM and INTERM components, where INTRAM components will be processed first as they reside in one sub-pattern and can be quickly processed. Processing INTERM components are more complex as they involves the processing of two LLPs at the same time. Processing the INTRAM components first will have a high impact on reducing the search space when manipulating the INTERM components. The matching process continue as long as there are more unprocessed component and as long as all LLPs are non-empty. As soon as any of LLPs is empty, this implies that a match for this pattern cannot be found for these sequences.

In the next two subsections, we will describe how to update and use the LLP structures to efficiently localize structured patterns in given sequences. The update will be shown, using an example, for the case of processing the INTERM components and then for the case of processing the INTRAM components. The matching process ends up with LLPs that have only subsequences with possible matches to the given pattern.

5.1   Updating the LLP Structure for INTRAM Components

INTRAM components can only resides in one sub-pattern, and that is why processing will be fast and easy. Processing them first will have a high impact on pruning the search space that is required to match other INTERM components. The algorithm can be explained in the following steps, for a given component $INTRAM_i = (OBoffset, CBoffset, len)$:

1. Start by processing internal opening and closing brackets: $OBlevel = OBffset + len - 1$ and $CBlevel = CBoffset$, where OBlevel and CBlevel are the corresponding levels in the corresponding LLP_i for $INTRAM_i$.
2. Match each node, $node_{OB}$, at level $OBlevel$ with all its successor nodes, $node_{CB}$, at level $CBlevel$. A successor node can be easily recognized by having a link to $node_{OB}$ in its parent links.
3. If no match is found, prune the sub-Trie rooted at $node_{CB}$. Notice that the sub-Trie rooted at $node_{OB}$ will be automatically pruned if all its successors are not matched. The pruning goes up to the highest parent that has only one child.
4. Repeat steps 2 and 3 until all nodes at both levels are processed, or LLP is empty, where a no match flag is returned.
5. Decrement $OBlevel$ and increment $CBlevel$.
6. Repeat from step 2 until all brackets are processed, where $OBlevel < OBoffset$.

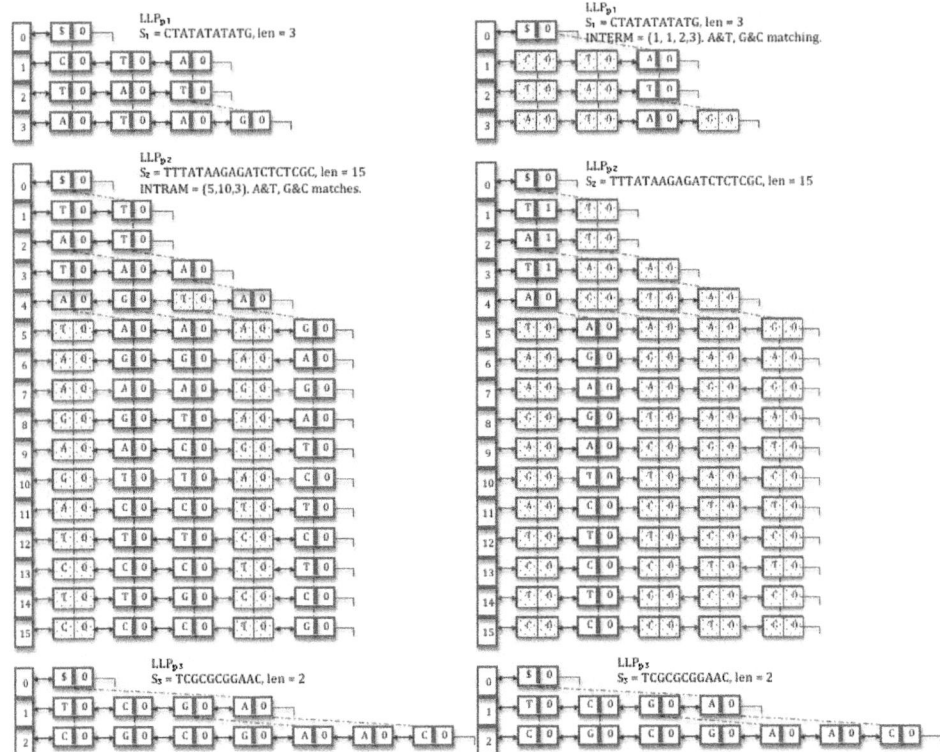

Fig. 4. Updating the LLP structure in Figure 3 for (left) the component $INTRAM = (5, 10, 3)$ in p_2 and (right) the component $INTERM = (1, 1, 2, 3)$ in p_1, for the pattern component-based structure that is shown in Figure 1 (b). Red nodes are nodes that are pruned from the search space and will not be considered any more in future processing.

5.2 Updating the LLP Structure for INTERM Components

Matching in an INTERM component, involves processing two LLPs structures. On one hand, matching INTERM components is more complex. The process involves matching each subsequence that corresponds to the OB in the first LLP, against all corresponding subsequences for the CB in the second LLP. On the other hand, processing INTERM components will result in progressively pruning the search space that is required for matching the next INTERM components. The algorithm can be explained in the following steps, for a given component $INTERM_j = (OBoffset, CBoffset, k, len)$ in p_j:

1. Start by processing internal opening and closing brackets: $OBlevel = OBffset + len - 1$ and $CBlevel = CBoffset$, where $OBlevel$ and $CBlevel$ are the corresponding levels in the corresponding LLP_{p_j} and LLP_{p_k} respectively for $INTERM_j$.
2. Start from the first node, $node_{OB}$, at level $OBlevel$ in LLP_{p_j}. $node_{OB}$, up to the parent, represents one possible subsequence in LLP_{p_j} for the OB.

3. Match $node_{OB}$ with each node, $node_{CB}$, at level $CBlevel$ in LLP_{p_k}. All $node_{CB}$, down to children, represent all possible subsequences in LLP_{p_k} for the CB. We have two cases:
   - If no match is found, prune $node_{OB}$ and Go to step 2.
   - If a match is found, increment the reference counters in $node_{CB}$.
4. increment $CBlevel$.
5. Change node $node_{OB}$ and make it point to its direct parent in LLP_{p_j}.
6. Match $node_{OB}$ with each node, $node_{CB}$ at level $CBlevel$ in LLP_{p_k} with a non-zero reference counter for its direct parent, reinitiating this counter to zero as soon as all its children are processed to be ready for next match process. We again have two match cases as in step 3.
7. Repeat from step 4 until all brackets are processed where $CBlevel > CBoffset + len - 1$.
8. Process next subsequence in LLP_{p_j} by making $node_{OB}$ point to the next node at level $OBlevel$ in LLP_{p_j} and repeat from step 2 until all nodes (subsequences) at level $OBlevel$ in LLP_{p_j} are processed.
9. Move through all nodes at level $CBoffset + len - 1$ in LLP_{p_k} and prune any node with a reference counter of a zero value. A node with a zero reference counter implies no possible match is found for a subsequence represented by that node n LLP_{p_j}.

Figure 4 (left) shows the resultant LLP_{s2} structure for s_2 in Figure 3, after processing $INTERM = (5, 10, 3)$ in p_2 shown in Figure 1 (b). This results in pruning some search space in string s_2 when manipulating other components. Figure 4 (right) shows updating the resultant structure after processing $INTERM1 = (1, 1, 2, 3)$ in p_1. This results in extensively pruning the search space in the LLP, leaving only a small portion of sequences for matching the last component $INTERM2 = (14, 1, 3, 2)$ in p_2, where a match is found. Snapshot for reference counter values is shown at the end of the matching process.

5.3 Pruning Search Space in LLP

Deleting a node in LLP, means deleting that node and all its descendants, which implies deleting all the subsequences represented by this node. Pruning the sub-Trie at a give node can be easily done in the corresponding LLP structure as follows, for a given node, $node_i$:

- Propagate up to the furtherest parent for $node_i$ with only one child, and update $node_i$ link to this parent.
- Let $nextNode$ and $prevNode$ point to the next node and previous nodes of $node_i$ respectively.
- Detach the sublist between $prevNode$ and $nextNode$ and update links information stored inside each of the two nodes accordingly.
- update $nextNode$ link to its last child node and $prevNode$ link to its last child node.
- Repeat from step 3 till the last level of the LLP structure.

There are also cases where $nextNode$ link or $prevNode$ link is null, when $node_i$ is at the beginning or the end of its linked list respectively. These cases can be easily manipulated by the pruning algorithm and are not shown to maintain the clarity of the algorithm. The time of the pruning operation is proportional to the corresponding pattern length.

6   Experimental Setup

We compared the performance of the backtracking (BT) and LLP algorithm using real-world data (Set 1) and synthetic data (Set 2). The patterns are defined by their components as described in Section 3 and sequences are indexed using the LLP data structures as described in Section 4.

Four patterns were used to compare the two approaches as follows:

- $P_1 = (p_{11}, p_{12}, p_{13})$ where $p_{11} = [[[[[[, p_{12} =]]]]]]]$ ....[[[[[, and $p_{13} =]]]]]]$.
  The sub-patterns contain only inter-molecular base pairs.
- $P_2 = (p_{21}, p_{22}, p_{23})$ where $p_{21} = [[[[[[, p_{22} =]]]]]].((((\ldots)))).[[[[[[, and $p_{23} =]]]]]]$.
  The sub-patterns contain both inter-molecular base pairs and intra-molecular base pairs.
- $P_3 = (p_{31}, p_{32}, p_{33})$ where $p_{31} = [[[[[[, p_{32} =]]]]]].((((.(((...))).(((...))).(((...)).)))).[[[[[, and $p_{33} =]]]]]]$.
  The sub-patterns contain both inter-molecular base pairs and intra-molecular base pairs.
- $P_4 = (p_{41}, p_{42}, p_{43})$ where $p_{41} = [[[[[[, p_{42} =]]]]]].((((\ldots)))).[[[[[[, and $p_{43} = ((((\ldots))))$.]]]]]].
  The sub-patterns contain both inter-molecular base-pairs and intra-molecular base pairs.

Two sets of sequences were used to represent databases as follows:

- Set 1: This benchmark is lifted from our ongoing collaborative research project to identify guide-like RNA elements in a recently sequenced diplonemid. In *Diplonema papillatum*, the mitochondrial genome consists of approximately one hundred circular chromosomes [11]. Each chromosome encodes a single gene fragment. The mechanism by which these fragments are joined together remains unknown. It has been hypothesized that RNA elements could guide the assembly process. For a gene consisting of n fragments, $n-1$ guide-like RNA elements are sought, such that each one consists of two segments that are the reverse complement of two consecutive gene fragments. The benchmark consists of finding guide-like RNA elements for the eight consecutive pairs of gene fragments that form the *cox1* gene. Each call to the pattern matcher comprises three sequences: fragment i, fragment $i + 1$, as well as a candidate sequence. The candidate sequence is selected from one of three pools: nuclear and mitochondrial EST sequences (TBest [13]), mitochondrial EST sequences (EST), or the mitochondrial genome (GENOME).
- Set 2: We generated synthetic data of sizes: 600, 1200, 2400, 4800, 9600, 19200, 38400. by carefully selecting one of the entries from the TBest database, where at least one match can be found for p_{i2} and two entries

from fragments i and $i + 1$, where a lot of matches can be found for p_{i1} and p_{i3}. We repeated the data to achieve the required sizes and also to show the speed up with respect to the data size, where approximately the same LLP structure will be generated for each data set when using the same pattern.

All the runs reported here were done on a Sun Fire E2900 server (24×1.8 GHz UltraSPARC-IV+ processors, 192 GB RAM) running Sun OS 5.10. The programs were developed using the Java programming language.

7   Results and Discussion

This section compares the execution time and memory usage of the backtracking and LLP algorithms on real-word and synthetic data with progressively more complex RNA interaction patterns.

Figure 5 shows histograms of the execution time for three data sets. The performance of the backtracking algorithm is worse when the input pattern consists exclusively of inter-molecular base pairs (P1). The algorithm is then forced to consider all possible starting locations for the 5'-end and 3'-end of the stems. The addition of intra-molecular base pairs (P2–P4) cuts the execution time in half because of the early failures. This effect is less pronounced for the third data set (TBest) because the sequences are on average shorter. The backtracking algorithm uses two stacks to keep track of the starting location of the matched 5' ends of inter- and intra-molecular stems. Since the number of stems is much smaller than the size of the input sequences, the memory usage for the stacks is negligible. Similarly, backtracking is implemented through recursion, but since the number of components of a pattern is always small compared to the length of the input sequences, the size of the system stack is negligible.

The LLP algorithm decreases considerably the execution time for Set 1. The biggest improvement is seen for the pattern P2 on the TBest data set. The observed speedup ($t_{\mathrm{BT}}/t_{\mathrm{LLP}}$) is 23. Particularly for patterns P2–P4, the larger part of the execution time is spent building the Trie and LLP data structures. Figure 7 shows the maximum memory requirement for all the patterns when matched against Set 1. The data set TBest comprises sequences that are on average shorter. Consequently, this data set has the lowest memory requirement, 7 to 8 megabyte. The largest memory requirement is for the GENOME data set. The maximum memory requirement varies between 41 and 72 megabyte. The memory requirement for the LLP approach depends on the pattern length (which implies the subsequence length) and repetitions in subsequences with that length in the given database. Sequence database sizes will have an impact on increasing the memory requirements of the LLP structure untill no more new subsequences can be obtained for a given pattern length. In this case, no new nodes will be created for the repeated subsequences.

Figure 6 shows the logarithm of the execution time for both algorithms for Set 2, where input sizes ranging from 600 to $38,400$ nucleotides. The execution time of the backtracking algorithm increases exponentially. For the LLP approach, the total execution time is only marginally larger than the build time. The execution time of the LLP approach increases much more slowly. Increasing the size of the input by a factor 64 translates into an increase of the execution time by 5,177 for the backtracking approach, but only a factor 68 for the LLP approach. The increase in the execution time of the LLP approach, for the same pattern, is only due to the increase in the time required for building the LLP structure for a bigger database. The maximum memory requirement for the synthetic data was 7.1 megabyte. Lower memory requirements here is because of high repetition in subsequences of Set 2, regardless of the variation in the sequence sizes. This also shows the advantage of the Trie structure in simultaneously matching similar subsequences whatever their start location are in the original database.

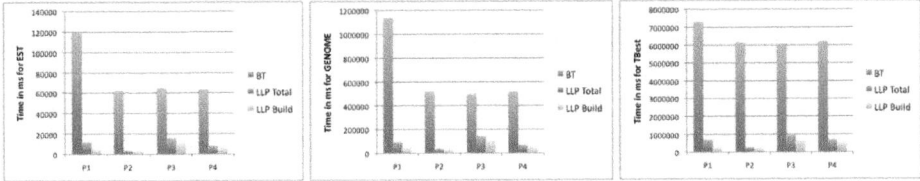

Fig. 5. Time in ms for databases in Set 1, when measured for localizing fours patterns P_1, P_2, P_3, and P_4, where: (left) the results for the EST database, (middle) the results for the GENOME database, and (right) the results for the TBest database. Both constructing time (LLP-Build) and total search time (LLP-Total), including the build time, are shown for the LLP approach. The build time is included in the total search time.

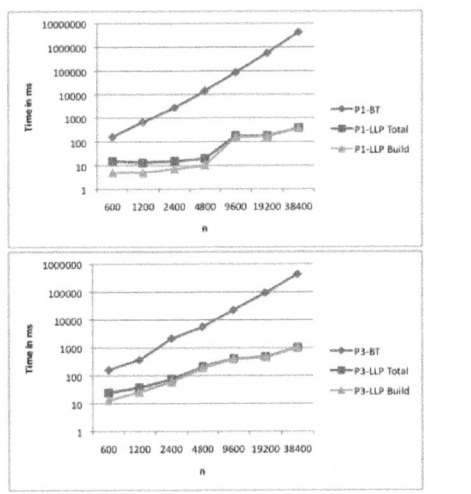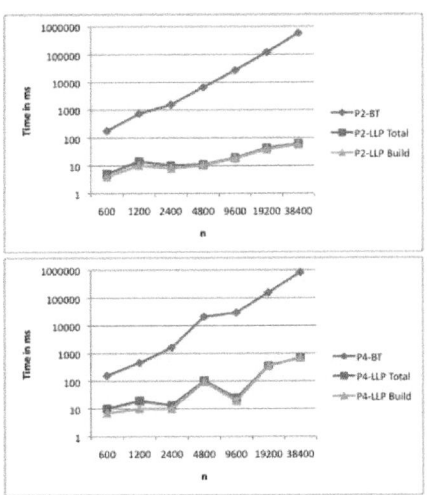

Fig. 6. Time in ms for the synthetic data set (Set 2) when measured for localizing: (top left) pattern P_1, (top right) pattern P_2, (bottom left) pattern P_3, and (bottom right) pattern P_4. Both constructing time (LLP-Build) and total search time (LLP-Total) are shown for the LLP approach. The build time is included in the total search time.

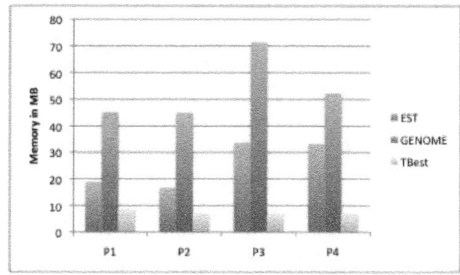

Fig. 7. Memory requirement in megabyte for the LLP approach for Set 1

8   Conclusion

We presented new algorithms and data structures for the localization of interaction patterns in multiple RNA sequences. An empirical study, using real-world and synthetic data, demonstrated that the proposed algorithm considerably improves the execution time while being practical in terms of memory usage.

Figures 5 and 6 show that the larger part of the execution time is spent on building the Trie structure. Therefore, further gain will be made whenever the Trie structure can be re-used or copied from one execution to the next. This would be the case if the matcher were used to implement a pattern discovery tool, where a fixed set of sequences is repeatedly searched.

Several extensions of this work are in progress. These include extensions to allow for small (bounded) variations of the length of the components. The development of this algorithm was prompted by a real-world application to identify RNA elements involved in splicing of mRNAs in a Diplonemid. We are also planning to use these data structures and algorithms to build a tool for the discovery of interaction patterns in Diplonemids. Free energy and p-values will be used for ranking the results.

Acknowledgement

The authors would like to thank Gertraud Burger for stimulating discussions related to RNA splicing and for sharing her data with us. The work is partially funded by the Natural Sciences and Engineering Research Council of Canada. The experiments were conducted at the High Performance Computing Virtual Laboratory.

References

1. Andronescu, M., Zhang, Z.C., Condon, A.: Secondary structure prediction of interacting RNA molecules. J. Mol. Biol. 345(5), 987–1001 (2005)
2. Anwar, M., Nguyen, T., Turcotte, M.: Identification of consensus RNA secondary structures using suffix arrays. BMC Bioinformatics 7, 244 (2006)

3. Badr, G.: Tries in information retrieval and syntactic pattern recognition. Ph.D. Thesis, School of Computer Science, Carleton University (June 2006)

4. Badr, G.: Optimized similarity measure over trie. In: MLDM Posters, pp. 236–150 (2007)

5. Bentley, J., Sedgewick, R.: Ternary search trees. Dr. Dobb's Journal (1998)

6. Carvalho, A.M., Freitas, A.T., Oliveira, A.L., Sagot, M.-F.: An efficient algorithm for the identification of structured motifs in DNA promoter sequences. IEEE/ACM Transactions on Computational Biology and Bioinformatics (TCBB) 3(2), 126–140 (2006)

7. Clement, J., Flajolet, P., Vallee, B.: The analysis of hybrid trie structures. In: Proc. Annual ACM-SIAM Symp. on Discrete Algorithms, San Francisco, California, pp. 531–539 (1998)

8. Gerlach, W., Giegerich, R.: GUUGle: a utility for fast exact matching under RNA complementary rules including G-U base pairing. Bioinformatics 22(6), 762–764 (2006)

9. Kato, Y., Akutsu, T., Seki, H.: A grammatical approach to RNA-RNA interaction prediction. Pattern Recognition 42(4), 531–538 (2009)

10. Kato, Y., Sato, K., Hamada, M., Watanabe, Y., Asai, K., Akutsu, T.: RactIP: fast and accurate prediction of RNA-RNA interaction using integer programming. Bioinformatics 26(18), i460–i466 (2010)

11. Marande, W., Burger, G.: Mitochondrial DNA as a genomic jigsaw puzzle. Science 318(5849), 415 (2007)

12. Marsan, L., Sagot, M.F.: Algorithms for extracting structured motifs using a suffix tree with an application to promoter and regulatory site consensus identification. J. Comput. Biol. 7(3-4), 345–362 (2000)

13. O'Brien, E.A., Koski, L.B., Zhang, Y., Yang, L.S., Wang, E., Gray, M.W., Burger, G., Lang, B.F.: TBestDB: a taxonomically broad database of expressed sequence tags (ESTs). Nucleic Acids Res. 35(Database issue), i445–i451 (2007)

14. Oommen, B.J., Badr, G.: Breadth-first search strategies for trie-based syntactic pattern recognition. Pattern Analysis and Applications 10(1), 1–13 (2007)

15. Pisanti, N., Carvalho, A.M., Marsan, L., Sagot, M.-F.: RISOTTO: Fast extraction of motifs with mismatches. In: Correa, J.R., Hevia, A., Kiwi, M. (eds.) LATIN 2006. LNCS, vol. 3887, pp. 757–768. Springer, Heidelberg (2006)

16. Salari, R., Backofen, R., Sahinalp, S.C.: Fast prediction of RNA-RNA interaction. Algorithms for Molecular Biology: AMB 5, 5 (2010)

17. Strothmann, D.: The affix array data structure and its applications to RNA secondary structure analysis. Theoretical Computer Science 389(1-2), 278–294 (2007)

A New Method for Identifying Essential Proteins Based on Edge Clustering Coefficient*

Huan Wang[1], Min Li[1,2], Jianxin Wang[1], and Yi Pan[1,2]

[1] School of Information Science and Engineering,
Central South University, Changsha 410083, P.R. China
[2] Department of Computer Science,
Georgia State University, Atlanta, GA 30302-4110, USA

Abstract. Identification of essential proteins is key to understanding the minimal requirements for cellular life and important for drug design. Rapid increasing of available protein-protein interaction data has made it possible to detect protein essentiality on network level. A series of centrality measures have been proposed to discover essential proteins based on network topology. However, most of them tended to focus only on topologies of single proteins, but ignored the relevance between interactions and protein essentiality. In this paper, a new method for identifying essential proteins based on edge clustering coefficient, named as SoECC, is proposed. This method binds characteristics of edges and nodes effectively. The experimental results on yeast protein interaction network show that the number of essential proteins discovered by SoECC universally exceeds that discovered by other six centrality measures. Especially, compared to BC and CC, SoECC is 20% higher in prediction accuracy. Moreover, the essential proteins discovered by SoECC show significant cluster effect.

Keywords: essential proteins, protein interaction network, centrality measures, edge clustering coefficient.

1 Introduction

Essential proteins are those proteins which are indispensable to the viability of an organism. The deletion of only one of these proteins is sufficient to cause lethality or infertility [1]. Thus, essential proteins (or genes) can be considered as potential drug targets for human pathogens. For example, the identification of conserved essential genes required for the growth of fungal pathogens offers an ideal strategy for elucidating novel antifungal drug targets [2]. Therefore, the identification of essential proteins is important not only for the understanding of

---

* This work is supported in part by the National Natural Science Foundation of China under Grant No.61003124 and No.61073036, the Ph.D. Programs Foundation of Ministry of Education of China No.20090162120073, the Freedom Explore Program of Central South University No.201012200124, the U.S. National Science Foundation under Grants CCF-0514750, CCF-0646102, and CNS-0831634.

J. Chen, J. Wang, and A. Zelikovsky (Eds.): ISBRA 2011, LNBI 6674, pp. 87–98, 2011.

the minimal requirements for cellular life, but also for practical purposes, such as drug design.

A variety of experimental procedures, such as single gene knockouts, RNA interference and conditional knockouts, have been applied to the prediction and discovery of essential proteins. However, these experimental techniques are generally laborious and time-consuming. Considering these experimental constraints, a highly accurate computational approach for identifying essential proteins would be of great value. With the development of high-throughput technology, such as yeast two-hybrid, tandem affinity purification and mass spectrometry, a wealth of protein-protein interaction (PPI) data have been produced, which open the door for us to lucubrate genomics and proteomics in network level [3].

Studies have shown that the topological properties of proteins in interaction networks could be strongly related to gene essentiality and cell robustness against mutations. In particular, a growing body of research focused on the centrality of a protein in an interaction network and suggested that a close relationship exists between gene essentiality and network centrality in protein interaction networks [4,5]. Consequently, A series of centrality measures have been used for discovering essential proteins based on network topological features.

The simplest of all centrality measures is the degree centrality which denotes the number of ties incident upon a node. It has been observed in several species, such as *E.coli*, *S.cerevisiae*, *C.elegans*, *D.melanogaster*, *M.musculus* and *H.sapiens* [4], that proteins with high degree (hubs) in the network are more likely to be essential. This phenomenon is identical with the centrality-lethality rule proposed by Jeong *et al.* [5]. Furthermore, many researchers have confirmed the correlation between degree centrality and protein essentiality. Besides the degree centrality, several other popular centrality measures, such as betweenness centrality [6], closeness centrality [7], subgraph centrality [8], eigenvector centrality [9], and information centrality [10], have also been proposed for discovering essential proteins. It has been proved that these centrality measures are significantly better than random selection in identifying essential proteins in recent reviews [11]. Furthermore, these centrality measures have also been used in weighted protein interaction networks and achieved a better result [12].

The current centrality measures only indicate the importance of nodes in the network but can not characterize the features of edges. In view of this, we propose a new essential proteins discovery method based on edge clustering coefficient, named as SoECC. we apply it in yeast protein interaction network (PIN). The experimental results show that the number of essential proteins predicted by SoECC is much more than that explored by degree centrality. For instance, among the top 100 proteins ranked by SoECC 78% are essential proteins. However, this percentage is only 57% for proteins ranked according to their number of interactions. Moreover, SoECC also outweigh another five measures of protein centrality in identifying essential proteins.

2 Materials and Methods

2.1 Experimental Data

Our analysis focuses on the yeast *Saccharomyces cerevisiae* because both the PPI and gene essentiality data of it are most complete and reliable among various species. The protein-protein interactions data of *S.cerevisiae* are downloaded from DIP database [13]. There are 4746 proteins and 15166 interactions in total after removing self-interactions and duplicate interactions.

A list of essential proteins of *S.cerevisiae* are integrated from the following databases: MIPS [14], SGD [15], DEG [16], and SGDP [17], which contains 1285 essential proteins altogether. In terms of the information about viability of gene disruptions cataloged by MIPS database, the 4746 proteins are grouped into three sets: essential protein set, non-essential protein set, and unknown set. By data matching, we find that out of all the 4746 proteins, 1130 proteins are essential which are derived from different types of experiment, such as systematic mutation set, classical genetics or large-scale survey, and 3328 proteins are non-essential. There are other 97 proteins whose essentiality are still unknown. The remaining 191 proteins in the yeast PIN neither covered by the essential protein set nor by the non-essential protein set. We incorporate this part of proteins into the unknown set.

2.2 Centrality Measures

Recently, many researchers found it is meaningful to predict essentiality by means of centrality measures. The notion of centrality comes from its use in social networks. Intuitively, it is related to the ability of a node to communicate directly with other nodes, or to its closeness to many other nodes, or to the quantity of pairs of nodes that need a specific node as intermediary in their communications. Here we will describe six classical centrality measures mentioned in the previous chapter, and all of which have been used for studies in biological networks.

A protein interaction network is conventionally regarded as an undirected graph $G = (V, E)$ which consists of a set of nodes V and a set of edges E. Each node $u \in V$ represents a unique protein, while each edge $(u, v) \in E$ represents an interaction between two proteins u and v. In order to facilitate description, we assign N as the total number of nodes in the network and A as the adjacency matrix of the network. The adjacency matrix $A = a_{u,v}$, which is a $N \times N$ symmetric matrix, whose element $a_{u,v}$ is 1 if there is a connecting edge between node u and node v, and 0 otherwise.

Degree Centrality (DC). The degree centrality $DC(u)$ of a node u is the number of its incident edges.

$$DC(u) = \sum_v a_{u,v} \tag{1}$$

Betweenness Centrality (BC). The betweenness centrality $BC(u)$ of a node u is defined as average fraction of shortest paths that pass through the node u.

$$BC(u) = \sum_s \sum_t \frac{\rho(s,u,t)}{\rho(s,t)}, s \neq t \neq u \tag{2}$$

Here $\rho(s,t)$ denotes the total number of shortest paths between s and t and $\rho(s,u,t)$ denotes the number of shortest paths between s and t that use u as an interior node. $BC(u)$ characterizes the degree of influence that the node u has in "communicating" between node pairs.

Closeness Centrality (CC). The closeness centrality $CC(u)$ of a node u is the reciprocal of the sum of graph-theoretic distances from the node u to all other nodes in the network.

$$CC(u) = \frac{N-1}{\sum_v d(u,v)} \tag{3}$$

Here $d(u,v)$ denotes the distance from node u to node v, i.e., the number of links in the shortest path between this pair of nodes. $CC(u)$ reflects the independency of the node u compared to all other nodes.

Subgraph Centrality (SC). The subgraph centrality $SC(u)$ of a node u measures the number of subgraphs of the overall network in which the node u participates, with more weight being given to small subgraphs. It is defined as:

$$SC(u) = \sum_{l=0}^{\infty} \frac{\mu_l(u)}{l!} = \sum_{v=1}^{N} [\alpha_v(u)]^2 e^{\lambda_v} \tag{4}$$

where $\mu_l(u)$ denotes the number of closed loops of length l which starts and ends at node u. $(\alpha_1,\alpha_2,...,\alpha_N)$ is an orthonormal basis of R^N composed by eigenvectors of A associated to the eigenvalues $\lambda_1,\lambda_2,...,\lambda_N$, where $\alpha_v(u)$ is the uth component of α_v.

Eigenvector Centrality (EC). The eigenvector centrality $EC(u)$ of a node u is defined as the uth component of the principal eigenvector of A.

$$EC(u) = \alpha_{max}(u) \tag{5}$$

Here α_{max} denotes the eigenvector corresponding to the largest eigenvalue of A. $\alpha_{max}(u)$ is the uth component of α_{max}. The core idea of EC is that an important node is usually connected to important neighbors.

Information Centrality (IC). The information centrality $IC(u)$ of a node u measures the harmonic mean lengths of paths ending at the node u. It is given by the following expression:

$$IC(u) = [\frac{1}{N} \sum_v \frac{1}{I_{uv}}]^{-1} \tag{6}$$

where $I_{uv} = (c_{uu} + c_{vv} - c_{uv})^{-1}$. Let D be a diagonal matrix of the degree of each node and J be a matrix with all its elements equal to one. Then we obtain the matrix $C = (c_{uv}) = [D - A + J]^{-1}$. For computational purposes, I_{uu} is defined as infinite. Thus, $\frac{1}{I_{uu}} = 0$.

2.3   Edge Clustering Coefficient

The current centrality measures only indicate the importance of nodes in the network but can not characterize how important the edges are. More recently, some researchers challenged the traditional explanation of the centrality-lethality rule and different points of view have been proposed [18,19]. He *et al.* [18] restudied the reason why hubs tend to be essential and advanced the idea of *essential protein-protein interactions*. They argued that the majority of proteins are essential due to their involvement in one or more *essential protein-protein interactions* that are distributed uniformly at random along the network edges. In another words, they concluded that the essentiality of proteins depend on *essential protein-protein interactions*. Inspired by this idea, we introduce the concept of edge clustering coefficient to measure the importance of edges in protein interaction network and apply it to discovery of essential proteins.

For an edge $E_{u,v}$ connecting node u and node v, we pay attention to how many other nodes that adjoin both u and v. The edge clustering coefficient of $E_{u,v}$ can be defined by the following expression:

$$ECC(u, v) = \frac{z_{u,v}}{min(d_u - 1, d_v - 1)} \tag{7}$$

where $z_{u,v}$ denotes the number of triangles that include the edge actually in the network, d_u and d_v are degrees of node u and node v, respectively. Then the meaning of $min(d_u - 1, d_v - 1)$ is the number of triangles in which the edge $E_{u,v}$ may possibly participate at most. For instance, in Fig. 1, the degrees of two end nodes n_1 and n_4 of edge $E_{1,4}$ are both 4. Therefore, this edge could constitute $min(4 - 1, 4 - 1) = 3$ triangles at most in theory. But in fact, there are only 2 triangles $\triangle_{145}$, $\triangle_{146}$, so $ECC(1, 4) = 2/3 = 0.67$.

Edge clustering coefficient characterizes the closeness between an edge's two connecting nodes and other nodes around them. The edges with higher clustering coefficient tend to involve in the community structure in network. It has been

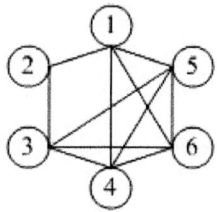

Fig. 1. Example of solving the edge clustering coefficient

proved that there is significant cluster effect in many large-scale complex networks including biological networks. Specifically, in protein interaction networks there exist many protein complexes which are considered to play a key role in carrying out biological functionality [3]. Furthermore, Hart *et al.* [20] examined the relationship between protein modularity and essentiality and indicated that essentiality is a product of the protein complex rather than the individual protein. On the basis of the existing studies, we believe it is significative to introduce edge clustering coefficient into identification of essential proteins.

Considering the factor of node's degree, we further define the sum of edge clustering coefficients of node u as:

$$SoECC(u) = \sum_{v \in N_u} ECC(u, v) \qquad (8)$$

where N_u denotes the set of all neighbor nodes of node u. Obviously, $SoECC(u)$ will be larger if node u that with higher degree. In this sense, SoECC has a integrated characteristic of edges and nodes. We take SoECC as a new measure for identifying essential proteins.

2.4   Evaluation Methods

For purpose of evaluating performance of SoECC, we compare SoECC and other six centrality measures in prediction accuracy by applying these methods to predict essential proteins in yeast PIN. A certain quantity of proteins with higher values computed by each measure are selected as candidates for essential proteins. Then we examine what proportion of them are essential. Moreover, several statistical measures, such as sensitivity (SN), specificity (SP), positive predictive value (PPV), negative predictive value (NPV), F-measure (F), and accuracy (ACC) are used to evaluate how effective the essential proteins identified by different methods. In advance, for notional convenience, we explain four frequently-used terms in statistics and their meanings in our work as follows:

- TP(true positives): essential proteins correctly predicted as essential.
- FP(false positives): non-essential proteins incorrectly predicted as essential.
- TN(true negatives):non-essential proteins correctly predicted as non-essential.
- FN(false negatives): essential proteins incorrectly predicted as non-essential.

Below are definitions of the six statistical measures mentioned above.

Sensitivity (SN). Sensitivity refers to the proportion of essential proteins which are correctly predicted as essential and the total essential proteins.

$$SN = \frac{TP}{TP + FN} \qquad (9)$$

Specificity (SP). Specificity refers to the proportion of non-essential proteins which are correctly eliminated and the total non-essential proteins.

$$SP = \frac{TN}{TN + FP} \qquad (10)$$

Positive Predictive Value (PPV). Positive predictive value is the proportion of essential proteins in the prediction which are correctly predicted as essential.

$$PPV = \frac{TP}{TP + FP} \tag{11}$$

Negative Predictive Value (NPV). Negative predictive value is the proportion of non-essential proteins in the prediction which are correctly eliminated.

$$NPV = \frac{TN}{TN + FN} \tag{12}$$

F-measure (F). F-measure is the harmonic mean of sensitivity and positive predictive value, which is given by:

$$F = \frac{2 * SN * PPV}{SN + PPV} \tag{13}$$

Accuracy (ACC). Accuracy is the proportion of correct prediction results of all the results, which is expressed as:

$$ACC = \frac{TP + TN}{P + N} \tag{14}$$

where P and N are the total number of essential proteins and non-essential proteins, respectively.

3 Results and Discussion

In order to measure the performance of SoECC, we carry out a comparison between SoECC and other six centrality measures. Similar to most experimental procedures [11], we firstly rank proteins in descending order according to their values of SoECC and centralities, then select the top 1%, top 5%, top 10%, top 15%, top 20%, top 25% proteins as essential candidates and determine how many of these are essential in the yeast PIN. We illustrate the number of essential proteins detected by SoECC and other six centrality measures in Fig. 2.

As can be seen in Fig. 2, SoECC performs significantly better than all centrality measures in identifying essential proteins in the yeast PIN. Especially, the improvement of SoECC compared to BC and CC are both more than 20% in any top percentage. Even comparing with the centrality measures with the best performance (SC, SC, EC, IC, IC, and DC) in each top percentage (top 1%, top 5%, top 10%, top 15%, top 20%, and top 25%), the number of essential proteins identified by SoECC are improved by 5.9%, 15.2%, 12.4%, 10.3%, 7.3%, and 3.0%, respectively.

For further assessing the performance of SoECC, we compare the several statistical indicators including sensitivity (SN), specificity (SP), positive predictive

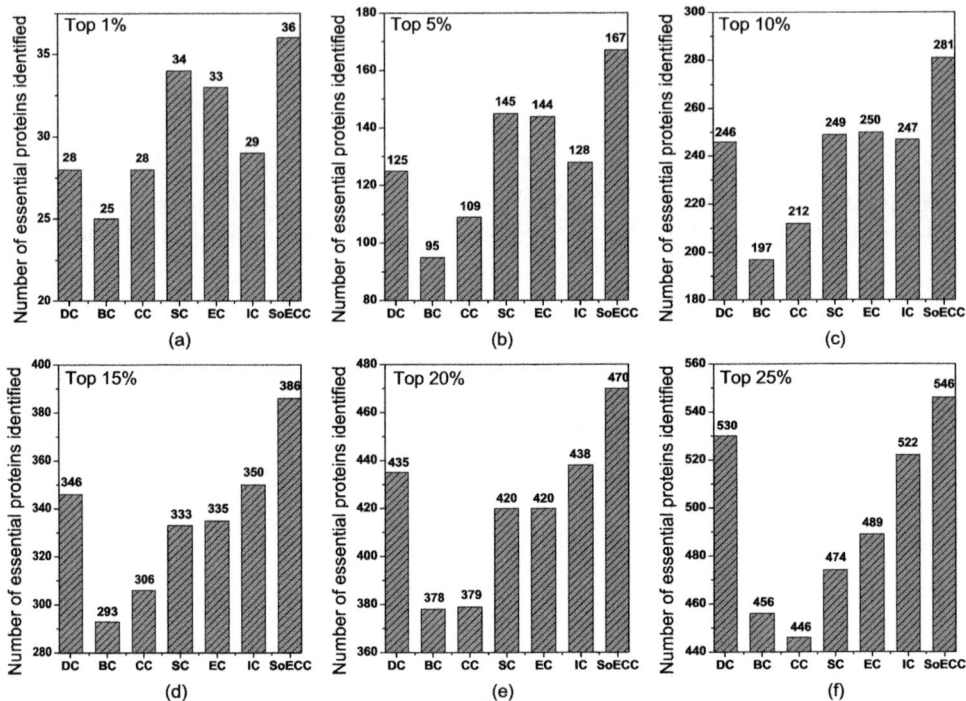

Fig. 2. Number of essential proteins detected by different methods

value (PPV), negative predictive value (NPV), F-measure (F), and accuracy (ACC) of SoECC and other six centrality measures. During the course of experimental data collection and processing previously, proteins were classified into three sets according to their essentiality. We have already known there are 1130 essential proteins, 3328 non-essential proteins and other 288 proteins whose essentiality are unknown contained in the yeast PIN. So we select the top 1130 proteins ranked by each method as essential proteins. The remaining 3616 proteins are considered as non-essential proteins. The values of six statistical indicators of each method is tabulated in Table 1.

From Table 1, we can see that the six statistical indicators (SN, SP, PPV, NPV, F, and ACC) of SoECC are consistently higher than that of any other centrality measure, which show that SoECC can identify essential proteins more accurately.

In addition, we make a comparison between the proteins predicted by SoEEC and that predicted by other six centrality measures. The number of overlaps in the top 100 proteins identified by any two different methods is shown in Table 2.

From Table 2, we can see that the common proteins identified by SoECC and each centrality measure are less than that identified by any pair of centrality measures. The overlapping rates between SoECC and each centrality measure are universally less than 50%. Especially, for BC and CC, the overlaps are both less than 30%. This indicates that there exist some difference in methodology

Table 1. Comparsion the values of sensitivity (SN), specificity (SP), positive predictive value (PPV), negative predictive value (NPV), F-measure (F), and accuracy (ACC) of SoECC and other six centrality measures

	SN	SP	PPV	NPV	F	ACC
DC	0.441	0.813	0.445	0.811	0.443	0.719
BC	0.388	0.798	0.395	0.793	0.392	0.694
CC	0.386	0.801	0.397	0.793	0.391	0.696
SC	0.414	0.811	0.427	0.803	0.420	0.710
EC	0.418	0.811	0.429	0.804	0.423	0.711
IC	0.443	0.814	0.447	0.812	0.445	0.720
SoECC	**0.468**	**0.824**	**0.475**	**0.820**	**0.471**	**0.734**

Table 2. Number of common proteins in the top 100 proteins identified by any two different methods

	DC	BC	CC	SC	EC	IC
DC	100	78	65	58	57	93
BC	78	100	72	48	47	73
CC	65	72	100	58	60	65
SC	58	48	58	100	97	62
EC	57	47	60	97	100	61
IC	93	73	65	62	61	100
SoECC	**44**	**27**	**27**	**46**	**44**	**44**

Table 3. Number of essential proteins in the different part of top 100 proteins identified by SoECC and other six centrality measures

	DC(56)	BC(73)	CC(73)	SC(54)	EC(56)	IC(56)
$CMs - SoECC$	24	27	28	32	31	22
$SoECC - CMs$	45	57	57	40	41	43

between SoECC and centrality measures. Actually, even comparing the top 100 proteins identified by SoECC with the union set of that identified by six centrality measures, there is only 57 common proteins.

To further demonstrate the efficiency of SoECC, we analyze the different proteins that identified by SoECC and other six centrality measures.

In Table 3, the first row $CMs - SoECC$ means essential proteins identified by centrality measures but ignored by SoECC. The second row $SoECC - CMs$ means essential proteins identified by SoECC but ignored by centrality measures. Obviously, $SoECC - CMs$ is generally more than $CMs - SoECC$, which means that SOECC is more effective and accurate in discovering essential proteins.

Table 4. Number and proportion of three types of interactions in the two networks composed by the top 100 proteins ranked by DC and SoECC respectively

		number	proportion
	$E_{ee}(u,v)$	107	0.405
DC	$E_{nn}(u,v)$	39	0.148
	$E_{en}(u,v)$	118	0.447
	$E_{ee}(u,v)$	243	0.666
SoECC	$E_{nn}(u,v)$	12	0.033
	$E_{en}(u,v)$	110	0.301

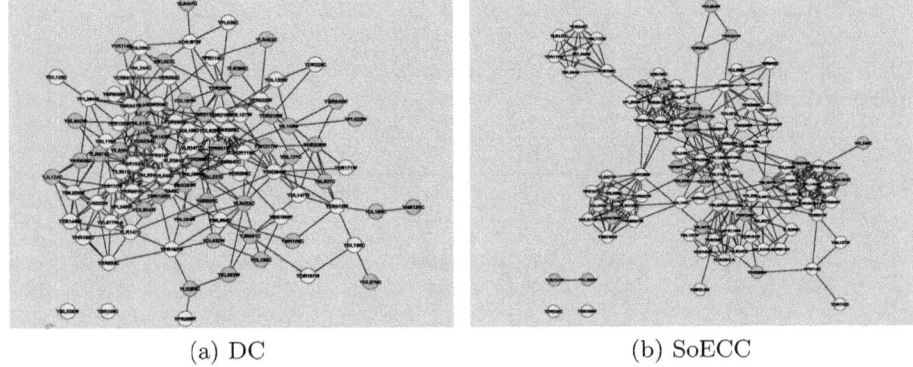

(a) DC   (b) SoECC

Fig. 3. The top 100 proteins in the yeast PIN identified by DC and SoECC respectively. The yellow nodes and red nodes represent essential proteins and non-essential proteins respectively.

It should be emphasized that, SoECC proposed in this paper has the dual characteristics of nodes and edges of the network. Some researchers suggested that the essentiality of proteins has correlation with essential interactions in PIN [18]. In order to validate that SoECC is involved in this viewpoint, we classify all interactions in the yeast PIN into three types according to the essentiality of each interacting protein pair as follows:

- $E_{ee}(u,v)$: u and v are both essential, i.e., essential interaction.
- $E_{nn}(u,v)$: u and v are both non-essential, i.e., non-essential interaction.
- $E_{en}(u,v)$: u and v are essential and non-essential respectively.

Then we examine the number and proportion of each type of interactions in the two networks composed by the top 100 proteins ranked by DC and SoECC respectively. As shown in Table 4, in terms of both quantity and proportion, the essential interactions generated by SoECC are obviously more than that generated by DC.

Moreover, we generate the two networks mentioned above using Cytoscape as Fig. 3.

As illustrated in Fig. 3, out of the top 100 proteins ranked by SoECC, there are 78 essential proteins, whereas this number is only 57 for DC. It also can be seen in Fig. 3 that, comparing with DC, there are more essential interactions in the subgraph induced by top 100 proteins ranked by SoECC. Furthermore, it is noteworthy that the essential proteins identified by SoECC (yellow nodes in Fig. 3(b)) show significant modularity. This cluster effect which is determined by the meaning of edge clustering coefficient is consistent with previous researches [19,20]. In this sense, our proposed method SoECC is meaningful in identifying essential proteins.

4   Conclusions

In the study of identification of essential proteins based on network topology, commonly used centrality measures are only reflect the importance of single proteins but can not characterize the essentiality of the interactions. However, recent studies have shown that there exist a relationship between protein-protein interactions and protein essentiality. In consideration of this idea, we propose a novel method for identifying essential proteins based on edge clustering coefficient, named as SoECC, which binds characteristics of edges and nodes effectively. The experimental results show that, the essential proteins detected by SoECC are universally more than that detected by other six centrality measures. Especially, compared to BC and CC, SoECC is 20% higher in prediction accuracy. Besides, by comparing the two networks composed by the top 100 proteins selected by DC and SoECC respectively, we found that there are more essential interactions in the latter. Significantly, essential proteins identified by SoECC show obvious modularity, which agreed with previous researches.

References

1. Winzeler, E.A., et al.: Functional characterization of the S. cerevisiae genome by gene deletion and parallel analysis. Science 285(5429), 901–906 (1999)
2. Hu, W., Sillaots, S., Lemieux, S., et al.: Essential Gene Identification and Drug Target Prioritization in Aspergillus fumigatus. PLoS Pathog. 3(3), e24 (2007)
3. Ho, Y., et al.: Systematic identification of protein complexes in *Saccharomyces cerevisiae* by mass spectrometry. Nature 415(6868), 180–183 (2002)
4. Hahn, M.W., Kern, A.D.: Comparative Genomics of Centrality and Essentiality in Three Eukaryotic Protein-Interaction Networks. Mol. Biol. Evol. 22(4), 803–806 (2005)
5. Jeong, H., Mason, S.P., Barabási, A.L., Oltvai, Z.N.: Lethality and centrality in protein networks. Nature 411(6833), 41–42 (2001)
6. Joy, M.P., Brock, A., Ingber, D.E., Huang, S.: High-betweenness proteins in the yeast protein interaction network. J. Biomed. Biotechnol. (2), 96–103 (2005)
7. Wuchty, S., Stadler, P.F.: Centers of complex networks. J. Theor. Biol. 223(1), 45–53 (2003)
8. Estrada, E., Rodríguez-Velázquez, J.A.: Subgraph centrality in complex networks. Phys. Rev. E. 71(5), 56103 (2005)

9. Bonacich, P.: Power and centrality: A family of measures. American Journal of Sociology 92(5), 1170–1182 (1987)
10. Stevenson, K., Zelen, M.: Rethinking centrality: Methods and examples. Social Networks 11(1), 1–37 (1989)
11. Estrada, E.: Virtual identification of essential proteins within the protein interaction network of yeast. Proteomics 6(1), 35–40 (2006)
12. Li, M., Wang, J., Wang, H., Pan, Y.: Essential Proteins Discovery from Weighted Protein Interaction Networks. In: Borodovsky, M., Gogarten, J.P., Przytycka, T.M., Rajasekaran, S. (eds.) ISBRA 2010. LNCS, vol. 6053, pp. 89–100. Springer, Heidelberg (2010)
13. Xenarios, I., Rice, D.W., Salwinski, L., et al.: DIP: the database of interacting proteins. Nucleic Acids Res. 28(1), 289–291 (2000)
14. Mewes, H.W., et al.: MIPS: analysis and annotation of proteins from whole genomes in 2005. Nucleic Acids Res. 34(Database issue), D169–D172 (2006)
15. Cherry, J.M., et al.: SGD: *Saccharomyces* Genome Database. Nucleic Acids Res. 26(1), 73–79 (1998)
16. Zhang, R., Lin, Y.: DEG 5.0, a database of essential genes in both prokaryotes and eukaryotes. Nucleic Acids Res. 37(Database issue), D455–D458 (2009)
17. *Saccharomyces* Genome Deletion Project, http://www-sequence.stanford.edu/group/yeast_deletion_project
18. He, X., Zhang, J.: Why do hubs tend to be essential in protein networks? PLoS Genet. 2(6), e88 (2006)
19. Zotenko, E., Mestre, J., O'Leary, D.P., et al.: Why do hubs in the yeast protein interaction network tend to be essential: reexamining the connection between the network topology and essentiality. PLoS Comput. Biol. 4(8), e1000140 (2008)
20. Hart, G.T., Lee, I., Marcotte, E.R.: A high-accuracy consensus map of yeast protein complexes reveals modular nature of gene essentiality. BMC Bioinformatics 8(1), 236 (2007)

Gene Order in Rosid Phylogeny, Inferred from Pairwise Syntenies among Extant Genomes

Chunfang Zheng[1] and David Sankoff[2]

[1] Département d'informatique et de recherche opérationnelle, Université de Montréal
[2] Department of Mathematics and Statistics, University of Ottawa

Abstract. Based on the gene order of four core eudicot genomes (cacao, castor bean, papaya and grapevine) that have escaped any recent whole genome duplication (WGD) events, and two others (poplar and cucumber) that descend from independent WGDs, we infer the ancestral gene order of the rosid clade and those of its main subgroups, the fabids and malvids. We use the gene order evidence to evaluate the hypothesis that the order Malpighiales belongs to the malvids rather than as traditionally assigned to the fabids. Our input data are pairwise synteny blocks derived from all 15 pairs of genomes. Our method involves the heuristic solutions of two hard combinatorial optimization problems, neither of which invokes any arbitrary thresholds, weights or other parameters. The first problem, based on the conflation of the pairwise syntenies, is the inference of disjoint sets of orthologous genes, at most one copy for each genome, and the second problem is the inference of the gene order at all ancestors simultaneously, minimizing the total number of genomic rearrangements over a given phylogeny.

1 Introduction

Despite a tradition of inferring common genomic structure among plants and despite plant biologists' interest in detecting synteny, e.g., [1,2], the automated ancestral genome reconstruction methods developed for animals [3,4,5,6] and yeasts [7,8,9,10,11] at the syntenic block or gene order levels, have yet to be applied to the recently sequenced plant genomes. Reasons for this include:

1. The relative recency of these data. Although almost twenty dicotyledon angiosperms have been sequenced and released, most of this has taken place in the last two years (at the time of writing) and the comparative genomics analysis has been reserved by the various sequencing consortia for their own first publication, often delayed for years following the initial data release.

2. Algorithms maximizing a well-defined objective function for reconstructing ancestors through the median constructions and other methods are computationally costly, increasing both with n, the number of genes orthologous across the genomes, and especially with $\frac{d}{n}$, where d is the number of rearrangements occurring along a branch of the tree.

3. Whole genome duplication (WGD), which is rife in the plant world, particularly among the angiosperms [12,13], sets up a comparability barrier between

J. Chen, J. Wang, and A. Zelikovsky (Eds.): ISBRA 2011, LNBI 6674, pp. 99–110, 2011.
© Springer-Verlag Berlin Heidelberg 2011

those species descending from a WGD event and species in all other lineages originating before the event [2]. This is largely due to the process of duplicate gene reduction, eventually affecting most pairs of duplicate genes created by the WGD, which distributes the surviving members of duplicate pairs between two homeologous chromosomal segments in an unpredictable way [14,15,16], thus scrambling gene order and disrupting the phylogenetic signal. This difficulty is compounded by the residual duplicate gene pairs created by the WGD, complicating orthology identification essential for gene order comparison between species descended from the doubling event and those outside it.

4. Global reconstruction methods are initially designed to work under the assumption of identical gene complement across the genomes, but if we look at dicotyledons, for example, each time we increase the set of genomes being studied by one, the number of genes common to the whole set is reduced by approximately $\frac{1}{3}$. Even comparing six genomes, retaining only the genes common to all six, removes 85 % of the genes from each genome, almost completely spoiling the study as far as local syntenies are concerned.

Motivated in part by these issues, we have been developing an ancestral gene order reconstruction algorithm PATHGROUPS, capable of handling large plant genomes, including descendants of WGD events, as soon as they are released, using global optimization criteria, approached heuristically, but with well-understood performance properties [9,10]. The approach responds to the difficulties enumerated above as follows:

1. The software has been developed and tested with all the released and annotated dicotyledon genome sequences, even though "ethical" claims by sequencing consortia leaders discourage the publication of the results on the majority of them at this time. In this enterprise, we benefit from the up-to-date and well organized CoGe platform [1,17], with its database of thousands of genome sequences and its sophisticated, user-friendly SynMap facility for extraction of synteny blocks.

2. PATHGROUPS aims to rapidly reconstruct ancestral genomes according to a minimum total rearrangement count (using the DCJ metric [18]) along all the branches of a phylogenetic tree. Its speed is due to its heuristic approach (greedy search with look-ahead), which allows it to return a solution for values of $\frac{d}{n}$ where exact methods are no longer feasible. The implementation first produces a rapid initial solution of the "small phylogeny" problem (i.e., where the tree topology is given and the ancestral genomes are to be constructed), followed by an iterative improvement treating each ancestral node as a median problem (one unknown genome to be constructed on the basis of the three given adjacent genomes).

3. The comparability barrier erected by a WGD event is not completely impenetrable, even though gene order fractionation is further confounded by genome rearrangement events. The WGD-origin duplicate pairs remaining in the modern genome will contain much information about gene order in the ancestral diploid, immediately before WGD. The gene order information is retrievable through the method of *genome halving* [19], which is incorporated in a natural way into PATHGROUPS.

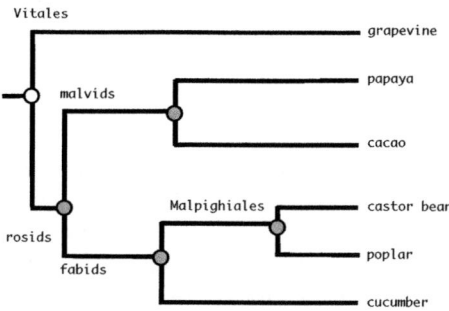

Fig. 1. Phylogenetic relationships among sequenced and non-embargoed eudicotyledon genomes (without regard for time scale). Poplar and cucumber each underwent WGD in their recent lineages. Shaded dots represent gene orders reconstructed here, including the rosid, fabid, malvid and Malpighiales ancestors.

4. One of the main technical contributions of this paper is the feature of PATH-GROUPS that allows the genome complement of the input genomes to vary. Where the restriction to equal gene complement would lead to reconstructions involving only about 15 % of the genes, the new feature allows close to 100% of the genes with orthologs in at least two genomes to appear in the reconstructions. The other key innovation we introduce here is our "orthologs for multiple genomes" (OMG) method for combining the genes in the synteny block sets output by SYNMAP for pairs of genomes, into orthology sets containing at most one gene from every genome in the phylogeny.

Both the PATHGROUPS and the OMG procedures are parameter-free. There are no thresholds or other arbitrary settings. We argue that the the appropriate moment to tinker with such parameters is during the synteny block construction and not during the orthology set construction nor the ancestral genome reconstruction. A well-tuned synteny block method goes a long way to attenuate genome alignment problems due to paralogy. It is also the appropriate point to incorporate thresholds for declaring homology, since these depend on evolutionary divergence time, which is specific to pairs of genomes. Finally, the natural criteria for constructing pairwise syntenies do not extend in obvious ways to three or more genomes.

2   Six Eudicotyledon Sequences

There are presently almost twenty eudicotyledon genome sequences released. Removing all those that are embargoed by the sequencing consortia, all those who have undergone more than one WGD since the divergence of the eudicots from the other angiosperms, such as *Arabidopsis*, and some for which the gene annotations are not easily accessible leaves us the six depicted in Fig. 1, namely cacao [20], castor bean [21], cucumber [22], grapevine [23,24], papaya [25] and poplar [26]. Of the two main eudicot clades, asterids and rosids, only the latter is represented, as well as the order Vitales, considered the closest relative of the rosids [12,27]. Poplar and cucumber are the only two to have undergone ancestral WGD since the divergence of the grapevine.

3   Formal Background

A genome is a set of chromosomes, each chromosome consisting of a number of genes linearly ordered. The genes are all distinct and each has positive or negative polarity, indicating on which of the two DNA strands the gene is located.

Genomes can be rearranged through the accumulated operation of number of processes: inversion, reciprocal translocation, transposition, chromosome fusion and fission. These can all be subsumed under a single operation called double-cut-and-join which we do not describe here. For our purposes all we need is a formula due to Yancopoulos *et al.* [18], stated in Section 3.1 below, that gives the genomic distance, or length of a branch in a phylogeny, in terms of the minimum number of rearrangement operations needed to transform one genome into another.

3.1   Rearrangement Distance

The genomic distance $d(G_1, G_2)$ is a metric counting the number of rearrangement operations necessary to transform one multichromosomal gene order G_1 into another G_2, where both contain the same n genes. To calculate D efficiently, we use the breakpoint graph of G_1 and G_2, constructed as follows: For each genome, each gene g with a positive polarity is replaced by two vertices representing its two ends, i.e., by a "tail" vertex and a "head" vertex in the order g_t, g_h; for $-g$ we would put g_h, g_t. Each pair of successive genes in the gene order defines an adjacency, namely the pair of vertices that are adjacent in the vertex order thus induced.

If there are m genes on a chromosome, there are $2m$ vertices at this stage. The first and the last of these vertices are called telomeres. We convert all the telomeres in genome G_1 and G_2 into adjacencies with additional vertices all labelled T_1 or T_2, respectively. The breakpoint graph has a blue edge connecting the vertices in each adjacency in G_1 and a red edge for each adjacency in G_2. We make a cycle of any path ending in two T_1 or two T_2 vertices, connecting them by a red or blue edge, respectively, while for a path ending in a T_1 and a T_2, we collapse them to a single vertex denoted "T".

Each vertex is now incident to exactly one blue and one red edge. This bi-coloured graph decomposes uniquely into κ alternating cycles. If n' is the number of blue edges, then [18]:

$$d(G_1, G_2) = n' - \kappa. \tag{1}$$

3.2   The Median Problem and Small Phylogeny Problem

Let G_1, G_2 and G_3 be three genomes on the same set of n genes. *The rearrangement median problem is to find a genome M such that $d(G_1, M) + d(G_2, M) + d(G_3, M)$ is minimal.*

For a given unrooted binary tree T on N given genomes $G_1, G_2, \cdots, G_N$ (and thus with $N - 2$ unknown ancestral genomes $M_1, M_2, \cdots, M_{N-2}$ and $2N - 3$ branches), *the small phylogeny problem is to infer the ancestral genomes so that the total edge length of T, namely $\sum_{XY \in E(T)} d(X, Y)$, is minimal.*

The computational complexity of the median problem, which is just the small phylogeny problem with $N = 3$, is known to be NP-hard and hence so is that of the general small phylogeny problem.

4 The OMG Problem

4.1 Pairwise Orthologies

As justified in the Introduction, we construct sets of orthologous genes across the set of genomes by first identifying pairwise synteny blocks of genes. In our study, genomic data were obtained and homologies identified within synteny blocks, using the SYNMAP tool in COGE [17,1]. This was applied to the six dicot genomes in COGE shown in Fig. 1, i.e., to 15 pairs of genomes. We repeated all the analyses to be described here using the default parameters of SYNMAP, with minimum block size 1, 2, 3 and 5 genes.

4.2 Multi-genome Orthology Sets

The pairwise homologies SYNMAP provides for all 15 pairs of genomes constitute the set of edges E of the *homology graph* $H = (V, E)$, where V is the set of genes in any of the genomes participating in at least one homology relation.

The understanding of orthologous genes in two genomes as originating in a single gene in the most recent common ancestor of the two species, leads logically to transitivity as a necessary consequence. If gene x in genome X is orthologous both to gene y in genome Y and gene z in genome Z, then y and z must also be orthologous, even if SYNMAP does not detect any homology between y and z.

Ideally, then, all the genes in a connected component of H should be orthologous. Insofar as SYNMAP resolves all relations of paralogy, we should expect *at most* one gene from each genome in such an orthology set, or two for genomes that descend from a WGD event. We refer to such a set as *clean*.

In practice, gene x in genome X may be identified as homologous to both y_1 and y_2 in genome Y. Or x in X is homologous both to gene y_1 in genome Y and gene z in genome Z, while z is also homologous to y_2. By transitivity, we again obtain that x is homologous to both y_1 and y_2 in the same genome. While one gene being homologous to several paralogs in another genome is commonplace and meaningful, this should be relatively rare in the output from SYNMAP, where syntenic correspondence is a criterion for resolving paralogy. Aside from tandem duplicates, which do not interfere with gene order calculations, and duplicates stemming from WGD events, we consider duplicate homologs in the same genome, inferred directly by SYNMAP or indirectly by being members of the same connected component, as evidence of error or noise.

To "clean" a connected component with duplicate homologs in the same genome (or more than two in the case of a WGD descendant), we delete a number of edges, so that it decomposes into smaller connected components, each one of which is clean. To decide which edges to change, we define an objective function

$$F(X) = \sum_i \sum_{G, i \notin G} C_X(i, G),$$
(2)

where i ranges over all genes and G ranges over all genomes, and $C_X(i, G) = 1$ if there is exactly one one edge connecting i to any of the genes j in G (possibly two such edges if G descends from a WGD event), otherwise $C_X(i, G) = 0$.

A global optimum, maximizing F, the exact solution of the "orthology for multiple genomes" (OMG) problem, would be hard to compute; instead we chose edges in H one at a time, to delete, so that the increase in F is maximized. We continue in this greedy way until we obtain a graph H^* with all clean connected components . Note that F is designed to penalize the decrease of $C_X(i, G) = 1$ to $C_X(i, G) = 0$ by the removal of the only homology relation between gene i and some gene in genome G, thus avoiding the trivial solution where $H*$ contains no edges.

Note that it is neither practical nor necessary to deal with H in its entirety, with its hundred thousand or so edges. It suffices to do the cleaning on each connected component independently. Typically, this will contain only a few genes and very rarely more than 100. The output of the cleaning is generally a decomposition of the homology set into two or more smaller, clean, sets. These we consider our orthology sets to input into the gene order reconstruction step.

5   PATHGROUPS

Once we have our solution to the OMG problem on the set of pairwise syntenies, we can proceed to reconstruct the ancestral genomes. First, we briefly review the PATHGROUPS approach (previously detailed in [9,10]) as it applies to the median problem with three given genomes and one ancestor to be reconstructed, *all having the same gene complement*. The same principles apply to the simultaneous reconstruction of all the ancestors in the small phylogeny problem, and to the incorporation of genomes having previously undergone WGD.

We redefine a path to be any connected subgraph of a breakpoint graph, namely any connected part of a cycle. Initially, each blue edge in the given genomes is a path. A *fragment* is any set of genes connected by red edges in a linear order. The set of fragments represents the current state of the reconstruction procedure. Initially the set of fragments contains all the genes, but no red edges, so each gene is a fragment by itself.

The objective function for the small phylogeny problem consists of the sum of a number of genomic distances, one distance for every branch in the phylogeny. Each of these distances corresponds to a breakpoint graph. A given genome determines blue edges in one breakpoint graph, while the red edges correspond to the ancestral genome being constructed. For each such ancestor, *the red edges are identical in all the breakpoint graphs corresponding to distances to that ancestor.*

A pathgroup is a set of three paths, all beginning with the same vertex, one path from each partial breakpoint graph currently being constructed. Initially, there is one pathgroup for each vertex.

Our main algorithm aims to construct three breakpoint graphs with a maximum aggregate number of cycles. At each step it adds an identical red edge to each path in the pathgroup, altering all three breakpoint graphs. It is always possible to create one cycle, at least, by adding a red edge between the two ends of any one of the paths. The strategy is to create as many cycles as possible. If alternate choices of steps create the same number of cycles, we choose one that sets up the best configuration for the next step. In the simplest formulation, the pathgroups are prioritized, 1. by the maximum number of cycles that can be created within the group, without giving rise to circular chromosomes, and
2. for those pathgroups allowing equal numbers of cycles, by considering the maximum number of cycles that could be created in the next iteration of step 1, in any one pathgroup affected by the current choice.

By maintaining a list of pathgroups for each priority level, and a list of fragment endpoint pairs (initial and final), together with appropriate pointers, the algorithm requires $O(n)$ running time.

In the current implementation of PATHGROUPS, much greater accuracy, with little additional computational cost, is achieved by designing a refined set of 163 priorities, based on a two-step look-ahead greedy algorithm.

5.1   Inferring the Gene Content of Ancestral Genomes

The assumption of equal gene content simplifies the mathematics of PATHGROUPS and allows for rapid computation. Unfortunately it also drastically reduces the number of genes available for ancestral reconstruction, so that the method loses its utility when more than a few genomes are involved.

Allowing unequal gene complements in the data genomes, we have to decide how to construct the gene complement of the ancestors.

Using dynamic programming on unrooted trees, our assignment of genes to ancestors simply assures that if a gene is in at least two of the three adjacent nodes of an ancestral genome, it will be in that ancestor. If it is in less than two of the adjacent nodes, it will be absent from the ancestor.

5.2   Median and Small Phylogeny Problems with Unequal Genomes

To generalize our construction of the three breakpoint graphs for the median problem to the case of three unequal genomes, we set up the pathgroups much as before, and we use the same priority structure. Each pathgroup, however, may have three paths, as before, or only two paths, if the initial vertex comes from a gene absent from one of the leaves. Moreover, when one or two cycles are completed by drawing a red edge, this edge must be left out of the third breakpoint graph if the corresponding gene is missing from the third genome.

The consequence of this strategy is that some of the paths in the breakpoint graph will never be completed into cycles, impeding the search for optimality.

We could continue to search for cycles, but this would be computationally costly, spoiling the linear run time property of the algorithm.

The small phylogeny problem can be formulated and solved using the same principles as the median problem, as with the case of equal genomes. The solution, however, only serves as an initialization. As in [10], the solution can be improved by applying the median algorithm to each ancestral node in turn, based on the three neighbour nodes, and iterating until convergence. The new median is accepted if the sum of the three branch lengths is no greater than the existing one. This strategy is effective in avoiding local minima.

6   Results on Rosid Evolution

In the process of reconstructing the ancestors, we can also graphically demonstrate the great spread in genome rearrangement rates among the species studied, in particular the well-known conservatism of the grapevine genome, as illustrated by the branch lengths in Fig. 2.

It has been suggested recently that the order Malpighiales should be assigned to the malvids rather than the fabids [28]. In our results, the tree supporting this suggestion is indeed more parsimonious than the more traditional one. However, based on the limited number of genomes at our disposal, this is not conclusive.

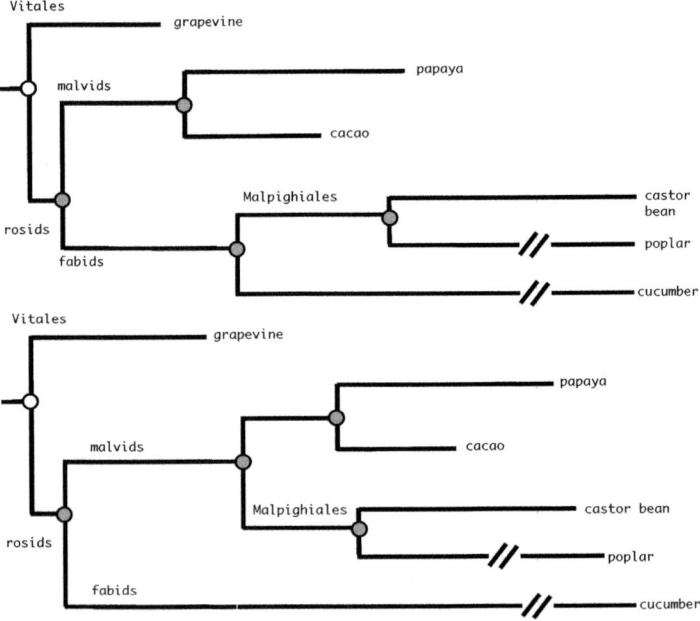

Fig. 2. Competing hypotheses for the phylogenetic assignment of the Malpighiales, with branch lengths proportional to genomic distances, following the reconstruction of the ancestral genomes with PATHGROUPS

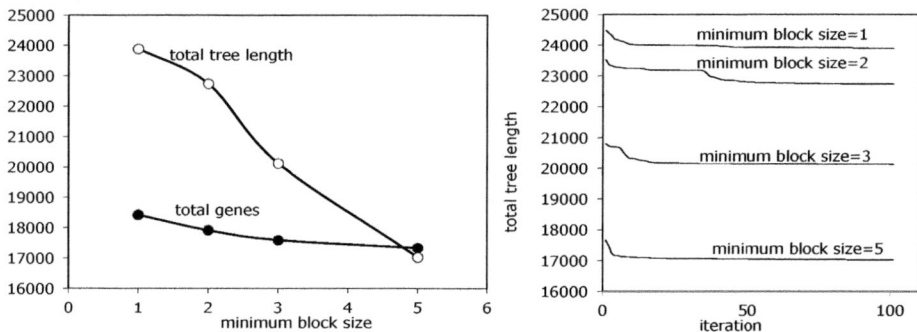

Fig. 3. Left: Effect of minimum block size on number of orthology sets and total tree length. Right: Convergence behaviour as a function of minimum block size.

6.1 Properties of the Solution as a Function of Synteny Block Size

To construct the trees in Fig. 2, from the 15 pairwise comparisons of the gene orders of the six dicot genomes, we identified some 18,000 sets of orthologs using SYNMAP and the OMG procedure. This varied surprisingly little as the minimum size for a synteny block was set to 1, 2, 3 or 5, as in Fig. 3. On the other hand, the total tree length was quite sensitive to minimum synteny block size. This can be interpreted in terms of risky orthology identifications for small block sizes.

Of the 18,000 orthology sets, the number of genes considered on each branch ranged from 12,000 to 15,000. When the minimum block size is 5, the typical branch length over the 11 branches of the tree (including one branch from each WGD descendant to its perfectly doubled ancestor plus one from that ancestor to a speciation node) is about 1600, so that $\frac{d}{n}$ is around 0.12, a low value for which simulations have shown PATHGROUPS to be rather accurate [10].

Fig. 3 shows the convergence behaviour as the set of medians algorithms is repeated at each ancestral node. Each iteration required about 8 minutes on a MacBook.

6.2 Block Validation

To what extent do the synteny blocks output by SYNMAP for a pair of genomes appear in the reconstructed ancestors on the path between these two genomes in the phylogeny? Answering this could validate the notion of syntenic conservation implicit in the block construction. Since our reconstructed ancestral genomes are not in the curated CoGE database (and are lacking the DNA sequence version required of items in the database), we cannot use SYNMAP to construct synteny blocks between modern and ancestor genomes. We can only see if the genes in the original pairwise syntenies tend to be colinear as well in the ancestor.

On the path connecting grapevine to cacao in the phylogeny in Fig. 1, there are two ancestors, the malvid ancestor and the rosid ancestor. There are 308 syntenic blocks containing at least 5 genes in the output of SYNMAP. A total of 11,229 genes are involved, of which 10,872 and 10,848 (97 %) are inferred to be in the malvid and rosid ancestor respectively.

Table 1. Integrity of cacao-grapevine syntenic blocks

synteny breaks	malvid ancestor		rosid ancestor	
	number	intra-block movement ≤ 1.0)	number	intra-block movement ≤ 1.0
0	140 (45%)	126 (90%)	153 (50%)	146 (95%)
1	66 (21%)	62 (94%)	64(21%)	58 (91%)
2	42 (14%)	39 (93%)	47(15%)	37 (79%)
> 2	60 (19%)	58 (97%)	44(14%)	38 (86%)

Table 1 shows that in each ancestor, roughly half of the blocks appear intact. This is indicated by the fact there are zero syntenic breaks in these blocks (no rearrangement breakpoints) and the average amount of relative movement of adjacent genes within these blocks is less than one gene to the left or right of its original position almost all of the time. Most of the other blocks are affected by one or two breaks, largely because the ancestors can be reconstructed with confidence by PATHGROUPS only in terms of a few hundred chromosomal fragments rather than intact chromosomes, for reasons given in Section 6.1. And it can be seen that the average shuffling of genes within these split blocks is little different from in the intact blocks.

7   Discussion and Future Work

We have developed a methodology for reconstructing ancestral gene orders in a phylogenetic tree, minimizing the number of genome rearrangements they imply over the entire tree. The input is the set of synteny blocks produced by SYNMAP for all pairs of genomes. The two steps in this method, OMG and PATHGROUPS, are parameter-free. Our method rapidly and accurately handles large data sets (tens of thousands of genes per genome, and potentially dozens of genomes). There is no requirement of equal gene complement.

For larger numbers of genomes, a problem would become the quadratic increase in the number of pairs of genomes, but this can be handled by SYNMAP only to pairs that are relatively close phylogenetically.

Future work will concentrate first on ways to complete cycles in the breakpoint graph which are currently left as paths, without substantially increasing computational complexity. This will increase the accuracy (optimality) of the results. Second, to increase the biological utility of the results, a post-processing component will be added to differentiate regions of confidence in the reconstructed genomes from regions of ambiguity.

Acknowledgments

Thanks to Victor A. Albert for advice, Eric Lyons for much help and Nadia El-Mabrouk for encouragement in this work. Research supported by a postdoctoral fellowship to CZ from the NSERC, and a Discovery grant to DS from the same agency. DS holds the Canada Research Chair in Mathematical Genomics.

References

1. Lyons, E., et al.: Finding and comparing syntenic regions among Arabidopsis and the outgroups papaya, poplar and grape: CoGe with rosids. Plant Phys. 148, 1772–1781 (2008)
2. Tang, H., et al.: Unraveling ancient hexaploidy through multiply-aligned angiosperm gene maps. Genome Res. 18, 1944–1954 (2008)
3. Murphy, W.J., et al.: Dynamics of mammalian chromosome evolution inferred from multispecies comparative maps. Science 309, 613–617 (2005)
4. Ma, J., et al.: Reconstructing contiguous regions of an ancestral genome. Genome Res. 16, 1557–1565 (2006)
5. Adam, Z., Sankoff, D.: The ABCs of MGR with DCJ. Evol. Bioinform. 4, 69–74 (2008)
6. Ouangraoua, A., Boyer, F., McPherson, A., Tannier, É., Chauve, C.: Prediction of Contiguous Regions in the Amniote Ancestral Genome. In: Salzberg, S.L., Warnow, T. (eds.) ISBRA 2009. LNCS, vol. 5542, pp. 173–185. Springer, Heidelberg (2009)
7. Gordon, J.L., Byrne, K.P., Wolfe, K.H.: Additions, losses, and rearrangements on the evolutionary route from a reconstructed ancestor to the modern *Saccharomyces cerevisiae* genome. PLoS Genet. 5, 1000485 (2009)
8. Tannier, E.: Yeast ancestral genome reconstructions: The possibilities of computational methods. In: Ciccarelli, F.D., Miklós, I. (eds.) RECOMB-CG 2009. LNCS, vol. 5817, pp. 1–12. Springer, Heidelberg (2009)
9. Zheng, C.: PATHGROUPS, a dynamic data structure for genome reconstruction problems. Bioinformatics 26, 1587–1594 (2010)
10. Zheng, C., Sankoff, D.: On the PATHGROUPS approach to rapid small phylogeny. BMC Bioinformatics 12(Suppl 1), S4 (2011)
11. Bertrand, D., et al.: Reconstruction of ancestral genome subject to whole genome duplication, speciation, rearrangement and loss. In: Moulton, V., Singh, M. (eds.) WABI 2010. LNCS, vol. 6293, pp. 78–89. Springer, Heidelberg (2010)
12. Soltis, D.E., et al.: Polyploidy and angiosperm diversification. Am. J. Bot. 96, 336–348 (2009)
13. Burleigh, J.G., et al.: Locating large-scale gene duplication events through reconciled trees: implications for identifying ancient polyploidy events in plants. J. Comp. Biol. 16, 1071–1083 (2009)
14. Langham, R.A., et al.: Genomic duplication, fractionation and the origin of regulatory novelty. Genetics 166, 935–945 (2004)
15. Thomas, B.C., Pedersen, B., Freeling, M.: Following tetraploidy in an *Arabidopsis* ancestor, genes were removed preferentially from one homeolog leaving clusters enriched in dose-sensitive genes. Genome Res. 16, 934–946 (2006)
16. Sankoff, D., Zheng, C., Zhu, Q.: The collapse of gene complement following whole genome duplication. BMC Genomics 11, 313 (2010)
17. Lyons, E., Freeling, M.: How to usefully compare homologous plant genes and chromosomes as DNA sequences. Plant J. 53, 661–673 (2008)
18. Yancopoulos, S., Attie, O., Friedberg, R.: Efficient sorting of genomic permutations by translocation, inversion, and block interchange. Bioinformatics 21, 3340–3346 (2005)

19. El-Mabrouk, N., Sankoff, D.: The reconstruction of doubled genomes. SIAM J. Comput. 32, 754–792 (2003)
20. Argout, X., et al.: The genome of *Theobroma cacao*. Nat. Genet. 43, 101–108 (2011)
21. Chan, A.P., et al.: Draft genome sequence of the oilseed species *Ricinus communis*. Nat. Biotechnol. 28, 951–956 (2010)
22. Haung, S., et al.: The genome of the cucumber, *Cucumis sativus* L. Nat. Genet. 41, 1275–1281 (2010)
23. Jaillon, O., et al.: The grapevine genome sequence suggests ancestral hexaploidization in major angiosperm phyla. Nature 449, 463–467 (2007)
24. Velasco, R., et al.: A high quality draft consensus sequence of the genome of a heterozygous grapevine variety. PLoS ONE 2, e1326 (2007)
25. Ming, R., et al.: The draft genome of the transgenic tropical fruit tree papaya (Carica papaya Linnaeus). Nature 452, 991–996 (2008)
26. Tuskan, G.A., et al.: The genome of black cottonwood, Populus trichocarpa (Torr. & Gray). Science 313, 1596–1604 (2006)
27. Forest, F., Chase, M.W.: Eudicots. In: Hedges, S.B., Kumar, S. (eds.) The Timetree of Life, pp. 169–176. Oxford University Press, Oxford (2009)
28. Shulaev, V., et al.: The genome of woodland strawberry (*Fragaria vesca*). Nat. Genet. 43, 109–116 (2011)

Algorithms to Detect Multiprotein Modularity Conserved during Evolution

Luqman Hodgkinson and Richard M. Karp

Computer Science Division, University of California, Berkeley,
Center for Computational Biology, University of California, Berkeley,
and the International Computer Science Institute
luqman@berkeley.edu, karp@icsi.berkeley.edu

Abstract. Detecting essential multiprotein modules that change infrequently during evolution is a challenging algorithmic task that is important for understanding the structure, function, and evolution of the biological cell. In this paper, we present a linear-time algorithm, Produles, that improves on the running time of previous algorithms. We present a biologically motivated graph theoretic set of algorithm goals complementary to previous evaluation measures, demonstrate that Produles attains these goals more comprehensively than previous algorithms, and exhibit certain recurrent anomalies in the performance of previous algorithms that are not detected by previous measures.

Keywords: modularity, interactomes, evolution, algorithms.

1 Introduction

Interactions between proteins in many organisms have been mapped, yielding large protein interaction networks, or interactomes [1]. The present paper continues a stream of scientific investigation focusing on conservation of modular structure of the cell, such as protein signaling pathways and multiprotein complexes, across organisms during evolution, with the premise that such structure can be described in terms of graph theoretic properties in the interactomes [2,3,4,5,6]. This stream of investigation has led to many successes, discovering conserved modularity across a wide range of evolutionary distances. However, there remain many challenges, such as running time, false positive predictions, coherence of predicted modules, and absence of a comprehensive collection of evaluation measures.

Evidence of conservation in the interaction data across organisms is essential for modules claimed by an algorithm to be conserved over a given evolutionary distance [7]. Due to the additivity of the scoring function for some previous algorithms in the interaction densities across organisms, a very dense network in one organism can be aligned with homologous proteins in another organism that have zero or few interactions among them. In this case, the interaction data does not support the claim of conservation across the given organisms.

J. Chen, J. Wang, and A. Zelikovsky (Eds.): ISBRA 2011, LNBI 6674, pp. 111–122, 2011.

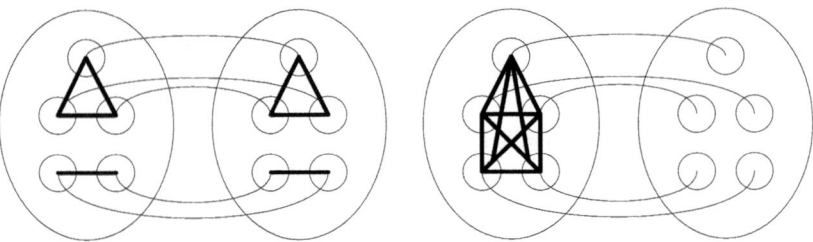

Fig. 1. Cartoons describing difficulties with additivity across data types and organisms. Organisms are represented by ovals. Proteins are represented by circles. Protein interactions are represented by thick lines. Proteins with high sequence similarity are connected with thin lines. Algorithms that are additive across the interaction and sequence data may predict the module on the left to be conserved due to high sequence similarity. In this case, the module boundaries are not well-defined, most likely containing portions of multiple modules that may have no relation with each other. Algorithms that are additive in the interactions across organisms may predict the module on the right to be conserved though there is no evidence for module conservation across the organisms in the protein interaction data.

Good boundaries are important for the modules that are returned by an algorithm. Some previous algorithms, such as NetworkBlast [3] and Graemlin [5], use a scoring function that is a sum of multiple scores: one score based on protein sequence similarity, and one score from each organism based on the density of interactions among the module proteins in the interactome for that organism. These algorithms then use a greedy search on this scoring function to find conserved modules. Due to the additivity, module pairs similar to the cartoons in Fig. 1 may receive high scores and be reported as conserved.

Produles is an important step to address these issues. It runs in linear time, scaling better than Match-and-Split [6] and MaWISh [4], and does not exhibit the recurrent anomalies that result from the additivity of the scoring function across organisms and data sources that forms the basis for NetworkBlast and Graemlin. Our objective is to initiate discussion of evaluation measures that are sensitive to these and similar issues by introducing the important algorithm goals described in Section 3.

2   Algorithms

Input Data

An interactome is an undirected graph $G = (V, E)$, where V is a set of proteins and $(v_1, v_2) \in E$ iff protein v_1 interacts with protein v_2. In this study the input is restricted to a pair of interactomes, $G_i = (V_i, E_i)$, for $i \in \{1, 2\}$, and protein sequence similarity values, $h : V_1 \times V_2 \to \mathbb{R}^+$, defined only for the most sequence similar pairs of proteins appearing in the interactomes. In this study, h is derived from BLAST [8] E-values. As BLAST E-values change when the order of the interactomes is reversed, h is defined with the rule

$$h(v_1, v_2) = h(v_2, v_1) = \frac{E(v_1, v_2) + E(v_2, v_1)}{2}$$

where $E(v_1, v_2)$ is the minimum BLAST E-value for $v_1 \in V_1, v_2 \in V_2$ when v_1 is tested for homology against the database formed by V_2. Any algorithm using only this data is a general tool as it can be easily applied to any pair of interactomes, including those for newly studied organisms.

Modularity, Conductance, and Degree Bounds

A modular system consists of parts organized in such a way that strong interactions occur within each group or module, but parts belonging to different modules interact only weakly [9]. Following this, a natural definition of multiprotein modularity is that proteins within a module are more likely to interact with each other than to interact with proteins outside of the module. Let $G = (V, E)$ be an interactome. A multiprotein module is a set of proteins $M \subset V$ such that $|M| \ll |V|$ and M has a large value of

$$\mu(M) = \frac{|E(M)|}{|\mathrm{cut}(M, V \backslash M)| + |E(M)|}$$

where $E(M)$ is the set of interactions with both interactants in M, and $\mathrm{cut}(M, V \backslash M)$ is the set of interactions spanning M and $V \backslash M$. Of the interactions involving proteins in M, the fraction contained entirely within M is given by $\mu(M)$. This is similar to the earlier definition of λ-module [10].

The conductance of a set of vertices in a graph is defined as

$$\Phi(M) = \frac{|\mathrm{cut}(M, V \backslash M)|}{|\mathrm{cut}(M, V \backslash M)| + 2\min(|E(M)|, |E(V \backslash M)|)}.$$

When $|E(M)| \leq |E(V \backslash M)|$, as for all applications in this study,

$$\Phi(M) = \frac{|\mathrm{cut}(M, V \backslash M)|}{|\mathrm{cut}(M, V \backslash M)| + 2|E(M)|} = \frac{1 - \mu(M)}{1 + \mu(M)}.$$

Thus, when searching for relatively small modules in a large interactome, minimizing conductance is equivalent to maximizing modularity. This relationship allows us to modify powerful algorithms from theoretical computer science designed for minimizing conductance [11]. It has been shown that conductance in protein interaction networks is negatively correlated with functional coherence, validating both this definition of modularity and the notion that biological systems consist of functional modules [12].

Assuming we are searching for modules of size at most b with modularity at least d, the vertices in any such module have bounded degree. Let $\delta(u)$ be the degree of u in G.

Theorem 1. *If $d > 0$, the objective function in the optimization problem*

$$\max_{G,M,u} \quad \delta(u)$$

$$s.t. \quad u \in M$$

$$|M| = b$$

$$\mu(M) \geq d$$

$$\mu(M) > \mu(M \backslash \{u\})$$

satisfies the bound $\delta(u) < (b-1)(1+d)/d$.

Proof. Let $M' \triangleq M \backslash \{u\}$. Let $y \triangleq |E(M')|$. Let $x \triangleq |\text{cut}(M', \{u\})|$.

$$\mu(M') = \frac{y}{|\text{cut}(M', V \backslash M')| + y} < \mu(M)$$

so

$$|\text{cut}(M', V \backslash M')| > \frac{y(1 - \mu(M))}{\mu(M)}$$

Thus,

$$\mu(M) = \frac{x+y}{[\delta(u) - x] + [|\text{cut}(M', V \backslash M')| - x] + [x+y]} < \frac{x+y}{\delta(u) - x + y + \frac{y(1-\mu(M))}{\mu(M)}}$$

which implies

$$\mu(M) < \frac{x}{\delta(u) - x}$$

As $\mu(M) \geq d$,

$$\delta(u) < \frac{x(1+d)}{d} \leq \frac{(b-1)(1+d)}{d} \qquad \qquad \square$$

The motivation for the restriction $\mu(M) > \mu(M \backslash \{u\})$ is that when searching for modules with high modularity, there may be proteins with such high degrees that it always improves the modularity to remove them from the module. It can be shown that this bound is tight and that neither requiring connectivity of M in the underlying graph nor requiring connectivity of $M \backslash \{u\}$ in the underlying graph can allow the bound to be further tightened.

Modularity Maximization Algorithm

PageRank-Nibble [11] is an algorithm for finding a module with low conductance in a graph $G = (V, E)$. Let A be the adjacency matrix for G. Let D be a diagonal matrix with diagonal entries $D_{ii} = \delta(i)$ where $\delta(i)$ is the degree of vertex i in G. Let $W = (AD^{-1} + I)/2$ where I is the identity matrix. W is a lazy random walk transition matrix that with probability $1/2$ remains at the current vertex and with probability $1/2$ randomly walks to an adjacent vertex. A PageRank vector is a row vector solution $\text{pr}(\alpha, s)$ to the equation

$$\text{pr}(\alpha, s) = \alpha s + (1 - \alpha)\text{pr}(\alpha, s)W^T$$

where $\alpha \in (0, 1]$ is a teleportation constant and s is a row vector distribution on the vertices of the graph called the preference vector. Define the distribution that places all mass at vertex v

$$\chi_v(u) = \begin{cases} 1 \text{ if } u = v \\ 0 \text{ otherwise} \end{cases}$$

Intuitively, when $s = \chi_v$, a PageRank vector can be viewed as a weighted sum of the probability distributions obtained by taking a sequence of lazy random walk steps starting from v, where the weight placed on the distribution obtained after t walk steps decreases exponentially in t [11].

Let p be a distribution on the vertices of G. Let the vertices be sorted in descending order by $p(\cdot)/\delta(\cdot)$ where ties are broken arbitrarily. Let $S_j(p)$ be the set of the first j vertices in this sorted list. For $j \in \{1, ..., |V|\}$, the set $S_j(p)$ is called a sweep set [11].

The PageRank-Nibble algorithm consists of computing an approximate Page-Rank vector with $s = \chi_v$, defined as $\text{apr}(\alpha, s, r) = \text{pr}(\alpha, s) - \text{pr}(\alpha, r)$, where r is called a residual vector, and then returning the sweep set $S_j(\text{apr}(\alpha, \chi_v, r))$ with minimum conductance [11].

From the definition, if p is a vector that satisfies $p + \text{pr}(\alpha, r) = \text{pr}(\alpha, \chi_v)$, then $p = \text{apr}(\alpha, \chi_v, r)$. Thus, $0 = \text{apr}(\alpha, \chi_v, \chi_v)$. After initializing $p_1 = 0$, $r_1 = \chi_v$, the solution is improved iteratively. Each iteration, called a push operation, chooses an arbitrary vertex u such that $r_i(u)/\delta(u) \geq \epsilon$. Then $p_{i+1} = p_i$ and $r_{i+1} = r_i$ except for the following changes:

1. $p_{i+1}(u) = p_i(u) + \alpha r_i(u)$
2. $r_{i+1}(u) = (1 - \alpha)r_i(u)/2$
3. For each v such that $(u, v) \in E$, $r_{i+1}(v) = r_i(v) + (1 - \alpha)r_i(u)/(2\delta(u))$

Intuitively, $\alpha r_i(u)$ probability is sent to $p_{i+1}(u)$, and the remaining $(1 - \alpha)r_i(u)$ probability is redistributed in r_{i+1} using a single lazy random walk step [11].

Each push operation maintains the invariant [11]

$$p_i + \text{pr}(\alpha, r_i) = \text{pr}(\alpha, \chi_v)$$

When no additional pushes can be performed, the final residual vector r satisfies

$$\max_{u \in V} \frac{r(u)}{\delta(u)} < \epsilon$$

The running time for computing $\text{apr}(\alpha, \chi_v, r)$ is $\mathcal{O}(1/(\epsilon\alpha))$ [11]. If we set ϵ and α to constants, which is reasonable given their meanings, and if we consider only the first b sweep sets, the algorithm runs in constant time. As we desire the degrees of the vertices in the final set to be bounded, we do not consider any sweep sets that contain vertices with degree $(b - 1)(1 + d)/d$ or greater, and we also require connectivity in the underlying graph.

Algorithm to Detect Conservation

The algorithm begins by finding a multiprotein module,

$$M \subset V_1$$

with high modularity in G_1 using the algorithm described previously. Let

$$\mathcal{H}_T(M) = \{v \mid \exists \, u \in M \text{ such that } h(u,v) \leq T\}$$

Modules corresponding to the connected components of the subgraph of G_2 induced by $\mathcal{H}_T(M)$ are candidates for conservation with M. Let these modules be $N_1, N_2, ..., N_k$. For $i = 1, ..., k$, let

$$\mathcal{R}_T(M, N_i) = \{u \in M \mid \exists \, v \in N_i \text{ such that } h(u,v) \leq T\}$$

If the following are true:

$$a \leq |\mathcal{R}_T(M, N_i)| \leq b$$
$$a \leq |N_i| \leq b$$
$$\frac{1}{c}|N_i| \leq |\mathcal{R}_T(M, N_i)| \leq c|N_i|$$
$$\mu(\mathcal{R}_T(M, N_i)) \geq d$$
$$\mu(N_i) \geq d$$

where a is a lower bound on size, b is an upper bound on size, c is a size balance parameter, and d is a lower bound on desired modularity, and if $\mathcal{R}_T(M, N_i)$ yields a connected induced subgraph of G_1, then we report the pair $(\mathcal{R}_T(M, N_i), N_i)$ as a conserved multiprotein module.

Each protein is used exactly once as a starting vertex for the modularity maximization algorithm. A counter is maintained for each protein in G_1. When a protein is placed in a module by the modularity maximization algorithm, the counter for the protein is incremented. Each counter has maximum value e for some constant e. If the modularity maximization algorithm returns a module containing any protein with counter value e, the entire module is ignored. If a protein in G_1 is reported to be in a conserved module, the counter for the protein is set to e in order to reduce module overlap. When all proteins in G_1 have been used as starting vertices, the roles of G_1 and G_2 are reversed, and the entire process is repeated.

Proof of Linear Running Time

Each value of $h(v, \cdot)$ for $v \in V$ is considered only when constructing $\mathcal{H}_T(M)$ for $\{M : v \in M\}$, so each value of $h(v, \cdot)$ is considered at most e times. If v is stored at each vertex in $\mathcal{H}_T(M)$ when constructing $\mathcal{H}_T(M)$, then constructing $\mathcal{R}_T(M, N_i)$ is a union of vertex lists and does not require additional considerations of $h(v, \cdot)$ values. As for all $v \in V_1$,

$$|\{M : v \in M\}| \leq e$$

the number of consideration of h values is

$$\sum_M \sum_{v \in M} |h(v, \cdot)| = \sum_v \sum_{M:v \in M} |h(v, \cdot)|$$

$$\leq e \sum_v |h(v, \cdot)|$$

$$= e|h(\cdot, \cdot)|$$

After finding $\mathcal{H}_T(M)$, it is necessary to compute $N_1, N_2, ..., N_k$. This can be problematic if any of the vertices in $\mathcal{H}_T(M)$ have large degree, which could conceivably be as large as $|V_2| - 1$. However, as we desire N_i such that $\mu(N_i) \geq d$ and $|N_i| \leq b$, which ideally do not contain any vertex u such that $\mu(N_i \setminus \{u\}) > \mu(N_i)$, we can discard, by Theorem 1, any vertex $v \in \mathcal{H}_T(M)$ with degree in G_2 of $(b-1)(1+d)/d$ or greater. A modified depth-first search that transitions only among vertices in $\mathcal{H}_T(M)$ is then used to compute $N_1, N_2, ..., N_k$. This requires time

$$\mathcal{O}((\frac{(b-1)(1+d)}{d})|\mathcal{H}_T(M)|) = \mathcal{O}(|\mathcal{H}_T(M)|)$$

As

$$|\mathcal{H}_T(M)| \leq \sum_{v \in M} |h(v, \cdot)|$$

all of these depth-first searches over the full run of the algorithm require time

$$\mathcal{O}(\sum_M |\mathcal{H}_T(M)|) = \mathcal{O}(\sum_M \sum_{v \in M} |h(v, \cdot)|) = \mathcal{O}(|h(\cdot, \cdot)|)$$

For a given M, constructing all $\mathcal{R}_T(M, N_i)$ by a union of lists stored at the vertices in the N_i requires time $\mathcal{O}(\sum_i |N_i| b \log b) = \mathcal{O}(|\mathcal{H}_T(M)|)$. Testing for connectivity of a single $\mathcal{R}_T(M, N_i)$ with a modified depth-first search that transitions only among vertices in $\mathcal{R}_T(M, N_i)$ requires constant time as $|\mathcal{R}_T(M, N_i)| \leq b$ and as each vertex in M has degree bounded by $(b-1)(1+d)/d$. All of these constructions and depth-first searches over the full run of the algorithm can be completed in time $\mathcal{O}(\sum_M |\mathcal{H}_T(M)|) = \mathcal{O}(|h(\cdot, \cdot)|)$.

Computing the modularity of module $U \in \{N_i, \mathcal{R}_T(M, N_i)\}$ requires computing the sum of degrees of the vertices in U and the number of edges with both endpoints in U. These can be computed in constant time when $|U| \leq b$ as each vertex in U has degree bounded by $(b-1)(1+d)/d$.

3   Biologically Motivated Algorithm Goals

These goals address the challenges described in the introduction. Goal 1 is a measure of how many sequence dissimilar proteins participate in the module. Goal 2 is a measure of quality of module boundaries. Goal 3 is a measure of evidence for the claim of conservation in the interaction data. Goal 4 measures fit to an evolutionary model that includes interaction formation and divergence, protein duplication and divergence, and protein loss. Goal 5 measures proteome coverage and module overlap. We now quantify these goals mathematically.

Definition 1. *(Algorithm output)* Let k pairs of conserved modules returned by an algorithm be $\mathcal{M} = \{(M_1^i, M_2^i) \mid i \in \{1, ..., k\}\}$. Let $(M_1, M_2) \in \mathcal{M}$. Let $M \in \{M_1, M_2\}$.

Definition 2. *(Filled module)* Let $G_{int}(M) = (M, E(M))$.

Definition 3. *(Module homology graph)* Let $G_{hom}(M_1, M_2) = (M_1 \cup M_2, H(M))$, where, for $p_1 \in M_1$, $p_2 \in M_2$, $(p_1, p_2) \in H(M)$ iff $h(p_1, p_2)$ is defined.

Definition 4. *(Module size)* Let $S(M) = |M|$.

Definition 5. *(Module density)* Let $\Delta(M) = |E(M)| / \binom{|M|}{2}$.

Definition 6. *(Interaction components)* Let $C(M)$ be the number of connected components in $G_{int}(M)$.

Definition 7. *(Module average)* Let $f_a(M_1, M_2) = (f(M_1) + f(M_2))/2$, where $f \in \{\mu, S, \Delta, C\}$.

Definition 8. *(Module difference)* Let $f_d(M_1, M_2) = |f(M_1) - f(M_2)|$, where $f \in \{\mu, S, \Delta, C\}$.

Definition 9. *(Module overlap)* Let $\mathcal{O}^i(M_1^i, M_2^i) = \max_{j \neq i} \min\{|M_1^j \cap M_1^i|/|M_1^i|, |M_2^j \cap M_2^i|/|M_2^i|\}$. A value of $\mathcal{O}^i = x$ implies that no module pair $j \neq i$ exists that covers more than fraction x of each module in module pair i.

Definition 10. *(Ancestral protein)* Let $p = (P_1, P_2)$, where $P_1 \subseteq M_1, P_2 \subseteq M_2$, and $G_{hom}(P_1, P_2)$ consists of a single connected component.

Definition 11. *(Ancestral protein projection)* For ancestral protein $p = (P_1, P_2)$, P_i is the projection of p on M_i for $i \in \{1, 2\}$.

Definition 12. *(Ancestral module)* Let $M_a(M_1, M_2)$ be the set of ancestral proteins for (M_1, M_2). The arguments, M_1, M_2, may be omitted for brevity when the context is clear.

Definition 13. *(Relationship disagreement)* Let $p, q \in M_a$, where $p = (P_1, P_2)$, $q = (Q_1, Q_2)$. For $i, j \in \{1, 2\}$, relationship disagreement means there is an interaction in G_i between some $p' \in P_i$ and some $q' \in Q_i$, but no interaction in G_j between any $p'' \in P_j$ with any $q'' \in Q_j$. Let $\mathcal{R}(M_1, M_2)$ be the number of relationship disagreements.

Definition 14. *(Relationship evolution)* Let $E_r(M_1, M_2) = \mathcal{R}(M_1, M_2)/\binom{|M_a|}{2}$, the fraction of possible relationship disagreements.

Definition 15. *(Ancestral module projection)* For $i \in \{1, 2\}$, let $\pi_i(M_a) = \{P_i \mid (P_1, P_2) \in M_a \wedge P_i \neq \emptyset\}$.

Definition 16. *(Number of protein duplications)* Let $\mathcal{D}(M_1, M_2) = |M_1| - |\pi_1(M_a)| + |M_2| - |\pi_2(M_a)|$.

Definition 17. *(Protein duplication evolution)* Let $E_d(M_1, M_2) = \mathcal{D}(M_1, M_2)/$ $(|M_1| + |M_2| - 2)$, the fraction of possible protein duplications.

Definition 18. *(Number of protein losses)* Let $\mathcal{L}(M_1, M_2) = 2|M_a| - |\pi_1(M_a)| - |\pi_2(M_a)|$.

Definition 19. *(Protein loss evolution)* Let $E_\ell(M_1, M_2) = \mathcal{L}(M_1, M_2)/(|M_2| + |M_1|)$, the fraction of possible protein losses.

Definition 20. *(Ancestral components)* Let $C(M_a)$ be the number of connected components in a graph with vertex set M_a, where an edge is defined between two ancestral proteins $p, q \in M_a$ if any protein in the projection of p on M_i interacts with any protein in the projection of q on M_i, for some $i \in \{1, 2\}$.

Definition 21. *(Proteome coverage)* Let $\mathcal{C}_i = |\mathcal{U}_i|/|V_i|$, where $\mathcal{U}_i$ is the set of proteins from V_i that are part of conserved modules. Let $\mathcal{C} = (\mathcal{C}_1 + \mathcal{C}_2)/2$.

Goal 1. $|M_a|$ is the number of ancestral proteins and should be reasonably large for significant multiprotein modules.

Goal 2. Any value of $C(M_a) > 1$ implies that the module pair is not well-defined as there is no evidence that the various connected components belong in the same module.

Goal 3. Δ_d, C_a, and C_d should be reasonably low to provide evidence for the claim of conservation across organisms. This may be problematic for models that are additive in the interaction densities across organisms.

Goal 4. E_r, E_d, and E_ℓ should be reasonably low for a good fit with evolution.

Goal 5. $\mathcal{C}$ and k should be in reasonable ranges with a low average value of $\mathcal{O}^i$.

4   Experiments and Results

Produles, NetworkBlast-M [13], Match-and-Split [6], and MaWISh [4] were applied to iRefIndex [14] binary interactions, Release 6.0, for *Homo sapiens* and *Drosophila melanogaster*, consisting of 74,554 interactions on 13,065 proteins for *H. sapiens* and 40,004 interactions on 10,050 proteins for *D. melanogaster*. The evaluation was performed on the module pairs returned that had 5-20 proteins per organism. This removes a large number of module pairs from Match-and-Split and MaWISh that consist of modules on two or three proteins, single edges or triangles, and the few huge modules from MaWISh with nearly a thousand proteins each and $C(M_a) > 1$, that likely have little information content. This has little effect on NetworkBlast-M for which nearly all of its modules have 13-15 proteins per organism [15]. All programs were run with varying h threshold, corresponding to varying numbers of homologous protein pairs: $h = 10^{-100}$: 5,675 pairs, $h = 10^{-40}$: 25,346 pairs, $h = 10^{-25}$: 50,831 pairs, and $h = 10^{-9}$: 138,824 pairs. Considering only the module pairs from NetworkBlast-M with highest NetworkBlast-M

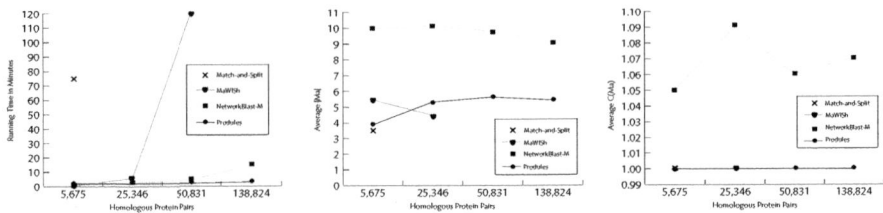

Fig. 2. Comparison of running time and performance on Goal 1 and Goal 2. The x-axis is the number of homologous protein pairs. The y-axis, from left to right, is the running time in minutes, the average $|M_a|$, and the average $C(M_a)$.

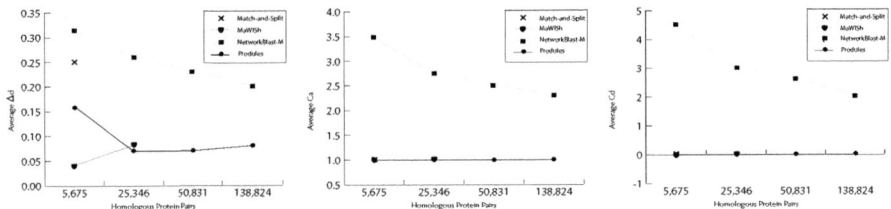

Fig. 3. Comparison of performance on Goal 3. The x-axis is as in Fig. 2. The y-axis, from left to right, is the average Δ_d, the average C_a, and the average C_d.

score does not significantly change the distributions [15]. Graemlin has nineteen network-specific parameters over a wide range of values. Together with the authors of Graemlin, we were unable to find settings that would yield results for the networks in this study.

As expected, Produles returns multiprotein modules with much higher values of μ than other approaches [15]. What is remarkable is that by focusing only on this measure, other desirable properties are attained. In Fig. 2, the linear running time of Produles is seen to be very desirable. Neither Match-and-Split nor MaWISh could complete on the data set with 50,831 homologous protein pairs. NetworkBlast-M has high average value of $|M_a|$ due mainly to its focus on modules with 13-15 proteins per organism. Both Match-and-Split and Produles guarantee that $C(M_a) = 1$. NetworkBlast-M has a high average value of $C(M_a)$ due to additivity across data types. MaWISh has a few module pairs with $C(M_a) > 1$ in a larger size range. Fig. 3 shows that NetworkBlast-M has difficulty with Goal 3 due to additivity of its scoring model in the interaction densities across organisms. NetworkBlast-M frequently aligns a dense module in one organism with a module that has zero or few interactions in the other organism. For all algorithms, average Δ_a is approximately 0.3 [15]. Fig. 4 shows that Produles performs comparably on the evolutionary model with algorithms that attempt to match topologies. By searching only for modularity, Produles detects conserved multiprotein module pairs in this data set that are consistent with evolution. Fig. 5 shows that NetworkBlast-M produces many overlapping module pairs. As indicated by Figs. 3-4, for many of these, the interaction data does not support the claim of conservation.

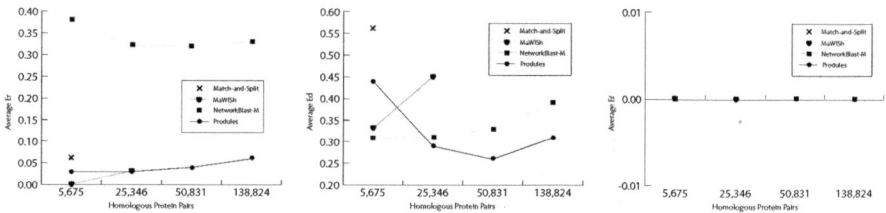

Fig. 4. Comparison of performance on Goal 4. The x-axis is as in Fig. 2. The y-axis, from left to right, is the average E_r, the average E_d, and the average E_ℓ.

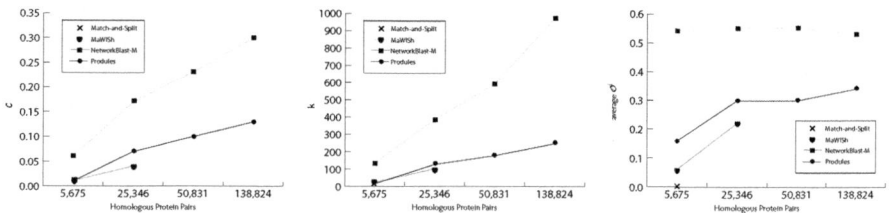

Fig. 5. Comparison of performance on Goal 5. The x-axis is as in Fig. 2. The y-axis, from left to right, is $\mathcal{C}$, k, and the average $\mathcal{O}^i$.

As a final test, we computed GO biological process enrichment [16] with Bonferroni correction at 0.05 significance level. All the algorithms in this study performed comparably for modules in each size range. More than 95% of Network-Blast-M modules were in the size range 13-15 proteins for which the percentage of modules enriched were: 100% for Produles, 98% for NetworkBlast-M, 100% for MaWISh, and 100% for Match-and-Split. All remaining NetworkBlast-M modules were in the size range 10-12 proteins for which the percentage of modules enriched were: 79% for Produles, 66% for NetworkBlast-M, 89% for MaWISh, with no modules in this size range for Match-and-Split.

5   Conclusion

We present a linear-time algorithm to detect conserved multiprotein modularity, and a new set of evaluation measures, comparing with leading algorithms and describing reasons for lower performance of earlier approaches. The measures introduced are sensitive to important issues not addressed by previous measures.

Acknowledgments. The authors would like to thank those who helped with the study: Maxim Kalaev with NetworkBlast-M, Jason Flannick and Antal Novak with Graemlin, Mehmet Koyutürk with MaWISh, Manikandan Narayanan with Match-and-Split, and Sabry Razick and Ian M. Donaldson with iRefIndex. This work is supported in part by NSF grant IIS-0803937. Bonnie Kirkpatrick graciously provided a critical reading of the manuscript for which the authors extend their warmest thanks.

References

1. Vidal, M.: Interactome modeling. FEBS Letters 579, 1834–1838 (2005)
2. Kelley, B.P., Sharan, R., Karp, R.M., Sittler, T., Root, D.E., Stockwell, B.R., Ideker, T.: Conserved pathways within bacteria and yeast as revealed by global protein network alignment. Proc. Natl. Acad. Sci. 100(20), 11394–11399 (2003)
3. Sharan, R., Suthram, S., Kelley, R.M., Kuhn, T., McCuine, S., Uetz, P., Sittler, T., Karp, R.M., Ideker, T.: Conserved patterns of protein interaction in multiple species. Proc. Natl. Acad. Sci. 102(6), 1947–1979 (2005)
4. Koyutürk, M., Kim, Y., Topkara, U., Subramaniam, S., Szpankowski, W., Grama, A.: Pairwise alignment of protein interaction networks. Journal of Computational Biology 13(2), 182–199 (2006)
5. Flannick, J., Novak, A., Srinivasan, B.S., McAdams, H.H., Batzoglou, S.: Graemlin: general and robust alignment of multiple large interaction networks. Genome Research 16, 1169–1181 (2006)
6. Narayanan, M., Karp, R.M.: Comparing protein interaction networks via a graph match-and-split algorithm. Journal of Computational Biology 14(7), 892–907 (2007)
7. Beltrao, P., Serrano, L.: Specificity and evolvability in eukaryotic protein interaction networks. PLoS Computational Biology 3(2), e25 (2007)
8. Altschul, S.F., Gish, W., Miller, W., Myers, E.W., Lipman, D.J.: Basic local alignment search tool. Journal of Molecular Biology 215(3), 403–410 (1990)
9. Simon, H.A.: The structure of complexity in an evolving world: the role of near decomposability. In: Callebaut, W., Rasskin-Gutman, D. (eds.) Modularity: Understanding the Development and Evolution of Natural Complex Systems. Vienna Series in Theoretical Biology. MIT Press, Cambridge (2005)
10. Li, M., Wang, J., Chen, J., Pan, Y.: Hierarchical organization of functional modules in weighted protein interaction networks using clustering coefficient. In: Măndoiu, I., Narasimhan, G., Zhang, Y. (eds.) ISBRA 2009. LNCS (LNBI), vol. 5542, pp. 75–86. Springer, Heidelberg (2009)
11. Andersen, R., Chung, F., Lang, K.: Local graph partitioning using PageRank vectors. In: 47th Annual IEEE Symposium on Foundations of Computer Science (FOCS 2006), pp. 475–486. IEEE Press, New York (2006)
12. Voevodski, K., Teng, S., Xia, Y.: Finding local communities in protein networks. BMC Bioinformatics 10, 297 (2009)
13. Kalaev, M., Bafna, V., Sharan, R.: Fast and accurate alignment of multiple protein networks. In: Vingron, M., Wong, L. (eds.) RECOMB 2008. LNCS (LNBI), vol. 4955, pp. 246–256. Springer, Heidelberg (2008)
14. Razick, S., Magklaras, G., Donaldson, I.M.: iRefIndex: a consolidated protein interaction database with provenance. BMC Bioinformatics 9, 405 (2008)
15. Hodgkinson, L., Karp, R.M.: Algorithms to detect multi-protein modularity conserved during evolution. EECS Department, University of California, Berkeley, Technical Report UCB/EECS-2011-7 (2011)
16. Boyle, E.I., Weng, S., Gollub, J., Jin, H., Botstein, D., Cherry, J.M., Sherlock, G.: Go:termfinder—open source software for accessing gene ontology information and finding significantly enriched gene ontology terms associated with a list of genes. Bioinformatics 20(18), 3710–3715 (2004)

The Kernel of Maximum Agreement Subtrees

Krister M. Swenson[1,3], Eric Chen[2],
Nicholas D. Pattengale[4], and David Sankoff[1]

[1]Department of Mathematics and Statistics, University of Ottawa, Ontario,
K1N 6N5, Canada
[2]Department of Biology, University of Ottawa, Ontario, K1N 6N5, Canada
[3]LaCIM, UQAM, Montréal Québec, H3C 3P8, Canada
[4]Sandia National Laboratories, Albuquerque, New Mexico

Abstract. A Maximum Agreement SubTree (MAST) is a largest subtree common to a set of trees and serves as a summary of common substructure in the trees. A single MAST can be misleading, however, since
there can be an exponential number of MASTs, and two MASTs for the
same tree set do not even necessarily share any leaves. In this paper we
introduce the notion of the Kernel Agreement SubTree (KAST), which is
the summary of the common substructure in all MASTs, and show that
it can be calculated in polynomial time (for trees with bounded degree).
Suppose the input trees represent competing hypotheses for a particular
phylogeny. We show the utility of the KAST as a method to discern the
common structure of confidence, and as a measure of how confident we
are in a given tree set.

1 Introduction

Phylogeny inference done on genetic data using maximum parsimony, maximum
likelihood, and Bayesian analyses usually yields a set of most likely trees (phylogenies). A typical approach used by biologists to discern the commonality of the
trees is to apply a consensus method which yields a single tree containing edges
that are well represented in the set. For example, the majority-rules consensus
tree contains only the edges (bipartitions of the leaf set) that exist in a majority
of input trees. Consensus methods are also commonly used for their original purpose [1], to summarize the information provided from *different* data sets (there
are other uses [32] but these are the two that we consider in this paper).

If one desires a more conservative summary, they may use the strict consensus
tree, which has an edge if and only if the edge exists in all of the input trees. Yet
even for this extremely conservative consensus method, there has been debate
as to its validity and the conditions under which it should be used [3,23,4]. In
particular, Barrett et al. [3] showed an example where a parsimony analysis of
two data sets yields a consensus tree that is at odds with the tree obtained by
combining the data. Nelson [23] replied with an argument that the error was not
the act of taking the consensus, but the act of pooling the data.

The issue at the heart of this debate is, essentially, that of *wandering* or *rogue*
leaves (taxa). Indeed, one or many leaves appearing in different locations of

J. Chen, J. Wang, and A. Zelikovsky (Eds.): ISBRA 2011, LNBI 6674, pp. 123–135, 2011.

otherwise identical trees have created the problems noticed by Barrett et al., and can also reduce the consensus tree to very few, if any, internal edges. On the other hand, Finden and Gordon [10] had already characterized Maximum Agreement SubTrees (MASTs): maximum cardinality subsets of the leaves for which all input trees agree. By calculating a MAST, one avoids Barrett's issue because all MASTs agree with the parsimonious tree they computed on the combined data. As we will see a single MAST can be misleading, however, as there can exist two MASTs (on a single set of trees) which share no leaves. Further, there are potentially an exponential (in the number of leaves) number of MASTs for a single set of trees [18]. For Barrett's example we will see that our new method appropriately excludes the contentious part of the tree, and so may be more fit than traditional consensus methods for comparing trees obtained from different analyses.

Wilkinson was the first to directly describe the issues surrounding rogue leaves and develop an approach to try to combat them [32]. Since then, a large body of work by Wilkinson and others has grown on the subjects of finding a single representative tree [32,33,34,31,11,24] or something other than a tree (forest, network, etc.) [2,17,9,15,26]. A full review of this work is out of the scope of this article so we refer the reader to the chapter of Bryant [8], the earlier work of Wilkinson [32,33], and Pattengale et al. [24]. Despite the myriad of options we notice a distinct lack of an efficiently computable base-line method for reporting subtrees of high confidence; a method analogous to the strict consensus, but less susceptible to rogue leaves. Thus, we introduce the Kernel Agreement SubTree (KAST) to summarize the information shared by all (potentially exponential) MASTs. Like the strict consensus, the KAST gives a summary of the common structure of high confidence, except that it excludes the rogue leaves that confound traditional consensus methods. The KAST has the benefits of having a simple definition, of summarizing the subtree of confidence by reporting a single tree, and unlike the other known subtree methods can be computed in polynomial time (when at least one input tree has bounded degree). Note that we do not use the term *kernel* in the machine learning sense (as in [28]).

When speaking of a reconstruction method that produces many most probable trees, Barrett et al.[3] called for "conservatism" and suggests the use of the strict consensus. In Section 5 we show the utility of the KAST as a means to get a conservative summary of many most probable trees. We then show the utility of the KAST in the original setting of consensus methods; on trees obtained through different analyses. In each setting we use the KAST not only to find subtrees of confidence, but as an indicator of randomness in the input trees.

The paper is organized as follows. We continue by formally defining the problem in Section 1.1 and showing properties of the MAST and KAST in Section 1.2. We then present Bryant's algorithm for computing the MAST in Section 2, on which our algorithm to compute the KAST (Section 3) is based on. Section 4 reports experimental values for the expected size of the KAST on various sets of trees generated at random while Section 5 shows how the KAST can be used to find subtrees of confidence, and report subsets of trees for which we are confident.

1.1 Definitions

Consider a set of trees $\mathcal{T} = \{T_1, T_2, \ldots, T_k\}$ and a set of labels L such that each $x \in L$ labels exactly one leaf of each T_i. We will restrict a tree to a subset L' of its leaf set L; $T_i|_{L'}$ is the minimum homeomorphic subtree of T_i which has leaves L'. An *agreement subtree* for $\mathcal{T}$ is a subset $L' \subseteq L$ such that $T_1|_{L'} = T_2|_{L'} = \cdots = T_k|_{L'}$. A *maximum agreement subtree* (MAST) is an agreement subtree of maximum size. The set of all maximum agreement subtrees is $\mathcal{M}$.

Definition 1. *The* Kernel Agreement SubTree (KAST) *is the intersection of all MASTs (i.e.* $\cap_{T \in \mathcal{M}} T$*).*

See Figure 1 for an example.

As usual, node a is an *ancestor* of b if the path from b to the root passes through a. b is a *descendant* of a. For nodes a and b, the least common ancestor $lca(a, b)$ is the ancestor of a and b that is a descendant of all ancestors of a and b.

1.2 Properties of a MAST and the KAST

In Section 3 we show that the KAST can be computed in the same time as the fastest known algorithm to compute the MAST, by a convenient use of dynamic programming. The current fastest know algorithms for the MAST problem are due to Farach et al. [13] and Bryant [7]. Let d_i be the maximum degree (number of children) of tree $T_i \in \mathcal{T}$. These algorithms run in $O(kn^3 + n^d)$ time where $n = |L|$, k is the number of trees in the input, and d is the minimum over all d_i, $1 \leq i \leq k$.

We devote this section to showing desirable properties of the KAST by contrasting it with the MAST. First we look at the role KAST can play in Barrett's example [3]. The rooted trees obtained by his parsimony analyses are $T_1 = (A, (B, (C, D)))$ and $T_2 = (A, ((B, C), D)))$ (written in Newick format). The set of maximum agreement subtrees for T_1 and T_2 is

$$\{(A, (B, D)), (A, (D, C)), (A, (B, C))\}.$$

Thus, the KAST has only a single leaf A, which indicates that there is not enough information to imply a subtree of confidence. This is the result we would prefer to see, given the circumstances. We see more examples in Section 5 that show a KAST which finds substantial common substructure, yet does not falter by including subtrees that are at odds with biological observation.

Take a tree set $\mathcal{T}$ with a MAST of size m. Adding a tree T to $\mathcal{T}$ cannot result in a MAST larger than m. This is due to the fact that an agreement subtree of $\mathcal{T} \cup \{T\}$ must also be an agreement subtree of $\mathcal{T}$. On the other hand, the signal for a particular kernel can become apparent when more trees that agree are added to the set.

Property 1. The KAST on tree set $\mathcal{T}$ can be smaller than that of $\mathcal{T} \cup T$, for some tree T.

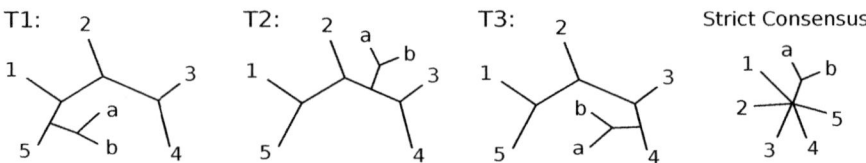

Fig. 1. The effect of adding a tree to the input set. The MASTs for $\{T_1, T_2\}$ are $\{1,2,3,4,5\}$, $\{a,b,1,3,4\}$, $\{a,b,2,3,4\}$, and $\{a,b,3,4,5\}$, yielding the KAST $\{3,4\}$. The MAST for $\{T_1, T_2, T_3\}$ is $\{1,2,3,4,5\}$, yielding the KAST $\{1,2,3,4,5\}$.

Figure 1 shows an example exhibiting this property. The KAST on input tree set $\{T_1, T_2\}$ has two leaves (is essentially empty) whereas the subtree on leaves $\{1,2,3,4,5\}$ is amplified by the addition of the tree T_3 to the set. We also see in Section 4 that the KAST size can often increase when adding somewhat similar trees to a set.

We finish with a few negative results about the MAST. The first shows that the MAST is not necessarily a good indicator of the common subtrees of confidence between two trees.

Property 2. There exists a family of tree sets that yields at least two MASTs, the intersection of which is size 2.

Take the caterpillar trees

$$(1, (2, (3, \ldots (n-1, n) \ldots))) \text{ and } (n/2, (n/2+1, \ldots, (n, (n/2-1, \ldots, (2,1) \ldots)) \ldots))$$

for even n. Two of the MASTs for these trees are $\{1, 2, \ldots, n/2, n/2 + 1\}$ and $\{n/2, \ldots, n-1, n\}$.

The second property shows that the number of MASTs and the size of them are not good indicators of their quality. We will see experimental evidence corroborating this fact in Section 4.

Property 3. There exists a family of tree sets that yields exactly two MASTs of size $\Omega(n)$, but the KAST is of size 4.

For this example we use trees that are nearly caterpillars. We write them as caterpillars, except S_1 denotes a subtree $(1a, 1b)$ while $S_{m/2+1}$ denotes a subtree $((m/2 + 1)a, (m/2 + 1)b)$. The first tree is then

$$(S_1, (2, (3, \ldots (S_{m/2+1}, \ldots (m-1, m) \ldots))))$$

and the second is

$$(n, (n-1, \ldots, (n/2 + 2, (S_1, (2, \ldots, (n/2, (S_{m/2+1})) \ldots))) \ldots))$$

where $m = n - 2$. The only two MASTs are now

$$\{1a, 1b, 2, \ldots, m/2, (m/2 + 1)a, (m/2 + 1)b\} \text{ and }$$
$$\{1a, 1b, (m/2 + 1)a, (m/2 + 1)b, m/2 + 2, \ldots, m-1, m\}.$$

2   A Dynamic Programming Algorithm to Find the MAST

While either of the fastest known algorithms [7,13] for finding a MAST can be adapted to compute the KAST, we find it instructive to describe the algorithm of Bryant. We are comprehensive in our description. However, we refer the reader to Bryant's dissertation [7] for a more precise description of the algorithm.

Take $a, b \in L$ and call $\mathcal{T}(a, b)$ the set of all agreement subtrees where the $lca(a, b)$ is the root of the tree. Let $\mathcal{M}(a, b) \subseteq \mathcal{T}(a, b)$ be the set of maximum agreement subtrees where $lca(a, b)$ is the root, and $MAST(a, b)$ be the number of leaves in any member of $\mathcal{M}(a, b)$. We devote the rest of this section to computing $MAST(a, b)$ since the size of the MAST is simply the maximum $MAST(a, b)$ over all possible a and b.

Take three leaves $a, b, c \in L$. $ac|b$ denotes a *rooted triple* where $lca(a, c)$ is a descendant of $lca(a, b)$. In this case we say that c is *on a's side of the root* with respect to b (when $lca(a, b)$ is the root). Leaves a,b, and c form a *fan triple*, written (abc), if $lca(a, b) = lca(a, c) = lca(b, c)$. Define R to be the set of rooted triples common to all trees in $\mathcal{T}$ and F to be the set of fan triples common to all trees in $\mathcal{T}$. Bryant showed that an agreement subtree in $\mathcal{T}$ is equivalent to a subset of the set of rooted and fan triples.

The algorithm to compute $MAST(a, b)$ hinges upon the fact that the triples on a's side of the root, and the triples on b's side of the root can be addressed independently. Consider the set $X = \{x : xa|b \in R\} \cup \{a\}$ such that $lca(a, b)$ is the root. In this case, X corresponds to the leaves in a subtree on a's side of the root. Define $MAST_a = max\{MAST(a, x) : x \in X\}$ to be the MAST of the leaves in a subtree on a's side of the root. $MAST_b$ is defined similarly, where $X = \{x : a|bx \in R\} \cup \{b\}$.

If F is empty (i.e. the root of every tree in $\mathcal{T}$ is binary), then we have simply,

$$MAST(a, b) = MAST_a + MAST_b.$$

Otherwise, consider the maximum size subset $C \subseteq F$ such that $(abc) \in F$ for $c \in C$. Again, $MAST_c$ is the MAST that considers only the vertices x such that $xc|b$. The triples corresponding to some $MAST_c$ are not the same as those for $MAST_a$ and $MAST_b$. However, $MAST_c$ and $MAST_{c'}$ for $c, c' \in C$ could correspond to the same triples. To avoid conflict we construct a graph $G(C)$ as follows: for each $c \in C$ create a vertex with weight $MAST_c$. Make an edge between v and w if and only if $(bvw) \in F$ (i.e. v and w have the potential to appear in a subtree from the root that does not include a or b). A maximum weight clique S in this graph is the MAST of all potential subtrees that do not include a or b. So $MAST(a, b)$ can be written

$$MAST(a, b) = MAST_a + MAST_b + \sum_{s \in S} MAST_s$$

where $MAST_s$ is defined similarly to $MAST_a$ but with $X = \{x : a|sx \in R\} \cup \{s\}$.

3   Finding the KAST

$KAST(a, b)$ is the intersection of all MASTs in $\mathcal{M}(a, b)$ (the MASTs where $lca(a, b)$ is the root). In this section we show how to compute $KAST(a, b)$ through a modification of the algorithm of section 2.

Let $\mathcal{M}_a$ be the set of all MASTs on the leaf set $\{x : xa|b \in R\}$. In other words, $\mathcal{M}_a$ is the collection of sets of leaves that correspond to some $MAST_a$. Call $L(\mathcal{M}_a)$ the set of leaves in any MAST in $\mathcal{M}_a$ (i.e. $L(\mathcal{M}_a) = \{z \in M : M \in \mathcal{M}_a\}$). Symmetrically, $\mathcal{M}_b = \{x : a|bx \in R\}$ and $L(\mathcal{M}_b) = \{z \in M : M \in \mathcal{M}_b\}$. We begin by showing how to find $KAST(a, b)$ for binary trees.

Theorem 1. *If the trees $T_1, T_2, \ldots, T_k$ are binary, then*

$$KAST(a, b) = (\cap_{T \in \mathcal{M}_a} T) \cup (\cap_{T \in \mathcal{M}_b} T)$$

Proof. If $a = b$ then this is trivially true. Assume by induction that $KAST(c, d)$ can be calculated where $lca(a, b)$ is an ancestor of $lca(c, d)$.

Recall that $MAST(a, b) = MAST_a + MAST_b$ when the trees in $\mathcal{T}$ are binary and that $\mathcal{M}_a$ is the set of MASTs that include only the leaves a and x such that $lca(a, b)$ is an ancestor of $lca(a, x)$. It follows that $\mathcal{M}_a$ and $\mathcal{M}_b$ have the following property:

$$L(\mathcal{M}_a) \cap L(\mathcal{M}_b) = \emptyset$$

So $KAST(a, b)$ depends on $\mathcal{M}_a$ and $\mathcal{M}_b$ independently.

Bryant showed that any leaf included in $\mathcal{M}_a$ or $\mathcal{M}_b$ will necessarily exist in some MAST for $\mathcal{T}$ (a corollary of theorem 6.8 in [7]). Since the KAST contains only the leaves that exist in every MAST, then $KAST(a, b)$ must be equal to the intersection of all MASTs in $\mathcal{M}_a$. □

So the algorithm to compute $KAST(a, b)$ takes the intersection over all sets $KAST(c, d)$ such that $ac|b, ad|b \in R$ and $MAST(c, d)$ is maximum. It does the same for b's side of the root, and then takes the union of the result.

The following theorem hints that the independence of subsolutions that gives rise to MAST dynamic programming algorithms will similarly give rise to a KAST algorithm.

Theorem 2. *If any MAST is such that two leaves x, y are on the same side of the root, it follows that every MAST containing both x and y will also have them on the same side of the root.*

Proof. A simple proof by contradiction suffices. Assume that the theorem does not hold, namely that there is another MAST containing both x and y where x occurs on the other side of the root from y. Since the second MAST has root $lca(x, y)$ (because x and y are on either side of the root), this implies that the second MAST is a valid subtree in the first MAST, a contradiction. □

The implication here is that in building KAST subsolutions for one side of the root under consideration, we need not worry about leaves that we exclude being candidates for inclusion on the other side of the root.

We now present the main result of this section. Recall from Section 2 that $MAST(a, b) = MAST_a + MAST_b + \sum_{s \in S} MAST_s$ and the graph $G(C)$ where C is the set of triples satisfying $(abc) \in F$.

Theorem 3. $KAST(a, b) = (\cap_{T \in \mathcal{M}_a} T) \cup (\cap_{T \in \mathcal{M}_b} T) \cup (\cap_{S \in \mathcal{K}} (\cup_{s \in S} (\cap_{T \in \mathcal{M}_s} T)))$ where $\mathcal{K}$ is the set of all maximum weight cliques on graph $G(C)$.

Proof. If $a = b$ then this is trivially true. Assume by induction that $KAST(c, d)$ can be calculated where $lca(a, b)$ is an ancestor of $lca(c, d)$.

Take any maximum weight clique $S \in \mathcal{K}$. Bryant showed that for $S = \{s_1, \ldots, s_m\}$, $\cup_{i=1}^{m} T_i$ where $T_i \in \mathcal{M}_{s_i}$, is a MAST on the set of leaves $\{c : (abc) \in C\}$. By the definition of $G(C)$ we know that $L(\mathcal{M}_{s_1}), L(\mathcal{M}_{s_2}), \ldots, L(\mathcal{M}_{s_m}), L(\mathcal{M}_a)$, and $L(\mathcal{M}_b)$ are pairwise disjoint. Further, any leaf in the sets $L(\mathcal{M}_{s_i})$, $L(\mathcal{M}_a)$, or $L(\mathcal{M}_b)$ are necessarily included in some MAST for $\mathcal{T}$ (a corollary of theorem 6.8 in [7]). So the leaves in a $KAST(a, b)$ could have only the leaves that are in every MAST in $\mathcal{M}_{s_i}$ (i.e. $(\cup_{s \in S} (\cap_{T \in \mathcal{M}_s} T)))$, for all $1 \leq i \leq m$. But each clique in $\mathcal{K}$ represents a different MAST, so only the leaves that are in every clique will be in the KAST. Finally, this set is disjoint from $L(\mathcal{M}_a)$ and $L(\mathcal{M}_b)$ for the same reason that $L(\mathcal{M}_a)$ and $L(\mathcal{M}_b)$ are disjoint from each other.   □

4   Experiments

We implemented the KAST from code that computes the MAST in the phylogenetic package RAxML [29]. In this section we report empirical evidence about the expected size of the KAST and MAST under two different models. The first model builds a tree set $\mathcal{T}$ of random trees constructed through a birth/death process, while the second starts with a random birth/death tree and then produces new trees by doing Nearest Neighbor Interchange (NNI) moves [27,22]. This way we see how the expected sizes react to adding drastically dissimilar, or fairly similar trees to the set $\mathcal{T}$. In Figure 2, we show that as the size of $\mathcal{T}$ (the tree set) increases, the size of the KAST decreases precipitously in the case of the birth/death model, whereas it decreases more gracefully in the case of the NNI model. Each plot is generated from an initial birth/death tree on 50 leaves, where new trees are added to the tree set according to the prescribed model. This process was repeated 10 times and the average is reported. Plots with various numbers of leaves are similar except that the curve is scaled on the "leaves" axis proportionately. In regards to the number of MASTs, the plots show an erratic curve, confirming that the phenomenon described in Property 3 is not a rarity.

5   Applications

We now demonstrate the application of the KAST in finding subtrees of confidence, as well as finding subsets of the input tree set of confidence. To do this we gleaned phylogenies from the literature that are known to have an agreed upon

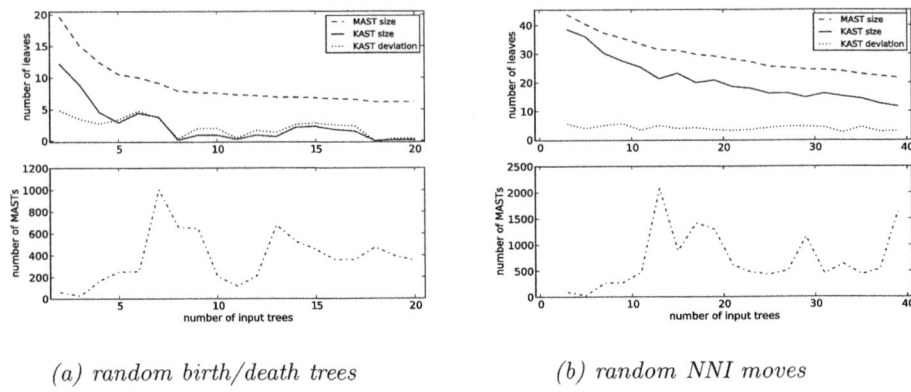

(a) random birth/death trees   (b) random NNI moves

Fig. 2. Expected values of the MAST and KAST sizes

structure, except for a few contentious leaves. Our intention is not to provide biological insight, but to confirm the utility of the KAST by comparing our results to familiar phylogenies. The real utility of the KAST will be on phylogenies that are much larger, so large as to make it difficult for a humans to process.

5.1   Analyses on Flatworm Phylogenies

In a recent publication by Philippe et al.[25], the proposed phylogeny describes the Acoel and the Nemertodermatids and Xenoturbellid as a sister-clade to Ambulacraria, which is vastly different from the previous publications. The competing hypotheses are depicted in Figure 3. In earlier publications both Nemertodermatids and Acoels are the outgroups with Xenoturbellid leaf grouping either with the Ambulacraria or with the Nemertodermatids and Acoels. Setting aside the interpretation and biological ramifications of the new proposed tree topology, it is a good real-world example for observing the effects of KAST on contentious trees.

There are two main objectives that we wish to explore through the use of this example. The first objective is to determine if the kernel of a set of phylogenetic trees can identify a subset that we are confident in. The second objective is to show the KAST as a measurement of how confident we are in the hypothesis of these trees.

Confidence in the phylogenetic tree. From their publication, Philippe et al.[25] presents three trees, two from prior publications and one from their own experimental result. With only 10 common leaves among the three trees it is very easy to identify the similarity between them by eye (Figure 3).

A quick observation can find that Ecdysozoans and Lophotrochozoans of Protostomia forms a clade, Vetebrates and Urochordates and Cephalochordates of Chordata forms a clade, and Hemichordates and Echinoderms of Ambulacraia forms a sister clade to Chordata; and that Xenoturbellid, Nemertodermatids, and Acoels are the rogue leaves. The KAST of these three trees agrees with this

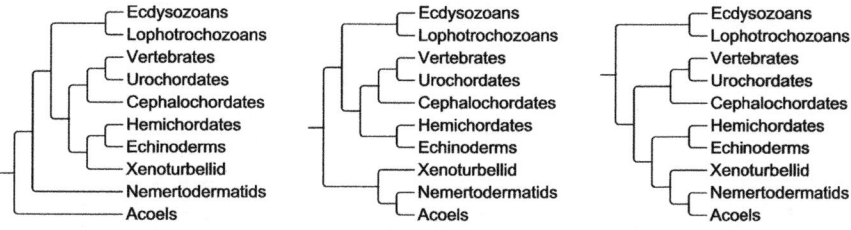

Fig. 3. Phylogenies from Figure 1 of Philippe et al.[25] Xenoturbellid, Nemertodermatids, and Acoels wander

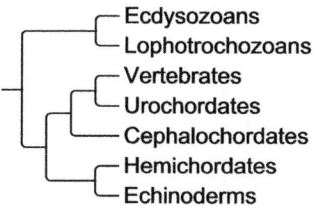

Fig. 4. The KAST of phylogenies of Figure 3

observation (Figure 4). This suggests that the KAST is able to find a subtree that is not only biologically obvious but also likely to have significant support.

With bigger trees it will be harder to identify the similarity. We would argue that the KAST can be an important tool in identifying or verifying these similarities.

Phylogeny reconstruction. To find if the kernel in a set of phylogenies could identify a subset of trees that we are confident of in the context of phylogeny reconstruction, we tried to replicate the analysis of Philippe et al.[25] The aligned mitochondrial gene set was taken from the supplementary material section and used as input for the Bayesian analysis that they used: PhyloBayes 3.2[20,19], with the CAT model[25,20] as the amino acid replacement model and default settings for everything else. We ran 10000 cycles and discarded the 1000 burn-ins, as they did. The consensus tree by majority rule was then obtained by CONSENSE [14] using all remaining 9000 trees. The consensus tree (Figure 5) we found is in agreement with the CAT + Γ model tree from their supplementary material Figure 1.

To test the validity of the conservative tree produced by the KAST, the kernel of the 9000 trees set is calculated. Of the 10 species in the KAST, three sponge species and two jellyfish species group together as predicted, the two annelida species group together as predicted, the three echinoderms also group together as predicted, and the topology of these phyla are also organized in the biologically obvious fashion (Figure 6). This corroborates the notion that the topology of KAST is the base-line topology.

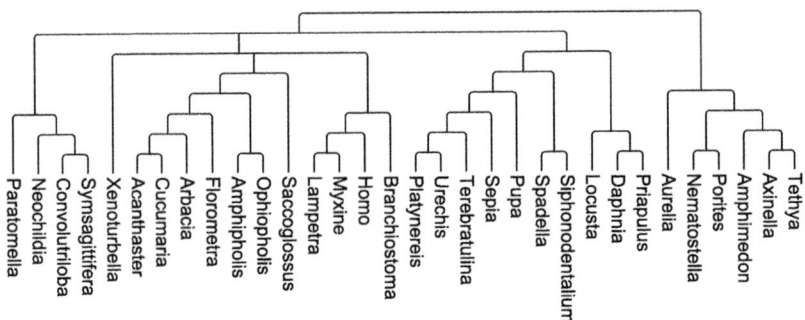

Fig. 5. The majority rule consensus tree using 9000 trees. The topology is essentially the same as Supplementary Figure 1 from Philippe et al.[25]

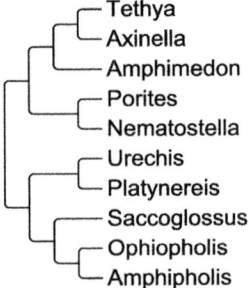

Fig. 6. The KAST of the 9000 trees from the Baysian analysis program PhyloBayes

Next we test the variability of the KAST within the tree set. Philippe et al.[25] sampled once every 10 cycles, to simulate this we sample 900 random trees in the 9000 tree set and calculate the KAST. We replicate this 1000 times and calculate the symmetric Robinson-Foulds distances (because the KAST is binary, we divide it by two) between every pair of KASTs generated. The average distance between these KASTs is 0.73 with an average KAST size of 10.73.

We also calculate the size of KAST with varying numbers of tree sets to test how sample size effects the KAST size. Samples of 5000, 2500, 1200, and 500 trees all have KAST size of 11. Starting with the samples of 250 trees the size of KAST start to increase, with samples of 30 trees having a KAST size of 16. While the KAST from the whole 9000 trees set is obviously more conservative, the KAST from the smaller samples agree with all known competing hypotheses while including up to half the leaves.

5.2   Analyses on γ-Proteobacteria Phylogenies

Finally, we test the KAST on the phylogeny of γ-proteobacteria that has been the subject of pains-taking study. We refer the reader to Herbeck et al. [16]

for a discussion of previous work. For our purposes, we concentrate on the studies related to 12 particular species used in Lerat et al. [21], who reconstructed a phylogeny based on hundreds of genes. Since then there have been other attempts to reconstruct the phylogeny based on the syntenic data of the whole genome[12,6,5,30].

We turn our attention to two studies that produced trees in discordance with that of Lerat. Belda et al. [5] produced two trees, one using Maximum Likelihood on amino acid sequences and the other using reversal distance on the syntenic information (they used the breakpoint distance as well, which produced the same tree as the inversion distance). The likelihood analysis gave a tree that agreed with Lerat's. The inversion distance gave a tree that has significant differences to that of Lerat; the KAST between the two has 9 of 12 leaves. However, we will see that when we add certain trees from the study of Blin et al., the KAST size is 10. Further, the leaves excluded are Wigglesworthia brevipalpis and Pseudomonas aeruginosa; the former identified by Herbeck et al. [16] as troublesome to place, and the latter being the outgroup that they used to root their trees.

Blin et. al [6] used model free distances (breakpoints, conserved intervals, and common intervals) on the syntenic data to reconstruct their phylogenies. They produced many trees with the various methods on two different data sets. The syntenic data that yielded the interesting phylogeny for our purposes was produced from coding genes along with ribosomal and transfer RNAs. Blin et al. noticed that their trees computed on this data, using conserved and common intervals, were more different from the Lerat tree than the others. The KAST confirms this: the KAST on the set of all published trees other than these two is 10 while the inclusion of either one (they are the same) yields a KAST of size 5. Our experimental data tells us that a sequence of six trees, each produced by a random NNI operation from the last, will yield a KAST of size 5 while six unrelated trees would produce a KAST of size 2. We conclude that we have higher confidence in the set of trees that don't include those two trees.

6   Conclusion

We claim that the utility of the KAST is two-fold. The first is that the KAST is a safe summary of the subtree of confidence for a set of trees. The second is that the size of the KAST is correlated with how related the set of trees is. The KAST is not as susceptible to rogue taxa as the very conservative strict consensus, and is not as misleading as the MAST can be. Furthermore, unlike the other methods that attempt to characterize structure in the presence of rogue taxa, our measure is computable in polynomial time.

Acknowledgments

The first author would like to thank Andre Aberer for use and discussions about his code to compute the MAST, and Bernard Moret for discussions about the kernel agreement subtree.

References

1. Adams, E.N.: Consensus techniques and the comparison of taxonomic trees. Syst. Zool. 21, 390–397 (1972)
2. Bandelt, H., Dress, A.: Split decomposition: A new and useful approach to phylogenetic analysis of distance data. Mol. Phyl. Evol. 1(3), 242–252 (1992)
3. Barrett, M., Donoghue, M.J., Sober, E.: Against consensus. Syst. Zool. 40(4), 486–493 (1991)
4. Barrett, M., Donoghue, M.J., Sober, E.: Crusade? a reply to Nelson. Syst. Biol. 42(2), 216–217 (1993)
5. Belda, E., Moya, A., Silva, F.J.: Genome rearrangement distances and gene order phylogeny in γ-proteobacteria. Mol. Biol. Evol. 22(6), 1456–1467 (2005)
6. Blin, G., Chauve, C., Fertin, G.: Genes order and phylogenetic reconstruction: Application to γ-proteobacteria. In: Lagergren, J. (ed.) RECOMB-WS 2004. LNCS (LNBI), vol. 3388, pp. 11–20. Springer, Heidelberg (2005)
7. Bryant, D.: Building trees, hunting for trees, and comparing trees. PhD dissertation, Department of Mathematics, University of Canterbury (1997)
8. Bryant, D.: A classification of consensus methods for phylogenetics. In: Bioconsensus. DIMACS Series in Discrete Mathematics and Theoretical Computer Science, vol. 61, pp. 163–184. AMS Press, New York (2002)
9. Bryant, D., Moulton, V.: Neighbor-net: an agglomerative method for the construction of phylogenetic networks. Mol. Biol. Evol. 21(2), 255–265 (2004)
10. Gordon, A.D., Finden, C.R.: Obtaining common pruned trees. J. Classification 2(1), 255–267 (1985)
11. Cranston, K.A., Rannala, B.: Summarizing a posterior distribution of trees using agreement subtrees. Syst. Biol. 56(4), 578–590 (2007)
12. Earnest-DeYoung, J.V., Lerat, E., Moret, B.M.E.: Reversing gene erosion – reconstructing ancestral bacterial genomes from gene-content and order data. In: Jonassen, I., Kim, J. (eds.) WABI 2004. LNCS (LNBI), vol. 3240, pp. 1–13. Springer, Heidelberg (2004)
13. Farach, M., Przytycka, T., Thorup, M.: On the agreement of many trees. Information Processing Letters, 297–301 (1995)
14. Felsenstein, J.: Phylogenetic Inference Package (PHYLIP), Version 3.5. University of Washington, Seattle (1993)
15. Gauthier, O., Lapointe, F.-J.: Seeing the trees for the network: consensus, information content, and superphylogenies. Syst. Biol. 56(2), 345–355 (2007)
16. Herbeck, J.T., Degnan, P.H., Wernegreen, J.J.: Nonhomogeneous model of sequence evolution indicates independent origins of primary endosymbionts within the enterobacteriales (gamma-proteobacteria). Mol. Biol. Evol. 22(3), 520–532 (2005)
17. Huson, D.H.: SplitsTree: analyzing and visualizing evolutionary data. Bioinformatics 14(1), 68–73 (1998)
18. Kubicka, E., Kubicki, G., McMorris, F.R.: On agreement subtrees of two binary trees. Congressus Numeratium 88, 217–224 (1992)
19. Lartillot, N., Brinkmann, H., Philippe, H.: Suppression of long-branch attraction artefacts in the animal phylogeny using a site-heterogeneous model. BMC Evol. Biol. 7(Suppl. 1) (2007); 1st International Conference on Phylogenomics, St Adele, CANADA, March 15-19 (2006)
20. Lartillot, N., Philippe, H.: A Bayesian mixture model for across-site heterogeneities in the amino-acid replacement process. Mol. Biol. Evol. 21(6), 1095–1109 (2004)

21. Lerat, E., Daubin, V., Moran, N.A.: From gene trees to organismal phylogeny in prokaryotes:the case of the γ-proteobacteria. PLoS Biol. 1(1), e19 (2003)
22. Moore, G.W., Goodman, M., Barnabas, J.: An iterative approach from the standpoint of the additive hypothesis to the dendrogram problem posed by molecular data sets. Journal of Theoretical Biology 38(3), 423–457 (1973)
23. Nelson, G.: Why crusade against consensus? a reply to Barret, Donoghue, and Sober. Syst. Biol. 42(2), 215–216 (1993)
24. Pattengale, N.D., Aberer, A.J., Swenson, K.M., Stamatakis, A., Moret, B.M.E.: Uncovering hidden phylogenetic consensus in large datasets. IEEE/ACM Transactions on Computational Biology and Bioinformatics 99(PrePrints) (2011)
25. Philippe, H., Brinkmann, H., Copley, R.R., Moroz, L.L., Nakano, H., Poustka, A.J., Wallberg, A., Peterson, K.J., Telford, M.J.: Acoelomorph flatworms are deuterostomes related to Xenoturbella. Nature 470(7333), 255–258 (2011)
26. Redelings, B.: Bayesian phylogenies unplugged: Majority consensus trees with wandering taxa, http://www.duke.edu/~br51/wandering.pdf
27. Robinson, D.F.: Comparison of labeled trees with valency three. Journal of Combinatorial Theory, Series B 11(2), 105–119 (1971)
28. Shin, K., Kuboyama, T.: Kernels based on distributions of agreement subtrees. In: Wobcke, W., Zhang, M. (eds.) AI 2008. LNCS (LNAI), vol. 5360, pp. 236–246. Springer, Heidelberg (2008)
29. Stamatakis, A.: RAxML-VI-HPC: maximum likelihood-based phylogenetic analyses with thousands of taxa and mixed models. Bioinformatics 22(21), 2688–2690 (2006)
30. Swenson, K.M., Arndt, W., Tang, J., Moret, B.M.E.: Phylogenetic reconstruction from complete gene orders of whole genomes. In: Proc. 6rd Asia Pacific Bioinformatics Conf. (APBC 2008), pp. 241–250 (2008)
31. Thorley, J.L., Wilkinson, M., Charleston, M.: The information content of consensus trees. In: Rizzi, A., Vichi, M., Bock, H. (eds.) Studies in Classification, Data Analysis, and Knowledge Organization, Advances in Data Science and Classification, pp. 91–98. Springer, Heidelberg (1998)
32. Wilkinson, M.: Common cladistic information and its consensus representation: reduced adams and reduced cladistic consensus trees and profiles. Syst. Biol. 43(3), 343–368 (1994)
33. Wilkinson, M.: More on reduced consensus methods. Syst. Biol. 44, 435–439 (1995)
34. Wilkinson, M.: Majority-rule reduced consensus trees and their use in bootstrapping. Mol. Biol. Evol. 13(3), 437–444 (1996)

A Consensus Approach to Predicting Protein Contact Map via Logistic Regression

Jian-Yi Yang* and Xin Chen

Division of Mathematical Sciences, School of Physical and Mathematical Sciences,
Nanyang Technological University, 21 Nanyang Link, Singapore, 637371
{yang0241,chenxin}@ntu.edu.sg

Abstract. Prediction of protein contact map is of great importance
since it can facilitate and improve the prediction of protein 3D structure. However, the prediction accuracy is notoriously known to be rather
low. In this paper, a consensus contact map prediction method called
LRcon is developed, which combines the prediction results from several
complementary predictors by using a logistic regression model. Tests on
the targets from the recent CASP9 experiment and a large dataset D856
consisting of 856 protein chains show that LRcon not only outperforms its
component predictors but also the simple averaging and voting schemes.
For example, LRcon achieves 41.5% accuracy on the D856 dataset for
the top $L/10$ long-range contact predictions, which is about 5% higher
than its best-performed component predictor. The improvements made
by LRcon are mainly attributed to the application of a consensus approach to complementary predictors and the logistic regression analysis
under the machine learning framework.

Keywords: Protein contact map; CASP; Logistic regression; Machine
learning.

1 Introduction

Protein contact map is a 2D description of protein structure, which presents the
residue-residue contact information of a protein. Two residues are considered to
be in contact if their distance in 3D space is less than a predefined threshold.
Prediction of protein contact map is of great importance because it can facilitate
and improve the computational prediction of protein 3D structure [21].

Many computational methods are already proposed to predict protein contact
map. These methods can be classified into two major categories: (i) methods
based on correlated mutations [20], [13], [10], [12], [17], and (ii) methods based
on machine learning [14], [15], [22], [4], [7], [19], [2], [23]. There also exist some
other methods, e.g., based on template-threading [18], [7] and integer linear
optimization [16]. However, the accuracy of contact prediction, especially for
long-range contact prediction, is still rather low [21], [11].

---

* Corresponding author.

J. Chen, J. Wang, and A. Zelikovsky (Eds.): ISBRA 2011, LNBI 6674, pp. 136–147, 2011.
© Springer-Verlag Berlin Heidelberg 2011

In this study, we intend to improve the accuracy of contact map prediction by using a consensus approach, which means that the prediction results from several existing predictors will be consolidated. To our best knowledge, not much effort has been made to develop a consensus contact prediction method except the following two approaches. Confuzz is a consensus approach based on the weighted average of the probability estimates from individual predictors (please refer to the website of CASP9). The other approach is based on integer linear programming [6]. We instead choose to tackle this problem in a different way. We consolidate the prediction results from individual predictors by using a logistic regression analysis under the machine learning framework. Tests on the CASP9 dataset as well as on another large dataset show that the proposed method not only outperforms its component predictors but also the simple averaging and voting schemes.

2   Materials and Methods

2.1   Datasets

In this study, two datasets are used to test the proposed method, which are downloadable at `http://www3.ntu.edu.sg/home2008/YANG0241/LRcon/`. The first one was collected from the targets in the recent CASP9 experiment. In CASP9, there are 28 participating groups in the contact prediction category. As one group might have several contact prediction models for the same target, here we selected the results only from the "model 1" of each predictor. In addition, we removed from further consideration those groups that made predictions for just a few targets and those targets that were predicted by just a few participating groups. As a result, we obtained 80 targets and 23 predictors. For the sake of convenience, we denote this dataset by D80. Finally, the true contact map for each target was derived from its 3D structure provided on the CASP9 website.

The second dataset was harvested from Protein Data Bank (PDB) [1] using the selected protein chains from the latest (May 2010) PDB_select 25% list [8]. Originally, there are 4869 protein chains in this list. A subset was extracted as follows. First, those chains with length less than 50 and/or coordinates information missing for some amino acids were removed. Second, those with pair-wise sequence identity higher than 25% and those with sequence identity to the NNcon training set [19] higher than 25% were further removed. This filtering process ends up with a total of 856 chains. We denote this dataset by D856.

2.2   Contact Definition

Two residues are defined to be in contact if the Euclidean distance between the 3D coordinates of their C_α atoms is less than or equal to 8 Å [4], [7], [19]. The CASP experiments [11], however, used C_β atoms instead of C_α atoms in determining two residues in contact. In this study, we choose the former definition because (i) it is a definition close to the one used in 3D structural modelling [24] and (ii) it was already used by two methods (i.e., [7] and [19]) that will be included in our consensus predictor.

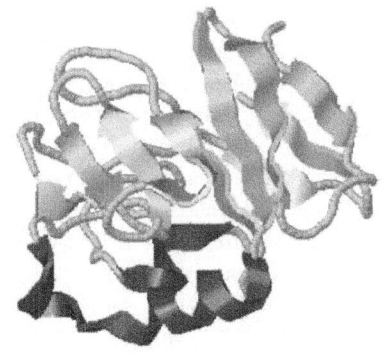 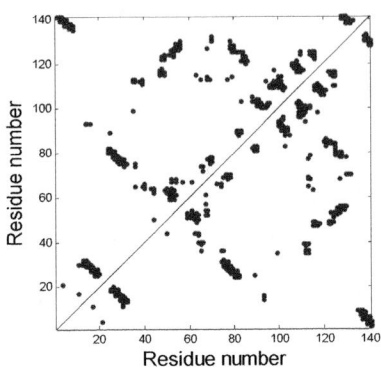

Fig. 1. An example of contact map at sequence separation $s \geq 6$. The left panel is a cartoon visualization of the 3D structure of the protein (PDB entry: 2NWF). The right panel is the contact map of this protein. A blue point in the figure indicates that the pair of residues are in contact. Note that the map is symmetrical with respect to the black main diagonal line.

For a protein with length L, the (true) contact information for all pairs of residues can be represented by a matrix $C = (c_{ij})_{L \times L}$, where $c_{ij} = 1$ if the residues i, j are in contact and $c_{ij} = 0$ otherwise. This matrix is often called a *contact map*. It is in fact a 2D description of protein structure, and a specific example of contact map is given in Figure 1.

Depending on the separation of two residues along the sequence, the residue-residue contact is classified into three classes: short-range contact (separation $6 \leq s < 12$), medium-range contact ($12 \leq s < 24$) and long-range contact ($s \geq 24$). Contacts for those residues too close along the sequence ($s < 6$) are omitted.

2.3   Performance Evaluation

The predicted contact map $PC = (pc_{ij})_{L \times L}$ is a matrix of probability estimates. The element pc_{ij} is the estimate for the contact probability of the residues i and j. In general, the top λL predictions (sorted by the probability estimates) are selected, which are then compared with the true contact map for evaluation. In the literature [11], [5], [19], [23], [15], the value of λ is usually set to be 0.1 or 0.2 and two metrics are used to evaluate the predictions: accuracy (Acc) and coverage (Cov).

$$\text{Acc} = \frac{\text{TP}}{\text{TP} + \text{FP}}, \quad \text{Cov} = \frac{\text{TP}}{\text{TP} + \text{FN}}, \tag{1}$$

where TP, FP, TN and FN, are true positive, false positive, true negative and false negative predictions, respectively. A residue pair is said to be a positive (resp., negative) pair if the two residues are (resp., are not) in contact.

In addition, a more robust metric called *F-measure* (Fm) is also used, which is basically a harmonic mean of precision and recall as defined below:

$$\text{Fm} = 2 \cdot \frac{\text{Acc} \times \text{Cov}}{\text{Acc} + \text{Cov}} \tag{2}$$

2.4 Consensus Prediction via Logistic Regression

Suppose there are p predictors, then we have p predicted contact maps for each protein. We attempt to combine these p maps to make a consensus prediction. The first difficulty appears that the output of some predictors (e.g., FragHMMent [2]) is not the whole map but part of the map. To overcome it, the probability estimates for those missing predictions are simply set to be 0 (i.e., not in contact).

A direct and simple way to combine the p predicted contact maps is to average over the p probability estimates for each residue pair and then select the top λL predictions. We call this method the *averaging* scheme. Another way is to first select the top λL predictions from each predicted map and then use these selected predictions to vote. The residue pairs with votes in the top λL positions are then output to be the top λL predictions. We call this method the *voting* scheme.

In this study, we propose to combine the p predicted contact maps via a logistic regression analysis. Logistic regression (LR) is a non-linear regression model in particular for a binary response variable [3]. It estimates the posterior probabilities by using the following formula:

$$P(Y_i = 1|P_i) = \frac{\exp(\alpha + \sum\limits_{j=1}^{p} \beta_j p_{ij})}{1 + \exp(\alpha + \sum\limits_{j=1}^{p} \beta_j p_{ij})} \tag{3}$$

where $P(Y_i = 1|P_i)$ is the posterior probability of the i-th residue pair being in contact given P_i. $P_i = (p_{i1}, p_{i2}, \cdots, p_{ip})$ is a probability vector for the i-th residue pair, of which each component p_{ij} is the probability estimate of the component predictor j on the i-th residue pair. The constants α and β_j $(j = 1, 2, \cdots, p)$ are the regression coefficients whose values can be estimated with a training set through Quasi-Newton optimization [3]. We used the implementation of LR in the software package Weka [9] (with default parameters) for our experiments.

2.5 Overall Architecture

Figure 2 depicts the overall architecture of our proposed method named LRcon. It comprises two major procedures: training and testing. In the training procedure, a logistic regression model (LR-Predictor) is built up with a training set of protein chains. In the testing procedure, a query amino acid sequence is first input into p individual predictors and, for each predictor, the top λL predictions are selected. Then, we take the union of all the selected residue pairs for further consideration (Please refer to Section 2.6 for more details). For each selected pair, the probability estimates of the p predictors are used to form a feature

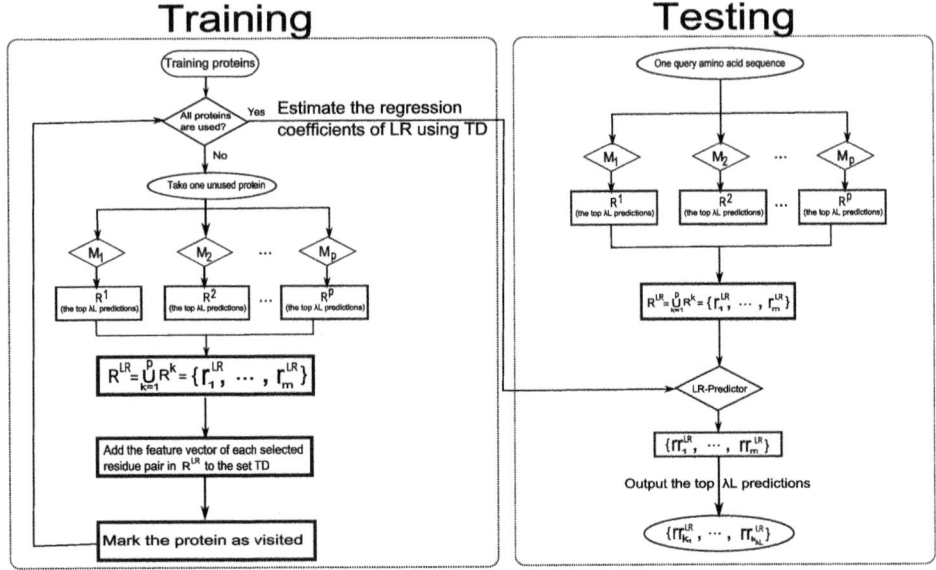

Fig. 2. The overall architecture of LRcon. In the training procedure, the consensus predictor is built upon p individual predictors $M_1, M_2, \cdots, M_p$. The set TD is used to store the training feature vectors of the selected residue pairs. During the testing, the prediction result is stored in $rr_i^{LR} = (r_{i,1}^{LR}, r_{i,2}^{LR}, p_i^{LR})$, where p_i^{LR} is the probability estimate for the residue pair i. $k_1, \cdots, k_{\lambda L}$ are the indices of the top λL predictions.

vector, which is then fed into the LR-Predictor for consensus contact prediction. Finally, the top λL contact predictions are selected as our consensus predictions.

For the D80 datasets, we use 23 predictors from CASP9 to build our consensus predictor. For the D856 dataset, only three predictors (SVM-SEQ (RR204) [7], NNcon (RR119) [19] and FragHMMent (RR158) [2]) are instead used, because there are no software available for the other predictors except SVMcon (RR002) [4]. SVMcon is excluded because it was developed based on the same classification algorithm as SVM-SEQ, so their predicted results are expected to have a large overlap.

2.6   Selection of Residue Pairs

Given a protein of length L, the total number of residue pairs is $(L + 1 - s) \times (L - s)/2$ for sequence separation at least s. If all these residue pairs are used, we would not be able to obtain a reliable LR-Predictor, due to at least two factors: (1) A large number of training samples does not allow to estimate the regression coefficients in a reasonable amount of computing time; (2) Most of the residue pairs belong to the negative class, so that a small proportion of positive samples make the predictions be severely biased against the positive class. This would inevitably discount the performance of LRcon if we choose to work this way.

Here we propose to use the *union* of the residue pairs corresponding to the top λL predictions from each of the p component predictors. For a protein of length L, we denote the set of the top λL residue pairs returned by the k-th predictor by $R^k = \{r_1^k, r_2^k, \cdots, r_{\lambda L}^k\}$, where $r_i^k = (r_{i,1}^k, r_{i,2}^k)$ represents a residue pair with $1 \leq r_{i,1}^k, r_{i,2}^k \leq L$. The residue pairs selected for this protein to train and test our LR-Predictor are then taken from the set

$$R^{LR} = \bigcup_{k=1}^{p} R^k \tag{4}$$

3   Results

In the following, the experimental results are evaluated on the top $0.1L$ and $0.2L$ predictions at sequence separations $6 \leq s < 12$, $12 \leq s < 24$ and $s \geq 24$.

3.1   Results on the CASP9 Dataset

In order to estimate the regression coefficients of LR and to assess the performance of LRcon, we applied 10-fold cross-validation to the CASP9 dataset D80. For the top $0.1L$ predictions, Figure 3 shows the average accuracy, coverage, and F-measure of LRcon, its component predictors, and the averaging and voting schemes (refer to Section 2.4) for a comprehensive comparison. It is evident from the figure that the averaging and voting schemes could perform better than most component predictors, but never all in any cases. On the contrary, the LRcon is able to outperform all the component predictors and the averaging and voting schemes as well. In addition, we can see from the F-measures that the prediction of long-range contact is much more challenging than the prediction of short-rang and medium-range contacts. For the top $0.2L$ predictions, LRcon also outperforms all the component predictors and the averaging and voting schemes. The detailed results are presented in Figure 1 of Supplementary Material, which is accessible at http://www3.ntu.edu.sg/home2008/YANG0241/LRcon/. A typical predicted CASP9 contact map where LRcon outperforms all the other predictors is depicted in Figure 2 of Supplementary Material.

We assess the statistical significance of the prediction differences between LRcon and each other predictor as follows. First, 80% protein chains are selected at random from the original dataset to construct a (sub)dataset. This is repeated 100 times so as to obtain 100 different datasets. Then, we collected the prediction results of all the tested predictors from these 100 datasets. Finally, the paired t-test is applied to assess their statistical significance on the F-measure differences. We summarized in Table 1 the experimental results tested on the CASP9 dataset.

We can observe from the above tests that: (1) The averaging scheme appears to perform better than the voting scheme, (2) Neither the averaging nor voting scheme achieve a better prediction than all the component predictors (in particular, e.g., RR391 and RR490), and (3) LRcon outperforms all the other predictors, including the averaging and voting schemes.

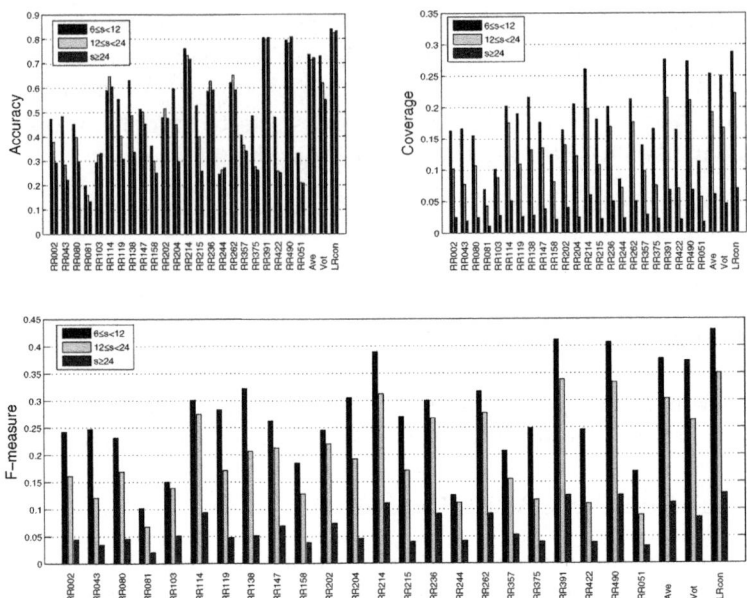

Fig. 3. Histogram of the accuracy, coverage and F-measure for the top $0.1L$ predictions of LRcon and other predictors on the CASP9 dataset D80. The predictor codes for the component predictors are directly taken from CASP9. Ave and Vot represent the averaging and voting schemes, respectively.

Table 1. The results of the statistical significance tests for the F-measures of LRcon and other predictors on the D80 dataset. The '+' /'-' indicates that the method in a given column is significantly better/worse than the method in a given row with p-value < 0.001, and '=' means that the results are not shown statistically different.

| Predictor | Top $0.1L$ predictions | | | | | | | | | Top $0.2L$ predictions | | | | | | | | |
| | $6 \leq s < 12$ | | | $12 \leq s < 24$ | | | $s \geq 24$ | | | $6 \leq s < 12$ | | | $12 \leq s < 24$ | | | $s \geq 24$ | | |
	Ave	Vot	LRcon	Ave	Vot	LRcon	Ave	Vot	LRcon	Ave	Vot	LRcon	Ave	Vot	LRcon	Ave	Vot	LRcon
RR002	+	+	+	+	+	+	+	+	+	+	+	+	+	+	+	+	+	+
RR043	+	+	+	+	+	+	+	+	+	+	+	+	+	+	+	+	+	+
RR080	+	+	+	+	+	+	+	+	+	+	+	+	+	+	+	+	+	+
RR081	+	+	+	+	+	+	+	+	+	+	+	+	+	+	+	+	+	+
RR103	+	+	+	+	+	+	+	+	+	+	+	+	+	+	+	+	+	+
RR114	+	+	+	+	-	+	+	-	+	+	+	+	+	-	+	+	-	+
RR119	+	+	+	+	+	+	+	+	+	+	+	+	+	+	+	+	+	+
RR138	+	+	+	+	+	+	+	+	+	+	+	+	+	+	+	+	+	+
RR147	+	+	+	+	+	+	+	+	+	+	+	+	+	+	+	+	+	+
RR158	+	+	+	+	+	+	+	+	+	+	+	+	+	+	+	+	+	+
RR202	+	+	+	+	+	+	+	+	+	+	+	+	+	+	+	+	+	+
RR204	+	+	+	+	+	+	+	+	+	+	+	+	+	+	+	+	+	+
RR214	-	-	+	-	-	+	=	-	+	-	-	+	-	-	+	-	-	+
RR215	+	+	+	+	+	+	+	+	+	+	+	+	+	+	+	+	+	+
RR236	+	+	+	+	-	+	+	-	+	+	+	+	+	-	+	+	-	+
RR244	+	+	+	+	+	+	+	+	+	+	+	+	+	+	+	+	+	+
RR262	+	+	+	+	-	+	+	-	+	+	+	+	+	-	+	+	-	+
RR357	+	+	+	+	+	+	+	+	+	+	+	+	+	+	+	+	+	+
RR375	+	+	+	+	+	+	+	+	+	+	+	+	+	+	+	+	+	+
RR391	-	-	+	-	-	+	-	-	+	-	-	+	-	-	+	-	-	+
RR422	+	+	+	+	+	+	+	-	+	+	+	+	+	+	+	+	+	+
RR490	-	-	+	-	-	+	-	-	+	-	-	+	-	-	+	-	-	+
RR051	+	+	+	+	+	+	+	+	+	+	+	+	+	+	+	+	+	+
Ave	=	-	+	=	-	+	=	-	+	=	-	+	=	-	+	=	-	+
Vot	+	=	+	+	=	+	+	=	+	+	=	+	+	=	+	+	=	+

Table 2. Comparison of accuracies, coverage and F-measures on the independent test (sub)dataset of the D856 dataset. The best results are shown in bold.

Predictor	Top 0.1L predictions									Top 0.2L predictions								
	$6 \leq s < 12$			$12 \leq s < 24$			$s \geq 24$			$6 \leq s < 12$			$12 \leq s < 24$			$s \geq 24$		
	Acc	Cov	Fm	Acc	Cov	Fm	Acc	Cov	Fm	Acc	Cov	Fm	Acc	Cov	Fm	Acc	Cov	Fm
FragHMMent	.365	.116	.176	.306	.079	.126	.275	.026	.047	.340	.219	.267	.275	.052	.088	.272	.143	.188
NNcon	.591	.187	.284	.455	.118	.187	.283	.026	.048	.478	.308	.374	.235	.045	.075	.376	.198	.260
SVM-SEQ	.610	.193	.293	.483	.125	.199	.366	.034	.062	.508	.327	.398	.323	.061	.103	.407	.215	.281
Ave	.529	.167	.254	.415	.108	.171	.365	.034	.062	.377	.243	.295	.288	.055	.092	.316	.167	.218
Vot	.563	.178	.271	.443	.115	.182	.314	.029	.053	.474	.305	.371	.289	.055	.092	.371	.195	.256
LRcon	**.650**	**.206**	**.313**	**.531**	**.138**	**.218**	**.415**	**.039**	**.071**	**.538**	**.346**	**.421**	**.355**	**.067**	**.113**	**.443**	**.234**	**.306**

Table 3. The results of the statistical significance tests for the F-measures of LRcon and other predictors on the independent test (sub)dataset of the D856 dataset

Predictor	Top 0.1L predictions									Top 0.2L predictions								
	$6 \leq s < 12$			$12 \leq s < 24$			$s \geq 24$			$6 \leq s < 12$			$12 \leq s < 24$			$s \geq 24$		
	Ave	Vot	LRcon	Ave	Vot	LRcon	Ave	Vot	LRcon	Ave	Vot	LRcon	Ave	Vot	LRcon	Ave	Vot	LRcon
FragHMMent	+	+	+	+	+	+	+	+	+	+	+	+	+	+	+	+	+	+
NNcon	-	-	+	-	-	+	+	+	+	-	-	+	-	-	+	+	+	+
SVM-SEQ	-	-	+	-	-	+	-	-	+	-	-	+	-	-	+	-	-	+
Ave	=	+	+	=	+	+	=	-	+	=	+	+	=	+	+	=	+	+
Vot	-	=	+	-	=	+	+	=	+	-	=	+	-	=	+	-	=	+

3.2 Results on the D856 Dataset

Because the size of the D856 dataset is significantly larger than that of the D80 dataset, we adopt here a different way, rather than using 10-fold cross-validation, to evaluate the prediction results of LRcon as follows. First, the 856 protein chains in the D856 dataset are partitioned at random into a training and a test dataset of equal size (i.e., 428 protein chains in each). Note that the training and test datasets are independent each other, so no sequence in the test dataset has over 25% sequence identity with any sequences in the training dataset. Second, three predictors, SVM-SEQ[7], NNcon[19], and FragHMMent[2], are used to make predictions on both the training and test datasets. Third, we use the predictions of these three predictors on the training dataset to estimate the regression coefficients of LR. Finally, the performance of LRcon is assessed on the test dataset.

The experimental results of LRcon and other predictors on the independent test dataset are listed in Table 2. We can see from the table that LRcon outperforms all the other predictors in terms of accuracy, coverage and F-measure. For example, LRcon achieves an average accuracy of 41.5% for the top $L/10$ long-range contact predictions, which is about 5% higher than its best-performed component predictor (i.e., SVM-SEQ). We also conduct a statistical significance test in the same way as we did earlier on the CASP9 dataset, and the results are shown in Table 3. It can be seen that the simple averaging and voting schemes perform better than NNcon and FragHMMent, but worse than SVM-SEQ. On the other hand, LRcon once again consistently outperforms all the other predictors, including the simple averaging and voting schemes.

4   Discussions

In this section, we discuss how the performance of LRcon is affected by the following three factors: the residue pair selection, the component predictors and the classification algorithm.

4.1   The Impact of Residue Pair Selection

For each protein chain, we have used formula (4) to select the residue pairs for training and testing LRcon. In order to demonstrate the effectiveness of this filtering process, we tested the performance of LRcon when all the residue pairs satisfying the sequence separation condition were used. Because it takes too much computing time and computer memory, we just tested the results of LR-con for the top $0.1L$ predictions at sequence separation $s \geq 24$ using the CASP9 dataset D80. In this case, the resulting average accuracy of LRcon decreases to 0.806, which is 0.026 lower than that obtained with the filtering process employed (see Figure 3). Therefore, it is necessary to train LRcon with a properly selected subset of residue pairs in order to achieve more accurate contact map prediction.

4.2   The Impact of Individual Predictors

The major reason of LRcon's superior performance is believed that its component predictors can make complementary predictions to each other. We say two predictors M_1 and M_2 are complementary if their correct predictions (denoted respectively by TP_1 and TP_2) among the top λL predictions are not completely the same. We have the following two observations. First, the sizes of TP_1 and TP_2 should be as large as possible ($\leq \lambda L$) for LRcon to be accurate enough. Second, in order to improve over M_1 and M_2 by combining them, TP_1 and TP_2 should not be the same. Otherwise, we would not be able to make any improvement by combining them; instead, the predictions might even become worse.

We conduct the experiments on the CASP9 dataset D80 to further confirm the above observations as follows. As mentioned in Section 2.1, when we selected component predictors, only "model 1" of each predictor was used in our previous experiments. For some participating groups there are several prediction models. For instance, the participating group RR114 in CASP9 have five prediction models for each target. We looked into these five models and found that their prediction results were in fact very similar, indicating that these models are not complementary to each other. When all models were used for each participating group in CASP9, we obtained 28 predictors in total. Using these predictors as component predictors of LRcon, we found that the predictions became worse than before. For example, the accuracy value decreased slightly from 0.832 to 0.827 for the case of the top $0.1L$ predictions at sequence separation $s \geq 24$. Therefore, one possible way to further improve the performance of LRcon is to select just those accurate while complementary predictors as component predictors. To this end, some metrics might be needed to quantify the complementary property among individual predictors.

Table 4. Comparison of accuracies, coverage and F-measures of five classification algorithms on the CASP9 dataset D80. The best results are shown in bold.

Algorithm	Top 0.1L predictions									Top 0.2L predictions								
	$6 \leq s < 12$			$12 \leq s < 24$			$s \geq 24$			$6 \leq s < 12$			$12 \leq s < 24$			$s \geq 24$		
	Acc	Cov	Fm	Acc	Cov	Fm	Acc	Cov	Fm	Acc	Cov	Fm	Acc	Cov	Fm	Acc	Cov	Fm
RF	.824	.283	.422	.810	.219	.345	.816	.069	.127	.689	.479	.565	.721	.394	.510	.786	.134	.228
NB	.769	.264	.393	.756	.204	.322	.780	.066	.122	.654	.454	.536	.656	.359	.464	.744	.126	.216
J48	.736	.253	.377	.724	.196	.308	.741	.061	.113	.658	.458	.540	.680	.372	.481	.741	.126	.215
KNN	.797	.274	.407	.777	.210	.330	.799	.067	.124	.685	.476	.562	.707	.386	.500	.773	.131	.225
LR	**.839**	**.288**	**.429**	**.822**	**.222**	**.350**	**.832**	**.070**	**.129**	**.710**	**.493**	**.582**	**.727**	**.398**	**.514**	**.799**	**.136**	**.232**

Table 5. Comparison of accuracies, coverage and F-measures of five classification algorithms on the independent test (sub)dataset of the D856 dataset. The best results are shown in bold.

Algorithm	Top 0.1L predictions									Top 0.2L predictions								
	$6 \leq s < 12$			$12 \leq s < 24$			$s \geq 24$			$6 \leq s < 12$			$12 \leq s < 24$			$s \geq 24$		
	Acc	Cov	Fm	Acc	Cov	Fm	Acc	Cov	Fm	Acc	Cov	Fm	Acc	Cov	Fm	Acc	Cov	Fm
RF	.594	.188	.286	.464	.120	.191	.352	.033	.060	.485	.313	.380	.388	.205	.268	.312	.059	.099
NB	.650	.206	.313	**.531**	**.138**	**.218**	.406	.038	.069	.538	.346	.421	**.443**	**.234**	**.306**	.349	.066	.111
J48	.621	.197	.299	.492	.127	.202	.394	.037	.067	.506	.326	.396	.414	.218	.286	.323	.061	.103
KNN	.623	.197	.299	.488	.126	.201	.383	.036	.065	.508	.327	.398	.404	.213	.279	.325	.062	.104
LR	**.650**	**.206**	**.313**	.530	.137	.218	**.415**	**.039**	**.071**	**.540**	**.348**	**.423**	.440	.232	.304	**.355**	**.067**	**.113**

4.3   The Impact of Classification Algorithm

Besides the logistic regression, the following four classification algorithms are experimented to explore the impact of a classification algorithm on the performance of LRcon. They are random forest (RF), k-nearest neighbor (k-NN), Naive Bayes (NB), and J48. The details about these algorithms can be obtained from Weka [9]. We also chose the algorithm implementations in Weka in our subsequent experiments. Except for k-NN, where k was set to be 10 to produce probability estimates, the parameters for RF, NB, and J48 were all set to be their respective default values.

The experimental results of LR and the other four algorithms on the datasets D80 and D856 are presented in Tables 4 and 5, respectively. We can see that the accuracies, coverage and F-measures of LR are consistently higher than those of any other algorithm on the D80 dataset. When tested on the the D856 dataset, LR and NB achieved comparable results and better than the other three algorithms. These observations lead to our selection of LR as the classification algorithm in this study.

5   Conclusions

Prediction of protein contact map plays an important role in the prediction of protein 3D structure. However, the accuracy of current computational methods is rather low. In this paper, we explored the possibility of improving the accuracy of an individual protein contact predictors by using a consensus approach.

Under the machine learning framework, an improved sequence-based protein contact map prediction method, named LRcon, has been developed based on logistic regression. LRcon is built upon the prediction results from its component contact map predictors. For each residue pair, the probability estimates of the component predictors are used to form a feature vector, which is then fed into the logistic regression-based algorithm to make a consensus prediction. Logistic regression models are trained and assessed under the machine learning framework by using independent training and test datasets. Experimental results on the CASP9 dataset and another large-sized dataset containing 856 protein chains show that LRcon can make statistically significant improvements over its component predictors and the simple averaging and voting schemes as well. We believe that these improvements made by LRcon are mainly attributed to the application of a consensus approach to the complementary predictors and the logistic regression analysis under the machine learning framework.

Acknowledgments

This work was partially supported by the Singapore NRF grant NRF2007IDM-IDM002-010 and MOE AcRF Tier 1 grant RG78/08.

References

1. Berman, H.M., Westbrook, J., Feng, Z., Gilliland, G., Bhat, T.N., Weissig, H., Shindyalov, I.N., Bourne, P.E.: The Protein Data Bank. Nucleic Acids Research 28, 235–242 (2000)
2. Björkholm, P., Daniluk, P., Kryshtafovych, A., Fidelis, K., Andersson, R., Hvidsten, T.R.: Using multi-data hidden Markov models trained on local neighborhoods of protein structure to predict residue-residue contacts. Bioinformatics 25, 1264–1270 (2009)
3. Cessie, L.S., van Houwelingen, J.C.: Ridge estimators in logistic regression. Applied Statistics 41, 191–201 (1992)
4. Cheng, J., Baldi, P.: Improved residue contact prediction using support vector machines and a large feature set. BMC Bioinformatics 8, 113 (2007)
5. Ezkurdia, I., Graña, O., Izarzugaza, J.M.G., Tress, M.L.: Assessment of domain boundary predictions and the prediction of intramolecular contacts in CASP8. Proteins 77, 196–209 (2009)
6. Gao, X., Bu, D., Xu, J., Li, M.: Improving consensus contact prediction via server correlation reduction. BMC Structural Biology 9, 28 (2009)
7. Wu, S., Zhang, Y.: A comprehensive assessment of sequence-based and template-based methods for protein contact prediction. Bioinformatics 24, 924–931 (2008)
8. Griep, S., Hobohm, U.: PDBselect 1992-2009 and PDBfilter-select. Nucleic Acids Research 38, D318–D319 (2009)
9. Hall, M., Frank, E., Holmes, G., Pfahringer, B., Reutemann, P., Witten, I.H.: The WEKA data mining software: an update. SIGKDD Explorations 11, 10–18 (2009)
10. Hamilton, N., Burrage, L., Ragan, M.A., Huber, T.: Protein contact prediction using patterns of correlation. Proteins 7, 679–684 (2004)
11. Izarzugaza, J.M.G., Graña, O., Tress, M.L., Valencia, A., Clarke, N.: Assessment of intramolecular contact predictions for CASP7. Proteins 69, 152–158 (2007)

12. Kundrotas, P.J., Alexov, E.G.: Predicting residue contacts using pragmatic corre-
    lated mutations method: reducing the false positives. BMC Bioinformatics 7, 503
    (2006)
13. Olmea, O., Valencia, A.: Improving contact predictions by the combination of cor-
    related mutations and other sources of sequence information. Folding & Design 2,
    S25–S32 (1997)
14. Pollastri, G., Baldi, P.: Prediction of contact maps by GIOHMMs and recurrent
    neural networks using lateral propagation from all four cardinal corners. Bioinfor-
    matics 70, S62–S70 (2002)
15. Punta, M., Rost, B.: PROFcon: novel prediction of long-range contacts. Bioinfor-
    matics 21, 2960–2968 (2005)
16. Rajgaria, R., Wei, Y., Floudas, C.A.: Contact prediction for beta and alpha-beta
    proteins using integer linear optimization and its impact on the first principles 3D
    structure prediction method ASTRO-FOLD. Proteins 78, 1825–1846 (2010)
17. Shackelford, G., Karplus, K.: Contact prediction using mutual information and
    neural nets. Proteins 69, 159–164 (2007)
18. Shao, Y., Bystroff, C.: Predicting interresidue contacts using templates and path-
    ways. Proteins 53, 497–502 (2003)
19. Tegge, A.N., Wang, Z., Eickholt, J., Cheng, J.: NNcon: improved protein contact
    map prediction using 2D-recursive neural networks. Nucleic Acids Research 37,
    W515–W518 (2009)
20. Thomas, D.J., Casari, G., Sander, C.: The prediction of protein contacts from
    multiple sequence alignments. Protein Engineering 9, 941–948 (1996)
21. Tress, M.L., Valencia, A.: Predicted residue-residue contacts can help the scoring
    of 3D models. Proteins 78, 1980–1991 (2010)
22. Vullo, A., Walsh, I., Pollastri, G.: A two-stage approach for improved prediction
    of residue contact maps. BMC Bioinformatics 7, 180 (2006)
23. Xue, B., Faraggi, E., Zhou, Y.: Predicting residue-residue contact maps by a two-
    layer, integrated neural-network method. Proteins 76, 176–183 (2009)
24. Zhang, Y., Kolinski, A., Skolnick, J.: TOUCHSTONE II: a new approach to ab
    initio protein structure prediction. Biophysical Journal 85, 1145–1164 (2003)

A Linear Time Algorithm for Error-Corrected Reconciliation of Unrooted Gene Trees

Paweł Górecki[1] and Oliver Eulenstein[2]

[1] Institute of Informatics, Warsaw University, Poland
gorecki@mimuw.edu.pl
[2] Department of Computer Science, Iowa State University, USA
oeulenst@cs.iastate.edu

Abstract. Evolutionary methods are increasingly challenged by the fast growing resources of genomic sequence information. Fundamental evolutionary events, like gene duplication, loss, and deep coalescence, account more then ever for incongruence between gene trees and the actual species tree. Gene tree reconciliation is addressing this fundamental problem by invoking the minimum number of gene-duplication and losses that reconcile a gene tree with a species tree. Despite its promise, gene tree reconciliation assumes the gene trees to be correctly rooted and free of error, which severely limits its application in practice. Here we present a novel linear time algorithm for error-corrected gene tree reconciliation of unrooted gene trees. Furthermore, in an empirical study on yeast genomes we successfully demonstrate the ability of our algorithm to (i) reconcile (cure) error-prone gene trees, and (ii) to improve on more advanced evolutionary applications that are based on gene tree reconciliation.

1 Introduction

The wealth of newly sequenced genomes has provided us with an unprecedented resource of information for phylogenetic studies that will have extensive implications for a host of issues in biology, ecology, and medicine, and promise even more. Yet, before such phylogenies can be reliably inferred, challenging problems that came along with the newly sequenced genomes have to be overcome. Evolutionary biologists have long realized that gene-duplication and subsequent loss, a fundamental evolutionary process [14], can largely obfuscate phylogenetic inference [20]. Gene-duplication can form complex evolutionary histories of genes, called gene trees, whose topologies are traditionally used to derive species trees. This approach relies on the assumption that the topologies from gene trees are consistent with the topology of the species tree. However, frequently genes that evolve from different copies of ancestral gene-duplications can become extinct and result in gene trees with correct topologies that are inconsistent with the topology of the actual species tree (see Fig. 1). In many such cases phylogenetic information from the gene trees is indispensable and may still be recovered using gene tree reconciliation.

Gene tree reconciliation is a well-studied method for resolving topological inconsistencies between a gene tree and a trusted species tree [5,9,11,18,20,22]. Inconsistencies are resolved by invoking gene-duplication and loss events that reconcile the gene tree

J. Chen, J. Wang, and A. Zelikovsky (Eds.): ISBRA 2011, LNBI 6674, pp. 148–159, 2011.

to be consistent with the actual species tree. Such events do not only reconcile gene trees, but also lay foundation for a variety of evolutionary applications including ortholog/paralog annotation of genes, locating episodes of gene-duplications in species trees [2,10,15], reconstructing domain decompositions [3], and species supertree construction [1,2,17,21].

A major problem in the application of gene tree reconciliation is its high sensitivity to error-prone gene trees. Even seemingly insignificant errors can largely mislead the reconciliation process and, typically undetected, infer incorrect phylogenies (e.g., [16,22]). Errors in gene trees are often topological errors and rooting errors. Topological error results in an incorrect topology of the gene tree that can be caused by the inference process (e.g. noise in the underlying sequence data) or the inference method itself (e.g. heuristic results). This problem has been addressed for rooted gene trees by 'correcting the error'; that is, editing the given tree such that the number of invoked gene-duplications and losses is minimized [6,7]. However, most inference methods used in practice return only un-rooted gene trees (e.g. parsimony and maximum likelihood based methods) that have to be rooted for the gene tree reconciliation process. Rooting error is a wrongly chosen root in an un-rooted gene tree. Whereas rooting can be typically achieved in species trees by outgroup analysis, this approach may not be possible for gene trees if there is a history of gene duplication and loss [22]. Other rooting approaches like midpoint rooting or molecular clock rooting assume a constant rate of evolution that is often unrealistic. However, rooting problems can be bypassed by identifying roots that minimize the invoked number of gene duplications and losses [6,7,12,22,24]. In summary, standard gene tree reconciliation requires gene trees that are free of error and correctly rooted [11]. Even small topological error or a slightly misplaced root can incorrectly identify large numbers of gene duplications and losses, and therefore mislead the reconciliation process. As previous work has incorporated topological error-correction separately from correctly rooting gene trees into the gene tree reconciliation process [6,12], this process can still be largely misled.

Our Contribution: We describe a novel linear time algorithm for error-corrected gene tree reconciliation that simultaneously incorporates the adjustment of topological error and rooting error. Given an un-rooted gene tree and a rooted species tree, the algorithm finds an 'error-corrected version' of the gene tree that requires the minimum number of gene duplication and loss events when reconciled with the species tree. Similar to the error-correcting approach from Durand et al. [7] we assume that the error-free gene tree can be found within the local search neighborhood consisting of all gene trees that are within at most one nearest neighbor interchange (NNI) operation (a standard tree edit operation) of the given unrooted gene tree. Under all rooted versions of the trees in this local neighborhood the algorithm identifies one that invokes the minimum number of gene duplications and losses. Our experiments on yeast genomes suggest that our algorithm can greatly improve on the accuracy of reconciling, and thus curating, error-prone gene trees. Furthermore, we show that our error-corrected reconciliations lead to improved predictions of invoked gene duplication and loss events that allow to infer more accurate yeast phylogenies.

Fig. 1. An lca-mapping M from the gene tree $\mathcal{G}$ into the species tree $\mathcal{S}$ and the corresponding embedding. M is shown for the internal nodes of $\mathcal{G}$.

2   Duplication-Loss Model

We introduce the fundamentals of the classical duplication-loss model. Our definitions are mostly adopted from [12]. For a more detailed introduction to the duplication-loss model we refer the interested reader to [11,15,20,8].

Let $\mathcal{I}$ be the set of species consisting of $N > 0$ elements. The *unrooted gene tree* is an undirected acyclic graph in which each node has degree 3 (internal nodes) or 1 (leaves), and the leaves are labeled by the elements from $\mathcal{I}$. A *species tree* is a rooted binary tree with N leaves uniquely labeled by the elements from $\mathcal{I}$. In some cases, a node of $\mathcal{S}$ will be referred by "cluster" of labels of its subtree leaves. For instance, a species tree $(a, (b, c))$ has 5 nodes denoted by: a, b, c, bc and abc. A *rooted gene tree* is a rooted binary tree with leaves labeled by the elements from $\mathcal{I}$. The internal nodes of a tree T we denote by $\text{int}(T)$.

Let $\mathcal{S} = \langle V_{\mathcal{S}}, E_{\mathcal{S}} \rangle$ be a *species tree*. $\mathcal{S}$ can be viewed as an upper semilattice with $+$ a binary least upper bound operation and $\top$ the top element, that is, the root. In particular for $a, b \in V_{\mathcal{S}}$, $a < b$ means that a and b are on the same path from the root, with b being closer to the root than a. We define the *comparability predicate* $D(a, b) = 1$, if $a \leq b$ or $b \leq a$ and $D(a, b) = 0$, when a and b are incomparable. The *distance function* $\rho(a, b)$ is used to denote the number of edges on the unique (non-directed) path connecting a and b.

We call distinct nodes $a, b \in V_{\mathcal{S}}$ *siblings* when $a + b$ is a parent of a and b. For $a, b \in V_{\mathcal{S}}$ let $\mathbf{Sb}(a, b)$ be the set of nodes defined by the following recurrent rule: **(i)** $\mathbf{Sb}(a, b) = \emptyset$, if $a = b$ or a and b are siblings, **(ii)** $\mathbf{Sb}(a, b) = \{c\} \cup \mathbf{Sb}(a + c, b)$, if $a < b$ or $a + c < a + b$; here c is the sibling of a, and **(iii)** $\mathbf{Sb}(a, b) = \mathbf{Sb}(b, a)$ otherwise.

By $L(a, b)$ we denote the number of elements in $\mathbf{Sb}(a, b)$. Observe that $L(a, b) = \rho(a, b) - 2 \cdot (1 - D(a, b))$. Let $M_{\mathcal{G}} \colon V_{\mathcal{G}} \to V_{\mathcal{S}}$ be the *least common ancestor (lca) mapping*, from rooted $\mathcal{G}$ into $\mathcal{S}$ that preserves the labeling of the leaves. Note, that if $a, b \in V_{\mathcal{G}}$ are children of v, then $M_{\mathcal{G}}(v) = M_{\mathcal{G}}(a) + M_{\mathcal{G}}(b)$. An example is depicted in Fig. 1.

In this general setting let us assume that we are given a *cost function* $\xi \colon V_{\mathcal{G}} \times V_{\mathcal{S}} \to \mathbf{R}$ which for all nodes $v \in V_{\mathcal{G}}, a \in V_{\mathcal{S}}$ assigns a real $\xi(v, a)$ representing a contribution to node a which comes from v when reconciling $\mathcal{G}$ with $\mathcal{S}$. Having ξ we can define $\kappa(v) = \sum_a \xi(v, a)$ to be a total contribution from v in the reconciliation of $\mathcal{G}$ with $\mathcal{S}$. We call κ a *contribution* function. Finally, $\sigma = \sum_v \kappa(v)$ is the total cost of reconciliation of $\mathcal{G}$ with $\mathcal{S}$.

Now we present examples of cost functions that are used in the duplication model. Let w_1 and w_2 the children of an internal node $v \in V_{\mathcal{G}}$. The *Duplication cost* function is defined as follows: $\xi^D(v, a) = 1$ if $v \in \text{int}(\mathcal{G})$ and $M(v) = M(w_i) = a$ for some i, and $\xi^D(v, a) = 0$ otherwise. Loss cost function: $\xi^L(v, a) = 1$ if $v \in \text{int}(\mathcal{G})$ and $a \in \mathbf{Sb}(M(w_1), M(w_2))$, and $\xi^L(v, a) = 0$ otherwise. It can be proved that if v is internal in $\mathcal{G}$, then $\kappa^D(v) = D(M(w_1), M(w_2))$ and $\kappa^L(v) = L(M(w_1), M(w_2))$ (in both cases 0 if v is a leaf).

Observe that a node $v \in V_{\mathcal{G}}$ is called a duplication [17,9] if $\kappa^D(v) = 1$. Moreover, $\kappa^L(v) = l(v)$, where $l(v)$ is the number of gene losses associated to v. It can be proved that σ^D and σ^L are the minimal number of gene duplications and gene losses (respectively) required to reconcile (or to embed) $\mathcal{G}$ with $\mathcal{S}$. Here details are omitted for brevity. The example of an embedding is depicted in Fig. 1.

2.1 Introduction to Unrooted Reconciliation

Here we highlight some results from [12] that are used for the design of our algorithm. From now on, we assume that $\mathcal{G} = \langle V_{\mathcal{G}}, E_{\mathcal{G}} \rangle$ is an unrooted gene tree. We define a rooting of $\mathcal{G}$ by selecting an edge $e \in E_{\mathcal{G}}$ on which the root is to be placed. Such a rooted tree will be denoted by $\mathcal{G}_e$, where v_* is a new node defining the root. To distinguish between rootings of $\mathcal{G}$, the symbols defined in previous section for rooted gene trees will be extended by inserting index e. Please observe, that the mapping of the root of $\mathcal{G}_e$ is independent of e. Without loss of generality the following is assumed: **(A1)** $\mathcal{S}$ and $\mathcal{G}$ have at least one internal node and **(A2)** $M_e(v_*) = \top$; that is, the root of every rooting is mapped into the root of $\mathcal{S}$ (we may always consider the subtree of the species tree rooted in $M_e(v_*)$ with no change of the cost).

First, we transform $\mathcal{G}$ into a directed graph $\widehat{\mathcal{G}} = \langle V_{\mathcal{G}}, \widehat{E_{\mathcal{G}}} \rangle$, where $\widehat{E_{\mathcal{G}}} = \{ \langle v, w \rangle \mid \{v, w\} \in E_{\mathcal{G}} \}$. In other words each edge $\{v, w\}$ in $\mathcal{G}$ is replaced in $\widehat{\mathcal{G}}$ by a pair of directed edges $\langle v, w \rangle$ and $\langle w, v \rangle$.

Edges in $\widehat{\mathcal{G}}$ are labeled by nodes of $\mathcal{S}$ as follows. If $v \in V_{\mathcal{G}}$ is a leaf labeled by a, then the edge $\langle v, w \rangle \in \widehat{E_{\mathcal{G}}}$ is labeled by a. When v is an internal node in $\widehat{\mathcal{G}}$ we assume that $\langle w_1, v \rangle$ and $\langle w_2, v \rangle$ are labeled by b_1 and b_2, respectively. Then the edge $\langle v, w_3 \rangle \in \widehat{E_{\mathcal{G}}}$, such that $w_3 \neq w_1$ and $w_3 \neq w_2$ is labeled by $b_1 + b_2$. Such labeling will be used to explore mappings of rootings of $\mathcal{G}$. An edge $\{v, w\}$ in $\mathcal{G}$ is called *asymmetric* if exactly one of the labels of $\langle v, w \rangle$ and $\langle w, v \rangle$ in $\widehat{\mathcal{G}}$ is equal to $\top$, otherwise it is called *symmetric*.

Every internal node v and its neighbors in $\widehat{\mathcal{G}}$ define a subtree of $\widehat{E_{\mathcal{G}}}$, called a *star* with a center v, as depicted in Fig. 2. The edges $\langle v, w_i \rangle$ are called *outgoing*, while the edges $\langle w_i, v \rangle$ are called *incoming*. We will refer to the undirected edge $\{v, w_i\}$ as e_i, for $i = 1, 2, 3$.

The are several types of possible star topologies based on the labeling (for proofs and details see [12]): (S1) a star has one incoming edge labeled by $\top$ and two outgoing edges labeled $\top$ and these edges are connected to the three siblings of the center, (S2) a star has exactly two outgoing edges labeled by $\top$, (S3) a star has all outgoing edges and exactly one incoming edgd labeled by $\top$, (S4) a star has all edges labelled by *top*, and (S5) a star has all outgoing edges and exactly two incoming edges labeled by $\top$. Figure 2 illustrates the star topologies.

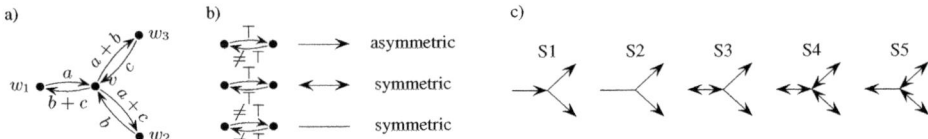

Fig. 2. a) A star in $\widehat{\mathcal{G}}$. b) Types of edges. c) All possible types of stars. We use simplified notation instead of the full topology.

In summary stars are basic 'puzzle-like' units that can be used to assemble them into unrooted gene trees. However, not all star compositions represent a gene tree. For instance, there is no gene tree with 3 stars of type S2. It follows from [12] (see Lemma 4) that we need the following additional condition: **(C1)** if a gene tree has two stars of type S2 then they share a common edge.

Now we overview the main result of [12] (see Theorem 1 for more details). Let $\mathcal{S}$ be a species tree and $\mathcal{G}$ be unrooted gene tree. The set of optimal edges[1] is defined as follows: $\mathbf{Min}_{\mathcal{G}} = \{e \in E_{\mathcal{G}} \mid \sigma_e^{M_{\alpha,\beta}} \text{ is minimal}\}$, where $\sigma_e^{M_{\alpha,\beta}}$ is the total cost for the weighted mutation cost defined by $\xi_e^{M_{\alpha,\beta}}(v,a) = \alpha \cdot \xi_e^D(v,a) + \beta \cdot \xi_e^L(v,a)$, e is an edge in $\mathcal{G}$ and α, β are two positive reals. Then **(M1)** if $|\mathbf{Min}_{\mathcal{G}}| > 1$, then $\mathbf{Min}_{\mathcal{G}}$ consists of all edges present in all stars of type S4 or S5, **(M2)** if $|\mathbf{Min}_{\mathcal{G}}| = 1$, then $\mathbf{Min}_{\mathcal{G}}$ contains exactly one symmetric edge that is present in star of type S2 or S3. From the above statements, (C1) and star topologies we can easily determine $\mathbf{Min}_{\mathcal{G}}$. More precisely, the star edges outside $\mathbf{Min}_{\mathcal{G}}$ are asymmetric and share the same direction. Thus, to find an optimal edge it is sufficient to follow the direction of non $\top$ edges in $\widehat{\mathcal{G}}$.

Now we summarize the time complexity of this procedure. It follows from [4] that a single lca-query (that, is $a + b$ for nodes a and b in $\mathcal{S}$) can be computed in constant time after an initial preprocessing step requiring $O(|\mathcal{S}|)$ time. Other structures like $\widehat{\mathcal{G}}$ with the labeling can be computed in $O(|\mathcal{G}|)$ time. The same complexity has the procedure of finding an optimal edge in $\mathcal{G}$. In summary an optimal edge/rooting and the minimal cost can be computed in linear time. See [12] for more details and other properties.

3   Algorithm

We describe the algorithm for computing the optimal cost and the set of optimal edges after one NNI performed on an unrooted gene tree. We show that a single NNI operation can be completed in constant time if all structures required for computing the optimal rootings are already constructed. First, let us assume that we are given: (a) two positive reals α and β, a species tree $\mathcal{S}$, (b) lca structure for $\mathcal{S}$ that allows to answer lca-queries in constant time, (c) an unrooted gene tree $\mathcal{G}$, (d) $\widehat{\mathcal{G}}$ with the labelling of edges, (e) $\mathbf{Min}_{\mathcal{G}}$ - the set of optimal edges, and (f) σ - the minimal total weighted mutation cost. As mentioned in the previous section (b),(d)-(f) can be computed in linear time $O(\max(|\mathcal{S}|, |\mathcal{G}|))$. Now we show that (c)-(f) can be computed in constant time after single NNI.

---

[1] Candidates for best rootings.

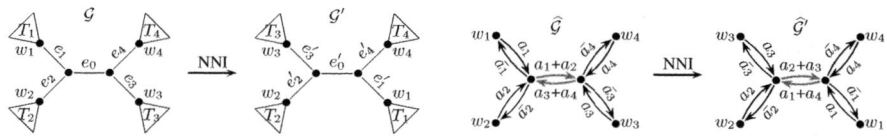

Fig. 3. A single NNI on $\mathcal{G}$ and $\widehat{\mathcal{G}}$. On the left e_i and e'_i (for $i = 0, \ldots, 4$) denote edges in $\mathcal{G}$ and its NNI-neighbor $\mathcal{G}'$, respectively. On the right each node a_i denote the labeling of edges in $\widehat{\mathcal{G}}$. Notation $\bar{a}_i$ denotes the lca-mapping of complementary subtrees, for instance, $\bar{a}_3 = a_1 + a_2 + a_4$, etc. For brevity, we omit each subtree T_i attached to w_i in the left diagram.

NNI operation (c) and the update of lca-mappings (d). We start with the definition of a single NNI operation.

Definition 1. *(Single NNI operation) NNI operation transforms a gene tree $\mathcal{G} = ((T_1, T_2), (T_3, T_4))$ into $\mathcal{G}' = ((T_2, T_3), (T_1, T_4))$, where T_i-s are (rooted) subtrees of $\mathcal{G}$. The edge that connects the roots of (T_1, T_2) and (T_3, T_4) in $\mathcal{G}$ is denoted by e_0 and called the center edge. For each $i = 1, 2, 3, 4$ we assume the following: w_i is the root of T_i, e_i is the edge connecting w_i with e_0 and a_i is the lca-mapping of T_i. Similarly, we define the center edge e'_0 and e'_i in $\mathcal{G}'$.*

The NNI operation is depicted in Fig. 3 with the transformation of $\widehat{\mathcal{G}}$ into $\widehat{\mathcal{G}'}$. The notation will be used from now on. Note that there is a second NNI operation, when $\mathcal{G}$ is replaced with $((T_1, T_3), (T_2, T_4))$. However, it can be easily defined and therefore it is omitted here for brevity. Observe that the NNI operation (without updating of lca-mappings) can be performed in constant time for both trees.

The right part of Fig. 3 depicts the transformation of $\widehat{\mathcal{G}}$. Please observe that the labels of the incoming and outgoing edges attached to each w_i in $\widehat{\mathcal{G}}$ do not change during this operation. Therefore, we can prove the following lemma.

Lemma 1. *NNI changes only the labels of the center edge.*

We conclude that updating $\widehat{\mathcal{G}}$ requires only two lca-queries and can be performed in constant time.

Reconstruction of optimal edges (e). We analyze the changes of the optimal set of edges $Min_{\mathcal{G}}$. To this end we consider a number of cases depending on the relation between the optimal set of edges and the set of edges, incident to the nodes of the center edge. Let $C_{\mathcal{G}} = \{e_i\}_{i=0,\ldots,4}$.

For convenience, assume that the NNI operation replaces e_i with e'_i as indicated in Fig.3. We call two disjoint edges from $C_{\mathcal{G}}$ *semi-alternating* if they will share a common node after the NNI operation. In Fig.3 $\{e_1, e_4\}$ and $\{e_2, e_3\}$ are semi-alternating. For two edges a and b sharing a common node let $\star(a, b)$ be the set of three edges defining the unique star that contains a and b.

Lemma 2. *Under assumption that e_i is replaced by e'_i after the NNI operation the set of optimal edges does not require additional changes iff one of the following conditions is satisfied:*

(EQ1) $\text{Min}_{\mathcal{G}} \cap C_{\mathcal{G}} = \emptyset$,
(EQ2) $\text{Min}_{\mathcal{G}} \supseteq C_{\mathcal{G}}$ *and each pair of semi-alternating edges contains at least one symmetric edge,*
(EQ3) $\text{Min}_{\mathcal{G}}$ *consists of only the center edge,*
(EQ4) $\text{Min}_{\mathcal{G}} \cap C_{\mathcal{G}} = \{e_i\}$ *for some $i > 0$ and the center is asymmetric after the NNI operation.*

Proof: (EQ1) All edges in $C_{\mathcal{G}}$ are asymmetric (2 stars S1). Then, after the NNI operation e_0' is asymmetric and ($C_{\mathcal{G}'}$ has 2 stars S1). (EQ2) $C_{\mathcal{G}}$ consists of 2 stars of type S4/S5 and at most two asymmetric edges. It follows from EQ2 that the asymmetric edges in $C_{\mathcal{G}'}$ cannot form a star of type other than S5. Together with M1 it follows $C_{\mathcal{G}'}$ is optimal. (EQ3) By M1 the center is symmetric in $\mathcal{G}$. It remains symmetric after NNI. From C1 and M2, $\text{Min}_{\mathcal{G}'}$ consists of the center edge. (EQ4) Note that the type of $\star(e_i', e_0')$ is S1, S2 or S3. □

Lemma 3 (NE1). *If $\text{Min}_{\mathcal{G}} \supseteq C_{\mathcal{G}}$ and there exists a pair $\{e_i, e_j\}$ of asymmetric semi-alternating edges, then $\text{Min}_{\mathcal{G}}' = \text{Min}_{\mathcal{G}} \setminus C_{\mathcal{G}} \cup (C_{\mathcal{G}'} \setminus \{e_i', e_j'\})$.*

Proof: The type of $\star(e_i', e_j')$ is S1 or S3 and the other star has type S4 or S5. By M2 e_i' and e_j' are not optimal. □

The proofs for the following lemmas are similar to the proof for Lemma 3 and are omitted for brevity.

Lemma 4 (NE2). *If $\text{Min}_{\mathcal{G}} \cap C_{\mathcal{G}} = \{e_i\}$ for some $i > 0$ and the center is symmetric after the NNI operation then $\text{Min}_{\mathcal{G}}' = \text{Min}_{\mathcal{G}} \setminus \{e_i\} \cup \star(e_0', e_i')$.*

Lemma 5. *Assume that $\text{Min}_{\mathcal{G}} \cap C_{\mathcal{G}} = \{e_0, e_i, e_j\}$, where $i \neq 0$,*

(NE3) *If both e_i and e_j are symmetric then $\text{Min}_{\mathcal{G}'} = \text{Min}_{\mathcal{G}} \setminus C_{\mathcal{G}} \cup C_{\mathcal{G}'}$,*
(NE4) *If e_j is asymmetric and e_0' is symmetric then $\text{Min}_{\mathcal{G}'} = \text{Min}_{\mathcal{G}} \setminus C_{\mathcal{G}} \cup \star(e_0', e_i')$.*
(NE5) *If both e_j and e_0' are asymmetric then $\text{Min}_{\mathcal{G}'} = \text{Min}_{\mathcal{G}} \setminus C_{\mathcal{G}} \cup \{e_i'\}$.*

Note that $\{e_0, e_i, e_j\}$ must be a star in $\mathcal{G}$, that is, $\{i, j\}$ equals either $\{1, 2\}$ or $\{3, 4\}$.

Computing the optimal cost (f). Observe that from Lemmas 2-5 at least one optimal edge remains optimal after the NNI operation. Therefore, to compute the difference in costs between optimal rootings of $\mathcal{G}$ and $\mathcal{G}'$ we start with the cost analysis for the rootings of such edge.

First, we introduce a function for computing the cost differences. Consider three nodes x, y, z of some rooted gene tree such that x and y are siblings and the parent of them (denoted by xy), is a sibling of z. In other words we can denote this subtree by $((x, y), z)$. Then, the partial contribution of $((x, y), z)$ to the total weighted mutation cost can be described as follows: $\sum_{a \in \mathcal{S}} \alpha * (\xi^D(xy, a) + \xi^D(xyz, a)) + \beta * (\xi^L(xy, a) + \xi^L(xyz, a))$. Assume that x, y and z are mapped into a, b and c (from the species tree), respectively. It can be proved from the definition of ξ^D and ξ^L that the above contribution equals: $\phi(a, b, c) = \alpha * (D(a, b) + D(a + b, c)) + \beta * (L(a, b) + L(a + b, c))$. Now, assume that a single NNI operation changes $((x, y), z))$ into $(x, (y, z))$. It should be clear that the cost difference is given by: $\Delta_3(a, b, c) = \phi(c, b, a) - \phi(a, b, c)$. Similarly, we can define a cost difference when a single NNI operation changes $((x, y), (z, v))$

into $((x,v),(y,z))$. Assume, that v is mapped into d. Then, the cost contribution of the first subtree is: $\phi'(a,b,c,d) = \phi(a,b,c+d) + \alpha * D(c,d) + \beta * L(c,d)$. The cost difference is given by: $\Delta_4(a,b,c,d) = \phi'(a,d,b,c) - \phi'(a,b,c,d)$.

Lemma 6. *If the center edge is optimal and remains optimal after the NNI operation then the cost difference equals $\Delta_4(a_1,a_2,a_3,a_4)$, where a_i (for $i = 1,2,3,4$) is the mapping as indicated in Fig.3.*

As mentioned the above lemma can be proved by comparing the rootings placed on the center edges in $\mathcal{G}$ and $\mathcal{G}'$. Lemma 6 gives a solution for cases: EQ2, EQ3, NE1 and NE3. The next lemma gives a solution for the remaining cases.

Lemma 7. *If for some $i > 0$ there exists an optimal edge in $T_i \cup \{e_i\}$ that remains optimal after the NNI operation (under assumption that e_i is replaced by e_i') then the cost difference is $\Delta_3(a_4,a_3,a_2)$ if $i = 1$, $\Delta_3(a_3,a_4,a_1)$ if $i = 2$, $\Delta_3(a_2,a_1,a_4)$ if $i = 3$ and $\Delta_3(a_1,a_2,a_3)$ if $i = 4$.*

Similarly to Lemma 6 we can prove Lemma 7 by comparing the rootings of e_i and e_i'.

Error correction algorithm. Finally, we can present the algorithm for computing the optimal weighted mutation cost for a given gene tree and its NNI neighborhood. See Alg.1 for details. It should be clear that the complexity of this algorithm is linear in time $O(\max(|\mathcal{G}|,|\mathcal{S}|))$. We write that a gene tree has *an error* if the optimal cost is computed for one of its NNI variants (line 7 of Alg.1). Otherwise, we write that a gene tree *does not require corrections*. Please note that it is straightforward to extend the algorithm to reconstruct the optimal variant of the input gene tree.

Algorithm 1. Optimal weighted cost for $\mathcal{G}$ and its NNI neighborhood

---

1. **Input** A species tree $\mathcal{S}$, an unrooted gene tree $\mathcal{G}$, $\alpha, \beta > 0$.
2. **Output** Optimal weighted cost for $\mathcal{G}$ and its NNI neighborhood.
3. Compute: the optimal weighted mutation cost σ, **Min**$_\mathcal{G}$, lca structure for $\mathcal{S}$ and $\widehat{\mathcal{G}}$ by the unrooted reconciliation algorithm [12]. Let $mincost := \sigma$.
4. **for** each internal edge e_0 in $\mathcal{G}$ **do**
5.   Transform $\mathcal{G}$ into $\mathcal{G}'$ and $\widehat{\mathcal{G}}$ into $\widehat{\mathcal{G}}'$ (in situ).
6.   Update **Min**$_\mathcal{G}$ according to cases NE1-NE5 and adjust the cost σ (Lemma 6, 7).
7. $mincost := min(mincost, \sigma)$
8.   Perform the reverse transformation to reconstruct the original $\mathcal{G}$, $\widehat{\mathcal{G}}$ and σ.
9.   Execute all steps 5-8 for the second NNI operation on e_0.
10. **return** $mincost$

3.1 General Reconstruction Problems

We present several approaches to problems of error correction and phylogeny reconstruction. Let us assume that $\sigma_{\alpha,\beta}(\mathcal{S},\mathcal{G})$ is the cost computed by Alg. 1, where $\alpha, \beta > 0$, $\mathcal{S}$ is a rooted species tree and $\mathcal{G}$ is an unrooted gene tree.

Problem 1 (NNIC). Given a rooted species tree $\mathcal{S}$ and a set of unrooted gene trees, G compute the total cost $\sum_{\mathcal{G} \in G} \sigma_{\alpha,\beta}(\mathcal{S},\mathcal{G})$.

The NNIC problem can be solved in polynomial time by an iterative application of Alg. 1. Additionally, we can reconstruct the optimal rootings as well as the correct topology of each gene tree.

Problem 2 (NNIST). Given a set of unrooted gene trees G find the species tree S that minimizes the total cost $\sum_{\mathcal{G} \in G} \sigma_{\alpha,\beta}(\mathcal{S}, \mathcal{G})$.

The complexity of the NNIST problem is unknown. However, similar problems for the duplication model are NP-hard [17]. Therefore we developed heuristics for the NNIST problem to use them in our experiments.

In applications there is typically no need to search over all NNI variants of a gene tree. For instance, a good candidate for an NNI operation is *a weak edge*. A weak edge is usually defined on the basis of its length, where short length indicates weakness. To formalize this property, let us assume that each edge in a gene tree $\mathcal{G}$ has length. We call an edge e in $\mathcal{G}$ *weak* if the length of e is smaller than ω, where ω is a non-negative real. Now we can define variants of NNI-C and NNI-ST denoted by ω-NNIC and ω-NNIST, respectively, where the NNI operations are performed on weak edges only. Formal definitions are omitted for brevity.

4 Experiments

We demonstrate the performance of our algorithms on empirical data sets.

Data preparation. First, we inferred 4133 unrooted gene trees with branch lengths from nine yeast genomes contained in the Genolevures 3 data set [23], which contains protein sequences from the following nine yeast species: *C. glabrata* (4957 protein sequences, abbreviation CAGL), *S. cerevisiae* (5396, SACE), *Z. rouxii* (4840, ZYRO), *S. kluyveri* (5074, SAKL), *K. thermotolerans* (4933, KLTH), *K. lactis* (4851, KLLA), *Y. lipolytica* (4781, YALI), *D. hansenii* (5006, DEHA) and *E. gossypii* (4527, ERGO).

We aligned the protein sequences of each gene family by using the program TCoffee [19] using the default parameter setting. Then maximum likelihood (unrooted) gene trees were computed from the alignments by using proml from the phylip software package. The original species tree of these yeasts [23], here denoted by G3, is shown in Fig. 4.

Software. The unrooted reconciliation algorithm [12] and its data structures are implemented in program URec [13]. Our algorithm partially depends on theses data structures and therefore was implemented as a significantly extended version of URec. Additionally, we implemented a hill climbing heuristic to solve NNIST and ω-NNIST.

Inferring optimal species trees. The optimal species tree reconstructed with error corrections (NNIST optimization problem) is depicted in Fig. 4 and denoted by FULLEC.

Tree	NNIST cost (rank)	NNIST Errors	No correct. cost (rank)
G3	50653 (5)	3680	66407 (4)
FULLEC	**48869 (1)**	3565	64665 (2)
NOEC	48909 (2)	3684	**64413 (1)**

Fig. 4. Species tree topologies. G3 - original phylogeny of Genolevures 3 data set [23]. FULLEC - optimal rooted species tree inferred from gene trees with all possible error corrections (no NNI restrictions, cost 48869 with 3565 corrections). NOEC - optimal species tree for the yeast gene trees with no NNI operations (cost 64413, no corrections). Rank denotes a position of a tree on the sorted list of the best trees.

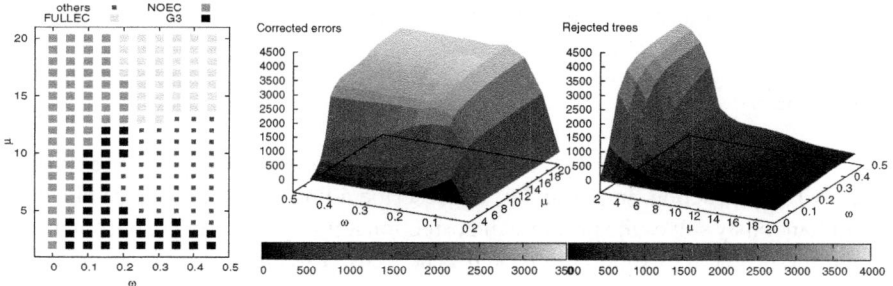

Fig. 5. ω-NNIST experiments for $\omega = 0, 0.05, 0.1, \ldots, 0.45$ and $\mu = 2, 3, \ldots, 20$. On the left - the diagram of the optimal ω-NNIST trees for all μ. In the center - the numbers of corrected errors. On the right - the numbers of rejected trees. Note that $\omega = 0$ does not represents weak edges, $\omega = 0.05$ sets 13.3% of all gene tree edges to be weak, $\omega = 0.1$ - approx. 28%, $\omega = 0.15$ - 41%, $\omega = 0.2$ - 49%, $\omega = 0.3$ - 62%, $\omega = 0.45$ - 77%. The longest edge has length 2.81898.

This tree differs from G3 in the rooting and in the middle clade with KLLA and ERGO. Additionally, we inferred by our ω-NNIST heuristic an optimal species tree, denoted here by NOEC, with no error corrections (that is, for $\omega = 0$). All the trees from this figure are highly scored in each of the optimization schemas.

From weak edges to species trees. In the previous experiment, the NNI operations were performed on almost every gene tree in the optimal solution and with no restrictions on the edges. In order to reconstruct the trees more accurately, we performed experiments for ω-NNIST optimization with various ω parameters and subsets of gene trees. The filtering of gene trees was determined by an integer $\mu > 0$ that defines the maximum number of allowed weak edges in a single gene tree. Each gene tree that did not satisfy such condition was rejected.

Fig. 5 depicts a summary of error correction experiments for weak edges. For each ω and μ we performed 10 runs of the ω-NNIST heuristic for finding the optimal species tree in the set of gene trees filtered by μ. The optimal species trees are depicted in the left diagram. We observed that G3, FULLEC and NOEC are significantly well represented in the set of optimal species trees. Note that the original yeast phylogeny is reconstructed for $\omega = 0.1$-0.15 (in other words approx. 30-40% of edges are weak) and $\mu \leq 10$. In particular for $\omega = 0.15$ and $\mu = 10$, 364 gene trees were rejected and 3164 errors were corrected.

From trusted species tree to weak edges in gene trees - automated and manual curation. Assume that the set of unrooted gene trees and the rooted (trusted) species tree S are given. Then we can state the following problem: find ω and μ such that S is the optimal species tree in ω-NNIST problem for the set of gene trees filtered by μ. For instance in our dataset, if we assume that G3 is a given correct phylogeny of yeasts, then from the left diagram in Fig. 4 one can determine appropriate values of ω and μ that yield G3 as optimal. In other words we can automatically determine weak edges by ω and filter gene trees by μ. This approach can be applied in tree curation procedures to correct errors in automated way as well as to find candidates (rejected trees) for further manual curation. For instance, in the previous case, when $\omega = 0.1$ and $\mu = 10$, we have

3164 trees that can be corrected and rooted by our algorithm, while the 364 rejected trees could be candidates for further manual correction.

5   Discussion

We present novel theoretical and practical results on the problem of error correction and phylogeny reconstruction. In particular, we describe a linear time and space algorithm that simultaneously solves the problem of correction topological errors in unrooted gene trees and the problem of rooting unrooted gene trees. The algorithm allows us to perform efficiently experiments on truly large-scale datasets available for yeast genomes. Our experiments suggest that our algorithm can be used to (i) detect errors, (ii) to infer a correct phylogeny of species under the presence of weak edges in gene trees, and (iii) to help in tree curation procedures. Software, datasets and documentation are freely available from http://bioputer.mimuw.edu.pl/~gorecki/ec.

Acknowledgment

The reviewers have provided several valuable comments that have improved the presentation. This work was conducted in parts with support from the Gene Tree Reconciliation Working Group at NIMBioS through NSF award #EF-0832858, with additional support from the University of Tennessee. PG was partially supported by the grant of MNiSW (N N301 065236) and OE was supported in part by NSF awards #0830012 and #10117189.

References

1. Bansal, M.S., Burleigh, J.G., Eulenstein, O., Wehe, A.: Heuristics for the gene-duplication problem: A $\Theta(n)$ speed-up for the local search. In: Speed, T., Huang, H. (eds.) RECOMB 2007. LNCS (LNBI), vol. 4453, pp. 238–252. Springer, Heidelberg (2007)
2. Bansal, M.S., Eulenstein, O.: The multiple gene duplication problem revisited. Bioinformatics 24(13), i132–i138 (2008)
3. Behzadi, B., Vingron, M.: Reconstructing domain compositions of ancestral multi-domain proteins. In: Bourque, G., El-Mabrouk, N. (eds.) RECOMB-CG 2006. LNCS (LNBI), vol. 4205, pp. 1–10. Springer, Heidelberg (2006)
4. Bender, M.A., Farach-Colton, M.: The lca problem revisited. In: Gonnet, G.H., Panario, D., Viola, A. (eds.) LATIN 2000. LNCS, vol. 1776, pp. 88–94. Springer, Heidelberg (2000)
5. Bonizzoni, P., Della Vedova, G., Dondi, R.: Reconciling a gene tree to a species tree under the duplication cost model. Theoretical Computer Science 347(1-2), 36–53 (2005)
6. Chen, K., Durand, D., Farach-Colton, M.: NOTUNG: a program for dating gene duplications and optimizing gene family trees. J. Comput. Biol. 7(3-4), 429–447 (2000)
7. Durand, D., Halldorsson, B.V., Vernot, B.: A hybrid micro-macroevolutionary approach to gene tree reconstruction. J. Comput. Biol. 13(2), 320–335 (2006)
8. Eulenstein, O., Huzurbazar, S., Liberles, D.A.: Reconciling phylogenetic trees. In: Dittmar, Liberles (eds.) Evolution After Gene Duplication. Wiley, Chichester (2010)
9. Eulenstein, O., Mirkin, B., Vingron, M.: Duplication-based measures of difference between gene and species trees. J. Comput. Biol. 5(1), 135–148 (1998)

10. Fellows, M.R., Hallett, M.T., Stege, U.: On the multiple gene duplication problem. In: Chwa, K.-Y., Ibarra, O.H. (eds.) ISAAC 1998. LNCS, vol. 1533, pp. 347–356. Springer, Heidelberg (1998)
11. Goodman, M., Czelusniak, J., Moore, G.W., Romero-Herrera, A.E., Matsuda, G.: Fitting the gene lineage into its species lineage, a parsimony strategy illustrated by cladograms constructed from globin sequences. Systematic Zoology 28(2), 132–163 (1979)
12. Górecki, P., Tiuryn, J.: Inferring phylogeny from whole genomes. Bioinformatics 23(2), e116–e122 (2007)
13. Górecki, P., Tiuryn, J.: Urec: a system for unrooted reconciliation. Bioinformatics 23(4), 511–512 (2007)
14. Graur, D., Li, W.-H.: Fundamentals of Molecular Evolution. Sinauer Associates, 2 sub edition (2000)
15. Guigó, R., Muchnik, I.B., Smith, T.F.: Reconstruction of ancient molecular phylogeny. Molecular Phylogenetics and Evolution 6(2), 189–213 (1996)
16. Hahn, M.W.: Bias in phylogenetic tree reconciliation methods: implications for vertebrate genome evolution. Genome Biology 8(7), R141+ (2007)
17. Ma, B., Li, M., Zhang, L.: From gene trees to species trees. SIAM Journal on Computing 30(3), 729–752 (2000)
18. Mirkin, B., Muchnik, I.B., Smith, T.F.: A biologically consistent model for comparing molecular phylogenies. J. Comput. Biol. 2(4), 493–507 (1995)
19. Notredame, C., Higgins, D.G., Jaap, H.: T-coffee: a novel method for fast and accurate multiple sequence alignment. J. Mol. Biol. 302(1), 205–217 (2000)
20. Page, R.D.M.: Maps between trees and cladistic analysis of historical associations among genes, organisms, and areas. Systematic Biology 43(1), 58–77 (1994)
21. Page, R.D.M.: GeneTree: comparing gene and species phylogenies using reconciled trees. Bioinformatics 14(9), 819–820 (1998)
22. Sanderson, M.J., McMahon, M.M.: Inferring angiosperm phylogeny from EST data with widespread gene duplication. BMC Evolutionary Biology 7(Suppl 1), S3 (2007)
23. Sherman, D.J., Martin, T., Nikolski, M., Cayla, C., Souciet, J.-L., Durrens, P.: Gènolevures: protein families and synteny among complete hemiascomycetous yeast proteomes and genomes. Nucleic Acids Research 37(suppl 1), D550–D554 (2009)
24. Wehe, A., Bansal, M.S., Burleigh, G.J., Eulenstein, O.: DupTree: a program for large-scale phylogenetic analyses using gene tree parsimony. Bioinformatics 24(13), 1540–1541 (2008)

Comprehensive Pharmacogenomic Pathway Screening by Data Assimilation

Takanori Hasegawa, Rui Yamaguchi, Masao Nagasaki,
Seiya Imoto, and Satoru Miyano

Human Genome Center, Institute of Medical Science, University of Tokyo,
4-6-1 Shirokanedai, Minato-ku, Tokyo 108-8639, Japan
{t-hasegw,ruiy,masao,imoto,miyano}@ims.u-tokyo.ac.jp

Abstract. We propose a computational method to comprehensively screen for pharmacogenomic pathway simulation models. A systematic model generation strategy is developed; candidate pharmacogenomic models are automatically generated from some prototype models constructed from existing literature. The parameters in the model are automatically estimated based on time-course observed gene expression data by data assimilation technique. The candidate simulation models are also ranked based on their prediction power measured by Bayesian information criterion. We generated 53 pharmacogenomic simulation models from five prototypes and applied the proposed method to microarray gene expression data of rat liver cells treated with corticosteroid. We found that some extended simulation models have higher prediction power for some genes than the original models.

1 Introduction

Construction and simulation of biological pathways are crucial steps in understanding complex networks of biological elements in cells [4, 7, 8, 9, 13, 15, 16]. To construct simulatable models, structures of networks and chemical reactions are collected from existing literature and the values of parameters in the model are set based on the results of biological experiments or estimated based on observed data by some computational method [9]. However, it is possible that there are some missing relationships or elements in the literature-based networks. Therefore, we need to develop a computational strategy to improve a prototype model and create better ones that can predict biological phenomena.

To propose novel networks of genes, statistical graphical models including Bayesian networks [3] and vector autoregressive models [5, 11] have been applied to gene expression data. An advantage of these methods is that we can find networks with a large number of genes and analyze them by a viewpoint of systems. However, due to the noise and the limited amount of the data, some parts of the networks estimated by these methods are not biologically reasonable and cannot be validated. In this paper, we focus on another strategy. Unlike the

J. Chen, J. Wang, and A. Zelikovsky (Eds.): ISBRA 2011, LNBI 6674, pp. 160–171, 2011.
© Springer-Verlag Berlin Heidelberg 2011

statistical methods, our method can create a set of extended simulatable models from prototype literature-based models.

There are two key points in our proposed strategy: One is that various structures of candidate simulation models are systematically generated from the prototypes. The other is that, for each created model, the values of parameters are automatically estimated by data assimilation technique [9, 16]; the values of parameters will be determined by maximizing the prediction capability of the model. For each of simulation models, by using data assimilation technique, we can discover that which genes are appropriately predicted their temporal expression patterns by the candidate model. Since we consider pharmacogenomic pathways, these genes are possibly placed on the mode-of-action of target chemical compound. The results obtained by our proposed strategy could be essential to create a larger and more comprehensive simulation model and systems biology driven pharmacology.

To show the effectiveness of the proposed strategy, we analyze time-course microarray data of rat liver cells treated with corticosteroid [2]. In the previous study, differential equation-based simulation models, named fifth generation model [12], were used and predictable expression patterns by this model were discussed for 197 genes selected by clustering analysis [2]. In this paper, we systematically generated 53 simulatable models from five prototypes and determined which 58 models suitably predict expression pattern of each gene. Finally, we show a comprehensive pharmacogenomics pathway screening that elucidates associations between genes and simulation models.

The paper is organized as follows: In Section 2, we elucidate a systematic method to create extended simulation models from prototype ones. The parameter estimation based on data assimilation technique with particle filter [9, 16] and a model selection method [6, 10] are also presented. We apply the proposed pharmacogenomic pathway screening strategy to constructed 58 models and time-course gene expression data of rat liver cells with corticosteroid in Section 3. Discussions are given in Section 4.

2   Method

2.1   Corticosteroid Pharmacokinetic and Pharmacogenomics Models

We first introduce a framework of pharmacokinetic and pharmacogenomic models employed in Jin et al. [2]. Under this framework, a pharmacokinetic model that represents a plasma concentration of methylprednisolone (MPL) in nanograms per milliliter, C_{MPL}, is given by

$$C_{\mathrm{MPL}} = C_1 \cdot e^{-\lambda_1 t} + C_2 \cdot e^{-\lambda_2 t}, \tag{1}$$

where C_1, C_2, λ_1 and λ_2 are coefficients for the intercepts and slopes and Jin et al. [2] set by $C_1 = 39,130$ (ng/ml), $C_2 = 12,670$ (ng/ml), $\lambda_1 = 7.54$ (h^{-1}) and $\lambda_2 = 1.20$ (h^{-1}). These values are obtained from other biological experiments than gene expression profilings that we will use for parameter estimation of

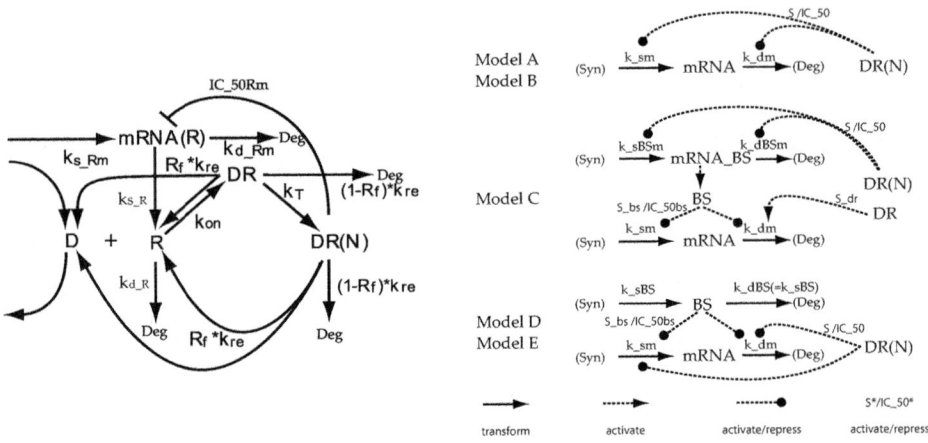

Fig. 1. The left figure and right figure shows core model for corticosteroid pharmacokinetics and prototype pharmacogenomic models with extensions respectively. In the right figure, the dashed lines with circle are the candidate relations to be extended and BS is the intermediate biosignal.

pharmacogenomic models described in the latter section. We thus use these four values for the corticosteroid pharmacokinetics.

In existing literature, corticosteroid pharmacogenomic pathways were investigated [2]. We show the core part of the pathway that includes corticosteroid, represented by D, and its receptor, R, in Figure 1 (left). Here, mRNA(R) denotes the mRNA of the receptor, DR is cytosolic drug-receptor complex and DR(N) is drug-receptor complex in nucleus. The reaction parameters in Figure 1 (left) were set according to Sun et al. [12] and summarized in Table 1 (left). The dynamics of the pathway can be represented by four differential equations given by

$$\frac{d\text{mRNA(R)}}{dt} = k_{s\_Rm} \cdot \left\{ 1 - \frac{\text{DR(N)}}{\text{IC}_{50\_Rm} + \text{DR(N)}} \right\}$$
$$- k_{d\_Rm} \cdot \text{mRNA(R)}, \tag{2}$$

$$\frac{d\text{R}}{dt} = k_{s\_R} \cdot \text{mRNA(R)} + R_f \cdot k_{re} \cdot \{\text{DR(N)} + \text{DR}\}$$
$$- k_{on} \cdot \text{D} \cdot \text{R} - k_{d\_R} \cdot \text{R}, \tag{3}$$

$$\frac{d\text{DR}}{dt} = k_{on} \cdot \text{D} \cdot \text{R} - (k_T + k_{re}) \cdot \text{DR}, \tag{4}$$

$$\frac{d\text{DR(N)}}{dt} = k_T \cdot \text{DR} - k_{re} \cdot \text{DR(N)}. \tag{5}$$

Based on the fundamental model represented in Figure 1 (left), we want to know how DR and DR(N) affect other genes in transcriptional level. As a basic pharmacogenomic model for finding relationship between drug-receptor complex and other genes, we consider extending five pharmacogenomic models [2] shown

Table 1. Parameter Setting for the core model and for the constructed pharmacogenomic models

Fixed Parameter	Value	Unit
k_{s_Rm}	2.90	fmol/g/h
k_{d_Rm}	0.1124	fmol/g/h
IC_{50_Rm}	26.2	fmol/mg
k_{on}	0.00329	1/nmol/h
k_T	0.63	h^{-1}
k_{re}	0.0572	h^{-1}
R_f	0.49	
k_{s_R}	1.2	h^{-1}
k_{d_R}	0.0572	h^{-1}
$mRNA_R^0$	25.8	fmol/g
R^0	540.7	fmol/mg

Estimated Parameter	Model	Unit
k_sm	All	1/nmol/h
k_dm	All	1/nmol/h
S or IC_50	All	1/nmol/h or fmol/mg
k_sBSm	C	1/nmol/h
k_dBSm	C	1/nmol/h
k_sBS	C, DE	1/nmol/h
k_dBS	C	1/nmol/h
S_bs or IC_50bs	C, DE	1/nmol/h or fmol/mg
S_dr	C	1/nmol/h
$mRNA\_BS^0$	C	fmol/mg
BS^0	DE	fmol/mg

in Figure 1 (right). The original five pharmacogenomic pathways [2] have the same elements as the core pharmacokinetic pathway, DR and DR(N), and represent relationships between corticosteroid and its downstream genes. However, more variations can be considered as candidates of pharmacogenomic pathway of corticoid. Therefore, from these five models, we automatically constructed 53 models with the following three rules.

(i) If a regulator, DR(N), DR or BS, activates (represses) the synthesis (degradation) of mRNA, a revised model tests to repress (activate) the degradation (synthesis) of mRNA. However, we do not consider combination effects of them.

(ii) If two regulators regulate the same element, we also consider either two regulator model or one regulator model that is defined by removing one of two edges.

(iii) If two regulators regulate the same element, we consider either independent regulation model that employs additive form or cooperative regulation model with the product of the regulators.

We create these rules for generating simulation models that covers all patterns of regulations when we do not change the number of elements such that mRNAs and proteins in each simulation model.

From Model A: One model with three parameters ("k_sm", "k_dm" and "S or IC_50") was generated by applying the rule (i). These models include only mRNA and can simply represent activation of mRNA expression.

From Model B: One model with three parameters ("k_sm", "k_dm" and "S or IC_50") was generated by applying the rule (i). These models include only mRNA and can simply represent repression of mRNA expression.

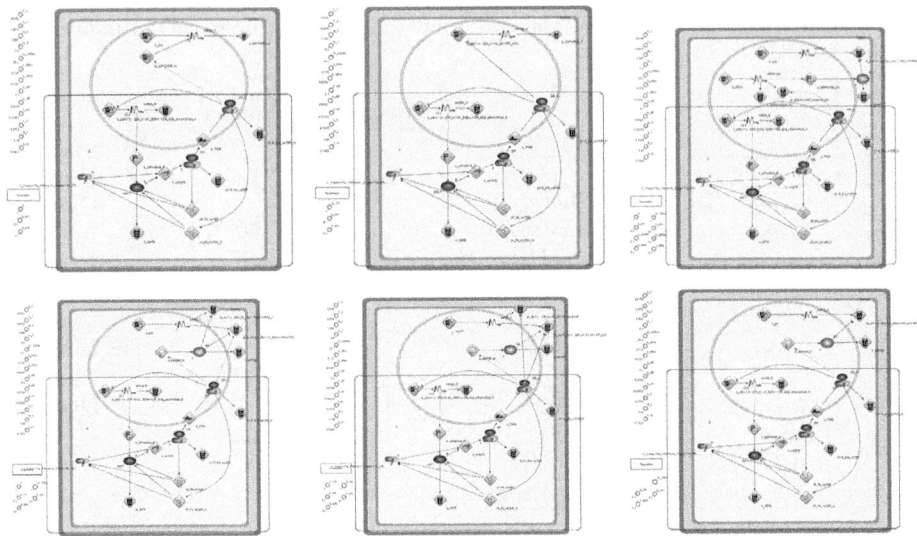

Fig. 2. Six representative pharmacogenomic simulation models (From top left to right, Model A, B, C12, DE10, DE12 and DE20). These models have high predictive power for many of 8799 rat liver genes. These models are described by Cell Illustrator 5.0.

From Model C: First, 15 models with 11 or 10 parameters ("k_sm", "k_dm", "S or IC_50", "k_sBSm", "k_dBSm", "k_sBS", "k_dBS", "S_bs or IC_50bs", "S_dr", and "initial values of mRNA_BS" and "BS") were generated by applying the rule (i) and (ii). These models include mRNA, BS, and mRNA_BS. Since DR is included only in Model C, we evaluate the necessity of the presence of DR by creating models without DR (rule (ii)). Therefore, 16 models that do not have DR were additionally created and finally we have 31 models from Model C.

From Model DE: 20 models with 5 or 6 parameters ("k_sm", "k_dm", "S or IC_50", "k_sBS", "S_bs" or "IC50_bs", and "initial value of BS") were generated by applying the rules (i), (ii) and (iii). These models include mRNA and BS. We unified the notation of Model D and E, because these two models are similar and the extended models are hard to be separated. We constructed 16 models, 4 models and 2 models according to rule (i), (ii) and (iii) respectively. In these simulation models, the parameters, "k_sBSm", "k_sBS", "k_sm", "BS^0 (initial concentration of BS)" and "$mRNA^0_{BS}$ (initial concentration of $mRNA_{BS}$)" were fixed in the original work [2], but we estimate these five parameters together with the other parameters.

For these 53 and original 5 pharmacogenomic models, we estimate the values of parameters by using time-course microarray gene expression data from liver cells of rats received glucocorticoid. We also evaluate which models can predict the expression profiles of each gene; it enables us to find better pharmacogenomic models for each gene. For this purpose, a mathematical technique called data assimilation for parameter estimation and model selection is described in the next section.

2.2   Data Assimilation for Parameter Estimation and Model Selection

To perform simulations by the pharmacogenomic models described in the previous section, we implemented them using Cell Illustrator [8], a software for biological pathway simulation based on hybrid functional Petri net with extensions. Six representative models in Cell Illustrator are shown in Figure 2.

Let $y_j[t]$ be the expression value of jth gene at time t and let $f(x, \theta)$ be a simulation model, where x is a vector of variables in the simulation model and θ is a parameter vector described in the previous section. For example, x includes the concentration of drug-recepter complex, DR. The simulation variable x will be updated by a system model:

$$x_t = f(x_{t-1}, \theta) + v_t, \quad t \in \mathcal{N}, \tag{6}$$

where x_t is the vector of values for the simulation variables at time t, v_t represents innovation noise and N is the set of simulation time points and set $N = \{1, ..., T\}$. To connect the simulation model with the observed data, we formulate an observation model:

$$y_j[t] = h(x_t) + w_t, \quad t \in \mathcal{N}_{obs}, \tag{7}$$

where h is a function that maps simulation variables to the observation and w_t is an observation noise. Here, N_{obs} is the set of time points that we measured gene expression data. We should note that N_{obs} is a subset of N. In our case, since x_t contains a variable representing the abundance of mRNA of the gene, i.e., the jth gene in Eq. (7), the function h takes out the element of x_t corresponding to $y_j[t]$. The model constructed by combining Eq.s (6) and (7) is called a nonlinear state space model. To simplify the notation, we assume $N_{obs} = N$, however, it is easy to generalize the theory described below to the case of $N_{obs} \subset N$.

The parameter vector θ is estimated by the maximum likelihood method that chooses the values of θ that maximize the likelihood

$$L(\theta|Y_{jT}) = \int p(x_0) \prod_{t=1}^{T} p(y_j[t]|x_t)p(x_t|x_{t-1}, \theta)dx_1 \cdots dx_T,$$

where $Y_{jT} = (y_j[1], ..., y_j[T])$. For the computation of the likelihood, we use the particle filter algorithm [9]. For details of the particle filter algorithm for biological pathway model, we refer Nagasaki et al. [8] and Koh et al. [4]. In the parameter estimation, we restricted the values of parameters so that they take positive and not so large from a biological point of view.

For the comparison of multiple simulation models $f_1, ..., f_M$, we employ Bayesian information criterion (BIC) [10]. For the mth model, f_m, BIC is defined by

$$\mathrm{BIC}(f_m) = -2 \log L(\hat{\theta}_m|Y_{jT}) + \nu_m \log T,$$

where $\hat{\boldsymbol{\theta}}_m$ is the maximum likelihood estimate of the vector of parameters in $\boldsymbol{f}_m$ and ν_m is the dimension of $\boldsymbol{\theta}_m$. Therefore, for the jth gene, the optimal simulation model, $\boldsymbol{f}^*$, can be obtained by

$$\boldsymbol{f}^* = \arg\min_{\boldsymbol{f}_m} \mathrm{BIC}(\boldsymbol{f}_m).$$

The model ranking for a gene can also be determined by the values of BIC.

3   Pharmacogenomic Pathway Screening for Corticosteroid 58 Models

3.1   Time-Course Gene Expressions

We analyze microarray time-course gene expression data of rat liver cells [2]. The microarray data were downloaded from GEO database (GSE487). The time-course gene expressions were measured at 0, 0.25, 0.5, 0.75, 1, 2, 4, 5, 5.5, 7, 8, 12, 18, 30, 48 and 72 hours (16 time-points) after receiving glucocorticoid. The data at time 0 hour are control (non-treated). The number of replicated observations is 2, 3 or 4 at a time point.

3.2   Results of Pathway Screening with Data Assimilation

First, we focused on 197 genes that were identified by the previous work [2] as the drug-affected genes by the clustering analysis. For the genes in each cluster, we explored which simulation models have better prediction power and the results are summarized in Figure 3. According to the results obtained previously [2, 12], the genes in the clusters 1, 2, 3, 4, 5 and 6 were reported to be well predicted by the Models "A", "A", "C", "D or E", "cell-cell interaction model" and "B, D or E", respectively. This result indicated that the genes in the cluster 1, 2 have almost same expression profiles. We should note that the cell-cell interaction model is not included in the five prototype models.

Figure 3 shows the results for each cluster and the gene expression profiles. We can summarize the results as follows:

Cluster 1: The previous research [2] suggested that these genes are well predicted by Model A. However, interestingly, in our results, Model A was selected few times. On the other hand, Models D and E and their extended models were selected many times. We presume the reason is that, particularly in the first part, the profiles of these genes are not so simple.

Cluster 2: These genes are also suggested to be suitably predicted with Model A. Like cluster 1, similar results, however, were obtained; for these genes, Model A was not selected in many times.

Cluster 3: The previous research [2] suggested that these genes fitted to Model C. However, in our results, not so many genes in cluster 3 are well predicted by Model C, but they fit to Models D and E and their extended models. We guess the reason is that Model C has more parameters than necessary. Therefore, in BIC, the second term, i.e., penalty for the number of parameters, takes large value and BIC cannot be small, so Model C and its extended versions were not selected. The same things can be said from the other works [1, 14].

Cluster 4: These genes were suggested to be fit with Models D or E. In our results, Model B and its extension and extension of Model A fit well, and Model E is especially fit, but Model D is not selected much. Instead, some extended versions of Models D and E fit well. The genes in cluster 4, we can see that some expression profiles do not vary widely. Such genes are well fit to Models A, B and its extensions, because of these simplicity. On the other hand, Models D and E and their extended models can follow complex behaviors and were selected in many times for other genes.

Cluster 5: Since these genes were judged to be fitted with the cell-cell interaction model that is not included in the five prototypes, these genes are not covered by our prepared models. However, in practice, the extended models of Model DE showed high predictive power for these genes. The expression profiles of these genes show sudden increasing patterns. Actually, our models can represent such dynamic patterns of gene expression profiles.

Cluster 6: These genes were suggested to be fit with Models B, D and E, but most genes were selected as the extended models of Models D and E. We presume the reason is that Models D and E are flexible and can follow various types of complex expression patterns.

We next illustrate the results of pharmacogenomic pathway screening for whole 8799 rat liver genes. Figure 4 shows the results with heatmap of the selected top 5 models for each gene and time-course expression profiles of genes that are specific for Models C6, C12, DE10 and DE12. For each gene, we test the significance of the top ranked simulation model by using Smirnov–Grubbs test. If the expression profile of a gene was predicted very well by several simulation models, we cannot find pharmacogenomic mechanism specific for the gene. However, if only one model could predict the behavior of a gene, the model is a strong candidate that represents corticosteroid's mode-of-action for the gene. In such a case, we say the gene is specific for the above model.

Unlike the genes from the clustering analysis, two prototype models, Models A and B, were selected as top 5 in many times. We presume the reason is that, in the whole gene, there are some genes whose expression patterns are somewhat flat (not show clear dynamic patterns) and Models A and B can follow them with a small number of parameters. Although the prototype D and E models were not selected many times, their extended models were frequently selected as top 5. This suggests that Models D and E can work well as the seed models for generating other simulation models with higher predictive power. The amount of genes obtained by this test varied widely depending on the models. From ModelA1, B1, C6, C12, C16, DE2, DE10, DE12 and DE20, we can obtained some specific genes. Interestingly, the number of genes fitting to Model C is relatively low, but many specific genes are obtained by Model C. It suggest that there are some expression profiles that can be represented by only the one of Model C. We then perform a functional analysis in order to reveal enriched gene

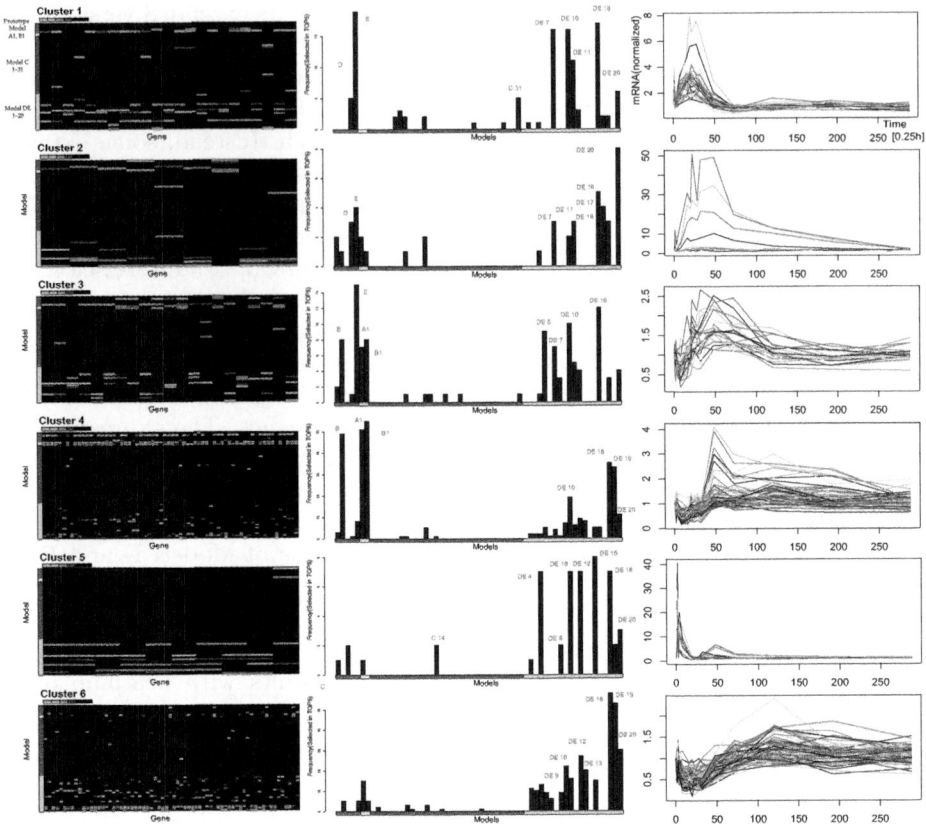

Fig. 3. Top 5 simulation models for each gene in a cluster defined by Jin et al. [2] are represented by a heat map. The green elements means that the model well fits to the gene expression profiles. The histograms of the frequencies of the models selected as top 5 are shown in the middle panels, and gene expression profiles are also shown in the right panels.

functions for each set of Model-specific genes. For the functional analysis, we used Ingenuity and the results can be summarized as follows:

ModelC_6: These genes have function of "Cellular Assembly and Organization" and "RNA Post–Transcriptional Modification" and relate to "Protein Ubiquitination Pathway". **ModelC_12:** These genes are most interesting genes. These have "Amino acid Metabolism", "Nucleic Acid Metabolizm", "Cell Death", "Cellular Grows and Proliferation", "Drug Metabolism" and "Lipid Metabolism" and so on. Additionally, these genes relate to "Aldosterone Signaling Epithelial Cells" and "Glucocorticoid Recepter Signaling". Beneficial effects of Corticosteroid is inhibition of immune system and adverse effect is numerous metabolic side effects, including osteoporosis, muscle wasting, steroid diabetes, and others. Therefore, these result in ModelC_12 is biologically significant because these genes may have a function concerning metabolic side effects. **ModelDE_10:** These genes are

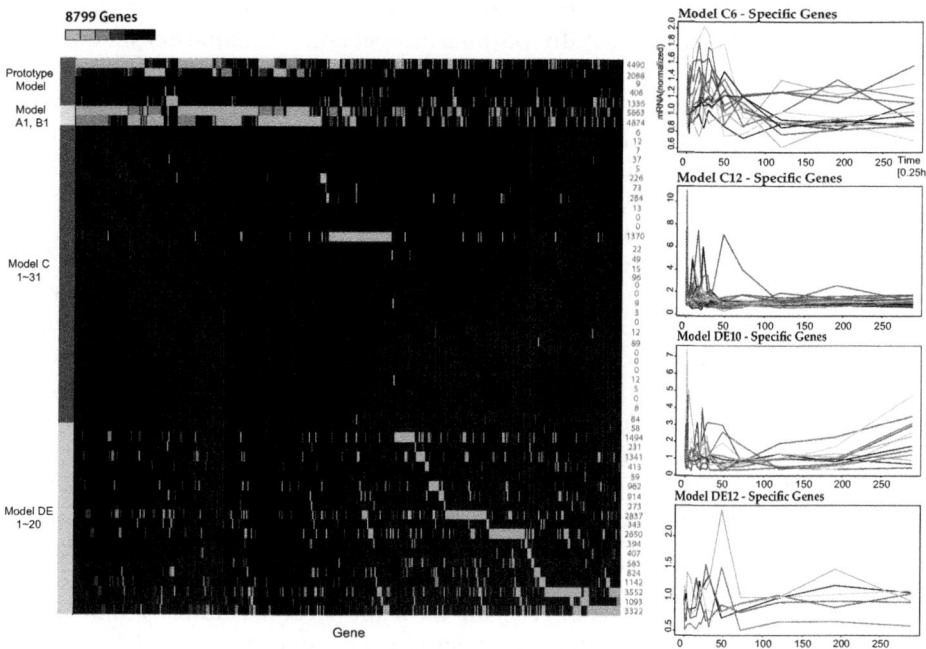

Fig. 4. The result of comprehensive pharmacogenomic pathway simulation model screening. Heat map for top 5 models is shown from 58 simulation models for 8799 rat liver genes. Time-course expression profiles are shown for genes that are specific for Models C6, C12, DE10 and DE12.

also interesting. The functions are "Neurological Disease", "Organismal Injury and Abnormalities" and "Immunological Disease", and are affected by "Graft–versus–Host Disease", "Autoimmune Thyroid Disease, "T Helper Cell Differenti-ation" and so on. Because of the above therapeutic and adverse effects of CS, the function of these genes are also significant concerning immune system function. **ModelDE_12:** The functions of these genes are "Cellular Development", "Car-diovascular Disease", and "Hematological Disease". These are also affected by "EIF2 signaling".

We consider that such genes are important among 8799 genes, because these were estimated to have a similar pathway and it may be difficult to collect these genes by clustering analysis simply using the gene expression profile.

4   Discussion

In this paper, we proposed a computational strategy for automatic generation of pharmacogenomic pathway simulation models from the prototype simulation models that are built based on literature information. The parameters in the con-structed simulation models were estimated based on the observed time-course gene expression data measured by dosing some chemical compound to the target cells. We constructed totally 58 pharmacogenomic simulation models on a pathway simulation software, Cell Illustrator, and used data assimilation technique for parameter estimation. For pathway screening, we introduce Bayesian information criterion for pathway model selection in the framework of data assimilation. We performed comprehensive pathway screening for constructed 58 pharmacogenoimc simulation models with gene expression data of rat liver cells treated with glucocorticoid.

The prototype five models fit to somewhat large number of genes well. However, there are more extended models that can predict the dynamic patterns of gene expressions better than the prototypes. This suggests that, from the prototype simulation models, we can automatically construct various extended simulation models and some of them could have higher prediction ability than the originals. Also, we performed a functional analysis to the sets of Model-specific genes identified by the Smirnov-Grubbs test. As shown above, some meaningful functions were found. We would like to discuss the relationship between Model-specific genes and enriched function in future paper with biological evidences.

We consider the followings as our future research topics. We simply use the pharmacokinetic model described in Section 2. However, we can generated many candidates and may construct true model from observed data by data assimilation technique. Also, we may combine multiple simulation models to create bigger one. As we mentioned before, data analysis based on statistical methods like Bayesian networks can produce network information that would be affected by a chemical compound. It should be useful if we combine the results from statistical data analysis with pharmacogenomic pathway simulations.

Acknowledgments. The computational resource was provided by the Super Computer System, Human Genome Center, Institute of Medical Science, University of Tokyo.

References

1. Hazra, A., Dubois, C.D., Almon, R.R., Snyder, H.G., Jusko, J.W.: Pharmacodynamic Modeling of Acute and Chronic Effects of Methylprednisolone on Hepatic Urea Cycle Genes in Rats. Gene Regulation and System Biology 2, 1–19 (2008)
2. Jin, Y.J., Almon, R.R., Dubois, D.C., Jusko, W.J.: Modeling of corticosteroid pharmacogenomics in rat liver using gene microarrays. The Journal of Pharmacology and Experimental Therapeutics 307(1), 93–107 (2003)
3. Kim, S., Imoto, S., Miyano, S.: Dynamic Bayesian network and nonparametric regression for nonlinear modeling of gene networks from time series gene expression data. Biosystems 75(1-3), 57–65 (2004)
4. Koh, C.H., Nagasaki, M., Saito, A., Wong, L., Miyano, S.: DA1.0: Parameter estimation of biological pathways using data assimilation approach. Bioinformatics 26(14), 1794–1796 (2010)
5. Kojima, K., Yamaguchi, R., Imoto, S., Yamauchi, M., Nagasaki, M., Yoshida, R., Shimamura, T., Ueno, K., Higuchi, T., Gotoh, N., Miyano, S.: A state space representative of VAR models with sparse learning for dynamic gene networks. Genome Informatics 22, 59–68 (2009)

6. Konishi, S., Ando, T., Imoto, S.: Bayesian information criteria and smoothing parameter selection in radial basis function networks. Biometrika 91(1), 27–43 (2004)
7. Matsuno, H., Inoue, S., Okitsu, Y., Fujii, Y.: A new regulatory interaction suggested by simulations for circadian genetic control mechanism in mammals. Journal of Bioinformatics and Computational Biology 4(1), 139–153 (2006)
8. Nagasaki, M., Yamaguchi, R., Yoshida, R., Imoto, S., Doi, A., Tamada, Y., Matsuno, H., Miyano, S., Higuchi, T.: Genomic data assimilation for estimating hybrid functional petri net from time-course gene expression data. Genome Informatics 17(1), 46–61 (2006)
9. Nakamura, K., Yoshida, R., Nagasaki, M., Miyano, S., Higuchi, T.: Parameter estimation of *In Silico* biological pathways with particle filtering toward a petascale computing. In: Pacific Symposium on Biocomputing, vol. 14, pp. 227–238 (2009)
10. Schwarz, G.: Estimating the dimension of a model. Ann. Statist. 6, 461–464 (1978)
11. Shimamura, T., Imoto, S., Yamaguchi, R., Fujita, A., Nagasaki, M., Miyano, S.: Recursive elastic net for inferring large-scale gene networks from time course microarray data. BMC Systems Biology 3, 41 (2009)
12. Sun, Y., Dubois, D.C., Almon, R.R., Jusko, W.J.: Fourth-generation model for corticosteroid pharmacodynamics: A model for methylprednisolone effects on receptor/gene-mediated glucocorticoid receptor down-regulation and tyrosine aminotransferase induction in rat liver. Journal of Pharmacokinetics and Biopharmaceutics 26(3), 289–317 (1998)
13. Tasaki, S., Nagasaki, M., Oyama, M., Hata, H., Ueno, K., Yoshida, R., Higuchi, T., Sugano, S., Miyano, S.: Modeling and estimation of dynamic EGFR pathway by data assimilation approach using time series protemic data. Genome Informatics 17(2), 226–238 (2006)
14. Yao, Z., Hoffman, P.E., Ghimbovschi, S., Dubois, C.D., Almon, R.R., Jusko, W.J.: Mathematical Modeling of Corticosteroid Pharmacogenomics in Rat Muscle following Acute and Chronic Methylprednisolone Dosing. Molecular Pharmaceutics 5(2), 328–339 (2007)
15. Yamaguchi, R., Imoto, S., Yamauchi, M., Nagasaki, M., Yoshida, R., Shimamura, T., Hatanaka, Y., Ueno, K., Higuchi, T., Gotoh, N., Miyano, S.: Predicting difference in gene regulatory systems by state space models. Genome Informatics 21, 101–113 (2008)
16. Yoshida, R., Nagasaki, M., Yamaguchi, R., Imoto, S., Miyano, S., Higuchi, T.: Bayesian learning of biological pathways on genomic data assimilation. Bioinformatics 24(22), 2592–2601 (2008)

The Deep Coalescence Consensus Tree Problem is Pareto on Clusters

Harris T. Lin[1], J. Gordon Burleigh[2], and Oliver Eulenstein[1]

[1] Department of Computer Science, Iowa State University, Ames, IA 50011, USA
{htlin,oeulenst}@iastate.edu
[2] National Evolutionary Synthesis Center, Durham, NC, USA,
University of Florida, Gainesville, FL, USA
gburleigh@ufl.edu

Abstract. Phylogenetic methods must account for the biological processes that create incongruence between gene trees and the species phylogeny. Deep coalescence, or incomplete lineage sorting creates discord among gene trees at the early stages of species divergence or in cases when the time between speciation events was short and the ancestral population sizes were large. The deep coalescence problem takes a collection of gene trees and seeks the species tree that implies the fewest deep coalescence events, or the smallest deep coalescence reconciliation cost. Although this approach can to be useful for phylogenetics, the consensus properties of this problem are largely uncharacterized, and the accuracy of heuristics is untested. We prove that the deep coalescence consensus tree problem satisfies the highly desirable Pareto property for clusters (clades). That is, in all instances, each cluster that is present in all of the input gene trees, called a consensus cluster, will also be found in every optimal solution. We introduce an efficient algorithm that, given a candidate species tree that does not display the consensus clusters, will modify the candidate tree so that it includes all of the clusters and has a lower (more optimal) deep coalescence cost. Simulation experiments demonstrate the efficacy of this algorithm, but they also indicate that even with large trees, most solutions returned by the recent efficient heuristic display the consensus clusters.

1 Introduction

The rapidly growing abundance of genomic sequence data has drawn attention to extensive incongruence among gene trees (e.g., [14, 15]) that may be caused by processes such as deep coalescence (incomplete lineage sorting), gene duplication and loss, or lateral gene transfer (see [5, 11]). Consequently, it is necessary to develop phylogenetic methods that account for the patterns of variation among gene trees, rather than simply assuming the gene tree topology reflects the relationships among species. One such phylogenetic approach is gene tree parsimony (GTP), which, given a collection of gene trees, seeks the species tree that implies the fewest evolutionary events causing incongruence among gene trees [5, 6, 11, 17]. One variation of GTP is the deep coalescence problem, which seeks a species tree that minimizes the number of deep coalescence events [11, 12]. Deep coalescence, or incomplete lineage sorting, may be present at the

J. Chen, J. Wang, and A. Zelikovsky (Eds.): ISBRA 2011, LNBI 6674, pp. 172–183, 2011.

early stages of speciation and whenever the time between speciation events was short and the ancestral population sizes were large. Consequently, there is much interest in phylogenetic approaches that account for coalescence (e.g., [4, 8]). Although the deep coalescence problem is NP-hard [22], recent algorithmic advances enable scientists to solve instances with a limited number of taxa [18] and efficiently compute heuristic solutions for larger data sets [1]. Still, little is known about the consensus properties of the deep coalescence problem or the accuracy of heuristics. In this study, we prove that the deep coalescence problem satisfies the Pareto consensus property. Furthermore, we introduce an efficient algorithm based on the Pareto property that can potentially improve heuristic solutions.

Related work. GTP approaches, including the deep coalescence problem, are examples of supertree problems, in which input trees with taxonomic overlap are combined to build a species tree that includes all of the taxa found in the input trees (see [2]). Although numerous supertree methods have been described, GTP methods are unique because they use a biologically based optimality criterion. One way of evaluating supertree methods is by characterizing their consensus properties (e.g., [3, 20]). The consensus tree problem is the special case of the supertree problem where all the input trees have the same taxa. Since all supertree problems generally seek to retain phylogenetic information from the input trees, one of the most desirable consensus properties is the Pareto property. A consensus tree problem satisfies the Pareto property on clusters (or triplets, quartets, etc.) if every cluster (or triplet, quartet, etc.) that is present in every input tree appears in the consensus tree [3, 20, 21]. Many, if not most, supertree problems in the consensus setting satisfy the Pareto property for clusters [3, 20]. However, this has not been shown for the deep coalescence problem.

Our contribution. We prove that the deep coalescence consensus tree problem satisfies the Pareto property for clusters. That is, for every instance of the problem, the consensus clusters appear in every resulting consensus tree. Consensus clusters are the clusters (or clades) that are present in all of the input trees. This result allows a major refinement of any given heuristic for the deep coalescence problem so that it will return only trees which (i) contain the consensus clusters of the input trees, and (ii) imply equal or fewer deep coalescence events than the tree found by the original heuristic. This follows directly from our construction, given in the proof of Theorem 1, which transforms every consensus tree that is not Pareto for clusters into one that displays this property, and implies fewer deep coalescence events. Furthermore, future heuristics for the deep coalescence problem may take advantage of the reduced search space that follows from our result.

2 Preliminaries

For brevity, proofs are omitted in the text but are available from the authors on request.

2.1 Basic Definitions

A *graph* G is an ordered pair (V, E) consisting of a non-empty set V of *nodes* and a set E of *edges*. We denote the set of nodes and edges of G by $V(G)$ and $E(G)$, respectively.

If $e = \{u, v\}$ is an edge of a graph G, then e is said to be *incident* with u and v. If v is a node of a graph G, then the *degree* of v in G, denoted $deg_G(v)$, is the number of edges in G that are incident with v. Let n be a natural number, the *degree inverse* of n in G, denoted $deg_G^-(n)$, is the set of all degree-n nodes in G.

A *tree* T is a connected graph with no cycles. T is *rooted* if it has exactly one distinguished node of degree one, called the *root*, and we denote it by $\text{Ro}(T)$. The unique edge incident with $\text{Ro}(T)$ is called the *root edge*.

Let T be a rooted tree. We define $\leq_T$ to be the partial order on $V(T)$ where $x \leq_T y$ if y is a node on the path between $\text{Ro}(T)$ and x. If $x \leq_T y$ we call x a *descendant* of y, and y an *ancestor* of x. We also define $x <_T y$ if $x \leq_T y$ and $x \neq y$, in this case we call x a *proper descendant* of y, and y a *proper ancestor* of x. The set of minima under $\leq_T$ is denoted by $\text{Le}(T)$ and its elements are called *leaves*. A node is *internal* if it is not a leaf. The set of all internal nodes of T is denoted by $I(T)$.

Let $X \subseteq \text{Le}(T)$, we write $\overline{X}$ to denote the *leaf complement* of X when the tree T is clear from the context, where $\overline{X} = \text{Le}(T) \setminus X$.

If $\{x, y\} \in E(T)$ and $x <_T y$ then we call y the *parent* of x denoted by $\text{Pa}_T(x)$ and we call x a *child* of y. The set of all children of y is denoted by $\text{Ch}_T(y)$. If two nodes in T have the same parent, they are called *siblings*. The *least common ancestor* (LCA) of a non-empty subset $X \subseteq V(T)$, denoted as $lca_T(X)$, is the unique smallest upper bound of X under $\leq_T$.

If $e \in E(T)$, we define T/e to be the tree obtained from T by identifying the ends of e and then deleting e. T/e is said to be obtained from T by *contracting* e. If v is a vertex of T with degree one or two, and e is an edge incident with v, the tree T/e is said to be obtained from T by *suppressing* v.

Examples of the following definitions are shown in Fig. 1. Let $X \subseteq V(T)$, the *subtree of T induced by* X, denoted $T(X)$, is the minimal connected subtree of T that contains $\text{Ro}(T)$ and X. The *restricted subtree* of T induced by X, denoted as $T|X$, is the tree obtained from $T(X)$ by suppressing all nodes with degree two. The *subtree of T rooted above node* $v \in V(T)$, denoted as T_v, is the restricted subtree induced by $\{u \in V(T) : u \leq_T v\}$.

T is *binary* if every node has degree one or three. Throughout this paper, the term tree refers to a rooted binary tree unless otherwise stated. Also, the subscript of a notation may be omitted when it is clear from the context.

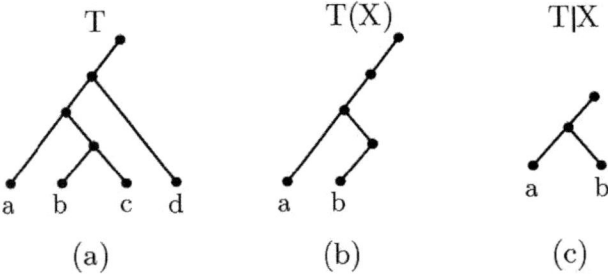

Fig. 1. (a) A rooted tree T with four leaves $\{a, b, c, d\}$. (b) The subtree of T induced by X where $X = \{a, b\}$. (c) The restricted subtree of T induced by X.

2.2   Deep Coalescence

We define the *deep coalescence* cost function as demonstrated in Fig. 2. Note, that our definition of the deep coalescence cost given in Def. 3, is somewhat different, but for our purposes equivalent, to its original definition also termed *extra lineage* given in Def. 6. The relation between both definitions is shown by Prop. 1.

Throughout this section we assume T and S are trees over the same leaf set.

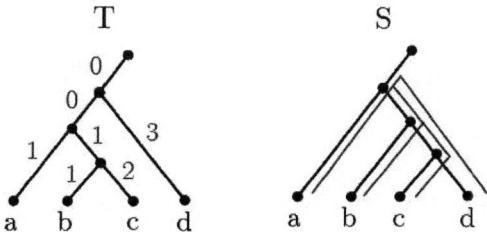

Fig. 2. Example showing the deep coalescence cost from T to S. Each edge of T is accompanied by its cost, and its corresponding path is shown on S.

Definition 1. *[Path Length] Suppose $x \leq_T y$, the* path length *from x to y, denoted $pl_T(x, y)$, is the number of edges in the path from x to y. Further, let $X \subseteq Y \subseteq \mathrm{Le}(T)$, we extend the path length function by $pl_T(X, Y) \triangleq pl_T(lca_T(X), lca_T(Y))$.*

Definition 2. *[LCA Mapping] Let $v \in V(T)$, the* LCA mapping *of v in S, denoted $M_{T \triangleright S}(v)$, is defined by $M_{T \triangleright S}(v) \triangleq lca_S(\mathrm{Le}(T_v))$.*

Definition 3. *[Deep Coalescence] The* deep coalescence cost *from T to S, denoted $DC(T, S)$, is*

$$DC(T, S) \triangleq \sum_{\substack{\{u,v\} \in E(T) \\ u < v}} pl_S(M_{T \triangleright S}(u), M_{T \triangleright S}(v))$$

Using the extended path lengths, the deep coalescence cost can be equivalently expressed as

$$DC(T, S) = \sum_{\substack{\{u,v\} \in E(T) \\ u < v}} pl_S(\mathrm{Le}(T_u), \mathrm{Le}(T_v))$$

Definition 4. *The* Boolean value *of a statement ϕ, denoted as $\llbracket \phi \rrbracket$, is 1 if ϕ is true, 0 otherwise.*

Definition 5. *[Edge Coverage] Let $\{u', v'\} \in E(S)$ and $u' < v'$, the* edge coverage *of $\{u', v'\}$ from T, denoted $C_{T \triangleright S}(u', v')$, is defined by*

$$C_{T \triangleright S}(u', v') \triangleq \sum_{\substack{\{u,v\} \in E(T) \\ u < v}} \llbracket M_{T \triangleright S}(u) \leq u' < v' \leq M_{T \triangleright S}(v) \rrbracket$$

Definition 6. *[Extra Lineage [11]] The* extra lineage cost *from T to S, denoted $EL(T, S)$, is*

$$EL(T, S) \triangleq \sum_{\substack{\{u', v'\} \in E(S) \\ u' < v' < \mathsf{Ro}(S)}} \left(C_{T \triangleright S}(u', v') - 1 \right)$$

Proposition 1. $EL(T, S) = DC(T, S) - |E(S)| + 1$

2.3   Consensus Tree

Definition 7. *[Consensus Tree Problem] Let $f\colon \mathcal{T}_X \times \mathcal{T}_X \to \Re$ be a cost function where X is a leaf set and $\mathcal{T}_X$ is the set of all trees over X. A consensus tree problem based on f is defined as follows.*
   Instance: A tuple of n trees $(T_1, \ldots, T_n)$ over X
   Find: The set of all trees that have the minimum aggregated cost with respect to f. Formally,

$$\operatorname*{argmin}_{S \in \mathcal{T}_X} \left(\sum_{i=1}^{n} f(T_i, S) \right)$$

This set is also called the solutions *for the consensus tree instance.*

Definition 8. *[Deep Coalescence Consensus Tree Problem] We define the* deep coalescence consensus tree problem *to be the consensus tree problem based on the deep coalescence cost function.*

2.4   Cluster and Pareto

Definition 9. *[Cluster] Let T be a tree, the* clusters *induced by T, denoted $\mathsf{Cl}(T)$, is the set of all leaves of some subtree in T. Formally, $\mathsf{Cl}(T) \triangleq \{\mathsf{Le}(T_v)\colon v \in V(T)\}$. Further, $X \in \mathsf{Cl}(T)$ is called a* trivial *cluster if $X = \mathsf{Le}(T)$ or $|X| = 1$, it is called* non-trivial *otherwise. Let $Y \subseteq \mathsf{Le}(T)$, we say that T* contains *(cluster) Y if $Y \in \mathsf{Cl}(T)$.*

Definition 10. *[Pareto on Clusters] Let P be a consensus tree problem based on some cost function. We say that P is* Pareto on clusters *if: for all instances $I = (T_1, \ldots, T_n)$ of P, for all solutions S of I, we have $\bigcap_{i=1}^{n} \mathsf{Cl}(T_i) \subseteq \mathsf{Cl}(S)$.*

3   Theorem Overview

We wish to show that the deep coalescence consensus tree problem is Pareto on clusters. We describe a high level structure of the proof in this section and provide necessary supporting lemmata in Sec. 4.

The proof proceeds by contradiction, assuming that the deep coalescence consensus tree problem is *not* Pareto on clusters. By Def. 10, the assumption implies that there exists an instance $I = (T_1, \ldots, T_n)$, a solution S for I, and a cluster $X \subseteq \mathsf{Le}(S)$ where $X \in \bigcap_{i=1}^{n} \mathsf{Cl}(T_i)$ but $X \notin \mathsf{Cl}(S)$. S being a solution for I, implies by Def. 7, that the aggregated deep coalescence cost, i.e. $\sum_{i=1}^{n} DC(T_i, S)$, is minimized. Then, based on the existence of the cluster X, we edit S and form a new tree R using a tree edit operation which will be introduced in Sec. 4. The properties of this new operation together with the properties of X (proved in Sec. 4), provides the key ingredients to calculate the changes in deep coalescence costs. With some further arithmetics, this allows us to conclude that R in fact has a smaller aggregated deep coalescence cost, i.e. $\sum_{i=1}^{n} DC(T_i, S) > \sum_{i=1}^{n} DC(T_i, R)$, hence contradicting the assumption that S is a solution for I.

4 Supporting Lemmata

4.1 Shallowest Regrouping Operation

In this section we formally define the new tree edit operation that forms the key part of the theorem. We begin with some useful definitions related to the depth of nodes. An example of this operation is shown in Fig. 3.

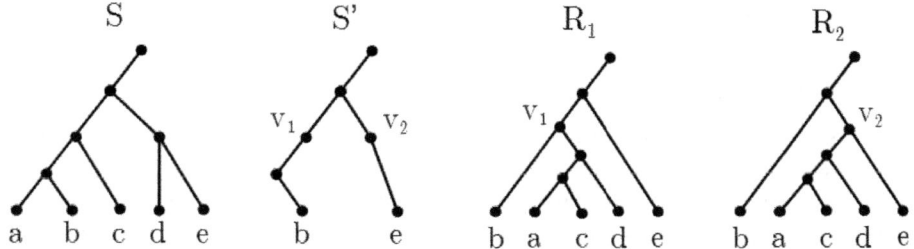

Fig. 3. Example of the shallowest regrouping operation of S by X where $X = \{a, c, d\}$. The intermediate tree $S' = S(\overline{X})$ shows its two shallowest degree-two nodes v_1 and v_2. R_1 and R_2 are the resulting trees of this operation. That is, $\widehat{\Gamma}(S, X) = \{R_1, R_2\}$ where $R_1 = \Gamma(S, X, v_1)$ and $R_2 = \Gamma(S, X, v_2)$.

Definition 11. *[Node Depth] The* depth *of a node* $v \in V(T)$, *denoted* $dep_T(v)$, *is* $pl(v, \text{Ro}(T))$.

Definition 12. *[Minimum Depth] Let* T *be a tree and* $X \subseteq V(T)$, *the* minimum depth *function, denoted* $mindep_T(X)$, *is the set of nodes in* X *which have the minimum depth among all nodes in* X. *Formally, we define* $mindep_T(X) \triangleq argmin_{v \in X} \left(dep_T(v) \right)$.

Now we have the necessary mechanics to define the new tree edit operation. In what follows, we assume S to be a tree, $\emptyset \subset X \subset \text{Le}(S)$, and $S' = S(\overline{X})$.

Definition 13. *[Regroup] Let* $v \in I(S')$. *The* regrouping *operation of* S *by* X *on* v, *denoted* $\Gamma(S, X, v)$, *is the (possibly non-binary) tree obtained from* S' *by*

1. *(R1) Identify* $\text{Ro}(S|X)$ *and* v.
2. *(R2) Suppress all nodes with degree two.*

Definition 14. *[Shallowest Regroup] The* shallowest *regrouping operation of* S *by* X, *denoted* $\widehat{\Gamma}(S, X)$, *defines a set of trees by* $\widehat{\Gamma}(S, X) \triangleq \{\Gamma(S, X, v): v \in mindep_{S'}(deg_{S'}^{-}(2))\}$.

As Fig. 3 shows, the shallowest regrouping operation pulls apart X from S and regroups X back onto each of the shallowest nodes in S.

4.2   Properties of the Shallowest Regrouping Operation

We examine some properties of the shallowest regrouping operation in this section. In general, these properties show that the path lengths defined by LCA's do not increase under several different assumptions. This preservation of path lengths would later assist in the calculation of deep coalescence costs. Throughout this section, we assume S to be a tree, $\emptyset \subset X \subset \text{Le}(S)$, and $A \subseteq B \subseteq \text{Le}(S)$. Further we let $S' = S(\overline{X})$ and $R \in \widehat{\Gamma}(S, X)$.

Lemma 1. *If $B \subseteq \overline{X}$, then $pl_S(A, B) = pl_{S'}(A, B)$.*

Lemma 2. *If $B \subseteq \overline{X}$, then $pl_S(A, B) \geq pl_R(A, B)$.*

Lemma 3. *If $B \subseteq X$, then $pl_S(A, B) \geq pl_R(A, B)$.*

Lemma 4. *If $A \subseteq \overline{X}$ and $X \subseteq B$, then $pl_S(A, B) \geq pl_R(A, B)$.*

4.3   Counting the Number of Degree-Two Nodes

In this section we present some important observations on the side effects of the shallowest regrouping operation. By definition the shallowest regrouping operation includes the step of suppressing nodes with degree two. Since this step affects path lengths and ultimately deep coalescence costs, we are required to count carefully the number of degree-two nodes under various conditions. Here we assume that T is a tree and $\{X, Y\}$ is a bipartition of $\text{Le}(T)$.

Lemma 5. *$deg^-_{T(X)}(2) \neq \emptyset$ and $deg^-_{T(Y)}(2) \neq \emptyset$.*

Lemma 6. *If $v \in deg^-_{T(X)}(2)$, then $\text{Le}(T_v) \cap X \neq \emptyset$ and $\text{Le}(T_v) \cap Y \neq \emptyset$.*

Lemma 7. *If $\text{Pa}(lca(X)) = \text{Ro}(T)$ and $v \in mindep(deg^-_{T(Y)}(2))$, then $dep(v) \leq |deg^-_{T(X)}(2)|$.*

5   Main Theorem

Theorem 1. *Deep coalescence consensus tree problem is Pareto on clusters.*

6   Algorithm for Improving a Candidate Solution

Algorithm 1 takes a consensus tree problem instance and a candidate solution as inputs. If the candidate solution does not display the consensus clusters, it is transformed into one that includes all of the consensus clusters and has a smaller (more optimal) deep coalescence cost.

The correctness of Algorithm 1 follows from the proof of Theorem 1. We now analyze its time complexity. Let m be the number of taxa present in the input trees. Line 3 takes $O(nm)$ time. Line 5, 6, and 7 each takes $O(m)$ time, and there are $O(m)$ iterations. Overall Algorithm 1 takes $O(nm + m^2)$ time.

Algorithm 1. Deep Coalescence Consensus Clusters Builder

1: **procedure** DCCONSENSUSCLUSTERSBUILDER(I,T)
   Input: A consensus tree problem instance $I = (T_1, \ldots, T_n)$, a candidate solution T for I
   Output: T, or an improved solution R that contains all consensus clusters of I
2:    $R \leftarrow T$
3:    $C \leftarrow$ Set of all consensus clusters of I
4:    **for all** cluster $X \in C$ **do**
5:        **if** R does not contain X **then**
6:            $v \leftarrow$ A node in $mindep(deg^-_{R(\overline{X})}(2))$ (shallowest degree-two node of $R(\overline{X})$)
7:            $R \leftarrow \Gamma(R, X, v)$ (regrouping operation of R by X on v)
8:        **end if**
9:    **end for**
10:    **return** R
11: **end procedure**

---

7 Experiment

We used a simulation experiment to test the efficacy of the Pareto property on our ability to estimate solutions to the deep coalescence consensus tree problem. Specifically, we test (i) if the solutions obtained from efficient heuristics presented in [1] display the Pareto property, and (ii) if the estimates can be improved based on our algorithm. We first generated a series of four 14-taxon trees that share a few clusters. To do this, we first generated random 11-taxon trees. Next, we generated random 4-taxon trees containing the species 11-14. We then replaced the one of the leaves in the 11-taxon tree with the random 4-taxon tree. This procedure produces gene trees that share a single 4-taxon cluster in common. Although this simulation does not reflect a biological process, it represents cases in which there is a high degree of error among gene trees. To generate more biologically plausible sets of input gene trees, we followed the general structure the coalescence simulation protocol described by Maddison and Knowles [12]. First, we generated 50 256-taxon species trees based on a Yule pure birth process using the r8s software package [16]. To transform the branch lengths from the Yule simulation to represent generations, we multiplied them all by 10^6. We then simulated coalescence within each species tree (assuming no migration or hybridization) using Mesquite [13]. All simulations produced a single gene copy from each species. For each species tree, we simulated 20 gene trees assuming a constant population size. The population size effects the number of deep coalescence events, with larger populations leading to more incomplete lineage sorting and consequently less agreement among the gene trees. Thus, to incorporate different levels of incomplete lineage sorting, for 25 of the species trees, we used a constant population size of 10,000, and for 25 we used a constant population size of 100,000. Thus, in total, we produced 50 sets of 20 gene trees, with each set simulated from a different 256-taxon species tree.

We performed a GTP phylogenetic analysis based on the deep coalescence consensus tree problem for each set of 20 gene trees using the fast local SPR search and software described by Bansal et al [1]. When the search returned its locally optimal tree, we examined the tree to see if it contained all of the consensus clusters from the input gene trees. If it did not, we used our algorithm to find a tree with a better deep coalescence score that includes all of the consensus clusters.

In three cases with the 14-taxon gene trees, we found that the SPR heuristic did not return a result that contained the consensus cluster. In these cases, our algorithm found a better solution that also contained the consensus cluster. The failure of the SPR heuristic in these cases appears to depend on the starting tree; these data sets did not fail with all starting trees. In contrast, the data sets produced under the 256-taxon coalescent simulations always returned species trees with all consensus clusters, even with relatively high levels of incongruence among gene trees. The average coalescence cost for the trees was 279 when the population size was 10,000 and 2038 when the population size was 100,000, and in all cases there existed simulation clusters. This suggests that the SPR heuristic may often perform well with biologically realistic data sets but may fail in cases of great conflict among gene trees.

8   Discussion

The Pareto property demonstrates that, in addition to offering a biologically informed optimality criterion to resolve incongruence among gene trees, the deep coalescence problem also is guaranteed to retain the phylogenetic clusters for which all gene trees agree. The Pareto property also has useful implications for heuristic estimates of the deep coalescence consensus tree problem. Since the deep coalescence problem is NP-hard [22], most meaningful instances will require heuristics to estimate a solution. While recently developed heuristics based on local search problems are efficient [1], it is difficult to evaluate their performance. The Pareto property suggests a simple method to diagnose suboptimal solutions. If the solution does not contain all the consensus from the input trees, not only is it not optimal, but also there must exist a better solution that contains the all the consensus clusters from the input trees. We further describe and implement an efficient algorithm to find better solutions that contain the consensus clusters given a proposed solution that does not contain the consensus clusters.

Our simulation experiments suggest that, in many cases, the SPR local search heuristic described by Bansal et al. [1] may return solutions that contain the consensus clusters. While this does not necessarily mean that the heuristic has found the optimal solution, it does mean, at the very least, that the heuristic estimates share many clusters with the optimal tree. We note that the size of the simulated data set, 256 taxa, exceeds the size of the largest published analysis of the deep coalescence consensus tree problem and is far beyond the largest instances (8 taxa) from which exact solutions have been calculated [18]. Although, the heuristics appear to perform well, it is likely that some estimated solutions will not contain all consensus clusters. Thus, we recommend always checking solutions for the Pareto property and, if necessary, improving the estimates with our algorithm. This requires very little additional computational cost, and can only improve the species tree estimate.

Although the Pareto property for the deep coalescence consensus tree problem is both desirable and useful, the implications for our results do have limitations. First, this property is limited to the consensus case, or, instances in which all of the input gene trees contain sequences from all of the species. Also, the Pareto property is only useful when all input trees share some clusters in common. If there are no consensus clusters among the input trees, then we cannot distinguish between any possible solutions based on the Pareto property. While this may seem like an extreme case, it is possible if there exists high levels of incomplete lineage sorting, or, perhaps more likely, much error in the gene tree estimates. Also, as we add more and more gene trees, we would expect more instances of conflict among the gene trees, potentially converging towards the elimination of consensus clusters.

Recently Than and Rosenberg [19] proved the existence of cases in which the deep coalescence problem is inconsistent, or converges on the wrong species tree estimate with increasing gene tree data. Although possible inconsistency is a concern for GTP analyses, the Pareto property provides some reassurance. Even in a worse case scenario in which the deep coalescence problem is misled, the optimal solutions will still contain all of the agreed upon clades from the gene trees. Still, perhaps the greatest advantage of the deep coalescence problem, especially compared to likelihood and Bayesian approaches that infer species trees based on coalescence models (e.g., [10, 9, 7]), is its computational speed and the feasibility of estimating a species tree from large-scale genomic data sets representing hundreds of taxa [1]. Here, the Pareto property also may help. Not only can our algorithm improve the performance of any existing heuristic, the Pareto property describes a limited subset of possible species trees that must contain the optimal solution. Future heuristics can greatly reduce the tree search by only focusing on trees that refine the strict consensus of the gene trees.

9   Conclusion and Future Work

We prove that the deep coalescence consensus tree problem satisfies the Pareto property for clusters and describe an efficient algorithm that, given a candidate solution that does not display the consensus clusters, transforms the solution so that it includes all the consensus clusters and has a lower deep coalescence cost. Simulation experiments demonstrate the efficacy of our algorithm. The simulations also suggest that existing heuristics developed for this problem may often perform well with biologically realistic data sets but may fail when there is much conflict among gene trees.

Our algorithm can be used to extend any given heuristic for the deep coalescence problem to obtain better solutions. It also suggests a new general approach to design phylogenetic algorithms. In most cases, heuristics to estimate solutions for phylogenetic inference problems are based on a few generic search strategies such as the local search heuristics based on NNI, SPR, or TBR branch swapping. Although these search strategies often appear to perform well, they are not connected to any specific phylogenetic problems or optimality criteria. Ideally, however, efficient and effective heuristics should be tailored to the properties of the phylogenetic problem. In the case of the deep coalescence consensus tree problem, the Pareto property provides an informative guiding constraint for the tree search. Specifically, when considering possible solutions, we need only consider solutions that contain all clusters from the input gene trees, or, in other words, that refine the strict consensus of the input gene trees. Our results show that generic local searches can be verified and improved based on the Pareto property, but future work will attempt to define more efficient strategies that directly search for the best Pareto solution.

Acknowledgments

The authors would like to thank our anonymous reviewers who have provided valuable comments. This work was conducted with support from the Gene Tree Reconciliation Working Group at NIMBioS through NSF award #EF-0832858, with additional support from the University of Tennessee. HL and OE were supported in parts by NSF awards #0830012 and #10117189.

References

1. Bansal, M., Burleigh, J.G., Eulenstein, O.: Efficient genome-scale phylogenetic analysis under the duplication-loss and deep coalescence cost models. BMC Bioinformatics 11(Suppl 1), S42 (2010)
2. Bininda-Emonds, O.R.P.: Phylogenetic supertrees: combining information to reveal the Tree of Life. Springer, Heidelberg (2004)
3. Bryant, D.: A classification of consensus methods for phylogenies. In: BioConsensus, DIMACS, pp. 163–184. AMS, Providence (2003)
4. Edwards, S.V.: Is a new and general theory of molecular systematics emerging? Evolution; International Journal of Organic Evolution 63(1), 1–19 (2009)
5. Goodman, M., Czelusniak, J., Moore, G.W., Romero-Herrera, A.E., Matsuda, G.: Fitting the gene lineage into its species lineage, a parsimony strategy illustrated by cladograms constructed from globin sequences. Systematic Zoology 28(2), 132–163 (1979)
6. Guigo, R., Muchnik, I., Smith, T.F.: Reconstruction of ancient molecular phylogeny. Mol. Phylogenet. Evol. 6(2), 189–213 (1996)
7. Heled, J., Drummond, A.J.: Bayesian inference of species trees from multilocus data. Molecular Biology and Evolution 27(3), 570–580 (2010)
8. Knowles, L.L.: Estimating species trees: Methods of phylogenetic analysis when there is incongruence across genes. Systematic Biology 58(5), 463–467 (2009)
9. Kubatko, L.S., Carstens, B.C., Knowles, L.L.: STEM: species tree estimation using maximum likelihood for gene trees under coalescence. Bioinformatics 25(7), 971–973 (2009)
10. Liu, L.: BEST: bayesian estimation of species trees under the coalescent model. Bioinformatics 24(21), 2542–2543 (2008)
11. Maddison, W.P.: Gene trees in species trees. Systematic Biology 46(3), 523–536 (1997)
12. Maddison, W.P., Knowles, L.L.: Inferring phylogeny despite incomplete lineage sorting. Systematic Biology 55(1), 21–30 (2006)
13. Maddison, W.P., Maddison, D.: Mesquite: a modular system for evolutionary analysis (2001), http://mesquiteproject.org
14. Pollard, D.A., Iyer, V.N., Moses, A.M., Eisen, M.B.: Widespread discordance of gene trees with species tree in drosophila: Evidence for incomplete lineage sorting. PLoS Genet. 2(10), e173 (2006)
15. Rokas, A., Williams, B.L., King, N., Carroll, S.B.: Genome-scale approaches to resolving incongruence in molecular phylogenies. Nature 425(6960), 798–804 (2003)
16. Sanderson, M.J.: r8s: inferring absolute rates of molecular evolution and divergence times in the absence of a molecular clock. Bioinformatics 19(2), 301–302 (2003)
17. Slowinski, J.B., Knight, A., Rooney, A.P.: Inferring species trees from gene trees: A phylogenetic analysis of the elapidae (Serpentes) based on the amino acid sequences of venom proteins. Molecular Phylogenetics and Evolution 8(3), 349–362 (1997)
18. Than, C., Nakhleh, L.: Species tree inference by minimizing deep coalescences. PLoS Computational Biology 5(9), e1000501 (2009)

19. Than, C.V., Rosenberg, N.A.: Consistency properties of species tree inference by minimizing deep coalescences. Journal of Computational Biology 18(1), 1–15 (2011)
20. Wilkinson, M., Cotton, J.A., Lapointe, F., Pisani, D.: Properties of supertree methods in the consensus setting. Systematic Biology 56(2), 330–337 (2007)
21. Wilkinson, M., Thorley, J., Pisani, D., Lapointe, F.-J., McInerney, J.: Some desiderata for liberal supertrees. In: Phylogenetic Supertrees: Combining Information to Reveal the Tree of Life, pp. 227–246. Springer, Dordrecht (2004)
22. Zhang, L.: From gene trees to species trees II: Species tree inference in the deep coalescence model. IEEE/ACM Trans. Comput. Biol. Bioinformatics (forthcoming, 2011)

Fast Local Search for Unrooted Robinson-Foulds Supertrees

Ruchi Chaudhary[1], J. Gordon Burleigh[2], and David Fernández-Baca[1]

[1] Department of Computer Science, Iowa State University, Ames, IA 50011, USA
[2] Department of Biology, University of Florida, Gainesville, FL 32611, USA

Abstract. A Robinson-Foulds (RF) supertree for a collection of input trees is a comprehensive species phylogeny that is at minimum total RF distance to the input trees. Thus, an RF supertree is consistent with the maximum number of splits in the input trees. Constructing rooted and unrooted RF supertrees is NP-hard. Nevertheless, effective local search heuristics have been developed for the restricted case where the input trees and the supertree are rooted. We describe new heuristics, based on the Edge Contract and Refine (ECR) operation, that remove this restriction, thereby expanding the utility of RF supertrees. We demonstrate that our local search algorithms yield supertrees with notably better scores than those obtained from rooted heuristics.

1 Introduction

Supertree techniques are widely used to combine multiple, usually conflicting, species trees for partially overlapping taxon sets into larger, comprehensive, phylogenies [7,13,23]. Matrix representation with parsimony (MRP) [3,25] is, by far, the most commonly used supertree method. While MRP often performs well [8,11,14], MRP supertrees may display biases and relationships that are not supported by any of the input trees [18,24,22]. Still, MRP remains popular because it can take advantage of fast and effective parsimony heuristics and use a broad range of input data, including rooted, unrooted, and non-binary trees [6]. In contrast to MRP, the Robinson-Foulds (RF) supertree method seeks a supertree that minimizes the total RF distance to the input phylogenies [2]. Thus, an RF supertree is consistent with the maximum number of splits in the input trees. Although the properties of the RF supertree method make it a desirable alternative to MRP, its use has been limited by existing heuristics. Bansal et al. [2] recently developed fast local search algorithms for the *rooted* RF problem, the special case where the input trees and the supertree are rooted. Here, we describe new local search algorithms for the *unrooted* RF problem. These are not only asymptotically as fast as the rooted RF heuristics, but they also allow more types of input data and improve the quality of supertree estimates, making the RF supertree method a viable alternative to MRP for nearly any data set.

The use of local search (hill-climbing) for constructing RF supertrees is motivated by the NP-hardness of the underlying optimization problem. Local search explores the space of possible supertrees in search of a *locally optimum* supertree, a tree whose score is minimum within its "neighborhood", where the neighborhood is defined by a *tree*

J. Chen, J. Wang, and A. Zelikovsky (Eds.): ISBRA 2011, LNBI 6674, pp. 184–196, 2011.

edit operation. The best known tree edit operations are Nearest Neighbor Interchange (NNI) [1], Subtree Prune and Regraft (SPR) [1,9], and Tree Bisection and Reconnection (TBR) [1]. The sizes of the respective neighborhoods are $\Theta(n)$, $\Theta(n^2)$, and $\Theta(n^3)$, where n is the number of taxa in the tree. Ganapathy et al. introduced p-Edge Contract and Refine (ECR) [15], which is based on selecting a set of p edges to contract, after which all possible refinements of the contracted tree are generated. The neighborhood of the 2-ECR operation has size $\Theta(n^2)$. Since the intersection of the TBR and 2-ECR neighborhoods has size $O(n)$ [16,15], a 2-ECR search can cover a significant part of the tree space left unexplored by TBR search. The effectiveness of combining TBR with ECR has been demonstrated for parsimony [17]. Further, the RF-distance between two trees is at most $2p$ if and only if they are one p-ECR move apart [16]. This suggests that ECR may be particularly well-suited for building RF supertrees.

We present fast NNI and 2-ECR local search algorithms for the unrooted RF supertree problem. To our knowledge, the only previous related work is [2] and the supertree analysis package Clann [12], which provides heuristics for maximizing the number of splits shared between the input trees and the supertree, but lacks any running time performance guarantees. Our NNI and 2-ECR search algorithms run in $\Theta(kn)$ and $\Theta(kn^2)$ time, where k is the number of input trees. They represent $\Theta(n)$ speed-ups over the naïve solutions for these problems. The algorithms produce binary supertrees, but the input trees are not required to be binary. The techniques used are, on the surface, similar to those used earlier for rooted trees [2]. In particular, we transform the unrooted problem into a rooted one and use an LCA mapping technique related to that of [2]. On the other hand, there are some important differences. For unrooted trees, we use LCA mappings from the supertree to each input tree, the opposite of what is done for rooted trees. This simplifies the algorithm considerably and allows us to compute RF distances without restricting the supertree to the leaf set of each input tree. It also enables us to handle multiple alternative rootings cleanly.

The results presented here are not only of algorithmic interest. It is often beneficial, if not necessary, to allow unrooted input. Identifying the root of a species tree is among the most difficult problems in phylogenetics (e.g., [28,29]), and conventional likelihood and parsimony-based phylogenetic methods typically produce unrooted trees. To root trees, most analyses include outgroup taxa that lie outside the clade of interest. However, in many cases, no useful outgroups exist, or the phylogenetic distance of available outgroups may contribute to systematic, or long-branch attraction, errors [29]. Methods for rooting trees in the absence of an outgroup also can be problematic. For example, rooting the tree by assuming a molecular clock, or similarly using mid-point rooting, may be misled by molecular rate variation throughout the tree [19,20], and the use of non-reversible models appears to perform well only when the substitution process is strongly asymmetric [20,30].

We examine the performance of our unrooted ECR-based RF supertree heuristic using several large data sets, and compare its performance with rooted RF supertrees obtained by SPR-based local search [2]. We demonstrate that the ability to handle unrooted trees allows us to construct, in a reasonable amount of time, higher-quality trees than those obtained by assuming fixed roots.

2 Preliminaries

2.1 Basic Notations and Problem Definition

A *phylogenetic tree* is an unrooted leaf labeled tree in which all the internal vertices have degree at least two [27]. We will use "phylogenetic tree" and "tree" interchangeably. The leaf set of a tree is denoted by $\mathcal{L}(T)$. The set of all vertices of a tree is denoted by $V(T)$ and set of all edges by $E(T)$. A tree is *binary* if every internal vertex has degree three. Let U be a subset of $V(T)$. We denote by $T(U)$ the minimum subtree of T that connects the elements in U. The *restriction* of T to U, denoted by $T_{|U}$, is the phylogenetic tree that is obtained from $T(U)$ by suppressing all vertices of degree two.

A *split* $A|B$ is a bipartition of the leaf set of a tree; A and B are the *parts* of split $A|B$. Order doesn't matter, so $A|B$ is identical to $B|A$. A split is *nontrivial* if each of A and B contains at least two elements. The set of all nontrivial splits of a tree T is denoted by $\Sigma(T)$.

Let T_1 and T_2 be two trees over the same leaf set. If an isomorphism exists between T_1, T_2, then we write $T_1 \simeq T_2$. The *Robinson-Foulds (RF) distance* [26] between T_1 and T_2, denoted by $RF(T_1, T_2)$, is defined as

$$RF(T_1, T_2) := |(\Sigma(T_1)\backslash\Sigma(T_2)) \cup (\Sigma(T_2)\backslash\Sigma(T_1))|.$$

We extend the notion of RF distance to the case where $\mathcal{L}(T_1) \subseteq \mathcal{L}(T_2)$ by letting $RF(T_1, T_2) := RF(T_1, T_{2|\mathcal{L}(T_1)})$.

A *profile* is a tuple of trees $\mathcal{P} := (T_1, T_2, ..., T_k)$, where each tree $T_i \in \mathcal{P}$ is called an *input tree*. A *supertree* on $\mathcal{P}$ is a phylogenetic tree S such that $\mathcal{L}(S) = \bigcup_{i=1}^{k} \mathcal{L}(T_i)$. We write n to denote $|\mathcal{L}(S)|$; i.e., n is the total number of distinct leaves in the profile.

We extend the notion of RF distance to profile and supertree as follows. Let $\mathcal{P}$ be a profile of unrooted trees and S be a supertree for $\mathcal{P}$. Then, the *RF distance* from $\mathcal{P}$ to S is $RF(\mathcal{P}, S) := \sum_{T \in \mathcal{P}} RF(T, S)$.

We now state our main problem. Let $\mathcal{B}(\mathcal{P})$ be the set of all binary supertrees for $\mathcal{P}$.

Problem 1 (Unrooted RF Supertree).
Input: A profile $\mathcal{P} = (T_1, T_2, ..., T_k)$ of unrooted trees.
Output: A supertree S^* for $\mathcal{P}$ such that $RF(\mathcal{P}, S^*) = \min_{S \in \mathcal{B}(\mathcal{P})} RF(\mathcal{P}, S)$.

The Unrooted RF Supertree problem is NP-hard even when all input trees have the same leaf set [21].

2.2 Local Search Problems

We shall consider local search based on two operations, NNI [1] and 2-ECR [15].

Definition 1 (NNI Operation). *Let e be an internal edge in a binary phylogenetic tree T_1. An NNI operation on T_1 consists of swapping one of the two subtrees on one side of e with one of the two subtrees on the other side of e (see Fig. 1).*

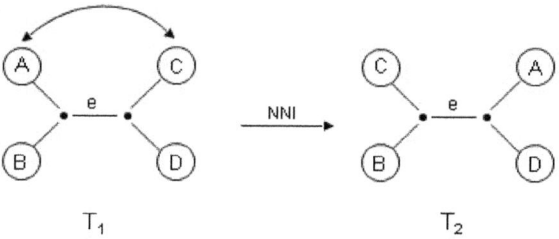

Fig. 1. An NNI operation. Tree T_2 results from T_1 after swapping subtree A with C.

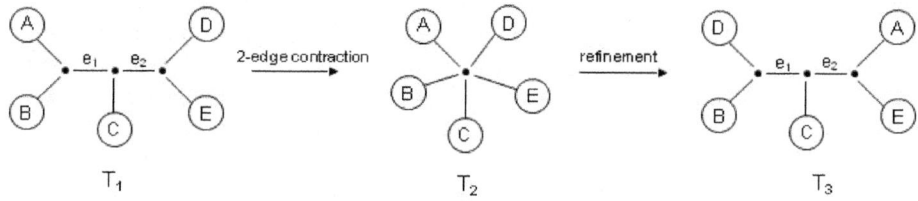

Fig. 2. A 2-ECR operation. Tree T_2 results from T_1 after contracting edge e_1 and e_2; T_2 is fully refined to build T_3. Observe the degree five vertex in T_2.

Definition 2 (2-ECR Operation). *Let T_1 be an unrooted binary tree. A 2-ECR operation on T_1 is the result of (i) choosing two internal edges e_1, e_2 of T_1, (ii) contracting e_1 and e_2; thereby creating one degree five vertex if edges were adjacent or two degree four vertices, otherwise, and (iii) refining the newly created vertex or vertices in some way; i.e., converting the non-binary tree into a binary tree (see Fig 2).*

For $\Delta \in \{\text{NNI}, 2\text{-ECR}\}$, let Δ_T denote the set of trees that can be obtained from a binary tree T by applying a single Δ operation.

Problem 2 (Δ Search).
Input: A profile $\mathcal{P} = (T_1, T_2, ..., T_k)$ of unrooted trees and a binary supertree S for $\mathcal{P}$.
Output: A tree $S^* \in \Delta_S$ such that $RF(\mathcal{P}, S^*) = \min_{S' \in \Delta_S} RF(\mathcal{P}, S')$.

We give algorithms that solve the NNI and 2-ECR search problems in time $\Theta(nk)$ and $\Theta(n^2 k)$, respectively. We achieve this by first executing a $O(kn)$-time preprocessing step (explained in Section 4), which is the same for both problems. After that, for each tree in the input profile, the RF distance from any tree in NNI_S or 2-ECR_S can be computed in constant time.

3   Structural Properties

In this section, we focus on the problem of obtaining the RF distance from an arbitrary input tree T to a supertree S. We solve the problem by exploiting its connection with its rooted version.

A *rooted* phylogenetic tree $\mathbb{T}$ has exactly one distinguished vertex $rt(\mathbb{T})$, called the *root*. A vertex v of $\mathbb{T}$ is *internal* if $v \in V(\mathbb{T}) \backslash (\mathcal{L}(\mathbb{T}) \cup rt(\mathbb{T}))$. The set of all internal

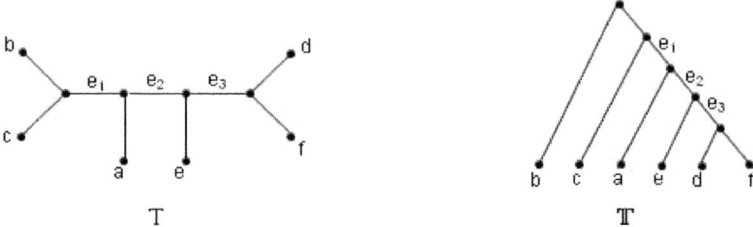

Fig. 3. Unrooted tree T with leaf set $\{a, b, c, d, e, f\}$. The rooted tree $\mathbb{T}$ with $r = b$ is also shown.

vertices of $\mathbb{T}$ is denoted by $I(\mathbb{T})$. We define $\leq_{\mathbb{T}}$ to be the partial order on $V(\mathbb{T})$ where $x \leq_{\mathbb{T}} y$ if y is a vertex on the path from $rt(\mathbb{T})$ to x. If $\{x, y\} \in E(\mathbb{T})$ and $x \leq_{\mathbb{T}} y$, then y is the *parent* of x and x is a *child* of y. Two vertices in $\mathbb{T}$ are *siblings* if they have the same parent. The *least common ancestor (LCA)* of a non-empty subset $L \subseteq V(\mathbb{T})$, denoted by $\mathrm{LCA}_{\mathbb{T}}(L)$, is the unique smallest upper bound of L under $\leq_{\mathbb{T}}$.

The subtree of $\mathbb{T}$ rooted at vertex $v \in V(\mathbb{T})$, denoted by $\mathbb{T}_v$, is the tree induced by $\{u \in V(\mathbb{T}) : u \leq v\}$. For each node $v \in I(\mathbb{T})$, $C_{\mathbb{T}}(v)$ is defined to be the set of all leaf nodes in $\mathbb{T}_v$. Set $C_{\mathbb{T}}(v)$ is called a *cluster*.

Let $\mathcal{H}(\mathbb{T})$ denote the set of all clusters of $\mathbb{T}$. The Robinson-Foulds distance between rooted trees $\mathbb{T}$, $\mathbb{S}$ over the same leaf set [26] is defined as

$$RF(\mathbb{T}, \mathbb{S}) := |(\mathcal{H}(\mathbb{T}) \backslash \mathcal{H}(\mathbb{S})) \cup (\mathcal{H}(\mathbb{S}) \backslash \mathcal{H}(\mathbb{T}))|.$$

Suppose S is a supertree for $\mathcal{P}$ and let T be a tree in $\mathcal{P}$. Throughout the rest of the paper, we assume that some arbitrary but fixed taxon $r \in \mathcal{L}(T) \cap \mathcal{L}(S)$ is chosen for T. We refer to r as the *outgroup*. Different outgroups may be used for different input trees. Let $\mathbb{T}$ and $\mathbb{S}$ be the trees that result from rooting T and S at the respective branches incident on r (see Fig. 3).

Lemma 1. *Let T and S be two unrooted phylogenetic trees with $\mathcal{L}(T) = \mathcal{L}(S)$, then,*

$$RF(T, S) = RF(\mathbb{T}, \mathbb{S}).$$

Proof. We will first show that $RF(T, S) \leq RF(\mathbb{T}, \mathbb{S})$. Recall that, $RF(T, S) := |(\Sigma(T) \backslash \Sigma(S)) \cup (\Sigma(S) \backslash \Sigma(T))|$. We will prove that for each unmatched split in the split set of T (respectively, S), there exists a unique unmatched cluster in the corresponding rooted tree $\mathbb{T}$ (respectively, $\mathbb{S}$). Let $A|B$ be a split such that $A|B \in \Sigma(T)$ but $A|B \notin \Sigma(S)$. Assume without loss of generality that $r \in A$. Then, $B \in \mathcal{H}(\mathbb{T})$ but $B \notin \mathcal{H}(\mathbb{S})$. The argument for S and $\mathbb{S}$ follows similarly. Thus $RF(T, S) \leq RF(\mathbb{T}, \mathbb{S})$ holds. The proof that $RF(T, S) \geq RF(\mathbb{T}, \mathbb{S})$ is similar. $\square$

We extend RF distance to the case where $\mathcal{L}(\mathbb{T}) \subseteq \mathcal{L}(\mathbb{S})$ in the same way as for unrooted trees. That is, $RF(\mathbb{T}, \mathbb{S}) := RF(\mathbb{T}, \mathbb{S}_{|\mathcal{L}(\mathbb{T})})$, where $\mathbb{S}_{|\mathcal{L}(\mathbb{T})}$ is the rooted phylogenetic tree obtained from $\mathbb{S}(\mathcal{L}(\mathbb{T}))$ by suppressing all non-root vertices of degree two. We now show how to compute the RF distance in this more general setting, without explicitly building $\mathbb{S}_{|\mathcal{L}(\mathbb{T})}$.

Definition 3 (Restricted Cluster). *Let* $v \in I(\mathbb{S})$. *The* restriction *of* $C_{\mathbb{S}}(v)$ *to* $\mathcal{L}(\mathbb{T})$ *is defined as*

$$\hat{C}_{\mathbb{T}}(v) := \{w \in \mathcal{L}(\mathbb{S}_v) : w \in \mathcal{L}(\mathbb{T})\}.$$

$\hat{C}_{\mathbb{T}}(v)$ *is called a* restricted cluster.

Definition 4 (Vertex Function). *The* vertex function $f_{\mathbb{S}}$ *assigns each* $u \in I(\mathbb{T})$ *the value* $f_{\mathbb{S}}(u) = |U|$, *where* $U := \{v \in I(\mathbb{S}) : C_{\mathbb{T}}(u) = \hat{C}_{\mathbb{T}}(v)\}$.

Observe that if $\mathcal{L}(\mathbb{S}) = \mathcal{L}(\mathbb{T})$, then for all $u \in I(\mathbb{T})$, $f_{\mathbb{S}}(u) \leq 1$.

We use $f_{\mathbb{S}}$ to define the following set, which will be used to compute $RF(\mathbb{T}, \mathbb{S})$.

$$\mathcal{F}_{\mathbb{S}} = \{u \in I(\mathbb{T}) : f_{\mathbb{S}}(u) = 0\}$$

We will drop the subscript from $f_{\mathbb{S}}$ and $F_{\mathbb{S}}$ when it is clear from the context.

Lemma 2. *Let* $\mathbb{S}' := \mathbb{S}_{|\mathcal{L}(\mathbb{T})}$. *Then* $RF(\mathbb{T}, \mathbb{S}) = |I(\mathbb{S}')| - |I(\mathbb{T})| + 2|\mathcal{F}_{\mathbb{S}'}|$.

Proof. Recall that $RF(\mathbb{T}, \mathbb{S}) := |(\mathcal{H}(\mathbb{T}) \backslash \mathcal{H}(\mathbb{S}')) \cup (\mathcal{H}(\mathbb{S}') \backslash \mathcal{H}(\mathbb{T}))|$. Let $\mathcal{G}_{\mathbb{S}'}$ be a set $\{u \in I(\mathbb{T}) : f_{\mathbb{S}'}(u) > 0\}$. Thus, $RF(\mathbb{T}, \mathbb{S}) = |I(\mathbb{S}')| + |I(\mathbb{T})| - 2|\mathcal{G}_{\mathbb{S}'}|$. Since $|\mathcal{G}_{\mathbb{S}'}| + |\mathcal{F}_{\mathbb{S}'}| = I(\mathbb{T})$ we have $RF(\mathbb{T}, \mathbb{S}) = |I(\mathbb{S}')| - |I(\mathbb{T})| + 2|\mathcal{F}_{\mathbb{S}'}|$.   □

Lemma 3. *Let* $\mathbb{S}' := \mathbb{S}_{|\mathcal{L}(\mathbb{T})}$. *Then* $|\mathcal{F}_{\mathbb{S}}| = |\mathcal{F}_{\mathbb{S}'}|$.

Proof. We prove the lemma by showing that for $u \in I(\mathbb{T})$, $f_{\mathbb{S}}(u) \neq 0$ iff $f_{\mathbb{S}'}(u) \neq 0$.
($\Rightarrow$) Since $f_{\mathbb{S}}(u) \neq 0$, there exists a vertex v in $\mathbb{S}$ such that $C_{\mathbb{T}}(u) = \hat{C}_{\mathbb{T}}(v)$. There are two cases.

Case 1: $\mathcal{L}(\mathbb{S}_v) = \hat{C}_{\mathbb{T}}(v)$. In this case v must exist in $\mathbb{S}'$, and so $f_{\mathbb{S}'}(u) \neq 0$.
Case 2: $\hat{C}_{\mathbb{T}}(v) \subset \mathcal{L}(\mathbb{S}_v)$. Let the children of v be v_1 and v_2. If $\mathcal{L}(\mathbb{T})$ is not disjoint with $\mathcal{L}(\mathbb{S}_{v_1})$ and $\mathcal{L}(\mathbb{S}_{v_2})$, then v exists in $\mathbb{S}'$. Otherwise, at most one of subtrees at these vertices, e.g v_1, may be absent in $\mathbb{S}'$ (if $\mathcal{L}(\mathbb{S}_{v_1})$ and $\mathcal{L}(\mathbb{T})$ are disjoint). In that case by applying the same argument inductively on v_2, we reach a vertex that stays in $\mathbb{S}'$. Thus we have a vertex with similar cluster present in $\mathbb{S}'$. Therefore, $f_{\mathbb{S}'}(u) \neq 0$.

($\Leftarrow$) Since $f_{\mathbb{S}'}(u) \neq 0$, there exists a vertex v in $\mathbb{S}'$ such that $C_{\mathbb{T}}(u) = C_{\mathbb{S}'}(v)$. Now we must have $v \in I(\mathbb{S})$, since restriction of $\mathbb{S}$ to $\mathcal{L}(\mathbb{T})$ does not introduce a new vertices in $\mathbb{S}'$. Thus, in $\mathbb{S}$, $\hat{C}_{\mathbb{T}}(v) = C_{\mathbb{T}}(u)$ (by the definition of $\mathbb{S}'$). Therefore, $f_{\mathbb{S}}(u) \neq 0$.   □

Corollary 1. $RF(\mathbb{T}, \mathbb{S}) = |\mathcal{L}(\mathbb{T})| + |I(\mathbb{T})| + 2|\mathcal{F}_{\mathbb{S}}| - 2$.

Proof. In Lemma 2, $|I(\mathbb{S}')| = |\mathcal{L}(\mathbb{T})| - 2$. Now the result is trivially true.   □

4   Preprocessing

We now describe a $O(n)$-time algorithm to compute the initial vertex function for a supertree $\mathbb{S}$ relative to input tree $\mathbb{T}$, along with the RF distance between these two trees.

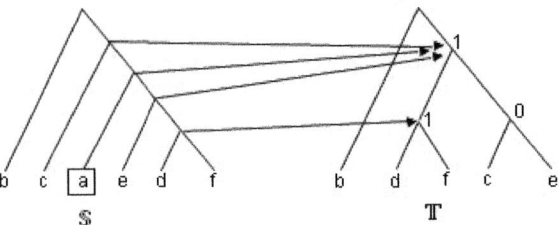

Fig. 4. The LCA mapping from $\mathbb{S}$ to $\mathbb{T}$. Vertex a in $\mathbb{S}$ is mapped to *null* as $a \notin \mathcal{L}(\mathbb{T})$. The internal vertices of $\mathbb{T}$ are labeled with the values of the vertex function.

Definition 5 (LCA Mapping). *For $\mathbb{S}$ and $\mathbb{T}$, the LCA mapping $\mathcal{M}_{\mathbb{S},\mathbb{T}} : V(\mathbb{S}) \to V(\mathbb{T})$ is defined as*

$$\mathcal{M}_{\mathbb{S},\mathbb{T}}(u) := \begin{cases} LCA_{\mathbb{T}}(\hat{C}_{\mathbb{T}}(u)), & \text{if } \hat{C}_{\mathbb{T}}(u) \neq \phi \, ; \\ null, & \text{otherwise.} \end{cases}$$

Fig. 4 illustrates LCA mappings.

Lemma 4. *For all $u \in I(\mathbb{T})$, $f(u) = |B|$, where $B := \{v \in I(\mathbb{S}) : \mathcal{M}_{\mathbb{S},\mathbb{T}}(v) = u$ and $|C_{\mathbb{T}}(u)| = |\hat{C}_{\mathbb{T}}(v)|\}$.*

Proof. By the definition of $f(u)$, it suffices to show that $B = U$, where $U := \{v \in I(\mathbb{S}) : C_{\mathbb{T}}(u) = \hat{C}_{\mathbb{T}}(v)\}$. If $v \in U$, then, by the definition of $\mathcal{M}_{\mathbb{S},\mathbb{T}}(v)$, $v \in B$. If $v \in B$, then $\mathcal{M}_{\mathbb{S},\mathbb{T}}(v) = u$ and $|C_{\mathbb{T}}(u)| = |\hat{C}_{\mathbb{T}}(v)|$ imply that $C_{\mathbb{T}}(u) = \hat{C}_{\mathbb{T}}(v)$. □

The LCA computation for $\mathbb{T}$ can be done in $O(n)$ time, and the LCA mapping from $\mathbb{S}$ to $\mathbb{T}$ can be done in $O(n)$ time [5] in bottom-up manner. Further, from Lemmas 2–4 we can compute the RF distance between $\mathbb{S}$ and $\mathbb{T}$ in $O(n)$ time as well. We assume that there is a distinct rooted copy $\mathbb{S}$ of S for each input tree T, and that $\mathbb{S}$ and $\mathbb{T}$ are rooted according to the outgroup chosen for T.

5   Solving the NNI Search Problem

Let T be an arbitrary tree in $\mathcal{P}$. We now show how to compute the RF distance from T to each tree in NNI$_S$ neighborhood in linear time of the size of neighborhood. The key idea is to simulate each NNI operation on unrooted tree S on its rooted version $\mathbb{S}$, using the LCA mapping from $\mathbb{S}$ to $\mathbb{T}$ to quickly compute the RF distance, This mapping changes at NNI operations are performed on S, but we show that it can be updated in constant time at each step.

Consider an NNI move across an edge $e = \{x, y\}$ of S. Let A and B be the two subtrees on x side of e, and C and D be the two subtrees on y side of e (Fig. 5).

Observation 1. *The tree obtained from swapping A and C is isomorphic to the tree obtained from swapping B and D.*

Proof. Both swaps produce the same sets of splits. Thus, by the Splits Equivalence Theorem [27], the trees are isomorphic. □

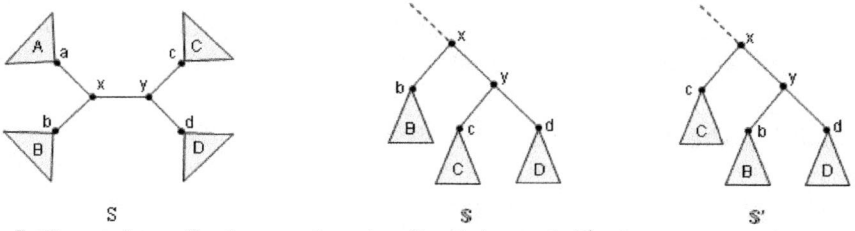

Fig. 5. Unrooted tree S, the rooted version $\mathbb{S}$ and the result $\mathbb{S}'$ of an NNI operation swapping subtrees B and C. The figure assumes that the outgroup lies in subtree A.

Without loss of generality, assume that the NNI move swaps B with C, resulting in tree S'. Also, assume that the subtrees A, B, C, and D connect with edge e through vertices a, b, c, and d, respectively. In $\mathbb{S}$, either x will be the parent of y or y will be the parent of x, depending on which side has the outgroup. In the first case, the children of y will be c and d. Further, if the sibling of y is b then the outgroup must be in subtree A (see Fig. 5), otherwise it is in subtree B. The other cases are analogous. Observe that the parent-child and sibling relationships can be checked in constant time.

Let the children of y in $\mathbb{S}$ be c and d, and the sibling of y be b. After the NNI operation the children of y will be b and d, and the sibling of y will be c. Let the resulting tree be called $\mathbb{S}'$. (Note that if outgroup was in B then we would have swapped A and D, since, from Observation 1 both operations produce the same result.)

Lemma 5. *(i) For all $u \in I(\mathbb{S}) \backslash \{y\}$, $\mathcal{M}_{\mathbb{S}',\mathbb{T}}(u) = \mathcal{M}_{\mathbb{S},\mathbb{T}}(u)$ and (ii) $\mathcal{M}_{\mathbb{S}',\mathbb{T}}(y) = lca(\mathcal{M}_{\mathbb{S},\mathbb{T}}(b), \mathcal{M}_{\mathbb{S},\mathbb{T}}(d))$.*

Proof. (i) For $v \in V(\mathbb{S}'_b) \bigcup V(\mathbb{S}'_c) \bigcup V(\mathbb{S}'_d)$, $\mathbb{S}_v \simeq \mathbb{S}'_v$. Thus, $\mathcal{M}_{\mathbb{S}',\mathbb{T}}(v) = \mathcal{M}_{\mathbb{S},\mathbb{T}}(v)$. Now, $\mathcal{L}(\mathbb{S}'_x) = \mathcal{L}(\mathbb{S}_x)$, thus $\mathcal{M}_{\mathbb{S}',\mathbb{T}}(x) = \mathcal{M}_{\mathbb{S},\mathbb{T}}(x)$. Also, except for subtree $\mathbb{S}_x$, the rest of the tree remains the same in $\mathbb{S}'$, thus for $v \in V(\mathbb{S}') \backslash V(\mathbb{S}'_x)$, $\mathcal{M}_{\mathbb{S}',\mathbb{T}}(v) = \mathcal{M}_{\mathbb{S},\mathbb{T}}(v)$. (ii) Observe that, b, d are children of y in $\mathbb{S}'$, and $\mathbb{S}'_b \simeq \mathbb{S}_b$, $\mathbb{S}'_d \simeq \mathbb{S}'_d$. So, $\mathcal{M}_{\mathbb{S}',T}(y) = $ LCA $(\mathcal{M}_{\mathbb{S}',T}(b), \mathcal{M}_{\mathbb{S}',T}(d)) = $ LCA$(\mathcal{M}_{\mathbb{S},T}(b), \mathcal{M}_{\mathbb{S},T}(d))$. □

Let $h := \mathcal{M}_{\mathbb{S},\mathbb{T}}(y)$ and $h' := \mathcal{M}_{\mathbb{S}',\mathbb{T}}(y)$. Note that h and h' may refer to the same vertex in $\mathbb{T}$. Let G denote the set $\{w \in \{h, h'\} : f_\mathbb{S}(w) = 0, \text{ but } f_{\mathbb{S}'}(w) \geq 1\}$, and L the set $\{w \in \{h, h'\} : f_\mathbb{S}(w) \geq 1, \text{ but } f_{\mathbb{S}'}(w) = 0\}$.

Lemma 6. $RF(\mathbb{S}', \mathbb{T}) = RF(\mathbb{S}, \mathbb{T}) - 2|G| + 2|L|$.

Proof. $RF(\mathbb{S}', \mathbb{T}) = 2|\mathcal{F}_{\mathbb{S}'}| = \{u \in I(\mathbb{T}) : f_{\mathbb{S}'}(u) = 0\} = 2|\mathcal{F}_\mathbb{S}| - 2|\{u \in \{h, h'\} : f_\mathbb{S}(u) = 0\}| + 2|\{u \in \{h, h'\} : f_{\mathbb{S}'}(u) = 0\}| = RF(\mathbb{S}, \mathbb{T}) - 2|G| + 2|L|$. □

Lemma 7. *The RF distance from T to any $S' \in NNI_S$ can be computed in $O(1)$ time.*

Proof. From Lemma 5, the LCA mapping of only one vertex y changes in $\mathbb{S}'$ and can be computed in constant time using the LCA pre-computation of $\mathbb{T}$. Also, the values of $f_{\mathbb{S}'}(h)$ and $f_{\mathbb{S}'}(h')$ can be updated in constant time. Finally, $RF(\mathbb{S}', \mathbb{T})$ is computed in constant time as shown in Lemma 6. Further, $RF(S', T) = RF(\mathbb{S}', \mathbb{T})$. □

Theorem 1. *The NNI Search problem can be solved in $\Theta(nk)$ time.*

Proof. There are $\Theta(n)$ edges in S. From Lemma 7, updating the RF distance after an NNI move takes constant time per input tree. Thus for k input trees it takes $\Theta(nk)$ time. Further, the pre-processing of Section 4 takes $\Theta(nk)$ time.   □

6   Solving the 2-ECR Search Problem

As seen in Section 2, a 2-ECR operation on a binary tree consists of contracting two edges e_1 and e_2, and then refining the contracted tree into a binary tree. These two edges may or may not be adjacent in the tree. Our algorithm for 2-ECR Search handles each case separately.

Case 1: The edges are not adjacent. We use the next result.

Lemma 8. *([15]) Let T be an unrooted leaf-labeled tree and let T' be a 2-ECR neighbor of T such that the 2-ECR move involves the contraction and refinement of two non-adjacent edges in T. Then T' can be reached from T through two NNI moves.*

Thus, when e_1, e_2 are not adjacent, the optimal 2-ECR neighbor can be obtained by computing an optimal NNI neighbor of an NNI neighbor of S. There are $\Theta(n)$ NNI neighbors of S and an optimal NNI neighbor of tree in it can be obtained in $\Theta(nk)$ time by Theorem 2. Therefore the optimal NNI neighbor of an NNI neighbor of S (i.e., the optimal 2-ECR neighbor of S) can be computed in $\Theta(n^2 k)$ time.

Case 2: The edges are adjacent. Note that there are $O(n)$ possible pairs of adjacent edges for a tree with n leaves. For a given pair (e_1, e_2) of edges, the 2-ECR operation contracts e_1 and e_2 and creates a degree-5 vertex. It then refines this vertex in one of the 15 possible ways to obtain a new binary tree. We will show that for each possible refinement the RF distance from an input tree can be computed in constant time.

Let $e_1 = \{x, y\}$ and $e_2 = \{y, z\}$ be the two edges in S chosen for the 2-ECR move. Let S' be the tree that results from the move. Let A and B be the subtree on x side of e_1, C be the subtree connected to y, and D and E be the subtree on z side of S. As in tree T_1 of Fig. 2. Also, assume that the subtrees A, B, C, D, and E connect with e_1 and e_2 through vertices a, b, c, d, and e, respectively.

Note that in tree S, any of the five subtrees A, B, C, D, E can contain the outgroup. We can easily check in constant time which case holds.

Now we divide the 15 possible S's into two categories for computing the RF distance from all possible S'.

Category 1: Subtree C doesn't change position. If C is fixed at the same place as S in S' then the remaining four subtrees can be arranged in three ways. Observe that one of them will be identical to S so we will not consider it. In the other two cases, we will be swapping a subtree on x side (A or B) with a subtree on z side (D or E). Notice that this move is similar to one NNI where the edge spans two edges e_1 and e_2. We show how to compute the RF distance of T from tree S', obtained by swapping A with D in S. The other case can be analyzed similarly.

First, we check which subtree among A, B, C, D, E contains the outgroup in S. If A or D contains the outgroup, then we swap B with E. The splits obtained from swapping

A and D are the same as the splits obtained from swapping B and E. Thus, by the Splits Equivalence Theorem [27], the trees are isomorphic.

Next, we find the vertices of $\mathbb{S}'$ with any change in LCA mapping in $\mathcal{M}_{\mathbb{S}',\mathbb{T}}$. Based on the topology of $\mathbb{S}$, there are three cases:

1. x *is parent of* y *and* y *is parent of* z. For all $t \in I(\mathbb{S}')\backslash\{y,z\}$, $\mathcal{M}_{\mathbb{S}',\mathbb{T}}(t) = \mathcal{M}_{\mathbb{S},\mathbb{T}}(t)$. Further, $\mathcal{M}_{\mathbb{S}',\mathbb{T}}(z) := lca(\mathcal{M}_{\mathbb{S},\mathbb{T}}(a), \mathcal{M}_{\mathbb{S},\mathbb{T}}(e))$, and $\mathcal{M}_{\mathbb{S}',\mathbb{T}}(y) := lca(\mathcal{M}_{\mathbb{S},\mathbb{T}}(c), \mathcal{M}_{\mathbb{S}',\mathbb{T}}(z))$.
2. y *is parent of* x *and* z. For all $t \in I(\mathbb{S}')\backslash\{x,z\}$, $\mathcal{M}_{\mathbb{S}',\mathbb{T}}(t) = \mathcal{M}_{\mathbb{S},\mathbb{T}}(t)$. Further, $\mathcal{M}_{\mathbb{S}',\mathbb{T}}(z) := lca(\mathcal{M}_{\mathbb{S},\mathbb{T}}(a), \mathcal{M}_{\mathbb{S},\mathbb{T}}(e)$, and $\mathcal{M}_{\mathbb{S}',\mathbb{T}}(x) := lca(\mathcal{M}_{\mathbb{S},\mathbb{T}}(d), \mathcal{M}_{\mathbb{S},\mathbb{T}}(b))$.
3. z *is parent of* y *and* y *is parent of* x. For all $t \in I(\mathbb{S}')\backslash\{y,x\}$, $\mathcal{M}_{\mathbb{S}',\mathbb{T}}(t) = \mathcal{M}_{\mathbb{S},\mathbb{T}}(t)$. Moreover, $\mathcal{M}_{\mathbb{S}',\mathbb{T}}(x) := lca(\mathcal{M}_{\mathbb{S},\mathbb{T}}(d), \mathcal{M}_{\mathbb{S},\mathbb{T}}(b))$, and $\mathcal{M}_{\mathbb{S}',\mathbb{T}}(y) := lca(\mathcal{M}_{\mathbb{S},\mathbb{T}}(c), \mathcal{M}_{\mathbb{S}',\mathbb{T}}(x))$.

It can be checked in constant time which one of the above three cases holds, and so the LCA mappings can be updated in constant time too. Let H be a set $\{u \in I(\mathbb{T}) : f_{\mathbb{S}'}(u) \neq f_{\mathbb{S}}(u)\}$. Set H can be computed in constant time. Observe that H will have at most four vertices. The new RF score is computed from the change in the f values of the vertices in H in the following way. For $t \in H$, if $f_{\mathbb{S}}(t) \geq 1$ and $f_{\mathbb{S}'}(t) = 0$ then the RF distance increases by 2 for t. Conversely, if $f_{\mathbb{S}}(t) = 0$ and $f_{\mathbb{S}'}(t) \geq 1$ then the RF distance decreases by 2 for t. Thus we have shown how the RF distance between a input tree and S', in Category 1, can be computed in constant time.

Category 2: Subtree C changes position. In this case the place of C in S' can be occupied by A, B, D, or E. Further, in each case rest of the four subtrees can be arranged at vertices x and z in three ways. Thus there are 12 possibilities in this Category. We will generate all S's in this in an order that will help us to compute RF distance easily. First, we will perform one NNI that swaps subtree C with a subtree from $\{A, B, D, E\}$ and compute the RF distance for the generated S'. For this S', we swap one subtree from x side with one subtree from z side to generate the other two S's. We will present our technique for one subtree, say A; the same can be done for the rest of the subtrees.

Once again, our algorithm first checks the topology of $\mathbb{S}$. If A or C has the outgroup, then we swap the subtrees other than A and C from x and y side of e_1. Observe that this is an NNI operation, and so the RF distance between T and S' can be computed in constant time from Lemma 7. The next two moves on S' are similar to Category 1. Thus, for each tree the RF distance can be computed in constant time.

We have shown how the RF distance between a input tree and S', in Categories 1 and 2 can be computed in constant time. This gives us our final result.

Theorem 2. *The 2-ECR Search problem can be solved in* $\Theta(n^2 k)$ *time.*

7   Experimental Results

We implemented our unrooted RF heuristic based on 2-ECR local search and ran it on a published supertree data set from Marsupials [10] as well as two unpublished data sets from the plant clades Gymnosperm and Saxifragales, where the trees were

Table 1. Experimental Results

Data Set	Supertree Method	RF Distance	Improvement
Saxifragales	Rooted RF	2220	
(959 taxa; 51 trees)	Unrooted RF	2152	3.06%
Marsupial	Rooted RF	1353	
(272 taxa; 158 trees)	Unrooted RF	1335	1.33%
Gymnosperm	Rooted RF	4112	
(950 taxa; 78 trees)	Unrooted RF	4050	1.50%

made with data assembled from GenBank. In our analyses, we first ran the SPR-based rooted RF supertree local search program [2] on each data set and then we selected the best supertree from the output as the starting supertree for our unrooted RF supertree program. We then compared the results of the unrooted RF supertree search with the original rooted RF supertrees (Table 1).

The rooted RF runs took between 53 minutes (for the Marsupial data set) and 32 hours (for the Gymnosperm data set). The unrooted RF runs required 19 search steps and 1.5 hours for the Saxifragales data set, 5 steps and 8 minutes for the Marsupial data set, and 13 steps and 1 hour for the Gymnosperm data set. In all three data sets, the unrooted RF program significantly improved the RF score of the starting supertree.

8   Conclusion

The RF supertree problem directly seeks a supertree that is most similar to input trees based on the RF distance, making it a desirable and potentially useful approach for building comprehensive phylogenies. Until now, the only existing heuristics for RF supertrees required rooted input [2]. However, nearly all recent supertree studies have included unrooted input trees (e.g., [4,7,10]). Thus, our new heuristics for the unrooted RF supertree problem greatly extend the utility of the RF supertree method. Further, our experiments show that they can easily handle data sets with nearly 1000 taxa and can notably improve upon the quality of rooted RF supertrees. This suggests that the RF supertree method is a viable alternative to MRP for nearly any data set. There are several directions for future development. In our experiments, the unrooted heuristic started from a high quality supertree (the rooted RF supertree). Although this strategy appears to be effective, it is also costly. Further tests are needed to examine the effects of the starting tree on the performance of the unrooted heuristic and to identify less costly strategies to build a starting tree. It is also important, and appears to be relatively straightforward, to incorporate uncertainty within the input trees into an RF supertree analysis by weighting the splits when calculating the RF distance.

Acknowledgements. R.C. and D.F.-B. were supported in part by NSF grant DEB-0829674. J.G.B. was supported in part by the NIMBioS Gene Tree Reconciliation Working Group, through NSF grant EF-0832858, with additional support from the University of Tennessee.

References

1. Allen, B.L., Steel, M.: Subtree transfer operations and their induced metrics on evolutionary trees. Annals of Combinatorics 5, 1–13 (2001)
2. Bansal, M.S., Burleigh, J.G., Eulenstein, O., Fernández-Baca, D.: Robinson-Foulds supertrees. Algorithms for Molecular Biology 5, 18 (2010)
3. Baum, B.R.: Combining trees as a way of combining data sets for phylogenetic inference, and the desirability of combining gene trees. Taxon 41, 3–10 (1992)
4. Beck, R.M.D., Bininda-Emonds, O.R.P., Cardillo, M., Liu, F.R., Purvis, A.: A higher-level MRP supertree of placental mammals. BMC Evolutionary Biology 6, 93 (2006)
5. Bender, M.A., Farach-Colton, M.: The LCA problem revisited. In: Gonnet, G.H., Viola, A. (eds.) LATIN 2000. LNCS, vol. 1776, pp. 88–94. Springer, Heidelberg (2000)
6. Bininda-Emonds, O.R.P., Beck, R.M.D., Purvis, A.: Getting to the roots of matrix representation. Syst. Biol. 54, 668–672 (2005)
7. Bininda-Emonds, O.R.P., Cardillo, M., Jones, K.E., MacPhee, R.D.E., Beck, R.M.D., Grenyer, R., Price, S.A., Vos, R.A., Gittleman, J.L., Purvis, A.: The delayed rise of present-day mammals. Nature 446, 507–512 (2007)
8. Bininda-Emonds, O.R.P., Sanderson, M.J.: Assessment of the accuracy of matrix representation with parsimony analysis supertree construction. Systematic Biology 50, 565–579 (2001)
9. Bordewich, M., Semple, C.: On the computational complexity of the rooted subtree prune and regraft distance. Annals of Combinatorics 8, 409–423 (2004)
10. Cardillo, M., Bininda-Emonds, O.R.P., Boakes, E., Purvis, A.: A species-level phylogenetic supertree of marsupials. Journal of Zoology 264, 11–31 (2004)
11. Chen, D., Eulenstein, O., Fernández-Baca, D., Burleigh, J.G.: Improved heuristics for minimum-flip supertree construction. Evolutionary Bioinformatics 2, 347–356 (2006)
12. Creevey, C.J., McInerney, J.O.: Clann: Investigating phylogenetic information through supertree analyses. Bioinformatics 21(3), 390–392 (2005)
13. Davies, T.J., Barraclough, T.G., Chase, M.W., Soltis, P.S., Soltis, D.E., Savolainen, V.: Darwin's abominable mystery: insights from a supertree of the angiosperms. Proceedings of the National Academy of Sciences of the United States of America 101, 1904–1909 (2004)
14. Eulenstein, O., Chen, D., Burleigh, J.G., Fernández-Baca, D., Sanderson, M.J.: Performance of flip supertree construction with a heuristic algorithm. Systematic Biology 53, 299–308 (2003)
15. Ganapathy, G., Ramachandran, V., Warnow, T.: Better hill-climbing searches for parsimony. In: Benson, G., Page, R.D.M. (eds.) WABI 2003. LNCS (LNBI), vol. 2812, pp. 245–258. Springer, Heidelberg (2003)
16. Ganapathy, G., Ramachandran, V., Warnow, T.: On contract-and-refine transformations between phylogenetic trees. In: SODA, pp. 900–909 (2004)
17. Goloboff, P.A.: Analyzing large data sets in reasonable times: Solutions for composite optima. Cladistics 15, 415–428 (1999)
18. Goloboff, P.A.: Minority rule supertrees? MRP, compatibility, and minimum flip display the least frequent groups. Cladistics 21, 282–294 (2005)
19. Holland, B., Penny, D., Hendy, M.: Outgroup misplacement and phylogenetic inaccuracy under a molecular clock — a simulation study. Syst. Biol. 52, 229–238 (2003)
20. Huelsenbeck, J., Bollback, J., Levine, A.: Inferring the root of a phylogenetic tree. Syst. Biol. 51, 32–43 (2002)
21. McMorris, F.R., Steel, M.A.: The complexity of the median procedure for binary trees. In: Proceedings of the International Federation of Classification Societies (1993)
22. Pisani, D., Wilkinson, M.: MRP, taxonomic congruence and total evidence. Systematic Biology 51, 151–155 (2002)

23. Pisani, D., Yates, A.M., Langer, M.C., Benton, M.J.: A genus-level supertree of the Di-
nosauria. Proceedings of the Royal Society of London 269, 915–921 (2002)
24. Purvis, A.: A modification to Baum and Ragan's method for combining phylogenetic trees.
Systematic Biology 44, 251–255 (1995)
25. Ragan, M.A.: Phylogenetic inference based on matrix representation of trees. Molecular Phylogenetics and Evolution 1, 53–58 (1992)
26. Robinson, D.F., Foulds, L.R.: Comparison of phylogenetic trees. Mathematical Biosciences 53, 131–147 (1981)
27. Semple, C., Steel, M.: Phylogenetics. Oxford University Press, Oxford (2003)
28. Smith, A.: Rooting molecular trees: problems and strategies. Biol. J. Linn. Soc. 51, 279–292
(1994)
29. Wheeler, W.: Nucleic acid sequence phylogeny and random outgroups. Cladistics 6, 363–368
(1990)
30. Yap, V., Speed, T.: Rooting a phylogenetic tree with nonreversible substitution models. BMC
Evol. Biol. 5, 2 (2005)

A Metric for Phylogenetic Trees Based on Matching

Yu Lin, Vaibhav Rajan, and Bernard M.E. Moret

Laboratory for Computational Biology and Bioinformatics,
Swiss Federal Institute of Technology (EPFL),
EPFL-IC-LCBB, INJ 230, Station 14, CH-1015 Lausanne, Switzerland
{yu.lin,vaibhav.rajan,bernard.moret}@epfl.ch

Abstract. Comparing two or more phylogenetic trees is a fundamental task in computational biology. The simplest outcome of such a comparison is a pairwise measure of similarity, dissimilarity, or distance. A large number of such measures have been proposed, but so far all suffer from problems varying from computational cost to lack of robustness; many can be shown to behave unexpectedly under certain plausible inputs. For instance, similarity measures based on maximum agreement are too strict, while measures based on the elimination of rogue taxa work poorly when the proportion of rogue taxa is significant; distance measures based on edit distances under simple tree operations (such as nearest-neighbor interchange or subtree pruning and regrafting) are NP-hard; and the widely used Robinson-Foulds distance is poorly distributed and thus affords little discrimination, while also lacking robustness in the face of very small changes—reattaching a single leaf elsewhere in a tree of any size can instantly maximize the distance.

In this paper, we introduce an entirely new pairwise distance measure, based on matching, for phylogenetic trees. We prove that our measure induces a metric on the space of trees, show how to compute it in low polynomial time, verify through statistical testing that it is robust, and finally note that it does not exhibit unexpected behavior under the same inputs that cause problems with other measures. We also illustrate its usefulness in clustering trees, demonstrating significant improvements in the quality of hierarchical clustering as compared to the same collections of trees clustered using the Robinson-Foulds distance.

1 Introduction

Comparing two or more phylogenetic trees is a fundamental task in computational biology. The simplest outcome of such a comparison is a pairwise measure of similarity, dissimilarity, or distance. A large number of such measures have been proposed, but so far all suffer from problems varying from computational cost to lack of robustness; many can be shown to behave unexpectedly under certain plausible inputs. For instance, similarity measures based on maximum agreement are too strict, while measures based on the elimination of rogue taxa work poorly when the proportion of rogue taxa is significant; distance measures based on edit distances under simple tree operations (such as nearest-neighbor interchange or subtree pruning and regrafting) are NP-hard; and the widely used Robinson-Foulds distance is poorly distributed and thus affords little discrimination, while also lacking robustness in the face of very small changes—reattaching a single leaf elsewhere in a tree of any size can instantly maximize the distance.

J. Chen, J. Wang, and A. Zelikovsky (Eds.): ISBRA 2011, LNBI 6674, pp. 197–208, 2011.
© Springer-Verlag Berlin Heidelberg 2011

In this paper, we introduce an entirely new pairwise distance measure, based on matching, for phylogenetic trees. We prove that our measure induces a metric on the space of trees, show how to compute it in low polynomial time, verify through statistical testing that it is robust, and finally note that it does not exhibit unexpected behavior under the same inputs that cause problems with other measures. Our matching metric can be viewed as a weighted extension of the Robinson-Foulds distance, but can also be interpreted in the context of tree editing, thus bridging two types of tree-to-tree measures.

We illustrate the use of our tree metric in clustering trees; we obtain significant improvements in the quality of hierarchical clustering as compared to the same collections of trees clustered using the Robinson-Foulds distance.

2   Background

2.1   Similarity, Editing, and Distance

Phylogenetic trees are leaf-labelled trees, most often unrooted. Perhaps the simplest way to quantify the similarity of a set of phylogenetic trees is to determine the smallest collection of leaves that, when removed, induce the same tree (on the remaining leaves) from each tree in the set. Such an induced tree is called the *Maximum Agreement SubTree (MAST)*. Several variations have been proposed on this theme, all seeking to identify a tree structure that is common, in exact or approximate form, to all trees in the given set. For a pair of trees, most such measures are fairly easy to compute. Trees can also be transformed through various operations that disconnect and reconnect subpieces; given any collection of such operations, and assuming that the operations are sufficiently powerful to enable us to transform any tree on n leaves into any other tree on n leaves, we can define an *edit distance* between two trees as the smallest number of allowed operations that will tranform one tree into the other. Computing such edit distances is typically NP-hard, however, nor is it clear which set of operations should be used in the characterization. Finally, we can focus on the characteristics of two trees to determine the number of differences and thus induce a distance measure based on outcomes rather than on transformations. The *Robinson-Foulds (RF) distance*, the most commonly used distance measure for trees, counts the number of edges (or, equivalently, bipartitions of the leaves) present in one tree, but not the other; it can be computed in linear time. We now look at each of these three approaches in turn.

2.2   Tree Similarity Measures

The MAST problem has been well studied [5,9,12]. While the general problem of finding the MAST of three or more trees is NP-hard [2], it can be solved in $O(n \log n)$ time for two binary trees [18]. Since requiring exact agreement may prove too demanding and lead to poor results, several authors proposed variations on this formulation, among them the *maximum information subtree (MIST)* [3] and the *maximum information subtree consensus (MISC)* [15], variations that are more robust than MAST in the presence of "rogue" taxa (taxa whose placement in the tree is unclear and highly variable). These methods work well in the presence of a small number of rogue taxa, but poorly (both in terms of running time and of quality of results) when rogues are numerous; they also work only on sizeable collections of trees, not on pairs of trees.

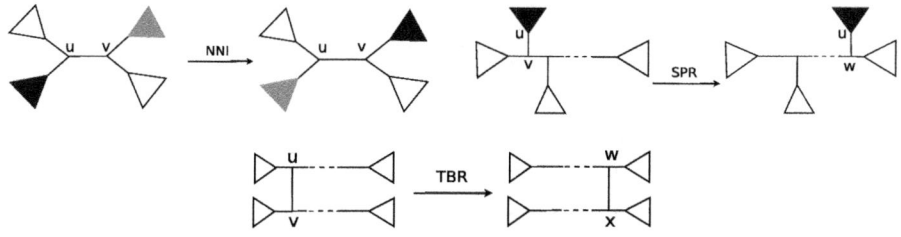

Fig. 1. NNI, SPR and TBR operations

2.3   Tree Editing

Editing operations are commonly used to explore tree space in phylogenetic inference, but also for comparing phylogenetic trees. We briefly describe the three most common operations, in increasing order of generality.

Nearest Neighbor Interchange (NNI). Let $e = \{u,v\}$ be an internal edge of a tree T and S_u and S_v be the set of subtrees connected to u and v respectively. A single NNI operation interchanges two subtrees across e: it disconnects one of the subtrees from S_u and connects it to vertex v, then disconnects one of the subtrees from S_v and connects it to vertex u, as illustrated in Fig. 1.

Subtree Prune and Regraft (SPR). An SPR operation disconnects a subtree from the larger tree by removing some edge $\{u,v\}$; the pruned subtree has vertex u, while the larger tree has vertex v. If the larger tree was binary, then v now has degree 2 and is eliminated by merging its two incident edges. Then the subtree is reconnected to the larger tree by creating a new vertex w on some edge of the larger tree and connecting it to the pruned subtree by a new edge $\{u,w\}$, as illustrated in Fig. 1. The *Leaf Prune and Regraft (LPR)* operation is the simplified version in which the subtree pruned always consists of a single leaf.

Tree Bisection and Reconnection (TBR). Let $e = \{u,v\}$ be an internal edge of a tree T and let C_1 and C_2 be the components of the tree formed by removing e and (if the tree was binary) suppressing vertices u and v. Form tree T' by choosing one edge in C_1 and adding a vertex w along that edge, choosing one edge in C_2 and adding a vertex x along that edge, and finally adding the edge $\{w,x\}$, as illustrated in Fig. 1. (If any of the components is just a single vertex, then the newly added edge is attached to the vertex.)

Any tree operation can be used to define an edit distance between trees: the minimum number of such operations needed to transform one tree into the other. Regrettably, computing the edit distance for each of the above three operations is NP-hard [1,6,11]. The NNI edit distance between two trees is $O(n\log n)$ [13] and can be approximated within a ratio of $O(\log n)$ [6]. The edit distances between two trees for SPR and TBR are $O(n)$ [1] and there is a 3-approximation algorithm to compute the TBR edit distance [20]. The LPR edit distance between two trees on n leaves is just n minus the number of leaves in the MAST of those two trees and so can be computed in polynomial time for two binary trees.

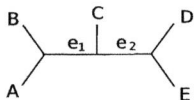

Fig. 2. An unrooted tree with 5 leaves

2.4   The Robinson-Foulds Distance

The Robinson-Foulds (RF) distance [16] is by far the most widely used measure of dissimilarity between trees. One of its main advantages is its independence from any model of tree editing: it does not infer any series of editing operations, but relies only on the current characteristics of the two trees.

Every internal edge e in a leaf-labeled tree T defines a non-trivial bipartition π_e on the leaves, and hence the tree T is uniquely represented by the set of bipartitions $\Gamma(T) = \{\pi_e \mid e \in E(T)\}$, where $E(T)$ is the set of internal edges in T. For example the unrooted tree in Fig. 2 is represented by two nontrivial bipartitions $\{AB|CDE, ABC|DE\}$ induced by edges e_1 and e_2, respectively. Given two unrooted leaf-labeled trees T_1 and T_2 on the same set of leaf labels, the Robinson-Foulds (RF) distance between them is the normalized count of the bipartitions induced by one tree and not the other, that is,

$$\mathcal{D}_{RF}(T_1, T_2) = \frac{1}{2}\big((|\Gamma(T_1) - \Gamma(T_2)|) + (|\Gamma(T_2) - \Gamma(T_1)|)\big).$$

Since there are at most $n-3$ non-trivial bipartitions in a tree on n leaves, the largest possible RF distance between two trees is $n-3$. The RF distance between two trees can be computed in linear time [7], while the RF distance matrix for a collection of trees can be computed in sublinear time [14]. However, the RF distance is overly sensitive to some small changes in the tree. For example, just moving a leaf at the end of a caterpillar tree (a single spine to which all leaves are attached) to the other end will create a tree with the maximum possible RF distance to the original tree, yet this change takes a single LPR operation. The RF distance between two random binary trees has a very skewed distribution [17,4] in which most values equal $n-3$ (also see section 4.1 for details).

3   Our Matching Distance

A tree T is uniquely represented by the set of bipartitions $\Gamma(T) = \{\pi_e \mid e \in E(T)\}$, where $E(T)$ is the set of internal edges in T. Given two trees, T_1 and T_2 on the same set of leaf labels, we define a complete weighted bipartite graph $B(X, Y, E)$ with $X = \Gamma(T_1)$ and $Y = \Gamma(T_2)$, that is, every bipartition is represented by a vertex in B. We denote this graph by $B(T_1, T_2)$. An edge (u, v) has weight 0 if the bipartitions $u \in \Gamma(T_1)$ and $v \in \Gamma(T_2)$ are the same, otherwise it has weight 1. We can then rephrase the RF distance between T_1 and T_2 as the weight of the minimum-weight matching in $B(T_1, T_2)$.

The binary weighting scheme does not make full use of the information in the bipartitions. Each bipartition π_e can be represented by a binary vector V_e of length n, where n is the number of leaves in T_1 (or T_2). For any leaf i, we set $V_e[i] = 1$ if leaf i and leaf 1 are on the same side of the bipartition π_e and set $V_e[i] = 0$ otherwise. We set the weight of each edge $e = \{u, v\}$ in $B(T_1, T_2)$ (where vertices u, v in B represent internal edges in T_1 and T_2 respectively) to:

$$W(u,v) = min\{\mathcal{D}_H(V_u, V_v), \mathcal{D}_H(V_u, \overline{V}_v)\},$$

where $\mathcal{D}_H$ is the Hamming distance between the two vectors and $\overline{V}$, the complement vector of V, is equal to $I - V$. The *matching distance* $\mathcal{D}_M(T_1, T_2)$ between trees T_1 and T_2 is the weight of the minimum-weight matching in $B(T_1, T_2)$ with the weighting scheme W. This definition is a natural choice since the Hamming distance between the two bipartitions represents the minimum number of leaves that must be moved in order to transform one into the other.

The minimum-weight matching problem can be solved in cubic time [8]. If the input weights are integers and the value of each weight is not greater than the number of leaves (as is the case for our matching problem), the running time of the algorithm can be improved to $O(n^{5/2}log(n))$ by cost scaling and blocking flow techniques [10] .

3.1 Basic Properties

First, we show that our distance measure is well defined: it is indeed a metric.

Lemma 1. *The matching distance $\mathcal{D}_M$ on binary leaf-labeled trees is a metric. For any binary trees T_i, T_j and T_k on n labeled leaves, we have*

1. *$\mathcal{D}_M(T_i, T_j) \geq 0$.*
2. *$\mathcal{D}_M(T_i, T_j) = 0$ if and only if $T_i = T_j$.*
3. *$\mathcal{D}_M(T_i, T_j) = \mathcal{D}_M(T_j, T_i)$.*
4. *$\mathcal{D}_M(T_i, T_j) + \mathcal{D}_M(T_j, T_k) \geq \mathcal{D}_M(T_i, T_k)$.*

Proof. Properties 1, 2 and 3 follow directly from the definition of the matching distance. We prove Property 4. Assume $M_{i,j}$ and $M_{j,k}$ are the minimum-weight matchings in $B(T_i, T_j)$ and $B(T_j, T_k)$. Construct a matching $M_{i,k} = \{(u,w) | (u,v) \in M_{i,j} \wedge (v,w) \in M_{i,j}\}$ in $B(T_i, T_k)$. Since $\mathcal{D}_M(T_i, T_k)$ is the minimum-weight matching in $B(T_i, T_k)$, we have

$$\mathcal{D}_M(T_i, T_k) \leq \sum_{(u,w) \in M_{i,k}} W(u,w)$$

$$\leq \sum_{(u,v) \in M_{i,j}, (v,w) \in M_{j,k}} (W(u,v) + W(v,w))$$

$$= \sum_{(u,v) \in M_{i,j}} W(u,v) + \sum_{(v,w) \in M_{j,k}} W(v,w)$$

$$= \mathcal{D}_M(T_i, T_j) + \mathcal{D}_M(T_j, T_k).$$

Next we investigate extremal properties of our matching distance.

Definition 1. *Let $T(n)$ be the space of all binary trees on n labeled leaves. The diameter (δ) of $T(n)$ with respect to a distance metric $\mathcal{D}$ on $T(n)$ is defined as*

$$\delta(T(n), \mathcal{D}) = max\{\mathcal{D}(T_1, T_2) \mid T_1, T_2 \in T(n)\}.$$

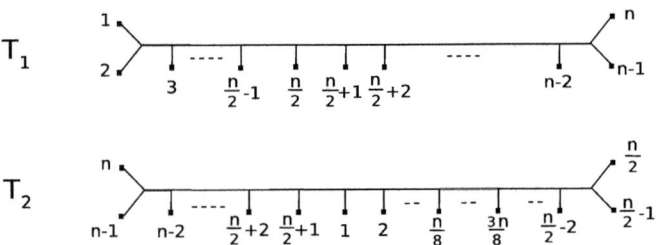

Fig. 3. Example for two trees on n leaves with Matching distance $\Theta(n^2)$

Theorem 1

$$\delta(T(n), \mathcal{D}_{RF}) = n - 3,$$
$$\delta(T(n), \mathcal{D}_M) = \Theta(n^2)$$

Proof. We prove the bounds on the diameter by explicitly constructing two trees T_1 and T_2. For the RF distance, choose T_1 and T_2 to be two caterpillar trees with different cherries, then no bipartition can appear both in $\Gamma(T_1)$ and $\Gamma(T_2)$, thus $(n-3)$ mismatches result in an RF distance of $(n-3)$. For the matching distance, construct two caterpillar trees T_1 and T_2 as shown in Fig. 3. The leaves in T_1 are ordered as $(1, \ldots, n)$, and the leaves in T_2 are ordered as $(n, \ldots, n/2 + 1, 1, 2, \ldots, n/2)$. It is easy to verify (by case analysis) that each bipartition corresponding to an internal edge along the path between leaf $n/8$ and leaf $3n/8$ in T_2 (marked in red) is at least $n/8$ away from every bipartition in T_1. Since there are $n/4$ such bipartitions in T_2, any matching between $\Gamma(T_1)$ and $\Gamma(T_2)$ will have a weight at least $(n/4) * (n/8) = \Omega(n^2)$. The upper bound is trivial.

3.2   Sensitivity to Tree Editing

We now study the change in the distance measures caused by a single tree editing operation. Let $\phi(T)$ be the set of trees derived by applying operation ϕ to a tree T, where ϕ can be one of NNI, SPR, TBR, LPR, or *Leaf Label Exchange (LLI)*, this last an operation that does not alter the tree structure, but simply exchanges the labels of two leaves.

Definition 2. *The gradient of a tree rearrangement operation ϕ with respect to a distance metric $\mathcal{D}$ on $T(n)$ is defined as*

$$\mathcal{G}(T(n), \mathcal{D}, \phi) = max\{\mathcal{D}(T_1, T_2) | T_1, T_2 \in T(n), T_2 \in \phi(T_1)\}.$$

Theorem 2

$$\mathcal{G}(T(n), \mathcal{D}_{RF}, NNI) = 1,$$
$$\mathcal{G}(T(n), \mathcal{D}_M, NNI) = \Theta(n).$$

Proof. Let T_2 be the tree obtained by applying one NNI operation on T_1. Every NNI operation changes only one bipartition in $\Gamma(T_1)$ into a new one in $\Gamma(T_2)$ (induced by the internal edge which is selected). Thus $\mathcal{G}(T(n), \mathcal{D}_{RF}, NNI) = 1$. Since $\Gamma(T_1)$ and $\Gamma(T_2)$ share $n - 4$ bipartitions, we can construct a matching $M_{1,2}$ in $B(T_1, T_2)$ that contains $n - 4$ matched pairs with weight zero and 1 matched pair with weight at most n. The sum of the weights for $M_{1,2}$ is upper bounded by n, and hence $\mathcal{D}_M(T_1, T_2) \leq n$. Let $e = (u, v)$ be an internal edge in T_1 connecting four rooted subtrees $\{S_1, S_2, S_3, S_4\}$ where S_1 and S_2 are attached to u and S_3 and S_4 are attached to v. Assume each of the four subtrees contains $n/4$ leaves and one NNI operation interchanges S_2 and S_3. The newly created bipartition by NNI in T_2 is now at least $\Theta(n)$ distance away from all possible bipartitions in T_1. So any matching in $B(T_1, T_2)$ will have weight at least $\Theta(n)$. From the upper and lower bounds, we have $\mathcal{G}(T(n), \mathcal{D}_M, NNI) = \Theta(n)$.

Theorem 3

$$\mathcal{G}(T(n), \mathcal{D}_{RF}, LPR) = n - 3$$
$$\mathcal{G}(T(n), \mathcal{D}_M, LPR) = \Theta(n).$$

Proof. The bound for $\mathcal{G}(T(n), \mathcal{D}_{RF}, LPR)$ is derived by applying LPR to a caterpillar tree T_1, where one leaf at one end of the tree is transposed to the other end of the tree. Let T_2 be the tree obtained by applying one LPR operation on T_1. T_2 shares no bipartitions with the tree T_1 and the RF distance between them is $n - 3$. The matching distance between T_1 and T_2 is $\Omega(n)$ since each pair of bipartitions from $\Gamma(T_1)$ and $\Gamma(T_2)$ contributes at least 1 to the matching weight and there are $n - 3$ pairs. Because every LPR operation only affects two internal edges in $\Gamma(T_1)$ (we remove an internal edge while pruning and create a new internal edge while regrafting), there are $n - 5$ internal edges left untouched and shared by T_1 and T_2. We can construct a matching $M_{1,2}$ in $B(T_1, T_2)$ that contains $n - 5$ matched pairs corresponding to the shared edges and another 2 matched pairs. For each matched pair for the shared edges, the weight is at most 1 since the corresponding bipartitions can only differ at the pruned leaf. For the other 2 matched pairs, the contribution to the total weight is at most $O(n)$. The weight for this matching $M_{1,2}$ is thus bounded by $O(n)$. From the upper and lower bounds, we have $\mathcal{G}(T(n), \mathcal{D}_M, LPR) = \Theta(n)$.

Theorem 4

$$\mathcal{G}(T(n), \mathcal{D}_{RF}, SPR) = n - 3,$$
$$\mathcal{G}(T(n), \mathcal{D}_M, SPR) = \Theta(n^2).$$

Proof. The bound for $\mathcal{G}(T(n), \mathcal{D}_{RF}, SPR)$ follows from Theorem 3 since LPR is a special case of SPR and $(n - 3)$ is already the maximum change in RF distance. The bound for $\mathcal{G}(T(n), \mathcal{D}_M, SPR)$ is obtained from the trees in Fig. 3, where one SPR operation on T_1 results in T_2 and $\mathcal{D}_M(T_1, T_2) = \Theta(n^2)$.

Theorem 5

$$\mathcal{G}(T(n), \mathcal{D}_{RF}, TBR) = n - 3,$$
$$\mathcal{G}(T(n), \mathcal{D}_M, TBR) = \Theta(n^2).$$

Proof. The results follow directly from Theorem 4 since SPR is a special case of TBR and both gradients have trivial upper bounds.

Theorem 6

$$G(T(n), \mathcal{D}_{RF}, LLI) = n - 3,$$
$$G(T(n), \mathcal{D}_M, LLI) = \Theta(n).$$

Proof. The bound for $G(T(n), \mathcal{D}_{RF}, LLI)$ is derived by applying LLI to a caterpillar tree T_1, where the labels of two leaves at two ends of the tree are interchanged. Let T_2 be the tree obtained by applying one LLI operation on T_1. T_2 shares no bipartitions with the tree T_1, and the RF distance between them is $n - 3$. The matching distance between T_1 and T_2 is $\Omega(n)$ since each pair of bipartitions from $\Gamma(T_1)$ and $\Gamma(T_2)$ contributes at least 1 to the matching weight and there are $n - 3$ pairs. Because every LLI operation only affects two leaves in T_1 and T_2, all $n - 5$ internal edges are left untouched. We can construct a matching $M_{1,2}$ in $B(T_1, T_2)$ that contains those $n - 3$ matched pairs corresponding to the shared edges. For each matched pair for the shared edges, the weight is at most 2 since the corresponding bipartitions can differ at not more than two leaves. The weight for this matching $M_{1,2}$ is thus bounded by $O(n)$. From the upper and lower bounds, we have $G(T(n), \mathcal{D}_M, LLI) = \Theta(n)$.

The ratio of the gradient to the diameter is an indication of the sensitivity of the distance measure. Our theorems indicate that the matching distance has the same asymptotic sensitivity as the RF distance with respect to NNI, SPR, and TBR, but is more sensitive than the RF distance with respect to LPR and LLI.

4   Experimental Results

The previous section gave extremal properties of our matching distance, but its main advantages are best seen by comparing its distribution of values to that of the RF distance. We have not derived an exact formula for the distribution, but present experimental results that show that our matching distance on random binary trees yields a distribution with a fairly broad bell curve, in sharp contrast to the highly skewed distribution of the RF distance.

4.1   Distribution of the Tree Distance Metrics

We first study the distribution of RF and matching distances by sampling pairs of random trees generated in two different ways. The first, uniformly sampled binary trees, are generated by the randomized leaf attachment process [17]), and the second, birth-death trees, are generated by a uniform, time-homogeneous birth-death process (birth rate = 0.1, death rate = 0). Fig. 4 shows the distribution of RF and matching distances for 100,000 pairs of uniformly sampled binary trees on 100 and 1,000 leaves each and birth-death trees on 100 leaves. The range of values for each distance is divided into 100 intervals and each point on the x axis represents an interval. Compared to RF (a very skewed distribution as shown in the figure and in [4]), our matching distance offers a larger range and is more broadly distributed, and thus also more discriminating.

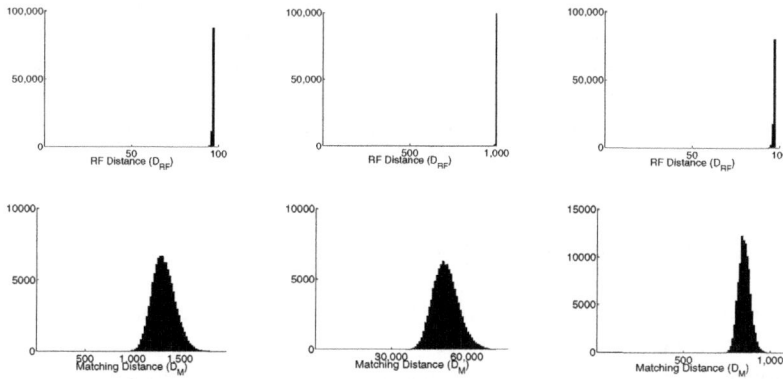

Fig. 4. Distribution of pairwise RF (above) and matching (below) distances between uniformly sampled binary trees on 100 leaves (left), on 1,000 leaves (middle), and between birth-death trees on 100 leaves (right)

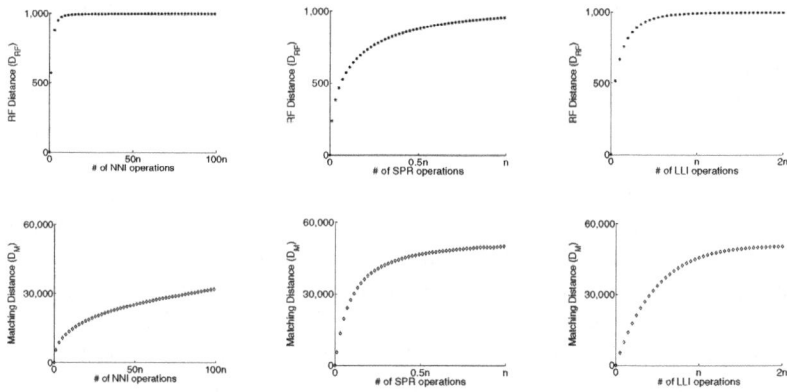

Fig. 5. RF (above) and matching (below) distances as a function of the number of NNI operations (left), SPR operations (middle), and LLI operations (right) for trees on 1,000 leaves (n = 1,000)

4.2   Tree Distance Metrics Under Tree Editing Operations

We study the behavior of both RF and matching metrics under various tree editing operations. For each operation, we study the change in the distance after successive applications of the operation. From the distributions of the two distance metrics seen in the previous section we expect the RF distance to saturate faster and the matching distance to have a better correlation with the number of tree rearrangement opertions which is indeed the case. Note that such a comparison with the edit distance (with respect to the rearrangement operations) is infeasible since it is NP-hard to compute the distances.

Our experiments start with 1,000 uniformly generated binary trees on 1,000 leaves each. We summarize the average pairwise RF and matching distances between the trees and the original as a function of the number of operations applied. Fig. 5 shows RF and matching distances as a function of the number of NNI operations. While the RF distance reaches saturation after $10,000$ ($10n$) operations ($n = 1,000$), our matching distance still shows an increasing trend; indeed, the average matching distance (~$30,000$) after $100,000$ ($100n$) operations is still far from the average matching distance (~$50,000$) between two randomly selected binary trees on $1,000$ leaves (as seen in Fig. 4). Similar results are shown in Fig. 5 for SPR and LLI operations.

5   Clustering Trees: An Application

In this section we provide a proof-of-concept study of the usefulness of the matching distance in clustering phylogenetic trees.

Phylogenetic analyses such as maximum-parsimony or maximum-likelihood analyses often produce many (possibly thousands) of candidate trees that are nearly optimal with respect to the defined objective function. To obtain a biologically relevant tree, postprocessing of these candidate trees is essential. Consensus tree methods are frequently used to extract the common structure from the candidate trees and summarize the output; however these methods often lose information and are sensitive to outliers. A different approach divides the set of candidate trees into several subsets using clustering methods, each cluster being characterized by its own consensus tree [19]. The authors of that approach demonstrate an improvement over traditional consensus methods by obtaining better resolved output trees and by providing details of the distribution of the candidate trees.

The efficacy of clustering relies on the dissimilarity measure used. We conducted a preliminary test on RF and matching metrics as dissimilarity measures in clustering. We generated 100 datasets, each of 200 random binary trees. The trees in each dataset were generated by a two-step process. We first sampled two binary trees on k leaves ($k < 100$) and used them as two different skeletons. Then from each of the two skeletons, a set of 100 trees is generated by adding the rest of the $(n - k)$ leaves one by one. To add a new leaf, an edge in the current tree is selected uniformly at random and the new leaf is attached to that edge. We vary k from 40 ($0.4n$) to 90 ($0.9n$). The 200 trees in each dataset is given as input to the clustering algorithm to check if the algorithm can distinguish the trees in the two clusters. Trees generated from the same skeleton are considered to be in the same cluster.

Notice that a MAST-based distance metric can easily distinguish the input trees into the correct cluster. We deliberately choose this experimental setup to provide a test case for the matching distance even in those settings where a MAST-based distance will perform better than RF.

We apply a standard hierarchical clustering approach (recommended for phylogenetic postprocessing in [19]) to the pairwise distance matrices generated by RF and matching distances. The similarity between clusters C_1 and C_2 is measured by the following three linkage criteria:

1. Complete linkage: $max\{\mathcal{D}(a,b)|a \in C_1, b \in C_2\}$.
2. Single linkage: $min\{\mathcal{D}(a,b)|a \in C_1, b \in C_2\}$.
3. Average linkage: $\frac{1}{|C_1||C_2|} \sum_{a \in C_1, b \in C_2} \mathcal{D}(a,b)$.

Table 1. Error rates for the clustering test

	0.4n		0.5n		0.6n		0.7n	
	$\mathcal{D}_{RF}$	$\mathcal{D}_M$	$\mathcal{D}_{RF}$	$\mathcal{D}_M$	$\mathcal{D}_{RF}$	$\mathcal{D}_M$	$\mathcal{D}_{RF}$	$\mathcal{D}_M$
Complete linkage	70%	0%	71%	0%	38%	0%	3%	0%
Single linkage	22%	8%	23%	0%	0%	0%	0%	0%
Average linkage	49%	0%	3%	0%	0%	0%	0%	0%

A run of the algorithm on a particular dataset is considered to err if it is unable to place every tree generated from the same skeleton in one cluster. We present the error rate obtained from 100 such datasets for each parameter in Table 1. For values of k higher than $0.7n$ both distance measures perform equally well, but, as expected, the matching distance has significantly better performance over a large range of input parameters.

6   Conclusion

We have introduced a new tree metric for phylogenetic analysis. This metric can be computed efficiently, in contrast to various edit distances, and offers better discrimination than the standard Robinson-Foulds distance, thanks to a much broader and less biased distribution of distance values. We have given extremal results as well as experimental results to characterize this new metric. Finally, we have demonstrated the use of this metric in clustering trees with an agglomerative hierarchical clustering method, where using our metric considerably improved over using the Robinson-Foulds metric.

Acknowledgement

We thank Nicholas D. Pattengale for many helpful discussions.

References

1. Allen, B.L., Steel, M.: Subtree transfer operations and their induced metrics on evolutionary trees. Annals of Combinatorics 5(1), 1–15 (2001)
2. Amir, A., Keselman, D.: Maximum agreement subtree in a set of evolutionary trees: Metrics and efficient algorithms. SIAM J. Computing 26(6), 1656–1669 (1997)
3. Bryant, D.: Hunting for trees, building trees and comparing trees: Theory and method in phylogenetic analysis. PhD thesis, University of Canterbury (1997)
4. Bryant, D., Steel, M.: Computing the distribution of a tree metric. ACM/IEEE Trans. on Comput. Biology and Bioinformatics 6(3), 420–426 (2009)
5. Cole, R., Farach-Colton, M., Hariharan, R., Przytycka, T., Thorup, M.: An O($n\log n$) algorithm for the maximum agreement subtree problem for binary trees. SIAM J. Computing 30(5), 1385–1404 (2000)
6. DasGupta, B., He, X., Jiang, T., Li, M., Tromp, J., Zhang, L.: On distances between phylogenetic trees. In: Proc. 8th ACM/SIAM Symp. Discrete Algs. (SODA 1997), pp. 427–436 (1997)
7. Day, W.H.E.: Optimal algorithms for comparing trees with labeled leaves. J. Classification 2(1), 7–28 (1985)

8. Edmonds, J., Karp, R.M.: Theoretical improvements in algorithmic efficiency for network flow problems. J. ACM 19(2), 248–264 (1972)
9. Farach, M., Przytycka, T.M., Thorup, M.: On the agreement of many trees. Inf. Process. Lett. 55(6), 297–301 (1995)
10. Gabow, H.N., Tarjan, R.E.: Faster scaling algorithms for network problems. SIAM J. Computing 18(5), 1013–1036 (1989)
11. Hickey, G., Dehne, F., Rau-Chaplin, A., Blouin, C.: SPR distance computation of unrooted trees. Evol. Bioinform. Online 4, 17–27 (2008)
12. Kao, M.Y.: Tree contractions and evolutionary trees. SIAM J. Computing 27(6), 1592–1616 (1998)
13. Li, M., Tromp, J., Zhang, L.: On the nearest-neighbour interchange distance between evolutionary trees. J. Theor. Biol. 182(4), 463–467 (1996)
14. Pattengale, N.D., Gottlieb, E.J., Moret, B.M.E.: Efficiently computing the Robinson-Foulds metric. J. Comput. Biol. 14(6), 724–735 (2007)
15. Pattengale, N.D., Swenson, K.M., Moret, B.M.E.: Uncovering hidden phylogenetic consensus. In: Borodovsky, M., Gogarten, J.P., Przytycka, T.M., Rajasekaran, S. (eds.) ISBRA 2010. LNCS, vol. 6053, pp. 128–139. Springer, Heidelberg (2010)
16. Robinson, D.R., Foulds, L.R.: Comparison of phylogenetic trees. Mathematical Biosciences 53, 131–147 (1981)
17. Steel, M., Penny, D.: Distributions of tree comparison metrics—some new results. Syst. Biol. 42(2), 126–141 (1993)
18. Steel, M., Warnow, T.: Kaikoura tree theorems: computing maximum agreement subtree problem. Information Processing Letters 48, 77–82 (1993)
19. Stockham, C., Wang, L.-S., Warnow, T.: Statistically-based postprocessing of phylogenetic analysis using clustering. In: Proc. 10th Conf. Intelligent Systems for Mol. Biol. (ISMB 2002). Bioinformatics, vol. 18, pp. S285–S293. Oxford U. Press, Oxford (2002)
20. Whidden, C., Zeh, N.: A unifying view on approximation and fpt of agreement forests. In: Bücher, P., Moret, B.M.E. (eds.) WABI 2006. LNCS (LNBI), vol. 4175, pp. 390–402. Springer, Heidelberg (2006)

Describing the Orthology Signal in a PPI Network at a Functional, Complex Level

Pavol Jancura[1], Eleftheria Mavridou[2], Beatriz Pontes[3], and Elena Marchiori[1]

[1] Institute for Computing and Information Sciences, Radboud University Nijmegen,
Postbus 9010, 6500 GL Nijmegen, The Netherlands
{jancura,elenam}@cs.ru.nl
[2] Department of Medical Microbiology, Radboud University Medical Center,
Postbus 9101, 6500 HB Nijmegen, The Netherlands
[3] Department of Computer Science, University of Seville,
Avda. Reina Mercedes s/n 41012 Seville, Spain

Abstract. In recent work, stable evolutionary signal induced by orthologous proteins has been observed in a Yeast protein-protein interaction (PPI) network. This finding suggests more connected subgraphs of a PPI network to be potential mediators of evolutionary information. Because protein complexes are also likely to be present in such subgraphs, it is interesting to characterize the bias of the orthology signal on the detection of putative protein complexes. To this aim, we propose a novel methodology for quantifying the functionality of the orthology signal in a PPI network at a protein complex level. The methodology performs a differential analysis between the functions of those complexes detected by clustering a PPI network using only proteins with orthologs in another given species, and the functions of complexes detected using the entire network or sub-networks generated by random sampling of proteins. We applied the proposed methodology to a Yeast PPI network using orthology information from a number of different organisms. The results indicated that the proposed method is capable to isolate functional categories that can be clearly attributed to the presence of an evolutionary (orthology) signal and quantify their distribution at a fine-grained protein level.

1 Introduction

In general, two proteins are orthologous if they originated from a common ancestor, having been separated in evolutionary time only by a speciation event. Orthologous proteins have high amino acid sequence similarity and usually retain the same or very similar function, which allows one to infer biological information between the proteins. Obviously, orthology as such is very important in studying evolution. Therefore, the problem of establishing proper orthology relations has been under the wide investigation in comparative genomics (see for instance [1]) and many databases and public resources of orthologs have been made available, such as Inparanoid [2] and OrthoMCL-DB[3].

J. Chen, J. Wang, and A. Zelikovsky (Eds.): ISBRA 2011, LNBI 6674, pp. 209–226, 2011.
© Springer-Verlag Berlin Heidelberg 2011

Recent studies used this form of evolutionary information to analyse protein modules and PPI networks, for instance [4,5,6,7,8,9,10,11,12]. In particular, in a study by Wutchy et al. [6] stable evolutionary signal was found to be present in a Yeast PPI network as examined by its pairwise orthologs with respect to various different species. They observed that a high local clustering around protein-protein interactions correlates with evolutionary conservation of the participating proteins. This means that highly connected proteins and protein pairs embedded in a well clustered neighbourhood tend to be evolutionary conserved and therefore retain their evolutionary signal. These findings suggest also that more connected areas of a PPI network are potential mediators of evolutionary information.

Because more connected regions of PPI networks contain protein modules or complexes, in this paper we focus on the explicit use of orthology to see whether there are functional complexes that can be clearly attributed to this evolutionary signal. To this aim, we try to characterize those functions of complexes predicted by clustering the subgraph of a PPI network induced by all proteins with orthologs in another given species, but not predicted (or predicted for a smaller fraction of proteins) when clustering the entire network. We consider these functions as strong characterization of the underlying evolutionary signal of orthologs, since they are suppressed or not observed when clustering using the entire network.

Specifically, we examine highly functionally coherent putative protein complexes as detected by two state-of-the-art clustering techniques in the Yeast PPI network using only proteins with orthologs in another given organism. Our target clusters should contain a function which can be genuinely attributed to the orthology signal and exclude the case that it could be attributed by chance. Therefore we consider three classes of clusters, consisting of putative complexes as detected by these clustering techniques applied to the Yeast PPI network with (1) all proteins, (2) only proteins with ortholog in the considered other organism, and (3) randomly selected proteins. The latter class of clusters is the collection of cluster sets produced by the application of clustering to the PPI network induced by a random selection of a set of proteins (of size equal to that of the set of proteins used to generate the class (2)) repeated for dozens times. For all clusters in each class we infer putative functions by measuring their gene ontology (GO) functional enrichment [13].

In general, protein functions belong to certain functional categories. Hence, we map all putative functions inferred from the clusters to these categories. For a set of clusters and a certain category, we compute the fraction of proteins contained in the clusters and having functions mapped to that category. This fraction quantifies (at protein level) the presence of that functional category in a given cluster set. This allows us to identify functional categories whose proteins' fraction is higher in clusters from the class (2) than in clusters from the other two classes. We consider the corresponding clusters in class (2) as describing the orthology signal (with respect to considered species). Furthermore, we analyse those clusters of class (2) having a predicted function for its proteins that is not inferred when using clusters of class (1). Finally we discuss the new meaningful functions for well-defined as well as for unknown proteins that are present in the compilation of putative complexes.

2   Other Related Work

In previous works on phylogenetic analysis of protein networks and complexes evolutionary information was usually used as a mean for evaluating the preservation of orthology information in functional modules [5,8,9,10]. Here, however, we incorporate evolutionary information beforehand for detecting evolutionary signal at complex, functional level. Our identification of protein complexes uses only the topology of the network of the considered species and orthology information from another species, without requiring knowledge on the interactome of the other species.

In general, our approach differs from comparative network methods [14], as the latter aim to find evolutionary conserved modules across species, and exploit both orthology and network topology of the considered organisms. The clusters we obtain are in one species and are related to the orthology signal with respect to another species, but are not required to be evolutionary conserved through species (we do not enforce any type of similarity at the graph-structure level). Furthermore, comparative methods mostly do not use 'known' orthologs in available databases but rather they rely on sequence similar proteins, where the level of required similarity is determined by a minimal similarity score threshold. Instead, our method exploits the orthology information available in existing databases.

3   Method

The following terminology is used in the sequel. A PPI network is represented by means of a graph $G(V, E)$, where V is the set of nodes (proteins) and E is the set of edges (binary interactions). Let X be a subset of nodes V (e.g. ortholog set). The set X induces a subgraph $G[X] = (X, E_X)$ of G, with set X of nodes and set E_X of those edges of E that join two nodes in X. For a set S, we denote by $|S|$ the number of its elements.

We are interested in quantifying the orthology signal by means of a set of functions of putative protein complexes detected by applying clustering to a PPI network. To this end, we directly exploit evolutionary information of proteins as described by the presence of orthologs in another, given species. We call these proteins 'true orthologs'. Specifically, we propose the following methodology.

Given a PPI network $G = (V, E)$ and a given species s, apply the following steps.

1. *Retrieve from a database the set O of 'true orthologs' of V with respect to s, with $|O| = n$.*
2. *Generate the following three classes of clusters, using a given clustering algorithm.*

(a) Class 1 clusters (GC). Apply clustering to the whole PPI network G.
(b) Class 2 clusters (OC). Apply clustering to the sub-network induced by O.
(c) Class 3 clusters (RC). Apply clustering to the sub-network induced by a randomly selected subset of V of size n. Repeat the process a number N of times. Consider all sets of clusters detected across these runs (RC = $\{RC_1, RC_2, \ldots, RC_N\}$).
3. *For each class of clusters,*
   (a) Infer putative complexes and their functional categories.
   (b) For each functional category, compute the fraction of those proteins in the detected complexes which have been assigned to that category.
4. *Select the set of those functional categories derived using clusters from class 2 and whose fraction are higher than those of the same category derived using clusters from class 1 or from class 3.*
5. *Output the set of clusters having at least one of the selected functional categories.*

In this study we consider as putative protein complex only a group of proteins of a higher complexity than just a single protein-protein interaction. Therefore, after applying any clustering method we retain only clusters of size greater or equal than 3.

In the sequel we describe in more detail the main steps of the proposed methodology.

Inferring Putative Complexes and their Functional Categories. In order to infer the the putative functions of a cluster, we measure the enrichment of functional annotations of the corresponding protein set, as entailed by the gene ontology (GO) annotation [13], using one of the well-established tools, the Ontologizer[1] [15]. The Ontologizer offers various algorithms for measuring GO enrichments. Here, we apply the standard statistical analysis method based on the one-sided Fisher's exact test [15], which measures the statistical significance of an enrichment and assigns to the cluster a p-value for each enriched function. The p-value is further corrected for multiple testing by means of a Bonferroni correction procedure.

The GO is known to have a hierarchical structure (directed acyclic graph) which can be used to define the level of an annotation. Specifically, the level of an annotation is equal to the length of the shortest path from the root of GO hierarchy to the annotation. The GO terms closer to the root of GO give more general description of biological functions while terms closer to the leaves of GO have granular and very specific biological definitions.

Each detected cluster is a potential protein complex. The quality of a protein cluster is given by the coherence of biological functions of proteins contained in the cluster. If a certain subset of proteins in a cluster has a significantly coherent function, a prediction of that function for all proteins in the cluster can be made. We may obtain more than one protein function prediction if we find

---

[1] http://compbio.charite.de/index.php/ontologizer2.html

more significantly coherent functions in the cluster. We say that proteins of a cluster have a *significantly coherent function* or functional GO annotation if the following criteria are satisfied:

1. the GO annotation is significantly enriched by the proteins in the cluster (p-value < 0.001).
2. more than half of the proteins in the cluster has this significant annotation.
3. the annotation is at least at the GO level four from the root of GO hierarchy.

In such a case the cluster can be used as protein function predictor and the significantly enriched GO annotation of the cluster is used to predict protein function of each of the proteins in that cluster. If a cluster does not satisfy the above conditions, no prediction can be made. Similar criteria were used by, e.g. [16,17]. The condition on GO hierarchy guarantees that the prediction about biological functions is sufficiently specific and informative [18]. Each cluster which is a predictor defines a putative protein complex and the set of significantly coherent functions defines the set of inferred functions.

Estimating the Frequency of a Functional Category. After identifying putative protein complexes, we use them to quantify, at a fine-grained, protein level, the frequency with which a functional category was detected: for each functional category inferred using the putative protein complexes, we count the fraction of those proteins in the putative complexes assigned to that category.

Specifically, functional categories are determined by GO slim functional terms, defined in the GO hierarchy as a subset of the higher level GO terms. Each GO slim characterizes a certain type of biological functions which have some features and tasks in common. As a result, each fine-grained term can be mapped to these GO slims' terms.

The GO also consists of three main independent domains, *biological process*, *molecular function* and *cellular component*, and each of them has its own GO slim terms and hence functional categories. Given a GO domain and proteins of a cluster group of interest one can map all inferred functions of each protein to their closest GO slims in the GO hierarchy. Then for every functional category we can count the number of proteins being mapped to the category. In this framework we define the frequency of a functional category as follows.

Let C be a set of putative complexes and $P(f)$ denote the set of proteins contained in C and being mapped to a functional category f. Let B be a set of background functional categories (functional background). Then *the frequency of a functional category f in C with respect to the background B* is

$$\phi_C(f) = \frac{|P(f)|}{|P(B)|}, \text{ where } P(B) = \bigcup_{\forall b \in B} P(b). \tag{1}$$

Notice that in our definition we consider an individual background for each GO domain.

The frequency of a functional category can be viewed as the expectation that a protein in a given set of putative complexes has that functional category. This results in a distribution of functional categories associated to a set of complexes.

Identifying Orthology-Related Categories. Since we are interested in analysing the evolutionary (orthology) signal, we use as the functional background the set of functional categories enriched by the class 2 putative complexes. Therefore we compute the functional frequencies for each class of putative complexes using this background. For class 3 (random sampling), for each functional category, we average the frequencies over all random simulations as follows

$$\overline{\phi_{RC}}(f) = \frac{\sum_{i=1}^{N} \phi_{RC_i}(f)}{N}. \tag{2}$$

Once the functional frequencies are computed, we isolate functional categories related to the orthology signal by the following simple rule:

A *functional category f* is *orthology-related* iff

$$\phi_{OC}(f) > \max\{\phi_{GC}(f), \overline{\phi_{RC}}(f)\}. \tag{3}$$

4   Experimental Settings

4.1   Data Collection

We chose to perform our analysis on the widely used and well-studied species Saccharomyces cerevisiae (yeast), which PPI network is one of the best established and information on functionality of its proteins are one of the most explored. This makes yeast as a good standard model species for protein network analysis.

We used the same yeast interaction data as in [19] which combines interaction data from DIP [20] and MPact [21], and interactions from the core datasets of the TAP mass spectrometry experiments [22,23]. This yeast interaction data are weighted by the method proposed by Jansen et al. [24] to measure the confidence of interactome. As a result, the low confidence interactions are ignored and the final yeast PPI network consists of 3545 proteins and 14354 interactions.

For obtaining orthology information we used the Inparanoid Database of Pairwise Ortholog[2] [2]. This database contains clusters of ortholog groups (COGs) constructed by the Inparanoid program, which is a fully automatic method for finding orthologs and in-paralogs between two species. Ortholog clusters in the Inparanoid are seeded with a two-way best pairwise match (the seed ortholog pair), after which an algorithm for adding in-paralogs is applied. The Inparanoid was found as one of the best performing algorithms for orthology detection with respect to its false negative and false positive rates [25].

Because in-paralogs are homologs that arise when duplication occurs after speciation, and the duplicated gene often still retains the function of the ortholog [26], they should be likely found in one protein complex. Therefore we consider all proteins present in COGs for inducing an orthology PPI sub-network and, for simplicity, we consider all proteins in a COG as orthologs. Specifically, further in this study we call orthologous protein or ortholog a protein which is a part of an orthologous cluster produced by the Inparanoid when comparing two species.

---

[2] http://inparanoid.sbc.su.se/

In our analysis, COGs were obtained for the following pairs of organisms:

- *Saccharomyces cerevisiae* vs. *Escherichia coli*
- *Saccharomyces cerevisiae* vs. *Caenorhabditis elegans*
- *Saccharomyces cerevisiae* vs. *Drosophila melanogaster*
- *Saccharomyces cerevisiae* vs. *Homo sapiens*

Escherichia coli (E.coli), Caenorhabditis elegans (worm), Drosophila melanogaster (fly) and Homo sapiens (human) are standard organisms used in protein network and genome comparative studies (e.g [27,28]) and represent the diverse life-forms from a prokaryote (E.coli) to the highly complex eukaryote (Human). Yeast proteins in the derived ortholog groups are called yeast orthologs. We considered the following 4 sets of yeast orthologs (present in the yeast PPI data), namely *Yeast-E.coli, Yeast-Worm, Yeast-Fly, Yeast-Human*, consisting of 451, 1664, 1724, and 1850 number of proteins.

4.2   Yeast Protein Function Annotations and Gene Ontology Files

In order to measure functional enrichments of clusters we used only experimentally verified annotations as reported in the yeast gene association file of Saccharomyces Genome Database[3] (SGD), available at the GO database[4]. We excluded all computationally assigned annotations to yeast proteins to avoid introducing a possible bias, because many of these techniques use protein structural or sequence similarity which may often refer to orthology. GO slims and terms are also available at GO database.

4.3   Clustering

In this study we used two clustering techniques: SiDeS and MCL. We briefly address their properties:

MCL [29] computes clusters based on simulation of stochastic flow in graphs and it is widely used on many domains. It is able to use information on weights of edges of a given network if available. A first successful application of this algorithm on biological networks was presented in [30]; MCL was also modified for detecting orthologous groups [31]. A recently published comparative study [32] indicated that MCL outperforms other algorithms for clustering PPI networks. The inflation parameter of the algorithm was set to 1.8 as suggested in [32].

SiDeS [33], in contrast, is not able to use information on weights of edges. However, the main advantage of SiDeS is that it directly addresses the problem of statistical significance of cluster density, based on the topological structure of a PPI network, during computation. Thus, all clusters isolated by SiDeS

---

[3] http://www.yeastgenome.org/
[4] http://www.geneontology.org/GO.downloads.shtml, SGD version: 1.1523 date: 11/13/2010, GO version: 1.1.1602 date: 16/11/2010, GO Slim version: 1.1.1543 date: 19/10/2010.

have statistically significant density and therefore the resulting clusters tend to be more biologically relevant than those produced by other methods, albeit fewer in number. SiDeS modifies an existing state-of-the-art graph clustering algorithm, HCS [34], based on recursive partitioning of a graph and incorporating the computation of statistical significance of clusters.

The above two clustering algorithms are different in their basic concepts and combining their results for identifying orthology-related functional categories should effectively minimize the possibility of finding an artefact. Therefore we applied both clustering algorithms on each yeast PPI sub-network induced by every set of yeast orthologs as well as on all yeast PPI sub-networks induced by repeated random protein selection of the same number of proteins as the protein count of a particular yeast ortholog set. We labelled each resulting cluster group as follows:

- OYC-E - yeast clusters found in the sub-network induced by the Yeast-E.coli ortholog set.
- OYC-W - yeast clusters found in the sub-network induced by the Yeast-Worm ortholog set.
- OYC-F - yeast clusters found in the sub-network induced by the Yeast-Fly ortholog set.
- OYC-H - yeast clusters found in the sub-network induced by the Yeast-Human ortholog set.

These groups are of the class (2) and we generally refer to them by the common name OYC. Analogically we also marked cluster groups induced by randomly sampled proteins as follows:

- RYC-E - yeast clusters found in the sub-network induced by random sampled proteins of the same number as the number of proteins in the Yeast-E.coli ortholog set.
- RYC-W - yeast clusters found in the sub-network induced by random sampled proteins of the same number as the number of proteins in the Yeast-Worm ortholog set.
- RYC-F - yeast clusters found in the sub-network induced by random sampled proteins of the same number as the number of proteins in the Yeast-Fly ortholog set.
- RYC-H - yeast clusters found in the sub-network induced by random sampled proteins of the same number as the number of proteins in the Yeast-Human ortholog set.

These groups belong to the class (3) and we generally refer to them by the common name RYC.

When MCL or SiDeS applied on the whole yeast network, we get clusters of the class (1) and we refer to them by the name GYC (general yeast clusters).

We randomly sampled proteins 1000 times for each given number of orthologs. Recall that every run produces one particular RYC group. In order to compare these clusters with GYC or OYC, we always report average values of RYC groups

Table 1. Numbers of Clusters

| Cluster Group | MCL | | SiDeS | |
	#Clusters	#Predictors	#Clusters	#Predictors
GYC	365	147	122	93
OYC-E	37	14	5	3
RYC-E	34.31 ($\pm$3.82)	12.69 ($\pm$2.96)	4.71 ($\pm$2.08)	3.8 ($\pm$1.76)
OYC-W	181	80	66	46
RYC-W	175.22 ($\pm$7.21)	67.85 ($\pm$5.87)	55.04 ($\pm$5.57)	40.32 ($\pm$4.54)
OYC-F	191	80	64	51
RYC-F	181.97 ($\pm$7.51)	70.32 ($\pm$6.01)	57.71 ($\pm$5.57)	42.25 ($\pm$4.49)
OYC-H	203	90	82	62
RYC-H	196.38 ($\pm$7.80)	75.71 ($\pm$6.21)	63.42 ($\pm$5.67)	46.12 ($\pm$4.68)

computed over all 1000 simulations according to a given ortholog set (average RYC values).

Tables 1 contains the number of all clusters and corresponding cluster predictors for GYC, all four OYC and average RYC, as identified by MCL and SiDeS.

5   Results

The detected cluster predictors are considered as putative protein complexes and used to identify orthology-related functional categories. For each cluster group of predictors we compute the functional frequencies with respect to the categories enriched by OYC, as explained in Section 3. Figure 1 shows the frequency distribution of GYC, OYC-W and RYC-W clusters as detected by SiDeS.

In Tables 4 and 5 (see the Appendix) frequencies are reported as measured by MCL and SiDeS OYC-W clusters, respectively. Observe that not all orthology-related functional categories are shared by MCL or SiDeS cluster groups. To minimize the possibility of false positives, we employed a conservative approach and considered as orthology-related functional categories only those identified by both clustering techniques. The results are listed in Table 2.

5.1   Orthology-Related Functional Categories

For Yeast-E.coli orthologs, the identified clusters have higher frequencies of ribosomal and mitochondrial proteins. Indeed, it has been shown that the ribosomes in the mitochondria of eukaryotic cells resemble those in bacteria, reflecting the likely evolutionary origin of this organelle [35].

Since worm, fly and human all belong to eukaryotes, we looked at which common functional categories have yeast clusters containing orthologs with respect to these species (reported in Table 2 in boldface). Considering molecular functions, we observed that protein binding proteins and kinases activity proteins are more frequently present in OYC clusters than in GYC clusters or in RYC

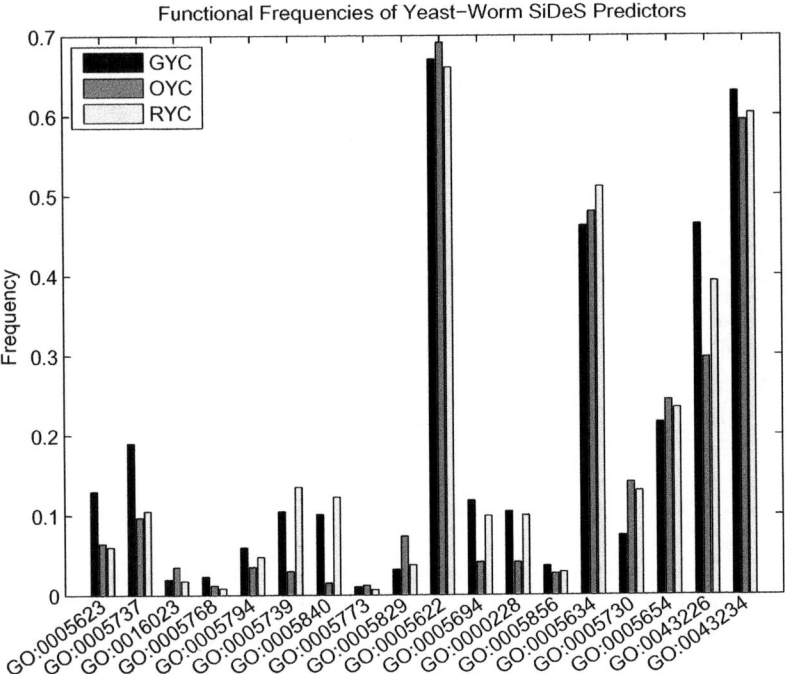

Fig. 1. Functional frequencies for Yeast-Worm orthologs as estimated by SiDeS predictors. On x-axis GO ids of GO slim functional categories are reported.

clusters. Thus these functional categories might be considered as orthology related. This is true in particular for proteins of protein kinase activity, which have been found conserved among eukaryotes: these kinase' functional conservations were investigated for yeast, worm, fly and human when studying their evolution [36]. Moreover, kinases' proteins are known to regulate the majority of cellular pathways, especially those involved in signal transduction. As we may see, signal transduction is also identified as orthology-related functional category. Regarding orthology-related protein binding, many functions of this category also showed high sequence conservation among eukaryotes (e.g [37,38]).

The next functional category which is orthology-related is translation. Many machineries involved with translation are expected to be evolutionary conserved as supported, e.g., by the evidence of finding a conserved protein family involved in translation [39], or by the presence of an evolutionary conserved mechanism for controlling the efficiency of protein translation [40].

Finally, we also observed OYC complexes containing vacuole proteins to be orthology-related. This is again supported by works which investigated yeast vacuole's proteins and function of their orthologs in other species. In particular, mammalian orthologs of yeast vacuolar protein sorting have been found to participate in early endosomal fusion and to interact with the cytoskeleton [41], and a very recent study of the same protein group revealed homologous genes and pathways that promote ageing in organisms ranging from yeast to mammals [42].

Table 2. Orthology-related functional categories. The spacing reflects the tree structure of GO slims in GO hierarchy. Functional categories in boldface are those shared by all OYC groups of eukaryotic orthologs.

Cluster Group	GO ID	Name	GO Domain
OYC-E	GO:0005739	mitochondrion	Cellular Component
	GO:0005840	ribosome	Cellular Component
OYC-W	GO:0005622	intracellular	Cellular Component
	GO:0005730	nucleolus	Cellular Component
	GO:0005773	**vacuole**	Cellular Component
	GO:0016023	cytoplasmic membrane-bounded vesicle	Cellular Component
	GO:0006412	**translation**	Biological Process
	GO:0007165	**signal transduction**	Biological Process
	GO:0009056	**catabolic process**	Biological Process
	GO:0019538	protein metabolic process	Biological Process
	GO:0005515	**protein binding**	Molecular Function
	GO:0003824	catalytic activity	Molecular Function
	GO:0004721	phosphoprotein phosphatase activity	Molecular Function
	GO:0016301	**kinase activity**	Molecular Function
	GO:0004672	**protein kinase activity**	Molecular Function
OYC-F	GO:0005730	nucleolus	Cellular Component
	GO:0005773	**vacuole**	Cellular Component
	GO:0043234	protein complex	Cellular Component
	GO:0006412	**translation**	Biological Process
	GO:0007165	**signal transduction**	Biological Process
	GO:0009056	**catabolic process**	Biological Process
	GO:0019538	protein metabolic process	Biological Process
	GO:0006139	nucleobase,-side,-tide and nucl. acid metab. proc.	Biological Process
	GO:0005515	**protein binding**	Molecular Function
	GO:0003677	DNA binding	Molecular Function
	GO:0008135	translation factor activity, nucleic acid binding	Molecular Function
	GO:0016301	**kinase activity**	Molecular Function
	GO:0004672	**protein kinase activity**	Molecular Function
OYC-H	GO:0005654	nucleoplasm	Cellular Component
	GO:0005773	**vacuole**	Cellular Component
	GO:0005829	cytosol	Cellular Component
	GO:0016023	cytoplasmic membrane-bounded vesicle	Cellular Component
	GO:0006412	**translation**	Biological Process
	GO:0007165	**signal transduction**	Biological Process
	GO:0009056	**catabolic process**	Biological Process
	GO:0005488	binding	Molecular Function
	GO:0005515	**protein binding**	Molecular Function
	GO:0016740	transferase activity	Molecular Function
	GO:0016301	**kinase activity**	Molecular Function
	GO:0004672	**protein kinase activity**	Molecular Function

5.2 Orthology-Related Putative Protein Complexes

We consider orthology-related clusters those clusters whose proteins perform at least one function of an orthology-related functional category. In Table 3 we report the number of orthology-related clusters found by the generated predictors. We call *unique* MCL *or* SiDeS *clusters* those orthology-related clusters whose proteins have a predicted function that is not inferred for those proteins by any GYC cluster identified by MCL or by SiDeS, respectively. These are the complexes that are new and derived using (the protein complex composition present in) the orthology sub-network, that is, uniquely linked to the orthology signal.

Given a unique cluster and its protein having a novel predicted function not inferred by any GYC cluster containing the protein. Then, if the function prediction is experimentally or computationally annotated in SGD, this prediction is verified. Analogously, if we find the novel predicted function has not been experimentally or computationally annotated in SGD, then this prediction is a new one. Observe that one cluster can have verified as well as new predictions at the same time. The number of clusters that produce verified and/or new protein predictions are reported in Table 3.

Examples of these novel complexes are given in the Appendix (Table 6): they demonstrate that by examining different set of orthologs we found specific putative complexes, most of them crucial for a living cell.

For instance, proteins of Cluster 1. are predicted to be involved in mitochondrial proton-transporting ATP synthase, catalytic core. While ATP1 and ATP2 are indeed the part of the catalytic core, ATP3 is part of the central stalk of mitochondrial proton-transporting ATP synthase. Cluster 1., however, gives a proper suggestion for the mechanism of the ATP3. Moreover, as ATP3 interacts with ATP2 it may be involved also in the catalytic core.

In Cluster 2. polyadenylation-dependent r-,t- and m-RNA catabolic process is newly predicted for NRD1 protein. This complies with recent findings that NRD1 is RNA-binding protein functioning in the poly(A) independent termination, in which binding to the combined and/or repetitive termination elements elicits efficient termination [43].

Cluster 3. is a predictor for INO80 complex. Three proteins, SWR1, IES6 and VPS72, have not yet been found to be part of this complex, however all of them associate with chromatin, where IES6 directly associates with the INO80 chromatin remodelling complex. This predictor has been found by both clustering methods independently.

In Cluster 4. ERR3 is a protein of unknown function, which has similarity to enolases. The predictor was found for Yeast-Worm as well as for Yeast-Fly orthologs, and it suggests that ERR3 is part of the ubiquitin conjugating enzyme complex.

Cluster 5. predicts COPII vesicle coat proteins. This cellular component was not predicted by any GYC predictor. Newly associated proteins with COPII are HIP1 and BUG1. These predictions seem to correctly suggest their functioning in a cell, as BUG1 is cis-golgi localized protein involved in endoplasmic reticulum to Golgi transport, and HIP1 is a high-affinity histidine permease, also involved in the transport of manganese ions.

Protein predictions for COPI vesicle coat are inferred by Cluster 6., where novel ones are for ARF1, ARF2 and ERV41 proteins. ARF1 and ARF2 are ADP-ribosylation factors involved in regulation of coated vesicle formation in intracellular trafficking within the Golgi. Because vesicles with COPI coats are

Table 3. Numbers of orthology-related clusters

Cluster Group	Method	#Predictors	#Ort-related.	#Unique	#Verified	#New
OYC-E	MCL	14	4	4	3	4
	SiDeS	3	2	1	0	1
OYC-W	MCL	80	37	32	29	31
	SiDeS	46	29	20	19	15
OYC-F	MCL	80	57	40	37	38
	SiDeS	51	44	34	32	28
OYC-H	MCL	90	33	24	20	23
	SiDeS	62	31	26	17	21

found associated with Golgi membranes at steady state [44], it suggests that these predictions might be correct. ERV41 is a protein localized to COPII-coated vesicles, but again our clusters at least properly predicts a possible role of protein in a cell.

Clusters 5. and 6. were partially also discovered by Yeast-Worm and Yeast-Human orthologs. Interestingly, each of them was discovered by a different clustering technique.

Cluster 7. consists of mostly DNA-directed RNA polymerase II proteins. Although proteins DST1, TFG2 and RPA135 have not been found to be directly part of this complex, the predictor properly associates these proteins with RNA polymerase system functioning. DST1 is a general transcription elongation factor TFIIS and it enables RNA polymerase II to read through blocks to elongation. TFG2 is a Transcription Factor II middle subunit involved in both transcription initiation and elongation of RNA polymerase II. Finally, RPA135 is RNA polymerase I second largest subunit A135. Thus, the protein is correctly associated with RNA polymerases and additionally our prediction also suggests that it may play a role in formation of RNA polymerase II.

6   Conclusions

We proposed a novel methodology for quantifying the functionality of the orthology signal in a PPI network at a protein complex level. The methodology performs a differential analysis between the functions of those complexes detected by clustering a PPI network using only proteins with orthologs in another given species, and the functions of complexes detected using the entire network or a sub-network generated by random sampling of proteins.

Results of our experimental analysis indicated the usefulness of the proposed methodology to identify functional categories clearly attributed to the presence of an evolutionary (orthology) signal. The distribution of these categories was described by means of protein functions inferred from those putative complexes detected by clustering a PPI network using an explicit orthology bias incorporated in the search space.

Acknowledgements. We are grateful to Elisabeth Georgii and Koji Tsuda for sharing the protein interaction data used in [19].

References

1. Kuzniar, A., van Ham, R.C., Pongor, S., Leunissen, J.A.: The quest for orthologs: finding the corresponding gene across genomes. Trends in Genetics 24(11), 539–551 (2008)
2. Remm, M., Storm, C.E., Sonnhammer, E.L.: Automatic clustering of orthologs and in-paralogs from pairwise species comparisons. Journal of Molecular Biology 314(5), 1041–1052 (2001)
3. Chen, F., Mackey, A.J., Stoeckert, C.J., Roos, D.S.: OrthoMCL-DB: querying a comprehensive multi-species collection of ortholog groups. Nucleic Acids Research 34(suppl 1) D363–D368
4. Vespignani, A.: Evolution thinks modular. Nature Genetics 35(2), 118–119 (2003)
5. Wuchty, S., Oltvai, Z.N., Barabási, A.L.: Evolutionary conservation of motif constituents in the yeast protein interaction network. Nature Genetics 35(2), 176–179 (2003)
6. Wuchty, S., Barabási, A.L., Ferdig, M.: Stable evolutionary signal in a yeast protein interaction network. BMC Evolutionary Biology 6(1), 8 (2006)
7. Brown, K., Jurisica, I.: Unequal evolutionary conservation of human protein interactions in interologous networks. Genome Biology 8(5), R95 (2007)
8. Campillos, M., von Mering, C., Jensen, L.J., Bork, P.: Identification and analysis of evolutionarily cohesive functional modules in protein networks. Genome Research 16(3), 374–382 (2006)
9. Fokkens, L., Snel, B.: Cohesive versus flexible evolution of functional modules in eukaryotes. PLoS Comput. Biol. 5(1), e1000276 (2009)
10. Erten, S., Li, X., Bebek, G., Li, J., Koyuturk, M.: Phylogenetic analysis of modularity in protein interaction networks. BMC Bioinformatics 10(1), 333 (2009)
11. Yosef, N., Kupiec, M., Ruppin, E., Sharan, R.: A complex-centric view of protein network evolution. Nucleic Acids Research 37(12), e88 (2009)
12. Woźniak, M., Tiuryn, J., Dutkowski, J.: MODEVO: exploring modularity and evolution of protein interaction networks. Bioinformatics 26(14), 1790–1791 (2010)
13. Ashburner, M., Ball, C.A., Blake, J.A., Botstein, D., Butler, H., Cherry, J.M., Davis, A.P., Dolinski, K., Dwight, S.S., Eppig, J.T., Harris, M.A., Hill, D.P., Issel-Tarver, L., Kasarskis, A., Lewis, S., Matese, J.C., Richardson, J.E., Ringwald, M., Rubin, G.M., Sherlock, G.: Gene ontology: tool for the unification of biology. the gene ontology consortium. Nature Genetics 25(1), 25–29 (2000)
14. Sharan, R., Ideker, T.: Modeling cellular machinery through biological network comparison. Nature Biotechnology 24(4), 427–433 (2006)
15. Bauer, S., Grossmann, S., Vingron, M., Robinson, P.N.: Ontologizer 2.0–a multifunctional tool for GO term enrichment analysis and data exploration. Bioinformatics 24(14), 1650–1651 (2008)
16. Liang, Z., Xu, M., Teng, M., Niu, L.: Comparison of protein interaction networks reveals species conservation and divergence. BMC Bioinformatics 7(1), 457 (2006)
17. Jancura, P., Marchiori, E.: Dividing protein interaction networks for modular network comparative analysis. Pattern Recognition Letters 31(14), 2083–2096 (2010)
18. Yon Rhee, S., Wood, V., Dolinski, K., Draghici, S.: Use and misuse of the gene ontology annotations. Nat. Rev. Genet. 9(7), 509–515 (2008)

19. Georgii, E., Dietmann, S., Uno, T., Pagel, P., Tsuda, K.: Enumeration of condition-dependent dense modules in protein interaction networks. Bioinformatics 25(7), 933–940 (2009)
20. Xenarios, I., Salwínski, Ł., Duan, X.J., Higney, P., Kim, S.M., Eisenberg, D.: Dip, the database of interacting proteins: a research tool for studying cellular networks of protein interactions. Nucleic Acids Research 30(1), 303–305 (2002)
21. Guldener, U., Munsterkotter, M., Oesterheld, M., Pagel, P., Ruepp, A., Mewes, H.W., Stumpflen, V.: MPact: the MIPS protein interaction resource on yeast. Nucl. Acids Res. 34(suppl_1), D436–D441 (2006)
22. Gavin, A.C., Bosche, M., Krause, R., Grandi, P., Marzioch, M., Bauer, A., Schultz, J., Rick, J.M., Michon, A.M., Cruciat, C.M., Remor, M., Hofert, C., Schelder, M., Brajenovic, M., Ruffner, H., Merino, A., Klein, K., Hudak, M., Dickson, D., Rudi, T., Gnau, V., Bauch, A., Bastuck, S., Huhse, B., Leutwein, C., Heurtier, M.A., Copley, R.R., Edelmann, A., Querfurth, E., Rybin, V., Drewes, G., Raida, M., Bouwmeester, T., Bork, P., Seraphin, B., Kuster, B., Neubauer, G., Superti-Furga, G.: Functional organization of the yeast proteome by systematic analysis of protein complexes. Nature 415, 141–147 (2002)
23. Krogan, N.J., Cagney, G., Yu, H., Zhong, G., Guo, X., Ignatchenko, A., Li, J., Pu, S., Datta, N., Tikuisis, A.P., Punna, T., Peregrín-Alvarez, J.M., Shales, M., Zhang, X., Davey, M., Robinson, M.D., Paccanaro, A., Bray, J.E., Sheung, A., Beattie, B., Richards, D.P., Canadien, V., Lalev, A., Mena, F., Wong, P., Starostine, A., Canete, M.M., Vlasblom, J., Wu, S., Orsi, C., Collins, S.R., Chandran, S., Haw, R., Rilstone, J.J., Gandi, K., Thompson, N.J., Musso, G., St Onge, P., Ghanny, S., Lam, M.H., Butland, G., Altaf-Ul, A.M., Kanaya, S., Shilatifard, A., O'Shea, E., Weissman, J.S., Ingles, C.J., Hughes, T.R., Parkinson, J., Gerstein, M., Wodak, S.J., Emili, A., Greenblatt, J.F.: Global landscape of protein complexes in the yeast saccharomyces cerevisiae. Nature 440(7084), 637–643 (2006)
24. Jansen, R., Yu, H., Greenbaum, D., Kluger, Y., Krogan, N.J., Chung, S., Emili, A., Snyder, M., Greenblatt, J.F., Gerstein, M.: A Bayesian Networks Approach for Predicting Protein-Protein Interactions from Genomic Data. Science 302(5644), 449–453 (2003)
25. Chen, F., Mackey, A.J., Vermunt, J.K., Roos, D.S.: Assessing performance of orthology detection strategies applied to eukaryotic genomes. PLoS ONE 2(4), e383 (2007)
26. Dolinski, K., Botstein, D.: Orthology and functional conservation in eukaryotes. Annual Review of Genetics 41(1), 465–507 (2007)
27. Bhardwaj, N., Lu, H.: Correlation between gene expression profiles and protein-protein interactions within and across genomes. Bioinformatics 21(11), 2730–2738
28. Sharan, R., Suthram, S., Kelley, R.M., Kuhn, T., McCuine, S., Uetz, P., Sittler, T., Karp, R.M., Ideker, T.: From the Cover: Conserved patterns of protein interaction in multiple species. Proceedings of the National Academy of Sciences 102(6), 1974–1979 (2005)
29. van Dongen, S.: Graph Clustering by Flow Simulation. PhD thesis, University of Utrecht (May 2000)
30. Enright, A.J., Van Dongen, S., Ouzounis, C.A.: An efficient algorithm for large-scale detection of protein families. Nucl. Acids Res. 30(7), 1575–1584 (2002)
31. Li, L., Stoeckert, C.J., Roos, D.S.: OrthoMCL: Identification of Ortholog Groups for Eukaryotic Genomes. Genome Research 13(9), 2178–2189 (2003)
32. Brohee, S., van Helden, J.: Evaluation of clustering algorithms for protein-protein interaction networks. BMC Bioinformatics 7(1), 488 (2006)

33. Koyuturk, M., Szpankowski, W., Grama, A.: Assessing significance of connectivity and conservation in protein interaction networks. Journal of Computational Biology 14(6), 747–764 (2007); PMID: 17691892
34. Hartuv, E., Shamir, R.: A clustering algorithm based on graph connectivity. Inf. Process. Lett. 76(4-6), 175–181 (2000)
35. Benne, R., Sloof, P.: Evolution of the mitochondrial protein synthetic machinery. Biosystems 21(1), 51–68 (1987)
36. Manning, G., Plowman, G.D., Hunter, T., Sudarsanam, S.: Evolution of protein kinase signaling from yeast to man. Trends in Biochemical Sciences 27(10), 514–520 (2002)
37. Sedeh, R.S., Fedorov, A.A., Fedorov, E.V., Ono, S., Matsumura, F., Almo, S.C., Bathe, M.: Structure, evolutionary conservation, and conformational dynamics of homo sapiens fascin-1, an f-actin crosslinking protein. Journal of Molecular Biology 400(3), 589–604 (2010)
38. Capra, J.A., Laskowski, R.A., Thornton, J.M., Singh, M., Funkhouser, T.A.: Predicting protein ligand binding sites by combining evolutionary sequence conservation and 3d structure. PLoS Comput. Biol. 5(12), e1000585 (2009)
39. Frolova, L., Le Goff, X., Rasmussen, H.H., Cheperegin, S., Drugeon, G., Kress, M., Arman, I., Haenni, A.L., Celis, J.E., Phllippe, M., Justesen, J., Kisselev, L.: A highly conserved eukaryotic protein family possessing properties of polypeptide chain release factor. Nature 372, 103–701 (1994)
40. Tuller, T., Carmi, A., Vestsigian, K., Navon, S., Dorfan, Y., Zaborske, J., Pan, T., Dahan, O., Furman, I., Pilpel, Y.: An evolutionarily conserved mechanism for controlling the efficiency of protein translation. Cell 141(2), 344–354 (2010)
41. Richardson, S.C.W., Winistorfer, S.C., Poupon, V., Luzio, J.P., Piper, R.C.: Mammalian late vacuole protein sorting orthologues participate in early endosomal fusion and interact with the cytoskeleton. Mol. Biol. Cell 15(3), 1197–1210 (2004)
42. Fabrizio, P., Hoon, S., Shamalnasab, M., Galbani, A., Wei, M., Giaever, G., Nislow, C., Longo, V.D.: Genome-wide screen in saccharomyces cerevisiae identifies vacuolar protein sorting, autophagy, biosynthetic, and trna methylation genes involved in life span regulation. PLoS Genet. 6(7), e1001024 (2010)
43. Hobor, F., Pergoli, R., Kubicek, K., Hrossova, D., Bacikova, V., Zimmermann, M., Pasulka, J., Hofr, C., Vanacova, S., Stefl, R.: Recognition of Transcription Termination Signal by the Nuclear Polyadenylated RNA-binding (NAB) 3 Protein. Journal of Biological Chemistry 286(5), 3645–3657 (2011)
44. Kirchhausen, T.: Three ways to make a vesicle. Nature Reviews. Molecular Cell Biology 1(3), 187–198 (2000)

Appendix

Table 4. Frequencies of functional categories for Yeast-Worm MCL predictors. Orthology-related functional categories are in boldface.

GO ID	GYC	OYC-W	RYC-W	Name
GO:0005623	0.162	0.083	0.103 (±0.032)	cell
GO:0005737	0.120	0.037	0.086 (±0.035)	cytoplasm
GO:0016023	0.026	**0.031**	0.025 (±0.015)	**cytoplasmic membrane-bounded vesicle**
GO:0005783	0.034	0.031	0.018 (±0.012)	endoplasmic reticulum
GO:0005768	0.034	**0.035**	0.014 (±0.010)	**endosome**
GO:0005794	0.052	**0.057**	0.050 (±0.017)	**Golgi apparatus**
GO:0005739	0.108	0.044	0.114 (±0.021)	mitochondrion
GO:0005773	0.008	**0.013**	0.008 (±0.006)	**vacuole**
GO:0005829	0.015	0.009	0.013 (±0.012)	cytosol
GO:0005622	0.629	**0.657**	0.587 (±0.056)	**intracellular**
GO:0005694	0.107	0.044	0.105 (±0.027)	chromosome
GO:0000228	0.077	0.031	0.098 (±0.025)	nuclear chromosome
GO:0005856	0.034	0.026	0.033 (±0.020)	cytoskeleton
GO:0005634	0.447	**0.510**	0.461 (±0.057)	**nucleus**
GO:0005730	0.059	**0.191**	0.111 (±0.044)	**nucleolus**
GO:0005815	0.017	0.006	0.007 (±0.008)	microtubule organizing center
GO:0005635	0.013	**0.020**	0.018 (±0.012)	**nuclear envelope**
GO:0005654	0.140	0.172	0.183 (±0.033)	nucleoplasm
GO:0043226	0.605	0.470	0.412 (±0.079)	organelle
GO:0005886	0.003	**0.009**	0.002 (±0.005)	**plasma membrane**
GO:0043234	0.418	**0.539**	0.536 (±0.053)	**protein complex**

Table 5. Frequencies of functional categories for Yeast-Worm SiDeS predictors. Orthology-related functional categories are in boldface.

GO ID	GYC	OYC-W	RYC-W	Name
GO:0005623	0.130	0.065	0.060 (±0.028)	cell
GO:0005737	0.190	0.097	0.105 (±0.045)	cytoplasm
GO:0016023	0.020	**0.035**	0.018 (±0.015)	**cytoplasmic membrane-bounded vesicle**
GO:0005768	0.023	0.012	0.008 (±0.009)	endosome
GO:0005794	0.060	0.035	0.048 (±0.021)	Golgi apparatus
GO:0005739	0.105	0.029	0.135 (±0.023)	mitochondrion
GO:0005840	0.101	0.015	0.123 (±0.029)	ribosome
GO:0005773	0.010	**0.012**	0.007 (±0.008)	**vacuole**
GO:0005829	0.032	**0.074**	0.038 (±0.030)	**cytosol**
GO:0005622	0.670	**0.691**	0.660 (±0.065)	**intracellular**
GO:0005694	0.118	0.041	0.100 (±0.031)	chromosome
GO:0000228	0.105	0.041	0.100 (±0.030)	nuclear chromosome
GO:0005856	0.037	0.026	0.028 (±0.020)	cytoskeleton
GO:0005634	0.462	0.479	0.511 (±0.054)	nucleus
GO:0005730	0.075	**0.141**	0.130 (±0.034)	**nucleolus**
GO:0005654	0.216	**0.244**	0.234 (±0.036)	**nucleoplasm**
GO:0043226	0.463	0.297	0.393 (±0.075)	organelle
GO:0043234	0.630	0.594	0.603 (±0.045)	protein complex

Table 6. Orthology-related clusters

Cluster ID	Proteins	Prediction	Cluster Group	Method
Cluster 1.	ATP1	mitochondrial proton-transporting ATP synthase, catalytic core	OYC-E	MCL
	ATP2		OYC-E	MCL
	ATP3		OYC-E	MCL
Cluster 2.	MTR4	nuclear polyadenylation-dependent r-,t-and m-RNA catabolic process	OYC-W	MCL
	TRF5		OYC-W	MCL
	PAP2		OYC-W	MCL
	NRD1		OYC-W	MCL
Cluster 3.	RVB1	INO80 chromatin remodelling complex	OYC-F	MCL, SiDeS
	RVB2		OYC-F	MCL, SiDeS
	ARP5		OYC-F	MCL, SiDeS
	ARP8		OYC-F	MCL, SiDeS
	INO80		OYC-F	MCL, SiDeS
	IES6		OYC-F	MCL, SiDeS
	SWR1		OYC-F	MCL, SiDeS
	VPS72		OYC-F	MCL, SiDeS
Cluster 4.	MMS2	ubiquitin conjugating enzyme complex	OYC-F,OYC-W	MCL
	UBC13		OYC-F,OYC-W	MCL
	ERR3		OYC-F,OYC-W	MCL
Cluster 5.	SEC23	COPII vesicle coat	OYC-F,OYC-W,OYC-H	MCL
	SEC24		OYC-F,OYC-W,OYC-H	MCL
	SFB2		OYC-F,OYC-W,OYC-H	MCL
	HIP1		OYC-F,OYC-W,OYC-H	MCL
	GRH1		OYC-F,OYC-W	MCL
	BUG1		OYC-F	MCL
Cluster 6.	RET2	COPI vesicle coat	OYC-F,OYC-H,OYC-W	SiDeS
	RET3		OYC-F,OYC-H,OYC-W	SiDeS
	SEC21		OYC-F,OYC-H,OYC-W	SiDeS
	SEC26		OYC-F,OYC-H,OYC-W	SiDeS
	SEC27		OYC-F,OYC-H,OYC-W	SiDeS
	ARF1		OYC-F,OYC-H,OYC-W	SiDeS
	ARF2		OYC-F,OYC-H,OYC-W	SiDeS
	COP1		OYC-F,OYC-H	SiDeS
	ERV41		OYC-F	SiDeS
Cluster 7.	SPT5	DNA-directed RNA polymerase II	OYC-H	SiDeS
	RPB2		OYC-H	SiDeS
	RPB3		OYC-H	SiDeS
	RPB4		OYC-H	SiDeS
	RPB7		OYC-H	SiDeS
	RPB8		OYC-H	SiDeS
	RPB9		OYC-H	SiDeS
	RPB11		OYC-H	SiDeS
	RPO21		OYC-H	SiDeS
	RPO26		OYC-H	SiDeS
	RPC10		OYC-H	SiDeS
	RPA135		OYC-H	SiDeS
	TFG2		OYC-H	SiDeS
	DST1		OYC-H	SiDeS

Algorithms for Rapid Error Correction for the Gene Duplication Problem

Ruchi Chaudhary[1], J. Gordon Burleigh[2], and Oliver Eulenstein[1]

[1] Department of Computer Science, Iowa State University, Ames, IA 50011, USA
{ruchic,oeulenst}@cs.iastate.edu
[2] Department of Biology, University of Florida, Gainesville, Florida 32611, USA
gburleigh@ufl.edu

Abstract. Gene tree - species tree reconciliation problems infer the patterns and processes of gene evolution within the context of an organismal phylogeny. In one example, the gene duplication problem seeks the evolutionary scenario that implies the minimum number of gene duplications needed to reconcile a gene tree and a species tree. While the gene duplication problem can effectively link gene and species evolution, error in gene trees can profoundly bias the results. We describe novel algorithms that rapidly search local Subtree Prune and Regraft (SPR) or Tree Bisection and Reconnection (TBR) neighborhoods of a gene tree to find a topology that implies the fewest duplications. These algorithms improve on the current solutions by a factor of n for searching SPR neighborhoods and n^2 for searching TBR neighborhoods, where n is the number of vertices in the given gene tree. They provide a fast error correction protocol for gene trees, in which we allow small gene tree rearrangements to improve the reconciliation cost. We tested the SPR tree rearrangement algorithm on a collection of 1201 plant gene trees, and in every case, the SPR algorithm identified an alternate topology that implied at least one fewer duplication. We also demonstrate a simple method to use the gene rearrangement algorithm to improve gene tree parsimony phylogenetic analyses, which infer a species tree based on the gene duplication problem.

1   Introduction

With the availability of large-scale genomic data from across a broad phylogenetic spectrum, scientists have an unprecedented opportunity to examine gene evolution. Gene tree - species tree (GT-ST) reconciliation seeks to map the history of gene trees into the context of species evolution and thus potentially link processes of gene evolution to phenotypic changes and diversification. One common approach is to infer the minimum number of evolutionary events (e.g., duplication, loss, coalescence, or lateral gene transfer) that are needed to reconcile a gene tree and species tree topology [21]. GT-ST reconciliation also can be extended to infer species phylogenies. For example, gene tree parsimony analyses take a collection of gene trees and seek a species tree that implies the fewest evolutionary events implied by the gene trees (e.g., [15,17,23,27]). Such approaches provide a truly genomic perspective on species relationships (e.g., [6,26]).

One underlying complication of all GT-ST reconciliation is uncertainty and error in the gene trees. Gene tree topologies often are estimated using heuristic methods from

J. Chen, J. Wang, and A. Zelikovsky (Eds.): ISBRA 2011, LNBI 6674, pp. 227–239, 2011.

short sequence alignments. Consequently, the inferred gene tree topologies likely differ from the true topologies. Error in gene tree estimates can have radical effects on GT-ST reconciliation and the interpretation of gene evolution. Rasmussen and Kellis [24] estimated that error in gene tree reconstruction can lead to 2-3 fold overestimates of gene duplications and losses. Gene tree error also produces biases in GT-ST reconciliations, often erroneously implying large numbers of duplications near the root of the species tree [7,18]. Furthermore, error in gene tree topologies can mislead gene tree parsimony phylogenetic analyses (e.g., [6,19,26]).

Several approaches have been proposed to address gene tree error in GT-ST reconciliation. First, questionable nodes in a gene tree or nodes with low support may be collapsed prior to gene tree reconciliation, and the resulting non-binary gene trees may be reconciled with species trees [4,9,29]. Similarly, GT-ST reconciliations can use a distribution of gene tree topologies, such as bootstrap gene trees, rather than a single gene tree estimate [7,11,20]. Both of these approaches may help account for stochastic error and uncertainty in gene tree topologies, but they do not explicitly confront gene tree error. Methods also exist to simultaneously infer the gene tree topology and the gene tree reconciliation with a fixed species tree [2,24]. While these sophisticated statistical approaches appear very promising, they are computationally intensive, and it is unclear if they will be tractable for large-scale analyses. Another, perhaps a more computationally feasible approach is to allow a limited number of, local rearrangements in the gene tree topology if they reduced the reconciliation cost [10,12]. For example, [10,12] described a method to allow NNI-branch swaps on selected branches of a gene tree to reduced the reconciliation cost.

Following [10,12], we address gene tree error in the reconciliation process by assuming that the correct gene tree can be found in a particular neighborhood of the given gene tree. Our approach is based on the gene duplication model, which identifies the fewest gene duplications implied from a given gene tree and given species tree. This *neighborhood* consists of all trees that are within one edit operation of the gene tree. While [10,12] use Nearest Neighbor Interchange (NNI) edit operations to define the neighborhood, we use the standard tree edit operations SPR [1,5] and TBR [1], which significantly extend on the search space of the NNI neighborhood. These extended search spaces may be more desirable to find the correct gene tree, if they can be efficiently searched. The *SPR and TBR local search problems* find a tree in the SPR and TBR neighborhood of a given gene tree, respectively, that has the minimum reconciliation cost when reconciled with a given species tree. Using the algorithm from Zhang [31] the best known (naïve) runtimes are $O(n^3)$ for the SPR local search problem and $O(n^4)$ for the TBR local search problem, where n is the number of vertices in the given gene tree. These runtimes typically forbid the computation of larger GT-ST reconciliations. We improve on these solutions by a factor of n for the SPR local search problem and a factor of n^2 for the TBR local search problem. This makes the local search under the TBR edit operation as efficient as under the SPR edit operation, and it provides a high speed gene tree error-correction protocol that is amenable to large-scale genomic data sets.

To evaluate the performance of our algorithms we implement the SPR local search algorithm for brevity. Note, that the SPR neighborhood is properly contained in the

TBR neighborhood for any given tree. Thus the performance of the SPR based program provides a conservative estimate of the performance of the TBR based program. We test our program on a collection of 1201 plant gene trees, some of which contain hundreds of leaves, and we demonstrate how it can be easily incorporated into large-scale gene tree parsimony phylogenetic analyses.

2   Basic Notations and Preliminaries

2.1   Basic Definitions and Notations

A *(phylogenetic) tree* is a leaf labeled tree in which the internal vertices have degree at least two. The set of all vertices of a tree is denoted by $V(T)$ and set of all edges by $E(T)$. T is *rooted* if it has exactly one distinguished vertex called *root*, which we denote by $Ro(T)$. Let T be a rooted tree. We define $\leq_T$ to be the *partial order* on $V(T)$, where $x \leq_T y$ if y is a vertex on the path from $Ro(T)$ to x. If $x \leq_T y$ we call x a *descendent* of y, and y an *ancestor* of x. We also define $x <_T y$ if $x \leq_T y$ and $x \neq y$, in this case x is called *proper descendent* of y and y *proper ancestor* of x. The set of minima under $\leq_T$ is denoted by $Le(T)$ and the contained elements are called *leaves*. If $\{x, y\} \in E(T)$ and $x \leq_T y$, then y is called *parent* of x and denoted by $Pa_T(x)$, and x a *child* of y. We write (y, x) to denote the edge $\{y, x\}$, where $y = Pa(x)$. The set of all children of a vertex y is denoted by $Ch_T(y)$. Two distinct vertices in T are called *siblings* if they have the same parent. We denote sibling of v in T by $Sb_T(v)$. The set of *internal vertices* of T, denoted $I(T)$, is defined to be $V(T) \backslash Le(T)$. The *least common ancestor* of a non-empty subset $L \subseteq V(T)$, denoted as $LCA_T(L)$, is the unique smallest upper bound of L under $\leq_T$. We define $T(U)$ the *minimum rooted subtree* of T that connects the elements in U for $U \subseteq V(T)$. Furthermore, the *restriction* of T to U, denoted by $T_{|U}$, is the rooted phylogenetic tree that is obtained from $T(U)$ by suppressing all non-root vertices of degree two. The *subtree* of T rooted at $v \in V(T)$, denoted T_v, is defined to be $T_{|U}$, for $U := \{u \in Le(T) : u \leq_T v\}$. T is *full binary* if every vertex has either zero or two children. Throughout this paper, the term tree refers to a rooted full binary tree. If an isomorphism exists between two trees T_1, T_2, then we write $T_1 \simeq T_2$.

2.2   The Gene Duplication Cost Model

A *species tree* and a *gene tree* are trees that represent the evolutionary relationships between species and genes (of a gene family) respectively. We assume that each leaf of the gene tree is labeled with the species from which that gene was sampled. Let G be a gene tree and S a species tree. In order to compare G with S, we require a mapping from each gene $g \in V(G)$ to the most recent species $s \in V(S)$ that could have contained it.

Definition 1 (Mapping). *The* leaf-mapping $\mathcal{L}_{G,S} \colon Le(G) \to Le(S)$ *maps a leaf* $g \in Le(G)$ *to that unique leaf* $s \in Le(S)$ *from which the gene* g *was sampled. The extension of* $\mathcal{L}_{G,S}$ *for all vertices of* G *is* $\mathcal{M}_{G,S} \colon V(G) \to V(S)$, *which is defined as* $\mathcal{M}_{G,S}(g) := LCA(\mathcal{L}_{G,S}(Le(G_g)))$.

Definition 2 (Comparability). G *is comparable to* S, *if for each* $g \in Le(G)$, *the leaf-mapping* $\mathcal{L}_{G,S}(g)$ *is well defined.*

Throughout this paper we assume that the gene tree G is comparable to the species tree S, and in addition to that, $Le(S) = \cup_{g \in Le(G)} \mathcal{L}_{G,S}(g)^1$. We also assume that n is the number of taxa present in both input trees.

Definition 3 (Duplication). *A vertex* $g \in I(G)$ *is a (gene) duplication (w.r.t. S) if* $\mathcal{M}_{G,S}(g) \in \mathcal{M}_{G,S}(Ch(g))^2$.

Definition 4 (Duplication Cost). *The duplication costs are defined as follows:*

(i) The duplication cost from $g \in G$ *to* S, $\Delta(G, S, g) := \begin{cases} 1, & \text{if } g \text{ is a gene duplication;} \\ 0, & \text{otherwise.} \end{cases}$

(ii) The duplication cost from G *to* S, $\Delta(G, S) := \sum_{g \in G} \Delta(G, S, g)$.

2.3   The Error-Correction Problems

Here we give definitions for tree rearrangement operations TBR [1] and SPR [1,5], and then formulate the Error-Correction problems that were motivated in the introduction.

Definition 5 (TBR operation). *Let T be a tree. For this definition, we regard the planted tree* $Pl(T)$ *as the tree obtained from adding root edge* $\{r, Ro(T)\}$ *to* $E(T)$, *where* $r \notin V(T)$. *Let* $e := (u, v) \in E(T)$, *and* X *and* Y *be the connected components that are obtained by removing edge e from T such that* $v \in X$ *and* $u \in Y$. *We define* $TBR_T(v, x, y)$ *for* $x \in X$ *and* $y \in Y$ *to be the tree that is obtained from* $Pl(T)$ *by first deleting edge e, and then adjoining a new edge f between X and Y as follows:*

1. *If* $x \neq Ro(X)$ *then suppress* $Ro(X)$ *and create new root by subdividing edge* $(Pa(x), x)$.
2. *Create a new vertex y' that subdivides the edge* $(Pa(y), y)$.
3. *Add edge f between vertices y' and* $Ro(X)$.
4. *Suppress the vertex u, and rename vertex y' as u.*
5. *Contract the root edge.*

We say that, the tree $TBR_T(v, x, y)$ is obtained from T by a *tree bisection and reconnection (TBR)* operation that bisects the tree T into the components X and Y, and reconnects them above the nodes x and y. We define the following *neighborhoods* for the TBR operation:

1. $TBR_G(v, x) := \cup_{y \in Y} TBR_G(v, x, y)$
2. $TBR_G(v) := \cup_{x \in X} TBR_G(v, x)$
3. $TBR_G := \cup_{(u,v) \in E(G)} TBR_G(v)$

Definition 6 (SPR operation). *The SPR operation is defined as a special case of TBR operation. Let* $e := (u, v) \in E(T)$, *and* X *and* Y *be the connected components that are obtained by removing edge e from T such that* $v \in X$ *and* $u \in Y$. *We define* $SPR_T(v, y)$ *for* $y \in Y$ *to be* $TBR_T(v, v, y)$. *We say that the tree* $SPR_T(v, y)$ *is obtained from* T *by performing subtree prune and regraft (SPR) operation that prunes subtree T_v and regrafts it above y. (See Fig. 1(a).)*

---

[1] Note that if $Le(G) \neq Le(S)$ then we can simply set the species tree to be $S_{|Le(G)}$. This takes $O(n)$ time and, consequently, does not affect the time complexity of our final algorithm.

[2] Gene duplication is actually a well studied theorem [22,14].

We define the following *neighborhoods* for the SPR operation:

1. $\mathrm{SPR}_G(v) := \cup_{y \in Y} \mathrm{SPR}_G(v, y)$
2. $\mathrm{SPR}_G := \cup_{(u,v) \in E(G)} \mathrm{SPR}_G(v)$

We now state the error-correction problems.

Problem 1 (SPR Based Error-Correction(SEC)).
Input: A gene tree G and a species tree S.
Output: A gene tree $G^* \in \mathrm{SPR}_G$ such that $\Delta(G^*, S) = \min_{G' \in \mathrm{SPR}_G} \Delta(G', S)$.

The TBR based error-correction (TEC) problem is defined analogously to the SPR based error-correction (SEC) problem.

3   Solving the SEC Problem

We first define a restricted version of SEC problem, called the restricted SPR based error-correction (R-SEC) problem.

Problem 2 (Restricted SPR Based Error-Correction (R-SEC)).
Input: A gene tree G, a species tree S, and $v \in V(G)$.
Output: A gene tree $G^* \in \mathrm{SPR}_G(v)$ such that $\Delta(G^*, S) = \min_{G' \in \mathrm{SPR}_G(v)} \Delta(G', S)$.

Observation 1. *SEC problem can be solved by solving R-SEC efficiently.*

Proof. Observe that there are $\Theta(n)$ different ways to select a subtree of G to be pruned. Let v be the root of pruned subtree. Furthermore, for each tuple $\langle G, S, v \rangle$ we call the solution of R-SEC problem. The tree with minimum duplication score among all R-SEC outputs is the solution of the SEC problem.   □

We now highlight how the R-SEC problem can be solved in $\Theta(n)$ time. In order to do so, it is sufficient to compute the value $\Delta(G', S)$ for each $G' \in \mathrm{SPR}_G(v)$. For a given gene tree G, the size of the $\mathrm{SPR}_G(v)$ neighborhood is $\Theta(n)$; the duplication cost for a gene tree and a species tree can be computed in $\Theta(n)$ time [31]. Thus the R-SEC problem can be solved in $\Theta(n^2)$ time. We describe a novel algorithm for R-SEC that gives the $\Theta(n)$ speed-up over this naïve solution. More precisely, we traverse the trees in $\mathrm{SPR}_G(v)$ neighborhood in such a special sequence in which the duplication score difference between two consecutive trees can be easily computed in constant time.

3.1   Structural Properties

In the following, we first define the NNI [1] operation, and then construct a graph on trees in $SPR_G(v)$, in which an edge exists between two trees if they are one NNI operation apart. We prove that such a graph is a rooted, binary tree. The tree topology yields an efficient algorithm for R-SEC problem.

Definition 7 (NNI operation). *We define the NNI operation as a special case of SPR operation. Let $e := (u, v) \in E(T)$, and X and Y be the connected components that are obtained by removing edge e from T such that $v \in X$ and $u \in Y$. We define $NNI_T(v)$ to be $SPR_T(v, y)$ for $y := Pa(u)$, and say that $NNI_T(v)$ is obtained from T by performing nearest neighbor interchange (NNI) operation that prunes subtree T_v and regrafts it above the parent of v's parent.*

Definition 8. *Let the NNI-distance $d_{NNI}(T_1, T_2)$ between two trees T_1 and T_2 be the minimum number of NNI operations required to transform T_1 into T_2.*

Definition 9. *The NNI-adjacency graph $\mathcal{X} = (V, E)$ is the graph with $V = SPR_G(v)$ and $\{G_u, G_v\} \in E \Longleftrightarrow d_{NNI}(G_u, G_v) = 1$.*

Lemma 1. $\mathcal{X}$ *is a tree.*

Proof. We prove it by showing that there exists a unique path between any two vertices in $\mathcal{X}$. Let $G', G'' \in V(\mathcal{X})$, thus $G', G'' \in SPR_G(v)$. Let $G' := SPR_G(v, x_1)$, $G'' := SPR_G(v, x_2)$, and let $d_G(x_1, x_2)$ be the distance between vertex x_1 and x_2 in G. We use induction on $d_G(x_1, x_2)$. Let $d_G(x_1, x_2) = 1$ and assume without loss of generality that $x_2 = Pa_G(x_1)$. Thus, $G' = NNI_{G''}(Sb(x_1))$. So the hypothesis holds for $d_G(x_1, x_2) = 1$. Assume now that the hypothesis is true for $d_G(x_1, x_2) \leq k$ and suppose $d_G(x_1, x_2) = k + 1$. Since G is a tree, there must be a unique path between x_1 and x_2; let y be a vertex on this path. Let $d_G(y, x_1) = 1$, and $G^n := SPR_G(v, y)$. If $y = Pa_G(x_1)$, then $G^n = NNI_{G'}(v)$; otherwise $G^n = NNI_{G'}(Sb(y))$. Since $d_G(y, x_2) = k$, thus (by induction hypothesis) the hypothesis is valid for $d_G(x_1, x_2) = k + 1$. □

Theorem 1. $\mathcal{X}$ *is a rooted full binary tree.*

Proof. In view of Lemma 1, it suffices to show that except a unique vertex of degree 2 all other vertices in $\mathcal{X}$ are of degree 1 or 3. Let $G' \in V(\mathcal{X})$, thus $G' = SPR_G(v, y)$ for some $y \in V(G)$. There are three cases.

Case 1: y is a root. Let $y_1 \in Ch_G(y)$. Let $G^1 := SPR_G(v, y_1)$, thus $G' = NNI_{G^1}(v)$. Hence $\{G^1, G'\} \in E(\mathcal{X})$. Since $|Ch_G(y)| = 2$, G' must be degree 2 vertex in $\mathcal{X}$.

Case 2: y is a leaf. Let $y_1 = Pa_G(y)$. Let $G^1 := SPR_G(v, y_1)$, thus $G^1 = NNI_{G'}(v)$. Hence $\{G^1, G'\} \in E(\mathcal{X})$, and consequently, G' is degree 1 vertex in $\mathcal{X}$.

Case 3: y is an internal vertex. Let $y_1 = Pa_G(y)$ and $y_2 \in Ch_G(y)$. Let $G^1 := SPR_G(v, y_1)$, thus $G^1 = NNI_{G'}(v)$. Let $G^2 := SPR_G(v, y_2)$, thus $G' = NNI_{G^2}(v)$. Since y has one parent and two children in G, thus G' is degree 3 vertex in $\mathcal{X}$. □

3.2   Characterizing Duplications

To solve the R-SEC problem we traverse tree $\mathcal{X}$. Two adjacent trees in $V(\mathcal{X})$ are one NNI operation apart. We show that duplication score of a tree can be computed in constant time from the LCA computation of its adjacent tree.

Let $e := (G', G'')$ be an edge in $\mathcal{X}$. Let $x := Pa(v)$, $y := Sb(v)$, and $z, z' \in Ch(y)$ in G' (see Fig. 1(b)). Without loss of generality, let $G'' := NNI_{G'}(z)$. (Observe G'' is similar to G'_r of Fig. 1(b).)

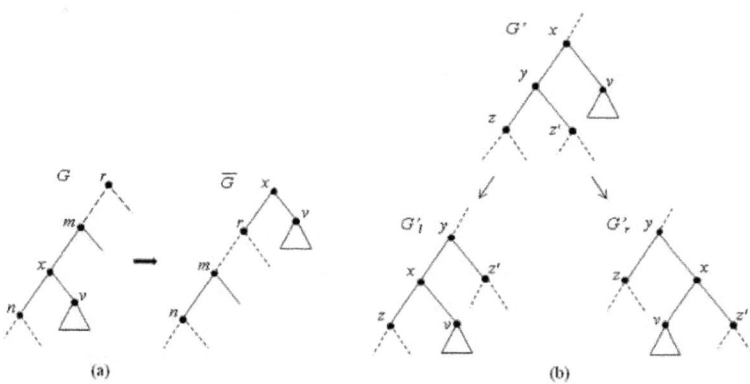

Fig. 1. (a) The tree $\overline{G}$ is obtained from G by pruning and regrafting subtree G_v to the root of G. The vertex $x \in V(G)$ is suppressed, and the new vertex above root in $\overline{G}$ is named x. (b) Two NNI operations $\text{NNI}_{G'}(z')$ and $\text{NNI}_{G'}(z)$ produce left-child G'_l and right-child G'_r of G' in $\mathcal{X}$.

Lemma 2. $\mathcal{M}_{G'',S}(y) = \mathcal{M}_{G',S}(x)$.

Proof. From NNI operation, $v, z' \in Ch_{G''}(x)$ and $z, x \in Ch_{G''}(y)$. Also, $G'_z \simeq G''_z$, $G'_{z'} \simeq G''_{z'}, G'_v \simeq G''_v$, so $Le(G'_y) = Le(G'_x)$. Thus, $\mathcal{M}_{G',S}(x) = \text{LCA}(\mathcal{L}_{G',S}(Le(G'_x)))$ $= \text{LCA}(\mathcal{L}_{G'',S}(Le(G''_y))) = \mathcal{M}_{G'',S}(y)$. □

Lemma 3. $\mathcal{M}_{G'',S}(w) = \mathcal{M}_{G',S}(w), \text{ for all } w \in V(G') \backslash \{x, y\}$.

Proof. For $g \in V(G'_v) \bigcup V(G'_z) \bigcup V(G'_{z'})$, since $G'_g \simeq G''_g$, therefore $\mathcal{M}_{G',S}(g) = \mathcal{M}_{G'',S}(g)$. Also, except for subtree G'_x, the rest of the tree remains the same in G''_x. Thus by Lemma 2, $\mathcal{M}_{G',S}(Pa_{G'}(x)) = \mathcal{M}_{G'',S}(Pa_{G''}(y))$. Inductively, $\mathcal{M}_{G',S}(g) = \mathcal{M}_{G'',S}(g)$, for all $g \in V(G') \backslash V(G'_x)$. □

Lemma 4. $\mathcal{M}_{G'',S}(x) = \text{LCA}(\mathcal{M}_{G',S}(v), \mathcal{M}_{G',S}(z'))$.

Proof. From Lemma 3, $\mathcal{M}_{G'',S}(v) = \mathcal{M}_{G',S}(v)$ and $\mathcal{M}_{G'',S}(z') = \mathcal{M}_{G',S}(z')$. Thus, $\mathcal{M}_{G'',S}(x) = \text{LCA}(\mathcal{M}_{G'',S}(v), \mathcal{M}_{G'',S}(z')) = \text{LCA}(\mathcal{M}_{G',S}(v), \mathcal{M}_{G',S}(z'))$. □

Lemma 5. $\Delta(G'', S, g) = \Delta(G', S, g), \text{ for all } g \in V(G'') \backslash \{x, y\}$.

Proof. The gene duplication status of a vertex in G' can change in G'' if its mapping or mapping of any of its children changes in $\mathcal{M}_{G'',S}$. From Lemma 3, and also, since $\mathcal{M}_{G'',S}(w) = \mathcal{M}_{G',S}(w)$, for $w \in Ch(Pa_{G'}(x))$, must have $\Delta(G'', S, Pa_{G'}(x)) = \Delta(G', S, Pa_{G'}(x))$. Thus the Lemma follows. □

Now, we define score $\Delta_e := \Delta(G'', S) - \Delta(G', S)$, for $e := (G', G'') \in E(\mathcal{X})$ and the given species tree S. Observe that this score can be negative too. We study how Δ_e can be computed efficiently for each edge e in $\mathcal{X}$.

Theorem 2. $\Delta_e = \sum_{g \in \{x, y\}} (\Delta(G'', S, g) - \Delta(G', S, g))$.

Proof. $\Delta_e = \Delta(G'', S) - \Delta(G', S) = \sum_{g \in V(G'')} (\Delta(G'', S, g) - \Delta(G', S, g))$

$$= \sum_{g \in V(G'') \setminus \{x,y\}} (\Delta(G'', S, g) - \Delta(G', S, g)) + \sum_{g \in \{x,y\}} (\Delta(G'', S, g) - \Delta(G', S, g))$$

$$= \sum_{g \in \{x,y\}} (\Delta(G'', S, g) - \Delta(G', S, g)) \qquad \square$$

Definition 10. *Let* $\overline{G} := SPR_G(v, Ro(G))$, *and let* $P_{G'}$ *be a path from* $\overline{G}$ *to* G' *in* $\mathcal{X}$. *For* G', *we define the* score-difference $\Delta_{\overline{G}, G'}$ *as* $\Delta_{\overline{G}, G'} := \sum_{e \in E(P_{G'})} \Delta_e$.

Theorem 3. *For given* S, G, *and* $v \in V(G)$, *the tree* $G' \in V(\mathcal{X})$ *is the output of R-SEC problem iff* $\Delta_{\overline{G}, G'} = min_{G'' \in V(\mathcal{X})} \Delta_{\overline{G}, G''}$.

Proof. Let $\Delta_{\overline{G}, G'} = min_{G'' \in V(\mathcal{X})} \Delta_{\overline{G}, G''}$. We prove that G' is the output of R-SEC problem. Since $\Delta_{\overline{G}, G'} = \sum_{e \in E(P_{G'})} \Delta_e = \Delta(G', S) - \Delta(\overline{G}, S)$, thus G' gives the minimum normalized duplication score over all trees in $V(\mathcal{X})$. Hence, G' must be the output of R-SEC problem. The other direction follows similarly. $\qquad \square$

3.3   The Algorithm

Following Lemmas 2-5 and Theorems 2-3, we now present our algorithm to solve the R-SEC problem. In our algorithm, we first regraft the subtree G_v at $Ro(G)$ and call it $\overline{G}$. We then compute LCA mapping and duplication score for $\overline{G}$. Further, we traverse the subtree of $\overline{G}$ rooted at sibling of v. For each traversed vertex k, we obtain the tree $G'' := SPR_G(v, k)$. The LCA mapping and duplication score for G'' can be computed in constant time from the LCA mapping of the tree $G' := SPR_G(v, k')$, where vertex k' was traversed right before k. Among all trees generated by regrafting G_v, the tree with the minimum normalized duplication score is returned as output. The algorithm is called Algo-R-SEC, and its description appears as Algorithm 1.

Lemma 6. *The R-SEC problem is correctly solved by Algo-R-SEC.*

Proof. It follows from Lemma 2-5 and Theorem 2-3. $\qquad \square$

Lemma 7. *The R-SEC and SEC problems can be solved in* $\Theta(n)$ *and* $\Theta(n^2)$ *time, respectively.*

Proof. In Algo-R-SEC, step 1 takes constant time. Step 2 precomputes LCA values for species tree [3], and so, finds LCA mapping and computes duplication score in $\Theta(n)$ time. Step 3 and 4 take constant time. The loop of step 5 runs for $\Theta(n)$ time. Inside the loop, step 10 and 17 run for constant time using pre-computed LCA values from step 2, resulting the execution of step 6-19 to take constant time. Hence, the R-SEC problem can be solved in $\Theta(n)$ time. From Observation 1, Algo-R-SEC is called $\Theta(n)$ times to solve SEC problem. Thus, the SEC problem can be solved in $\Theta(n^2)$ time. $\qquad \square$

Algorithm 1. Algo-R-SEC

Input: A gene tree G, a species tree S, and $v \in V(G)$
Output: A tree $G^* \in SPR_G(v)$ such that $\Delta(G^*, S) = \min_{G' \in \text{SPR}_G(v)} \Delta(G', S)$
1. Find $\overline{G}$ by pruning G_v and regrafting at Ro(G)
2. Compute $\mathcal{M}_{\overline{G}, S}$ and $\Delta(\overline{G}, S)$
3. Set BestTree := $\overline{G}$ and BestScore := 0
4. Set $G' := \overline{G}$, $\mathcal{M}_{G', S} := \mathcal{M}_{\overline{G}, S}$, $\Delta(G', S) := \Delta(\overline{G}, S)$, and $\Delta_{\overline{G}, G'} := 0$
5. **for** each node $k \neq Ro(\overline{G}_{Sb(v)})$ in preorder traversal of $\overline{G}_{Sb(v)}$ **do**
6.   **if** not backtracking, **then**
7.   Set $x := Pa_{G'}(v)$, $y := Sb_{G'}(v)$
8.   Compute $G'' := \text{NNI}_{G'}(Sb_{G'}(k))$
9.   Set $\mathcal{M}_{G'', S} := \mathcal{M}_{G', S}$ and $\mathcal{M}_{G'', S}(y) := \mathcal{M}_{G', S}(x)$
10. $\mathcal{M}_{G'', S}(x) := LCA(\mathcal{M}_{G', S}(k), \mathcal{M}_{G', S}(v))$
11. $\Delta_{\overline{G}, G''} := \Delta_{\overline{G}, G'} + \Delta_{(G', G'')}$
12.   **if** $\Delta_{\overline{G}, G''} <$ BestScore, **then** BestTree := G'', BestScore := $\Delta_{\overline{G}, G''}$
13.   **else,**
14.   Set $x := Pa_{G'}(v)$ and $y := Pa_{G'}(x)$
15.   Compute $G'' := \text{NNI}_{G'}(v)$
16.   Set $\mathcal{M}_{G'', S} := \mathcal{M}_{G', S}$, $\mathcal{M}_{G'', S}(x) := \mathcal{M}_{G', S}(y)$
17. $\mathcal{M}_{G'', S}(y) := LCA(\mathcal{M}_{G', S}(Sb_{G'}(v)), \mathcal{M}_{G', S}(k))$
18. $\Delta_{\overline{G}, G''} := \Delta_{\overline{G}, G'} - \Delta_{(G'', G')}$
19.   Set $G' := G''$, $\mathcal{M}_{G', S} := \mathcal{M}_{G'', S}$, $\Delta_{\overline{G}, G'} := \Delta_{\overline{G}, G''}$
20. **return** BestTree

---

4   Solving the TEC Problem

We extend our solution for the SEC problem to solve the TEC problem. A TBR operation can be viewed as an SPR operation except that the pruned subtree can be rerooted before it is regrafted. We define the R-TEC problem for the TEC Problem, as we defined the R-SEC problem for the SEC problem. We will show that the R-TEC problem can be solved by solving two smaller problems separately and combining their solutions.

Definition 11. *Let T be a tree and $x \in V(T)$. $RR(T, x)$ is defined to be the tree T, if $x = Ro(T)$ or $x \in Ch(Ro(T))$. Otherwise, $RR(T, x)$ is the tree obtained by suppressing $Ro(T)$, and subdividing the edge $(Pa(x), x)$ by the new root node.*

Lemma 8. *Given a tuple $\langle G, S, v \rangle$, and $G'' := TBR_G(v, x, y)$, for $x \in V(G_v)$, $y \in V(G) \backslash V(G_v)$. Then, $\Delta(G'', S) \leq_{G' \in TBR_G(v)} \Delta(G', S)$ iff $\Delta(RR(G_v, x), S) \leq_{x' \in V(G_v)} \Delta(RR(G_v, x'), S)$ and $\Delta(G'', S) \leq_{G' \in TBR_G(v, x)} \Delta(G', S)$.*

Proof. ($\Rightarrow$) Let $G^1 := TBR_G(v, x_1, y)$, for $x_1 \in V(G_v)$, and $x_1 \neq x$. Now observe that, $\forall g \in V(G) \backslash V(G_v)$, $\Delta(G'', S, g) = \Delta(G^1, S, g)$. Also, let $G^2 := TBR_G(v, x, y_1)$, for $y_1 \in V(G) \backslash V(G_v)$, and $y_1 \neq y$. Observe that, $\forall g \in V(G_v)$, $\Delta(G'', S, g) = \Delta(G^2, S, g)$. Thus, if G'' gives the minimum duplication score among all trees in $\text{TBR}_G(v)$, then the score contribution of vertices in $V(G_v)$ and $V(G) \backslash V(G_v)$ is independent. Now looking at vertices of G, the best score is achieved when G_v is rooted at x, i.e. $\Delta(RR(G_v, x), S) \leq_{x' \in V(G_v)} \Delta(RR(G_v, x'), S)$; also the best score is achieved when $RR(G_v, x)$ is regrafted at y, i.e., $\Delta(G'', S) \leq_{G' \in \text{TBR}_G(v,x)} \Delta(G', S)$. ($\Leftarrow$) This follows similarly.   □

Lemma 8 implies that a tree in $\text{TBR}_G(v)$ with the minimum duplication cost can be obtained by optimizing the rooting for the pruned subtree, and the regraft location, separately. A best rooting for the pruned subtree is linear time computable [16], and the solution to the R-SEC problem identifies a best regraft location in $\Theta(n)$ time. This allows to obtain a tree in $\text{TBR}_G(v)$ with the minimum duplication cost by evaluating only $\Theta(n)$ trees. Thus the R-TEC problem can be solved in $\Theta(n)$ time. The TEC problem can be solved by calling the solution of R-TEC problem $\Theta(n)$ times, and Theorem 4 follows.

Theorem 4. *The TEC problem can be solved in $\Theta(n^2)$ time.*

5   Experimental Results

We tested the performance of the gene tree rearrangement algorithms on a set of 1201 plant gene trees. Specifically, we wanted to examine how often and how much a single SPR rearrangement in the gene tree reduces the gene duplication score. We first downloaded sequences from the gene families from GreenPhyl, an online plant comparative genomic database [25]. We chose only gene families that had at least one sequence from all of the 13 land plant species included in the database. They each contained between 15 and 983 sequences (mean = 62.7; median = 47). We aligned the amino acid sequences from each gene family using MUSCLE [13] and a performed maximum likelihood (ML) phylogenetic analysis on each gene alignment using RAxML-VI-HPC version 7.0.4 [28]. We identified a root for each ML tree that implied the fewest gene duplications based on an accepted tree of the land plant relationships. The resulting 1201 rooted gene family trees implied between 5 and 667 gene duplications to reconcile with the species tree. The SPR gene tree rearrangement algorithm identified a new gene tree topology with a lower reconciliation cost (implied duplications) for all 1201 gene trees. The rearrangements reduced the reconciliation cost by between 1 and 6 duplications (ave. 2.2), which corresponded to between a 0.4% to 33% reduction in estimated duplications.

We also implemented a method to use the gene rearrangement algorithm to correct for gene tree error in gene tree parsimony phylogenetic analyses. We first took a collection of input gene trees and performed a SPR species tree search using Duptree [30]. After finding the locally optimal species tree, we used our SPR gene tree rearrangement algorithm to find gene tree topologies with a lower duplication cost. We then performed another SPR species tree search using Duptree, starting from the locally optimal species tree and using the new gene tree topologies. This search strategy is similar to re-rooting protocol in Duptree, which checks for better gene tree roots after a SPR species tree search [8,30]. We used this protocol on data set of 6084 genes (with a combined 81,525 leaves) from 14 seed plant taxa. This is the same data set used by [8], except that all gene tree clades containing sequences from a single species were collapsed to a single leaf. Our original SPR tree search found a species tree with 23,500 duplications. The SPR tree search after the gene tree rearrangements identified the same species tree, but the new gene trees had a reconciliation cost of only 18,213, a 22.5% reduction. The species tree was consistent with accepted seed plant phylogenetic hypotheses [8]. This tree search protocol took just under 4 hours on a Mac Powerbook with a 2 GHz Intel Core 2 Duo processor and 2 GB memory.

6 Conclusion

Gene tree - species tree reconciliation offers a powerful approach to study the patterns, processes, and effects of gene and genome evolution. Yet it can be thwarted by the error that is an inherent part of gene tree inference. Any reliable method for GT - ST reconciliation must account for gene tree error, and any useful method must be computationally tractable for large-scale genomic data. We introduce fast and effective algorithms to correct error in the gene trees based on the gene duplication problem. These algorithms, based on SPR and TBR rearrangements, greatly extend upon the range of possible errors in the gene tree from existing algorithms [10,12], while remaining fast enough to use on data sets with thousands of genes. Our analyses on 1201 plant gene trees demonstrates not only the feasibility of applying the algorithm to large-scale data but also the ubiquity of gene tree error, which necessitates an error correction protocol. The SPR gene rearrangement algorithm reduced the duplication cost in *every single gene tree*. The gene tree parsimony analysis emphasized how much gene tree error likely inflates the duplication scores while providing one possible protocol to address this.

While the results of the experiments are promising, they also suggest several directions for future research. First, further investigation is needed to characterize the effects of error on gene tree topologies. For example, it seems likely that in many cases gene tree errors may extend beyond a single SPR or TBR neighborhood; yet if we allow unlimited rearrangements, the gene trees will simply converge on the species tree topology. One simple improvement may be to weight the possible gene tree rearrangements based on support for different clades in the gene tree. Thus, well-supported clades may be rarely or never be subject to rearrangement, while poorly supported clades may be subject to extensive rearrangements. Finally, these approaches implicitly assume that all differences between gene trees and species trees are due to either duplications or errors. Future work will also incorporate other evolutionary processes that cause discord among trees, including gene losses, coalescence, and lateral transfer.

Acknowledgements. This work was conducted in parts with support from the Gene Tree Reconciliation Working Group at NIMBioS through NSF award EF-0832858, with additional support from the University of Tennessee. R.C. and O.E. were supported in parts by NSF awards #0830012 and #10117189.

References

1. Allen, B.L., Steel, M.: Subtree transfer operations and their induced metrics on evolutionary trees. Annals of Combinatorics 5, 1–13 (2001)
2. Arvestad, L., Berglund, A., Lagergren, J., Sennblad, B.: Gene tree reconstruction and orthology analysis based on an integrated model for duplications and sequence evolution. In: RECOMB, pp. 326–335 (2004)

3. Bender, M.A., Farach-Colton, M.: The LCA problem revisited. In: Gonnet, G.H., Viola, A. (eds.) LATIN 2000. LNCS, vol. 1776, pp. 88–94. Springer, Heidelberg (2000)
4. Berglund-Sonnhammer, A., Steffansson, P., Betts, M.J., Liberles, D.A.: Optimal gene trees from sequences and species trees using a soft interpretation of parsimony. Journal of Molecular Evolution 63, 240–250 (2006)
5. Bordewich, M., Semple, C.: On the computational complexity of the rooted subtree prune and regraft distance. Annals of Combinatorics 8, 409–423 (2004)
6. Burleigh, J.G., Bansal, M.S., Eulenstein, O., Hartmann, S., Wehe, A., Vision, T.J.: Genome-scale phylogenetics: inferring the plant tree of life from 18,896 discordant gene trees. Systematic Biology 60(2), 117–125 (2011)
7. Burleigh, J.G., Bansal, M.S., Wehe, A., Eulenstein, O.: Locating large-scale gene duplication events through reconciled trees: Implications for identifying ancient polyploidy events in plants. Journal of Computational Biology 16, 1071–1083 (2009)
8. Chang, W., Burleigh, J.G., Fernández-Baca, D., Eulenstein, O.: An ILP solution for the gene duplication problem. BMC Bioinformatics 12(Suppl 1), S14 (2011)
9. Chang, W., Eulenstein, O.: Reconciling gene trees with apparent polytomies. In: Chen, D.Z., Lee, D.T. (eds.) COCOON 2006. LNCS, vol. 4112, pp. 235–244. Springer, Heidelberg (2006)
10. Chen, K., Durand, D., Farach-Colton, M.: Notung: a program for dating gene duplications and optimizing gene family trees. Journal of Computational Biology 7, 429–447 (2000)
11. Cotton, J.A., Page, R.D.M.: Going nuclear: gene family evolution and vertebrate phylogeny reconciled. P. Roy. Soc. Lond. B Biol. 269, 1555–1561 (2002)
12. Durand, D., Halldórsson, B.V., Vernot, B.: A hybrid micro-macroevolutionary approach to gene tree reconstruction. Journal of Computational Biology 13(2), 320–335 (2006)
13. Edgar, R.C.: MUSCLE: multiple sequence alignment with high accuracy and high throughput. Nucleic Acids Research 32, 1792–1797 (2004)
14. Eulenstein, O.: Predictions of gene-duplications and their phylogenetic development, Ph.D. thesis, University of Bonn, Germany, 1998, GMD Research Series No. 20 / 1998 (1998) ISSN: 1435-2699
15. Goodman, M., Czelusniak, J., Moore, G.W., Romero-Herrera, A.E., Matsuda, G.: Fitting the gene lineage into its species lineage. a parsimony strategy illustrated by cladograms constructed from globin sequences. Systematic Zoology 28, 132–163 (1979)
16. Górecki, P., Tiuryn, J.: Inferring phylogeny from whole genomes. In: ECCB (Supplement of Bioinformatics), pp. 116–122 (2006)
17. Guigó, R., Muchnik, I., Smith, T.F.: Reconstruction of ancient molecular phylogeny. Molecular Phylogenetics and Evolution 6(2), 189–213 (1996)
18. Hahn, M.W.: Bias in phylogenetic tree reconciliation methods: implications for vertebrate genome evolution. Genome Biology 8, R141 (2007)
19. Huang, H., Knowles, L.L.: What is the danger of the anomaly zone for empirical phylogenetics? Systematic Biology 58, 527–536 (2009)
20. Joly, S., Bruneau, A.: Measuring branch support in species trees obtained by gene tree parsimony. Systematic Biology 58, 100–113 (2009)
21. Maddison, W.P.: Gene trees in species trees. Systematic Biology 46, 523–536 (1997)
22. Page, R.D.M.: Maps between trees and cladistic analysis of historical associations among genes, organisms, and areas. Systematic Biology 43(1), 58–77 (1994)
23. Page, R.D.M., Charleston, M.A.: From gene to organismal phylogeny: reconciled trees and the gene tree/species tree problem. Molec. Phyl. and Evol. 7, 231–240 (1997)
24. Rasmussen, M.D., Kellis, M.: A bayesian approach for fast and accurate gene tree reconstruction. Molecular Biology and Evolution 28, 273–290 (2011)
25. Rouard, M., Guignon, V., Aluome, C., Laporte, M., Droc, G., Walde, C., Zmasek, C.M., Périn, C., Conte, M.G.: Greenphyldb v2.0: comparative and functional genomics in plants. Nucleic Acids Research 39, D1095–D1102 (2010)

26. Sanderson, M.J., McMahon, M.M.: Inferring angiosperm phylogeny from EST data with widespread gene duplication. BMC Evolutionary Biology 7(suppl 1), S3 (2007)
27. Slowinski, J.B., Knight, A., Rooney, A.P.: Inferring species trees from gene trees: A phylogenetic analysis of the elapidae (serpentes) based on the amino acid sequences of venom proteins. Molecular Phylogenetics and Evolution 8, 349–362 (1997)
28. Stamatakis, A.: RAxML-VI-HPC: maximum likelihood-based phylogenetic analyses with thousands of taxa and mixed models. Bioinformatics 22(21), 2688–2690 (2006)
29. Vernot, B., Stolzer, M., Goldman, A., Durand, D.: Reconciliation with non-binary species trees. Computational Systems Bioinformatics 53, 441–452 (2007)
30. Wehe, A., Bansal, M.S., Burleigh, J.G., Eulenstein, O.: Duptree: a program for large-scale phylogenetic analyses using gene tree parsimony. Bioinformatics 24(13) (2008)
31. Zhang, L.: On a Mirkin-Muchnik-Smith conjecture for comparing molecular phylogenies. Journal of Computational Biology 4(2), 177–187 (1997)

TransDomain: A Transitive Domain-Based Method in Protein–Protein Interaction Prediction

Yi-Tsung Tang and Hung-Yu Kao

Department of Computer Science and Information Engineering,
National Cheng Kung University, Tainan, Taiwan
{p7895125,hykao}@mail.ncku.edu.tw

Abstract. The prediction of new protein–protein interactions is important due to many unknown functions of biological pathways. In addition, many protein–protein interaction databases contain different types of protein interactions, i.e., protein associations, physical protein associations and direct protein interactions. Moreover, discovering new crucial protein–protein interactions through biological experiments is still difficult. Therefore, there is increasing demand to discover not only protein associations but also direct protein interactions. Many studies have predicted protein–protein interactions by directly using biological features, such as Gene Ontology (GO) functions and domains of protein structure between two interacting proteins. In this article, we propose TransDomain, a new method of predicting potential protein–protein interactions by using a new strong transitive relationship between interacting protein domains. Our results demonstrate that TransDomain can effectively predict potential protein–protein interactions from existing identified protein interaction relationships. TransDomain achieved 90% precision rate and 91% accuracy in the prediction of all types of protein–protein interactions and outperformed the existing PPI prediction systems and simulated GO-based prediction methods.

Keywords: protein-protein interaction, transitive relationship, protein domain.

1 Introduction

Protein–protein interactions (PPIs) play a crucial role in most biological processes, and their identification is key to understanding global pathways. New PPIs are identified by pull-down experiments or other protein structure techniques. However, these experiments are usually too slow and expensive to identified new PPIs. Due to this reason, many previous works have proposed prediction methods to discover new PPIs from non-structure data, such as biological literature, biological ontology and protein domains.

The biological literature contains a diversity of PPIs, and various methods have been proposed to determine PPIs. Moreover, several bioinformatic tools have been developed to analyze gene and protein interactions from scientific articles [1]. For example, Protein Interaction information Extraction (PIE) methods used natural language processing and machine learning techniques to extract sentences containing PPIs from scientific articles [1]. Some methods use word patterns instead of

J. Chen, J. Wang, and A. Zelikovsky (Eds.): ISBRA 2011, LNBI 6674, pp. 240–252, 2011.

complicated natural language processing techniques to extract PPIs from scientific articles [2, 3]. Using patterns to discover specific biological information is a common and useful technique.

There are several public online resources that store identified PPIs, such as HPRD [4], DIP [5], IntAct [6], BioGRID [7], MINT [8] and Reactome [9]. These databases provide lists of several kinds of PPIs, including direct and physical interactions etc. Many of these databases contain protein interactions that are collected from not only biological techniques and experiments but also from scientific articles.

Each database has its own online interface to search identified PPIs from back-end databases. Searching for more interaction partners is one problem within PPI searches. A previous method proposed an ontology-based search engine, that is based on Gene Ontology (GO) hierarchy relationships, to discover PPIs [10, 11]. Extracting matches and extended matches from protein interaction database, this ontology-based method get more terms or concepts, that related to the query protein, for finding more protein interactions.

The prediction of PPIs remains an important issue in bioinformatics due to the large numbers of undiscovered protein interactions. The STRING database contains a text-mining technique based on two proteins described together in the literature [12, 13]. PPI Finder is an online tool for predicting unidentified PPIs of human proteins [14]. In PPI Finder, the IE module extracts candidate interactions from scientific articles and the IR module predicts positive PPI pairs using the GO sharing method. In this study, two proteins sharing common functions were hypothesized to have a higher probability of interaction. GO annotation can also be applied to other GO-based predicting methods [15]. This method focuses on two GOs (cellular component and biological process). This method was based on the hypothesis that a protein pair should have the same protein location and similar biological function.

In previous study, a combination of several orthogonal protein features within a probabilistic framework was proposed to increase the coverage of the human interaction map [16]. In this framework, a novel scoring function for local network topology, named "*Transitive Module*", was also investigated. This topology feature greatly enhanced the predictions. The transitive module investigates whether two proteins that share many common interactors and have few additional interactors that are not common to both proteins are more likely to interact than two proteins that share few common interactors. Based on this hypothesis, in this paper we evolved a new relationship of two proteins, called *the strong transitive relationship*, to inference the predicable interaction between two proteins with many common interactors. As shown in Fig. 1, there are two target proteins, A and C, their common interactors, $B_1 \sim B_n$, and known interactions between these proteins in this relationship. If these interactions between A, C and their common interactors can support to inference the interaction of A and C, we call this relationship a strong transitive relationship. We hypothesized that the use of the strong transitive relationship among two interaction proteins and their common interactors can make up for a deficiency of the prediction performance of the GO-based prediction methods.

Because the protein domain is a crucial factor of the protein interaction, protein domain is key to find pair domain relationship for PPIs prediction. Consider the example in Fig. 1. ER and p65 are known to interact with each other and MYC has not been identified to interact with ER. An example of the interaction between ER and

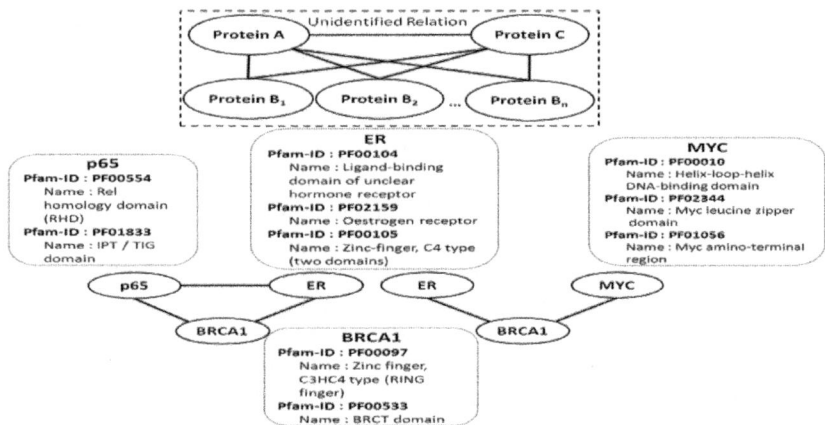

Fig. 1. Example of the strong transitive relationship

BRCA1, ER consists of Zinc-finger domain and Ligand-binding domain and BRCA1 also contains Zinc-finger domain. Due to the obvious relationships between common interactor and related protein domains, the interaction of ER and p65 can be identified. This is because that the relationship of Zinc-finger domain and Ligand-binding domain is a crucial factor within the transitive relationship for predicting protein interaction between ER and p65. In addition, an interaction between ER and MYC cannot be identified by their domains and the pair domain relationship because the pair domain relationship between these two proteins is not strong for protein interaction. Therefore, considering only pair domain relationship between a protein pair is also not sufficient for predicting protein interaction consequently. Nevertheless, the domains of a common interactor, BRCA1 between ER and MYC can support the pair domain relationship for the protein interaction prediction. For example, BRCA1 is known to interact with MYC and BRCA1 and ER also the identified interaction protein pair. The domain relationship among ER, BRCA1 and MYC can be extended by the pair domain relationship for protein interaction prediction because Zinc-finger domain and Ligand-binding domain are related domains in the protein interaction. The domain relationship in the transitive relationship can also support the protein interaction prediction with few common interactors. Therefore, we hypothesized that the use of relationships among protein domains can make up for a deficiency of fewer common interactors. As a result, the potential interactions including the relationships of hub-proteins and non-hub proteins will be possibly predictable by our proposed method. We aimed to use the common interactors of a potential protein pair and the relationships of their protein domains in the transitive relationship to better predict unidentified PPIs.

2   Method

2.1   Overall Architecture of TransDomain

The proposed PPI prediction method named TransDomain used the transitive relationship and the domain relationships within transitive relationships. In TransDomain,

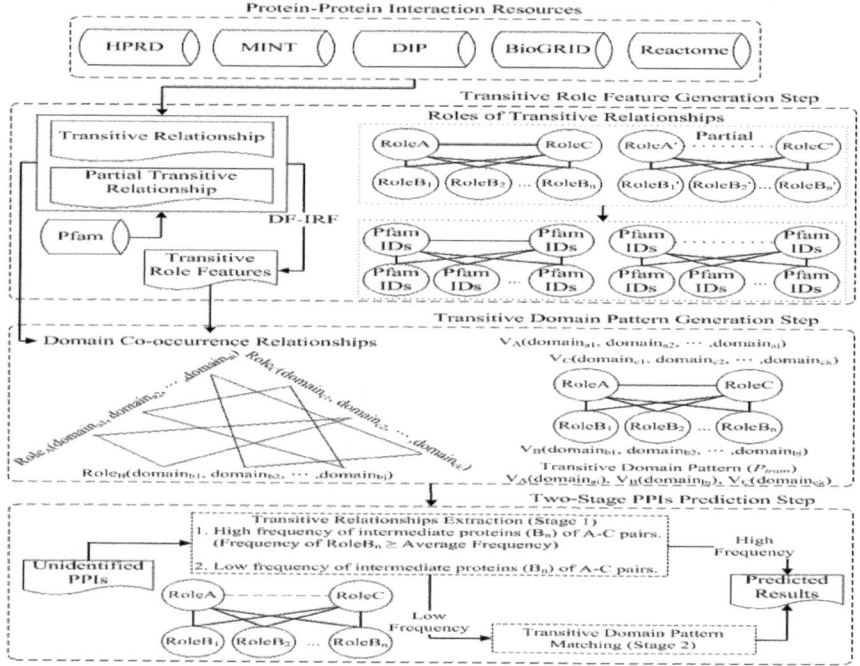

Fig. 2. Overall flowchart of TransDomain

the transitive relationship is an important concept for finding high confidence protein interacting pairs. We also applied domain co-occurrence relationships within transitive relationships to model the strength of the transitive relationship and filter out low-confidence protein interacting pairs. In this paper, for predicting unidentified protein interactions, we use the domain co-occurrence relationships to represent the transitive patterns to predict unidentified protein interactions. The domain co-occurrence patterns were used for predicting PPIs containing less common interactors. Therefore, TransDomain used both transitive relationships and transitive domain patterns to predict new protein interacting pairs.

The overall flowchart of TransDomain is shown in Fig. 2 and is composed of three major steps: the transitive role feature generation step, the transitive domain pattern generation step and two-stage PPIs prediction step.

2.2 Transitive Role Feature Generation Step

In the transitive role feature generation step, six roles (RoleA, RoleB, RoleC, RoleA', RoleB' and RoleC') were defined to represent three important roles in transitive relationships and partial transitive relationships. RoleA and RoleC are defined as two interacting proteins while RoleB is the common interactor of RoleA and RoleC. Two role pairs, i.e., RoleA-RoleB and RoleB-RoleC, are also the identified PPIs. The relationship among RoleA, RoleC and their common interactor RoleB is defined as the transitive relationship. In addition, RoleA' and RoleC' are defined as two non-interacting proteins. The relationship among RoleA', RoleC' and their common interactor, RoleB' is defined as the partial transitive relationship. The partial transitive relationship consisted of two identified PPIs, i.e., RoleA'-RoleB' and RoleB'-RoleC'. The examples of roles in the transitive relationship and the partial transitive relationship are shown in Fig. 3. RoleB and RoleB' are defined as an intermediate role in a transitive relationship. A protein in RoleB and RoleB' should be the intermediate protein between RoleA and RoleC. In addition, three roles (RoleA', RoleB' and RoleC') in partial transitive relationships indicate incomplete transitive relationships with one unidentified protein–protein interaction, such as RoleA'-RoleC' in Fig. 3.

The relationships among RoleA, RoleB and RoleC represent crucial information in transitive relationships. The protein domains in each role are extracted from Pfam database [17] and then used for generating features in the transitive role feature generation module.

We defined a new weighting scheme, *DF-IRF* to find transitive role features of RoleA, RoleB and RoleC. The transitive role feature is defined as a protein domain that plays a crucial role in the transitive relationship. Consider the example in Fig. 3, there are three domains, i.e., 8, 9, and 10, play RoleA in transitive relationship. There are four domains, i.e., 1, 3, 4 and 5, play RoleA in transitive relationship and RoleA' partial transitive relationship. In addition, protein domains 1, 3 and 15 in RoleB and protein domains 4, 7 and 11 in RoleC also play the crucial roles in transitive relationship. The goal of *DF-IRF* weighting scheme is to find the protein domains ((8, 9 and 10 in RoleA), (1, 3 and 15 in RoleB) and (4, 7 and 11 in RoleC)) that play the crucial roles in the transitive relationship.

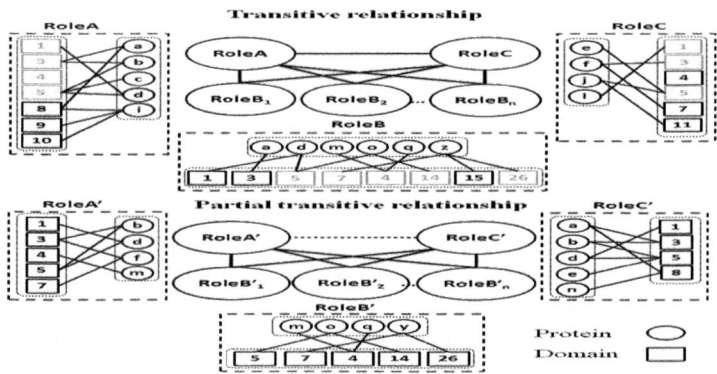

Fig. 3. Example of transitive role feature generation

The relationships between protein domains and three roles from transitive relationships and partial transitive relationships were also considered. Therefore, we download the SwissPfam data from the Pfam website to get Swissprot and Pfam IDs [17]. The transitive role features of RoleA, RoleB and RoleC of the transitive relationship were then extracted by the defined weighting scheme, *DF-IRF*.

The domain frequency (*DF*) is defined as the number of occurrence of protein domains in some role. *DF* of each domain is calculated from transitive relationships of defined training datasets. The domain frequency denotes the number of occurrences of protein domain i in role r for all transitive relationships. $Freq_{i,r}$ is the number of occurrences that domain i appears in role r. The domain frequency is normalized by the total frequency of all domains in role r.

$$DF_{i,r} = \frac{Freq_{i,r}}{\sum_j Freq_{j,r}}$$

In addition, the inverse role frequencies of each domain were calculated from the transitive relationships of the defined training dataset. The inverse role frequency (IRF) is defined as the inverse role frequency and has the equation:

$$IRF_i = \log \frac{N}{R_i}$$

N is the total number of roles in the training dataset and R_i denotes the number of roles that contain domain i. Finally, the weighting score of a domain i in role r of a transitive relationship was calculated by the DF-IRF weighting scheme according to the equation:

$$Weight_{i,j} = DF_{i,r} \times IRF_i$$

The transitive role features of three roles (RoleA, RoleB and RoleC) were generated by calculating DF-IRF weighting scores of domains from the training dataset. A domain with a high weighting score in a role indicates that the domain is a crucial transitive role feature. In the high-weighting score results, the ratio of overlapping domains among the three roles was very low. A few overlapped transitive role features appeared in the high-weighting score results, such as "7 transmembrane receptor (hodopsin family)" because it is a signaling transduction domain in many proteins. Many proteins can play a signaling transduction role in many pathways. In conclusion, the transitive role features produced by the transitive role feature generation are very specific.

In transitive role feature generation, the training dataset is needed for generating transitive role features. Before generate the training dataset, a global answer set of PPIs was generated firstly by integrating five exists PPI databases: HPRD [4, 18], MINT [8], DIP [5], BioGRID [7] and Reactome [9] with Swissprot ID and Uniprot ID. The training datasets were collected from the global answer set. In our experiments, we randomly selected 2,000 transitive relationships and 2,000 partial transitive relationships from the global answer set as the training datasets of TransDomain.

2.3 Transitive Domain Pattern Generation Step

To predict an interaction between two proteins using TransDomain, the transitive domain pattern (P_{trans}) must be generated after the transitive role feature generation step. In transitive domain pattern generation step, the generated transitive role features and defined transitive relationship were used to generate P_{trans}. The number of transitive role features of individual role (RoleA, RoleB and RoleC) in the transitive relationship is 20 due to the ratio of crucial domains. P_{trans} denotes the co-occurrence relationship of domains among RoleA, RoleB and RoleC. Therefore, the co-occurrence frequencies of triple domains were calculated. An example of transitive domain pattern generation step is shown in Fig. 4. Three vectors $V_A(domain_{a1}, domain_{a2}, ..., domain_{ai})$, $V_B(domain_{b1}, domain_{b2}, ..., domain_{bj})$ and $V_C(domain_{c1},$

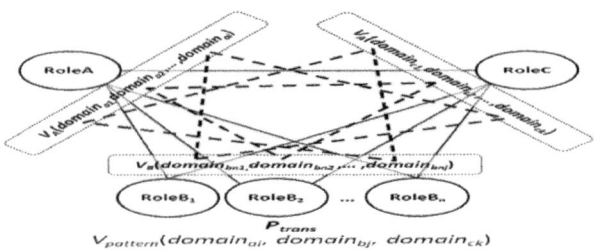

Fig. 4. Example of generating P_{trans}

$domain_{c2}$, ..., $domain_{ck}$) denote the domain vectors of RoleA, RoleB and RoleC, respectively. The triangle graph denotes the co-occurrence frequency of the protein domains with transitive role features among RoleA, RoleB and RoleC sharing the same transitive relationship within the training dataset. P_{trans} can also be defined as the domains co-existing in the same transitive relationship and can be denoted as a vector $V_{Pattern}(domain_{ai}, domain_{bj}, domain_{ck})$.

In our training dataset, the proteins with a 7-transmembrane-receptor domain will have a high probability to play RoleB in the transitive relationship, because these receptors usually play a signaling transduction role in protein interactions.

2.4 Two Stages PPIs Prediction Step

The PPI prediction module of TransDomain has two stages: the transitive relationships extraction stage and the transitive domain pattern matching stage to predict protein interactions.

The transitive relationships extraction stage can determine a protein pair that shares many intermediate proteins. The protein pairs will be predicted as interacting protein pairs according to the high frequency of intermediate proteins in RoleB. An example of the two-stage PPIs prediction process is shown in Fig. 5. The threshold value within the transitive relationships extraction stage is an average frequency, which is defined as the average count of number of unique intermediate proteins (RoleB) from all of the tested protein pairs. If the frequency of unique intermediate proteins (RoleB) of a protein pair is lower than the average frequency, this pair will be filtered out in stage one and put into the feature relationships extraction stage. By contrast, if the frequency of unique intermediate proteins (RoleB) of a protein pair is higher than the average frequency, this pair will be output into the predicted results. A protein pair with a low frequency of unique intermediate proteins (RoleB) is less likely to interact with each other.

In the feature relationships extraction stage, the domains of all of the testing protein pairs and all unique intermediate proteins (RoleB) were found first. A vector of a testing protein pair $V_{protein\_pair}(V_A(domain_{ai}), V_B(domain_{bj}), V_C(domain_{ck}))$ was used to match the P_{trans} vector $V_{pattern}(domain_i, domain_j, domain_k)$. If $V_{protein\_pair}$ of a protein pair matched $V_{pattern}$ in the feature relationship extraction stage, this pair is output into the predicted results. The matching example of feature relationships extraction is shown in Fig. 5. The matching process of feature relationships extraction consists of two matching strategies. In the first matching strategy, each domain in RoleA, RoleB

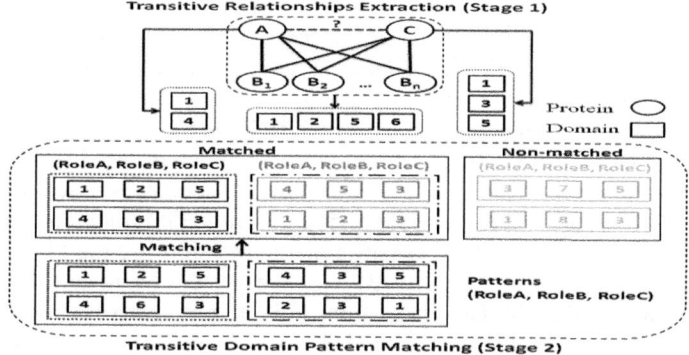

Fig. 5. Example of transitive relationships extraction

and RoleC in the testing protein pair matches domains in RoleA, RoleB and RoleC in learned P_{trans} respectively. Secondly, each domain in RoleA in a testing protein matches RoleB and RoleC in the P_{trans} while the first matching strategy is not exactly matched. RoleB and RoleC have the same processes in the second matching strategy after RoleA had matched a role within RoleB and RoleC in the P_{trans}.

3   Results and Discussions

3.1   Datasets Generation

Two datasets were randomly generated and served as the predefined training datasets to generate the transitive role features and the P_{trans}. Transitive relationships and partial transitive relationships were collected from five exists PPI databases that were described earlier in the training dataset generation. The dataset represented by RoleA, RoleB and RoleC contained 2,000 role pairs of transitive relationships. In addition, 2,000 role pairs of partial transitive relationships were used and are represented by RoleA', RoleB' and RoleC'. Two testing datasets, combined interaction (T_{mix}) and direct interaction only (T_{direct}), were used to evaluate the prediction performance of PPIs by TransDomain. T_{mix} consisted of three interactions types, i.e., direct interaction, physical association and association in positive dataset and non-interaction type in negative dataset. T_{direct} contained only direct interaction and non-direct interaction types in positive datasets and negative datasets respectively. These datasets each contained 200 interacting protein pairs (Positive) and 200 non-interacting protein pairs (Negative).

3.2   Performance Evaluation of Predicting Combined Protein Interactions

Two domain-based baseline methods were used for comparison with TransDomain. The domain-based baseline method one ($Domain_{m1}$) was defined as the comparison of domain similarity between one testing protein and all of the domains of the neighbor proteins of another testing protein. The neighbor protein was defined as proteins that have interaction with one of the testing proteins. In addition, the neighbor proteins of one testing protein not consist of another testing protein. Another domain-based baseline method (Domain$_{m2}$) was defined as the comparison of domain similarity between two testing proteins. The STRING database contains a text mining result that two proteins co-occur in scientific articles [12, 13]. Based on this idea, the co-occurrence concept was used as a simulated method of STRING in this comparison. PPI Finder is a recently developed PPI prediction system [14] and contains a protein pair extraction module and an interacting pair prediction module. PPI Finder is based on two proteins co-occurring in scientific articles and sharing the same GO function terms. We used the co-occurrence concept and GO sharing method to simulate the PPI Finder prediction method. Besides, GO Similarity BP CC is a protein localization- and function-based predicting method by GO annotations [15]. In our comparison, we simulated a GO annotation-based predicting method by two predicting strategies: sharing the same location (same cellular component) and functional similarity (similarity of biological process). The threshold of similarity score was 0.33 due to the coverage of interacting protein pairs. In addition, the threshold of similarity score was defined as 0 and the protein domains of two domain-based baseline methods were annotated by the Pfam database.

In addition, three previous methods, two domain-based baseline methods and TransDomain results were the average from T_{mix}. The comparative results are shown in Table 1. PPI Finder had a better precision rate than STRING because it compared protein functions. GO Similarity BP CC did not have a better precision rate than PPI Finder, even though it using all of the GO annotations to predict PPIs. Therefore, sharing a protein location or biological process had little effect in PPI prediction. Domain$_{B1}$ had a better precision among two domain-based baseline methods. Considering domains with the domains of neighbor proteins of another protein is nearly the biological concept of PPIs prediction. In other words, if a protein pair has a high probability to interact with each other, one protein of the protein pair also has high probability to be a neighbor of another protein of the protein pair. However, all of the domain-based baseline methods not perform good recall rate in T_{mix}. TransDomain had a better performance than the simulated methods and all of the domain-based baseline methods because the transitive relationships combined with the protein domains were successful in predicting PPIs.

Table 1. Performance comparison results in T_{mix}

Methods	Average performance in T_{mix}			
	Precision	Recall	F-measure	Accuracy
STRING	0.61	0.61	0.59	0.56
PPI Finder	0.70	0.52	0.59	0.63
GO Similarity BP CC	0.50	0.92	0.65	0.50
Domain$_{m1}$	0.68	0.14	0.23	0.54
Domain$_{m2}$	0.53	0.29	0.41	0.50
TransDomain	**0.90**	**0.93**	**0.91**	**0.91**

In the top-K precision evaluation, TransDomain was evaluated and compared with three previous methods and two domain-based baseline methods. The performance values were the average value of T_{mix}. The comparison result is shown in Fig. 6.

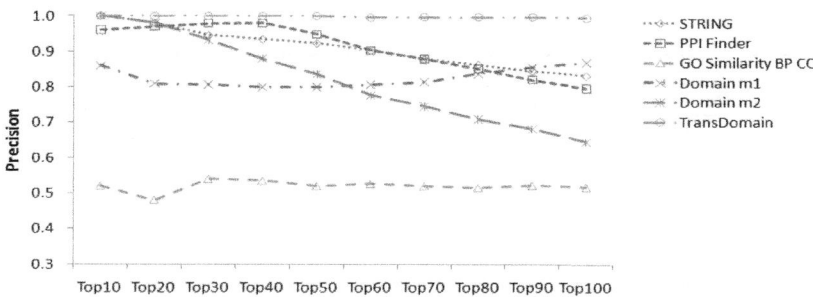

Fig. 6. Top-K average precision comparison results in T_{mix}

STRING and PPI Finder both had good precision rates with the top-100 predicting results. In addition, GO Similarity BP CC had lower precision rates with the top-100 prediction results. In two domain-based baseline methods, the precision rate of Domain$_{m1}$ was around 0.8 in each top-N prediction evaluation and the precision rate of Domain$_{m2}$ decreased while the prediction results increased. The result of Domain$_{B1}$ indicates that the similarity between protein domain and the protein domains of their neighbor proteins is the crucial evidence for protein interaction. In other words, if two proteins interact with each other, one protein of the two proteins is a neighbor protein of another protein of the two proteins.

The result of Domain$_{m2}$ indicates that two proteins will interact with each other while the domain similarity between their neighbor proteins is very high. However, TransDomain performed better than the simulated methods in the top-100 comparison of precision rate because the transitive relationship combined with protein domains can predict not only indirect protein interactions and physical associations but also all other protein interaction types. In addition, the AUC of TransDomain also outperformed simulated predicting methods. The AUC values of TransDomain were 0.891, 0.984, 0.978, 0.965 and 0.969 of T_{mix}.

3.3   Performance Evaluation of Predicting Direct Protein Interactions

Direct protein interactions are actual interactions between two interacting proteins and are crucial in protein interaction predictions. Therefore, the performance of predicting direct protein interactions was evaluated in the following experiments. T_{direct} was used as the performance evaluation dataset. Three previous methods and two domain-based baseline methods were also compared with TransDomain.

The comparison results for predicting direct interacting protein pairs are shown in Table 2. PPI Finder had a better precision rate than STRING due to the comparison of protein function. In addition, GO Similarity BP CC did not have better precision rate than PPI Finder by considering shared locations and biological processes. Domain$_{m1}$ and Domain$_{m2}$ had a lower recall rate than all of the other compared methods. This is because identified domains of two testing proteins and neighbor proteins are not sufficient for using neighbor proteins to predict unidentified direct protein interactions. TransDomain had a better performance than the simulated methods because the transitive relationship combined with protein domains performs well in this evaluation. In conclusion, TransDomain can effectively predict potential interacting protein pairs that are direct interacting protein pairs.

Table 2. Performance comparison results in T_{direct}

Methods	Average performance in T_{direct}			
	Precision	Recall	F-measure	Accuracy
STRING	0.50	0.98	0.66	0.51
PPI Finder	0.60	0.83	0.70	0.64
GO Similarity BP CC	0.50	0.95	0.66	0.50
Domain$_{m1}$	0.80	0.16	0.27	0.56
Domain$_{m2}$	0.58	0.36	0.44	0.55
TransDomain	**0.79**	**0.81**	**0.80**	**0.80**

3.4 Case Studies

In this evaluation, TransDomain was applied to predict PPIs not reported in the literature. Finding protein interactions of the ER receptor protein remains a crucial task in MCF7 cells used in breast cancer studies. Therefore, ER was selected as the query protein and TransDomain was used to predict interacting proteins. We used the HPRD protein interaction database, which contains 39,194 identified PPIs, as the predicting dataset [18].

First, the identified interacting proteins of ESR1 were filtered from the predicting dataset and the remaining protein entries were used as the predicting proteins. Top ten predicted interacting proteins of ER, (NR3C1, RB1, PPARG, STAT3, PPARA, VDR, HDAC1, PML, CSNK2A1 and MYC) were predicted by TransDomain. After evaluating the predicted results by two experts, MYC was selected because it is a crucial protein in breast cancer studies. Zinc finger proteins and tubulin alpha ubiquitous play RoleB have transitive relationships and have been found in pull-down experiments with ER and MYC. Previous studies reported an association between ER and MYC [19] and MYC was found as an interacting protein of ER by pull-down experiments. These results show that TransDomain is an effective method for predicting unidentified PPIs.

4   Conclusion

The developed PPI prediction system TransDomain was able to effectively and precisely predict interacting protein pairs from protein pairs that have not been identified using transitive role features and P_{trans}. These protein pairs were successfully predicted by the transitive role features and P_{trans}. In addition, not only the interactions between hub proteins and it interaction partners but also the interactions between hub proteins and non-hub proteins can be predicted by TransDomain. Our proposed method can also be applied to predict other relationships between biological entities, such as protein–DNA interactions or gene–protein interactions. The predicted results were sorted according to the frequency of transitive relationships, indicating that our method can achieve high accuracies for different kinds of interactions, such as protein associations, physical protein associations or direct protein interactions. In the future, the extended transitive relationship of roles will be considered in detail for the prediction of different types of interactions, such as protein–DNA interactions and gene–protein interactions.

References

1. Kim, S., Shin, S.Y., Lee, I.H., Kim, S.J., Sriram, R., Zhang, B.T.: PIE: an online prediction system for protein-protein interactions from text. Nucleic Acids Res. 36, W411–W415 (2008)
2. Huang, T.W., Tien, A.C., Huang, W.S., Lee, Y.C., Peng, C.L., Tseng, H.H., Kao, C.Y., Huang, C.Y.: POINT: a database for the prediction of protein-protein interactions based on the orthologous interactome. Bioinformatics 20, 3273–3276 (2004)
3. Ono, T., Hishigaki, H., Tanigami, A., Takagi, T.: Automated extraction of information on protein-protein interactions from the biological literature. Bioinformatics 17, 155–161 (2001)
4. Peri, S., Navarro, J.D., Amanchy, R., Kristiansen, T.Z., Jonnalagadda, C.K., Surendranath, V., Niranjan, V., Muthusamy, B., Gandhi, T.K., Gronborg, M., Ibarrola, N., Deshpande, N., Shanker, K., Shivashankar, H.N., Rashmi, B.P., Ramya, M.A., Zhao, Z., Chandrika, K.N., Padma, N., Harsha, H.C., Yatish, A.J., Kavitha, M.P., Menezes, M., Choudhury, D.R., Suresh, S., Ghosh, N., Saravana, R., Chandran, S., Krishna, S., Joy, M., Anand, S.K., Madavan, V., Joseph, A., Wong, G.W., Schiemann, W.P., Constantinescu, S.N., Huang, L., Khosravi-Far, R., Steen, H., Tewari, M., Ghaffari, S., Blobe, G.C., Dang, C.V., Garcia, J.G., Pevsner, J., Jensen, O.N., Roepstorff, P., Deshpande, K.S., Chinnaiyan, A.M., Hamosh, A., Chakravarti, A., Pandey, A.: Development of human protein reference database as an initial platform for approaching systems biology in humans. Genome Res. 13, 2363–2371 (2003)
5. Salwinski, L., Miller, C.S., Smith, A.J., Pettit, F.K., Bowie, J.U., Eisenberg, D.: The Database of Interacting Proteins: 2004 update. Nucleic Acids Res. 32, D449–D451 (2004)
6. Kerrien, S., Alam-Faruque, Y., Aranda, B., Bancarz, I., Bridge, A., Derow, C., Dimmer, E., Feuermann, M., Friedrichsen, A., Huntley, R., Kohler, C., Khadake, J., Leroy, C., Liban, A., Lieftink, C., Montecchi-Palazzi, L., Orchard, S., Risse, J., Robbe, K., Roechert, B., Thorneycroft, D., Zhang, Y., Apweiler, R., Hermjakob, H.: IntAct–open source resource for molecular interaction data. Nucleic Acids Res. 35, D561–D565 (2007)
7. Breitkreutz, B.J., Stark, C., Reguly, T., Boucher, L., Breitkreutz, A., Livstone, M., Oughtred, R., Lackner, D.H., Bahler, J., Wood, V., Dolinski, K., Tyers, M.: The BioGRID Interaction Database: 2008 update. Nucleic Acids Res. 36, D637–D640 (2008)
8. Chatr-aryamontri, A., Ceol, A., Palazzi, L.M., Nardelli, G., Schneider, M.V., Castagnoli, L., Cesareni, G.: MINT: the Molecular INTeraction database. Nucleic Acids Res. 35, D572–D574 (2007)
9. Vastrik, I., D'Eustachio, P., Schmidt, E., Gopinath, G., Croft, D., de Bono, B., Gillespie, M., Jassal, B., Lewis, S., Matthews, L., Wu, G., Birney, E., Stein, L.: Reactome: a knowledge base of biologic pathways and processes. Genome Biol. 8, R39 (2007)
10. Ashburner, M., Ball, C.A., Blake, J.A., Botstein, D., Butler, H., Cherry, J.M., Davis, A.P., Dolinski, K., Dwight, S.S., Eppig, J.T., Harris, M.A., Hill, D.P., Issel-Tarver, L., Kasarskis, A., Lewis, S., Matese, J.C., Richardson, J.E., Ringwald, M., Rubin, G.M., Sherlock, G.: Gene ontology: tool for the unification of biology. The Gene Ontology Consortium. Nat. Genet. 25, 25–29 (2000)
11. Park, B., Han, K.: An ontology-based search engine for protein-protein interactions. BMC Bioinformatics 11(Suppl 1), S23 (2010)
12. von Mering, C., Huynen, M., Jaeggi, D., Schmidt, S., Bork, P., Snel, B.: STRING: a database of predicted functional associations between proteins. Nucleic Acids Res. 31, 258–261 (2003)

13. von Mering, C., Jensen, L.J., Kuhn, M., Chaffron, S., Doerks, T., Kruger, B., Snel, B., Bork, P.: STRING 7–recent developments in the integration and prediction of protein interactions. Nucleic Acids Res. 35, D358–D362 (2007)

14. He, M., Wang, Y., Li, W.: PPI finder: a mining tool for human protein-protein interactions. PLoS One 4, e4554 (2009)

15. De Bodt, S., Proost, S., Vandepoele, K., Rouze, P., Van de Peer, Y.: Predicting protein-protein interactions in Arabidopsis thaliana through integration of orthology, gene ontology and co-expression. BMC Genomics 10, 288 (2009)

16. Scott, M.S., Barton, G.J.: Probabilistic prediction and ranking of human protein-protein interactions. BMC Bioinformatics 8, 239 (2007)

17. Finn, R.D., Mistry, J., Tate, J., Coggill, P., Heger, A., Pollington, J.E., Gavin, O.L., Gunasekaran, P., Ceric, G., Forslund, K., Holm, L., Sonnhammer, E.L., Eddy, S.R., Bateman, A.: The Pfam protein families database. Nucleic Acids Res. 38, D211–D222 (2010)

18. Mishra, G.R., Suresh, M., Kumaran, K., Kannabiran, N., Suresh, S., Bala, P., Shivakumar, K., Anuradha, N., Reddy, R., Raghavan, T.M., Menon, S., Hanumanthu, G., Gupta, M., Upendran, S., Gupta, S., Mahesh, M., Jacob, B., Mathew, P., Chatterjee, P., Arun, K.S., Sharma, S., Chandrika, K.N., Deshpande, N., Palvankar, K., Raghavnath, R., Krishnakanth, R., Karathia, H., Rekha, B., Nayak, R., Vishnupriya, G., Kumar, H.G., Nagini, M., Kumar, G.S., Jose, R., Deepthi, P., Mohan, S.S., Gandhi, T.K., Harsha, H.C., Deshpande, K.S., Sarker, M., Prasad, T.S., Pandey, A.: Human protein reference database–2006 update. Nucleic Acids Res. 34, D411–D414 (2006)

19. Cheng, A.S., Jin, V.X., Fan, M., Smith, L.T., Liyanarachchi, S., Yan, P.S., Leu, Y.W., Chan, M.W., Plass, C., Nephew, K.P., Davuluri, R.V., Huang, T.H.: Combinatorial analysis of transcription factor partners reveals recruitment of c-MYC to estrogen receptor-alpha responsive promoters. Mol. Cell 21, 393–404 (2006)

Rapid and Accurate Generation of Peptide Sequence Tags with a Graph Search Approach

Hui Li[1], Lauren Scott[1], Chunmei Liu[1,*], Mugizi Rwebangira[1],
Legand Burge[1], and William Southerland[2]

[1] Department of Systems and Computer Science
chunmei@scs.howard.edu
[2] Department of Biochemistry,
Howard University
2400 Sixth Street, NW
Washington, DC 20059
United States

Abstract. Protein peptide identification from a tandem mass spectrum (MS/MS) is a challenging task. Previous approaches for peptide identification with database search are time consuming due to huge search space. De novo sequencing approaches which derive a peptide sequence directly from a MS/MS spectrum usually are of high complexities and the accuracies of the approaches highly depend on the quality of the spectra. In this paper, we developed an accurate and efficient algorithm for peptide identification. Our work consisted of the following steps. Firstly, we found a pair of complementary mass peaks that are b-ion and y-ion, respectively. We then used the two mass peaks as two tree nodes and extend the trees such that in the end the nodes of the trees are elements of a b-ion set and a y-ion set, respectively. Secondly, we applied breadth first search to the trees to generate peptide sequence tags. Finally, we designed a weight function to evaluate the reliabilities of the tags and rank the tags. Our experiment on 2620 experimental MS/MS spectra with one PTM showed that our algorithm achieved better accuracy than other approaches with higher efficiency.

Keywords: Tandem mass spectrum, Post-translational modification (PTM), Peptide sequence tags.

1 Introduction

Protein identification from a tandem mass spectrum is an important but challenging problem in proteomics. The problem becomes more difficult if post-translational modifications (PTMs) present in the spectrum. The presence of post-translation modifications in proteins is very common and most proteins contain one or more PTMs. Currently, algorithms for protein identification using tandem mass spectrometry (MS/MS) for proteins that contain PTMs have been extensively developed [1-5, 17].

Existing approaches for protein identification fall into two popular categories. One category contains database based approaches and the other contains de novo peptide

---

* Corresponding author.

J. Chen, J. Wang, and A. Zelikovsky (Eds.): ISBRA 2011, LNBI 6674, pp. 253–261, 2011.
© Springer-Verlag Berlin Heidelberg 2011

sequencing approaches. Database search based approaches usually pre-specify a set of common PTM types and search databases [5-7]. The approaches align the query spectrum with a peptide database and use an evaluation function to evaluate the alignments. If a peptide sequence gives the best alignment score, it will be selected as the peptide sequence of the query spectrum. However, due to the existence of natural or artificial modifications in hundreds of species, database search based approaches may need to explore a huge search space. Moreover, if the peptide sequence of the query spectrum is not contained in the database, this kind of approaches will fail.

In contrast, de novo sequencing approaches directly find a peptide sequence for a spectrum [8-12, 19]. De novo sequencing approaches could achieve a good performance without looking up a peptide database. However, because a spectrum usually is not complete and contains a lot of noise peaks and PTMs, it is hard to derive a correct peptide sequence of full length. In other words, the accuracy of a de novo sequencing approach highly depends on the quality of the query spectra. Recently, approaches for blind PTM identification have been proposed [5,13,16,18]. For example, in [5], it employs a point process model to align a query spectrum with the theoretical spectrum of each peptide sequence in a peptide database. Alternatively, in [13], it proposes a dynamic programming algorithm to derive a peptide sequence for each spectrum. The major obstacle of the approaches is that it is very time consuming for those approaches to do spectral alignments due to the large size of search space.

Recently, approaches that combine both de novo sequencing and database search have been proposed. The approaches derive short peptide sequences (peptide sequence tags) instead of full-length peptide sequences and use the tags to filter a peptide database [13-15, 18]. The peptide sequence tags can filter out most peptide sequences from the database such that only a small fraction of candidate peptide sequences are left for further spectral alignments. The approaches can reduce much search space and thus significantly speed up the search. For example, PepNovo [14] uses a score function to evaluate the reliability of sequence tags obtained from de novo sequencing with a graph based dynamic programming approach. Using a fragmentation of a peptide, GutenTag [15] creates many short sequence tags which are contained in the peptide and uses the tags to find the full-length sequence in a peptide database. SeqTag [18] uses a tree-decomposition based approach to derive peptide sequence tags and then use a point process model to align the query spectrum with the theoretical spectrum of each candidate peptide sequence after database filtration with the tags. The approaches reduce significant amount of search space and thus can speed up the search process. It has become a promising direction for protein identification.

In this paper, we derive peptide sequence tags with a novel graph search approach. The experimental results on a dataset consisting of spectra with one PTM show that our algorithm achieves better or comparable accuracies on tag lengths of 3 and 4 with less amount of time than PepNovo [14] and SeqTag [18].

2   Algorithms

Let $S = \{s_1, s_2, s_k \ldots, s_n\}$ be a tandem mass spectrum, where $s_i = \{P_i, I_j\}$, P_i is the mass value of the ith peak of the spectrum, and I_j is the intensity of the peak. We have the following important observations:

1) The mass difference between two ions of the same type always is equal to   the mass of a single amino acid or the total mass of some amino acids;

2) If the mass difference between two peaks is not equal to the mass of any amino acid or the total mass of some amino acids, the two peaks must be of different ion types or at least one peak must be a noise peak; and

3) For each peak, its complementary peak should also be included in the spectrum. If this is not the case, the complementary ion should be added to the spectrum to make the spectrum more complete.

For each spectrum, a graph $G = (V, E)$ is constructed. In the graph, each vertex represents a peak of the spectrum. If the corresponding mass value difference of two vertices V_i and V_j is the mass of a single amino acid or the total mass of some amino acids, there is a directed edge from V_i to V_j if the corresponding mass value of V_i is less than the mass value of V_j.

Extensive graph based algorithms for PTM identification have been developed [8-10]. In the graph algorithms, each path of the graphs is labeled with an amino acid sequence. Although graph based algorithms may successfully identify some PTMs, the following challenges still exist and need to be attacked:

1) Many noise peaks exist in MS/MS spectral data. In our dataset, the percentage of noise peaks is more than 75%;
2) There are missing peaks;
3) The computing complexity is usually high due to huge search space; and
4) The existence of PTMs will break the chain of amino acid sequences.

In order to solve the problems above, we propose a novel graph search based approach. The approach consists of the following stages. Firstly, we find a pair of complementary mass peaks that are b-ion and y-ion, respectively and the pair has the highest intensity. We then use the two mass peaks as two tree nodes and extend the trees such that in the end the nodes of the trees are the elements of a b-ion set G_B and a y-ion set G_Y, respectively. Secondly, we apply breadth first search to the trees to generate peptide sequence tags. Finally, we design a weight function to evaluate the reliabilities of the tags. From the observation, the complementary ions that have the highest intensity are always located in around the middle part of a MS/MS spectrum. Identifying a correct pair of complementary ions in the first step plays a fundamental role in identifying correct tags. If noise peaks are identified as the starting nodes to extend the trees, the identified amino acid sequences will be wrong. In this paper, the constraint conditions for choosing the complementary pair of mass peaks $s_i = (P_i, I_i)$ and $s_j = (P_j, I_j)$ are as the following:

1) P_i is complementary to P_j;
2) The sum of the intensities I_i and I_j should be the highest among all complementary pairs; and
3) The mass difference between P_{i-1} and P_i or the mass difference between P_i and P_{i+1} should be equal to the mass of a single amino acid or the combination of multiple amino acids; the mass difference between P_{j-1} and P_j or the mass difference between P_j and P_{j+1} should be equal to the mass of a single amino acid or the total mass of multiple amino acids.

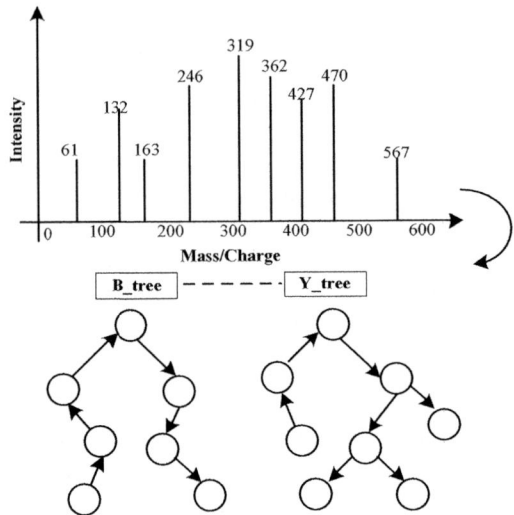

Fig. 1. An example MS/MS spectrum and its two trees

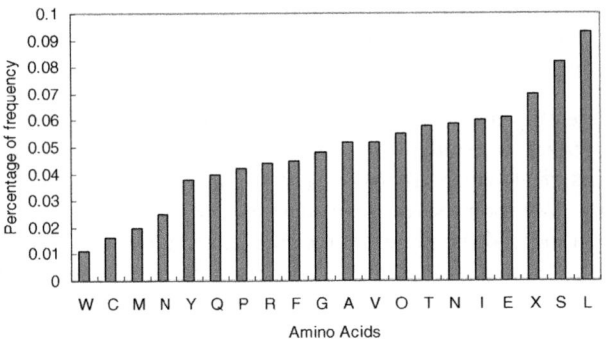

Fig. 2. Amino acid frequencies that are calculated from a yeast peptide database

After we find the pair of complementary ions, we use their corresponding vertices as the starting nodes of two trees, respectively, and extend the trees such that in the end, the nodes of the two trees are the elements of a b-ion set G_B and a y-ion set G_Y, respectively. Figure 1 shows an example MS/MS spectrum and its two trees.

In peptide sequences, the frequency of each amino acid is different. We calculate the frequency of each amino acid in a yeast peptide database.

In order to reduce graph search space, we introduce a heuristic search to find the best path in the tree graphs. We design a function to estimate the score of an edge. For an edge (V_i, V_j) labeled with amino acid a, the estimation function $F(V_i, V_j)$ is defined as follows:

$$F(V_i,V_j) = f(a)(\ln(\frac{r_i}{r_i-1}) + \ln(\frac{r_i}{r_i+1}) + \ln(\frac{r_j}{r_j-1}) + \ln(\frac{r_j}{r_j+1})) \quad (1)$$

Create the OpenList and CloseList.
// All the child nodes of the current parent will be inserted into OpenList and all the extended nodes will be inserted into CloseList.
For $i = 0$ To L do (L denotes the layer of the tree graph)

Rank the Openlist by the function in Equation 1;

Add node i to OpenList;
Change the parent index of node i;
If the node is a leaf, record the whole path and re-
    cursively search the path; delete node i from
    OpenList, and add it to CloseList;
If node i belongs to B_tree, set ArrayMark[i]=1;
If node i belongs to Y_tree and ArrayMark[i]=0, set
    ArrayMark[i] to be 2;
If node i belongs to Y_tree and ArrayMark[i]=1, set
    ArrayMark[i] to be 3;
Go back to the root;
    If the OpenList in null, break the loop;
End For
For each i, if ArrayMark[i]=1, put the node into Bset
else if ArrayMark[i]=2, put the node into Yset
else if ArrayMark[i]=3, put the node into Cset

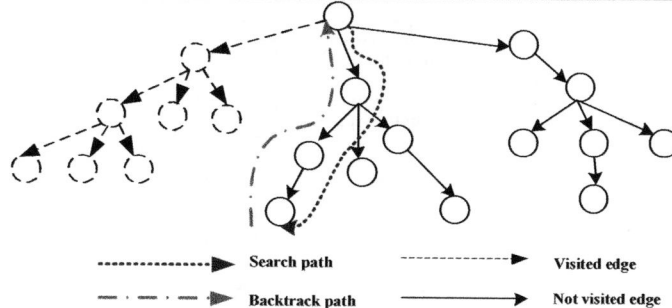

- - - - - - ▶ Search path   - - - - - - ▶ Visited edge
— · — · — ▶ Backtrack path   ————————▶ Not visited edge

Fig. 3. A heuristic search on a tree for identifying a b-ion set and a y-ion set

Where $f(a)$ is the frequency of amino acid a, and r_i and r_j are the relative intensities of mass peaks s_i and s_j. We use the following heuristic search algorithm to generate the b-ion set and y-ion set.

After getting the Bset and Yset as above, we process them as follows:

1) SBset= Bset - CSet
2) SYset= YSet - CSet
3) CSBset is a set after adding the complementary peaks to SBset
4) CSYset is a set after adding the complementary peaks to SYset.

The following pseudcode describes the entire algorithm:

1) Choose a pair of complementary mass peaks RootB and RootY that are *b*-ion and y-ion and has the highest sum of the intensities among complementary pairs;
2) Create Openlist and Closelist;
3) Use the heuristic search method to get BSet and YSet;
4) Get the sets CSBset and CSYset;
5) BSeqSet =BFS(CSBset) (BSeqSet is the set of sequences obtained from the breadth first search on CSBset);
6) YSeqSet =BFS(CSYsets) (YSeqSet is the set of sequences obtained from the breadth first search on CSYset);
7) Generate the tags from the BSeqSet and YSeqSet and use Equation 1 to rank the tags.

The whole algorithm consists of two components: heuristic search for a *b*-ion set and a *y*-ion set and breath first search for peptide sequence tags. The complexities of the two components are $O(n)$ and $O(e+n)$, respectively, where *n* is the number of vertices and *e* is the number of edges in the tree graph. Figure 4 shows the flowchart of the whole algorithm.

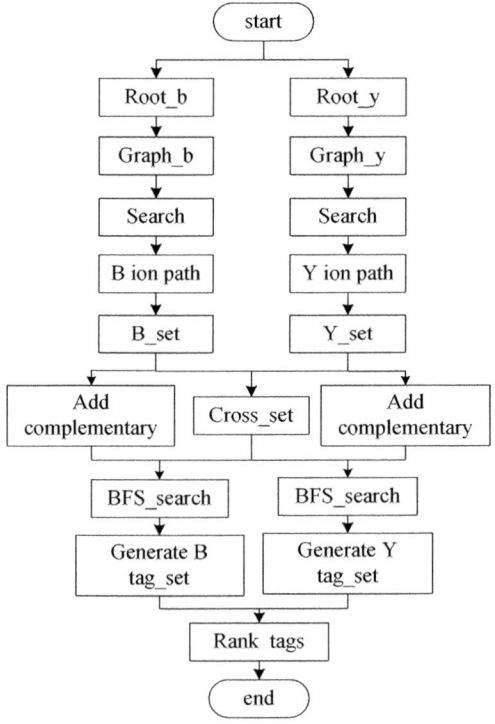

Fig. 4. The flowchart of the whole algorithm

3 Experiments and Results

We tested our algorithm on the same datasets of 2657 annotated yeast ion trap tandem mass spectra that contain no PTMs and 2620 spectra where each spectrum contains one artificially added PTM from a common PTM set as in [5]. We have implemented our algorithm using C++ programming language and run the program on Dell Power Edge 1950 server with 4 CPUs. Table 1 shows the accuracies of the selections of complementary pairs on the two datasets. It is shown from the table that our program is able to correctly identify a pair of complementary ions in 95.3% and 95.2% spectra in the two datasets, respectively. Table 2 shows the comparison of the results of peptide sequence tag selections of our algorithm with SeqTag [18] and PepNovo [14]. From the table, we can see that our program is able to identify more correct tags as rank 1 tags for both tag lengths of 3 and 4 and achieves better or comparable accuracies for other ranks while use less time than the other two programs.

Table 1. The accuracies of the selections of a pair of complementary ions

No PTM (2657)	One PTM (2620)
95.3%	95.2%

Table 2. A comparison between the performance of our tag selection program and that of SeqTag and PepNovo at different tag lengths

Tag length	Algorithm	R=1(%)	R=3(%)	R=5(%)	R=10(%)	R=25(%)	T(s)
3	Ours	69.7	84.2	89.5	92.3	95.3	0.27
	SeqTag	68.1	84.8	90.3	94.8	97.1	0.32
	PepNovo	62.8	83.7	89.7	94.9	97.8	3.59
4	Ours	55.4	74.1	74.6	85.2	91.2	0.31
	SeqTag	53.5	71.2	78.6	84.8	90.0	0.32
	PepNovo	51.1	71.7	79.3	85.8	91.4	3.65

The table shows the accuracies of peptide sequence tag selections on 2620 experimental spectra with one artificially added PTM. Columns for R = 1, 3, 5, 10, 25 are the percentages of spectra that have at least one correct tag in top 1, 3, 5, 10, 25 tags generated by our program, SeqTag, and PepNovo, respectively; T is the average time in seconds used for generating peptide sequence tags for one spectrum.

4 Conclusions

In this paper, we propose a novel graph search approach for identifying peptide sequence tags. Considering the fact that an experimental MS/MS spectrum contains a large amount of noise peaks and missing peaks, we first identify a pair of complementary ions. The experiments show that our program not only gains better or comparable accuracy of peptide sequence tag selection than SeqTag and PepNovo but also our program uses less amount of computing time. In future work, we will use the sequence tags the program generates to search a peptide database for candidate peptide sequences. We then will design an alignment algorithm such that for each query experimental spectrum, we align it with the theoretical spectrum of each candidate peptide sequence and evaluate each alignment with a score function. The peptide sequence that gives the highest alignment score will be chosen as the peptide sequence of the query spectrum.

Acknowledgement

This work was supported by NSF CAREER (CCF-0845888) (Li, Scott, Liu), NSF Science & Technology Center grant CCF-0939370 (Li, Scott, Liu, Rwebangira, and Burge), and 2 G12 RR003048 from the RCMI program, Division of Research Infrastructure, National Center for Research Resources, NIH (Rwebangira and Southerland).

References

1. Wilkins, M.R., Gasteiger, E., Gooley, A.A., Herbert, B.R., Molloy, M.P., Binz, P.A., Ou, K., Sanchez, J.C., Bairoch, A., Williams, K.L., Hochstrasser, D.F.: High-throughput Mass Spectrometric Discovery of Protein Post-Translational Modifications. Journal of Molecular Biology 289, 645–657 (1999)
2. Walsh, C.T.: Posttranslational Modification of Proteins: Expanding Nature's Inventory. Roberts & Company Publishers, Englewood, Colorado (2005)
3. Mann, M., Jensen, O.N.: Proteomic Analysis of Post-Translational Modifications. Nat. Biotechnol. 21, 255–261 (2003)
4. Han, Y., Ma, B., Zhang, K.: SPIDER: Software for Protein Identification from Sequence Tags Containing Sequencing Error. Journal of Bioinformatics and Computational Biology, 97–716 (2005)
5. Yan, B., Zhou, T., Wang, P., Liu, Z., Emanuele II, V.A., Olman, V., Xu, Y.: A Point-Process Model for Rapid Identification of Post-Translational Modifications. In: Proceedings of 2006 Pacific Symposium on Biocomputing, pp. 327–338 (2006)
6. Liu, C., Yan, B., Song, Y., Xu, Y., Cai, L.: Fast De Novo Peptide Sequencing and Spectral Alignment via Tree Decomposition. In: Proceedings of the 11th International Pacific Symposium on Biocomputing (PSB 2006), pp. 255–266 (2006)
7. Fu, Y., Jia, W., Lu, Z., Wang, H., Yuan, Z., Chi, H., Li, Y., Xiu, L., Wang, W., Liu, C., et al.: Efficient Discovery of Abundant Post-Translational Modifications and Spectral Pairs Using Peptide Mass and Retention Time Differences. BMC Bioinformatics 10(Suppl 1), S50 (2009)
8. Ma, B., Zhang, K., Hendrie, C., Liang, C., Li, M., Doherty-Kirby, A., Lajoie, G.: PEAKS: Powerful Software for Peptide De Novo Sequencing by Tandem Mass Spectrometry. Rapid Communication in Mass Spectrometry 17, 2337–2342 (2003)
9. Searle, B.C., Dasari, S., Turner, M., Reddy, A.P., Choi, D., Wilmarth, P.A., McCormack, A.L., David, L.L., Nagalla, S.R.: High-Throughput Identification of Proteins and Unanticipated Sequence Modifications Using a Mass-Based Alignment Algorithm for MS/MS De Novo Sequencing Results. Anal. Chem. 76, 2220–2230 (2004)

10. Yan, B., Pan, C., Olman, V.N., Hettich, R.L., Xu, Y.: A Graph-Theoretic Approach for the Separation of b and y Ions in Tandem Mass Spectrometry. Bioinformatics 21, 563–574 (2005)
11. Perkins, D.N., Pappin, D.J., Creasy, D.M., Cottrell, J.S.: Probability-based Protein Identification by Searching Sequence Databases Using Mass Spectrometry Data. Electrophoresis 20, 3551–3567 (1999)
12. Eng, J.K., McCormack, A.L., Yates III, J.R.: An Approach to Correlate Tandem Mass Spectral Data of Peptides with Amino Acid Sequences in A Protein Database. Journal of the American Society of Mass Spectrometry 5, 976–989 (1994)
13. Frank, A., Tanner, S., Pevzner, P.: Peptide Sequence Tags for Fast Database Search in Mass-Spectrometry. Journal of Proteome Research 4, 1287–1295 (2005)
14. Frank, A., Pevzner, P.: PepNovo: De Novo Peptide Sequencing via Probabilistic Network Modeling. Anal. Chem. 77, 964–973 (2005)
15. Tabb, D.L., Saraf, A., Ates, J.R.: GutenTag: High-Throughput Sequence Tagging via an Empirically Derived Fragmentation Model. Analytical Chemistry 75, 6415–6421 (2003)
16. Tsur, D., Tanner, S., Zandi, E., Bafna, V., Pevzner, P.: Identification of Post-translational Modifications by Blind Search of Mass Spectra. Nature Biotechnology 23, 1562–1567 (2005)
17. Tanner, S., Shu, H., Frank, A., Wang, L.C., Zandi, E., Mumby, M., Pevzner, P.A., Bafna, V.: InsPecT: Identification of Posttranslationally Modified Peptides from Tandem Mass Spectra. Analytical Chemistry 77, 4626–4639 (2005)
18. Liu, C., Yan, B., Song, Y., Xu, Y., Cai, L.: Peptide Sequence Tag-Based Blind Identification of Post-Translational Modifications with Point Process Model. In: The 14th International Conference on Intelligent Systems for Molecular Biology (ISMB 2006), Fortaleza, Brazil (2006)
19. Liu, C., Yan, B., Song, Y., Xu, Y., Cai, L.: Graph Tree Decomposition Based Fast Peptide Sequencing and Spectral Alignment. International Journal of Computational Science, Special Issue on Bioinformatics and Computational Biology 2, 632–645 (2008)

In Silico Evolution of Multi-scale Microbial Systems in the Presence of Mobile Genetic Elements and Horizontal Gene Transfer

Vadim Mozhayskiy and Ilias Tagkopoulos[*]

Department of Computer Science and Genome Center
University of California Davis
One Shields Avenue, Davis, CA 95616, USA
itagkopoulos@ucdavis.edu

Abstract. Recent phylogenetic studies reveal that Horizontal Gene Transfer (HGT) events are likely ubiquitous in the Tree of Life. However, our knowledge of HGT's role in evolution and biological organization is very limited, mainly due to the difficulty tracing HGT events experimentally, and lack of computational models that can capture its dynamics. Here, we present a novel, multi-scale model of microbial populations with the capacity to study the effect of HGT on complex traits and regulatory network evolution. We describe a parallel load-balancing framework, which was developed to overcome the innate challenges of simulating evolving populations of such magnitude and complexity. Supercomputer simulations of *in silico* cells that mutate, compete, and evolve, show that HGT can significantly accelerate, but also disrupt, the emergence of advantageous traits in microbial populations. We show that HGT leaves a lasting imprint to gene regulatory networks when it comes to their size and sparsity. In any given experiment, we observed phenotypic variability that can be explained by individual gain and loss of function during evolution. Analysis of the fossil mutational and HGT event record, both for evolved and non-evolved populations, reveals that the distribution of fitness effect for HGT has different characteristics in terms of symmetry, shape and bias from its mutational counterpart. Interestingly, we observed that evolution can be accelerated when populations are exposed in correlated environments of increased complexity, especially in the presence of HGT.

Keywords: Horizontal Gene Transfer, Microbial Evolution, Biological Networks, Simulation, Multi-scale Modeling, High Performance Computing.

1 Introduction

Horizontal Gene Transfer (HGT) is the process of horizontal transfer of genetic material within and across species. It is a mechanism of genetic exchange complementary to vertical transfer, which occurs through cell division and results in the transfer of genetic information from an ancestor to its offspring cells. Although largely ignored in the past, recent phylogenetic evidence suggests that its impact on bacterial evolution is significant and should be investigated more thoroughly [1, 2].

---

[*] Corresponding author.

J. Chen, J. Wang, and A. Zelikovsky (Eds.): ISBRA 2011, LNBI 6674, pp. 262–273, 2011.
© Springer-Verlag Berlin Heidelberg 2011

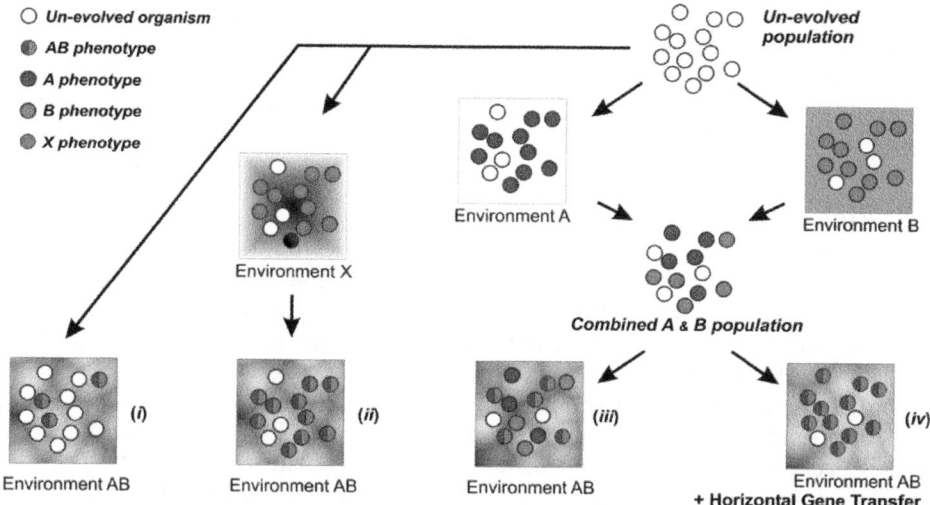

Fig. 1. General overview of the simulated ecological setting: Microbial populations evolve under environment AB either directly (*i*), or indirectly through an intermediate environment X of lower complexity (*ii*). In addition, we test whether pre-exposing to environments A and B and then merging the respective populations without (*iii*) or with (*iv*) HGT changes the rate and characteristics of evolution.

For instance, it has been estimated that up to a 32% of the bacterial genome is acquired by HGT [3]. However, even this number is a lower bound of the HGT events that take place through bacterial evolution, since only a small fraction of transferred material is positively selected, fixed and, consequently, observable through phylogenetic analysis [4].

Due to our limited ability to observe HGT dynamics in an experimental setting, theoretical models have been traditionally employed to elucidate the impact of HGT on evolution. Continuous kinetic [5, 6] and stochastic models [7-9] were developed to fit experimentally observed short-term dynamics of HGT in twenty-four-hour experiments [9] and to analyze the interplay between rates of HGT, mutations, and selection pressure parameters. It was shown that transferred genes can be successfully fixed in a population if the HGT rate is comparable to the mutational inactivation rate [7] and that high rates of HGT may affect evolution rate in a simple population model [8]. Previous models, although insightful, have a limited scope as they lack any notion of gene regulation, cellular networks and processes, multi-scale structures, and temporal expression dynamics. To address these issues, we extended our previous work [10] to develop a multi-scale simulation framework, capable of simulating the evolution of unicellular organisms in the presence of HGT, which is described in Sec. 2 and 3.

An overview of the general simulation setting discussed in this paper is illustrated in Fig. 1. We start with a random initial population of cells and three dynamic environments, namely A, B and AB, where the latter is the combination of the first two, and hence of higher complexity. The un-evolved initial population is exposed in one of following three settings: (a) it is directly placed into environment AB, (b) it is first placed in environment X, which is of lower complexity, (c) it is initially evolved in environments A and B, which leads to two distinct populations, that are subsequently randomly sampled (keeping the same effective size) to form a final population that is then placed in the AB environment with and without HGT. This setting allows us to investigate questions related to HGT and evolution in environments that are both correlated and increasingly more complex.

2 Biological Model

In our model, each *in silico* organism encompass functions and parameters that model basic biological phenomena, while its core consists of a gene regulatory and biochemical network with abstract molecular representations. The model has been extended to incorporate Horizontal Gene Transfer in addition to the other cellular (transcription, translation, modification, growth, death, etc.) and evolutionary (mutation and natural selection) processes. In a simulation, a fixed-size population of cells mutates, competes and evolves in well-defined, temporal, multivariate environments. Each cell comprises three types of nodes: Gene/mRNA, Protein, and Modified Protein (Fig. 2a). The Promoter/Gene/RNA node captures gene regulation and transcription, while the Protein and Modified Protein nodes capture translation and post-translational modification (acetylation, phosphorylation, etc.), respectively.

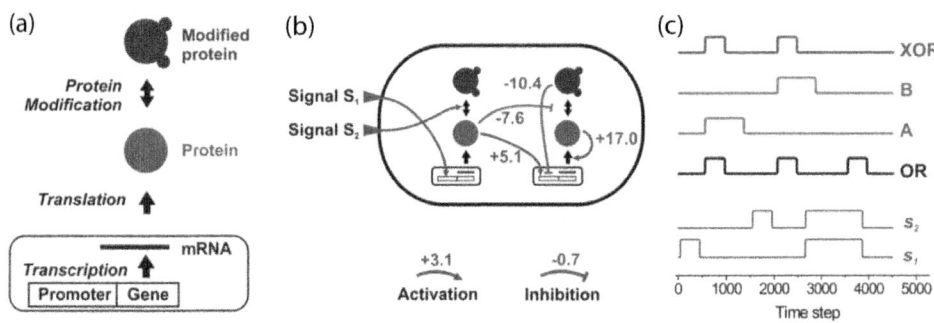

Fig. 2. (a) Basic cellular modeling in our simulation framework; a "triplet" captures processes of transcription, translation, and post-translational modification. (b) Example of a gene regulatory and biochemical network in an organism where environmental signals (e.g. oxygen, temperature, etc.) regulate the expression of certain genes/proteins. (c) Environmental signals (s_1 and s_2) and nutrient abundance for four environments (bottom to top: OR, A, B, XOR) shown as a function of time steps within one epoch. Nutrient presence is a delayed function of the two signals.

A "triplet" consists of a specific gene node and its products, i.e. the corresponding protein and modified protein node, and generally captures the "central dogma" of molecular biology (Fig. 2a). Each organism has its own distinct gene regulatory and biochemical network (i.e. a collection of various triplets and weighted regulatory edges) that can be depicted as a directed weighted graph (see Fig. 2b). There exists a set of "special triplets", which are common in all cells, and encode physiological responses. It is important to note that we do not impose any objective function or arbitrary selection. Instead, we model the environment in which synthetic organisms live and evolve, which consists of signals, nutrients and other chemicals (e.g. toxic compounds) with concentrations that can fluctuate over time. In this work, every environment has only one nutrient type and each organism possess one special triplet, whose expression allows the organism to metabolize the nutrients that are present. Since nutrients are present for a short duration, organisms that evolve the capacity to infer their presence and be prepared (e.g. express the metabolic triplet) have a selective advantage, in analogy to real microbial systems. We utilize this framework to address questions regarding the impact of HGT on trait evolution, and gene regulatory network organization.

In the setting discussed here, two signals s_1 and s_2 carry information regarding the presence of nutrients in the environment (Fig. 2c). For example, in a XOR environment, the "Nutrients Presence [XOR]" = Delayed (s_1 XOR s_2). Similarly, the correlation-structure of environments A and B is "Nutrients Presence [A]" = Delayed (s_1 AND NOT(s_2)) and "Nutrients Presence [B]" = Delayed (NOT(s_1) AND s_2) respectively. Despite the fact that the combined AB environment (delayed XOR) is a simple combination of the A and B environments, its complexity is significantly higher when compared to the other two (A and B) as it is not linearly separable [11]; in contrast to both A and B environments that can be separated linearly. To assess the fitness level of each organism, we report the Pearson correlation between nutrient abundance and response protein expression level over a predefined interval of time, which we call an "epoch" (4,500 time units in our simulations). We stress that this similarity measure is used for visualization purposes as a proxy to each organism's fitness, and at no point participates or interferes with the selection or evolutionary trajectory of cells during the simulation.

The probability of molecule creation at each node and at each time step is a function of the regulatory effect of other nodes on that specific node, and the availability of substrate molecules. We model the molecule production probability as a two-level sigmoid function that captures saturation effects for any given regulator and for the expression of any given node. As such, the molecule production probability of node i is given by:

$$G_i = basal_i + (1 - basal_i) \cdot tanh\left(\frac{\sum_{j=1}^{n}\left(w_{ij} \cdot f_{ij}(v_j, \tilde{m}_{ij}, \tilde{s}_{ij})\right) - m_i}{s_i}\right), \tag{1}$$

where the sigmoid function f_{ij} describes the regulatory effect of node j on node i:

$$f_{ij}(v_j, \tilde{m}_{ij}, \tilde{s}_{ij}) = \frac{1}{2} \cdot \left[1 + tanh\left(\frac{v_j - \tilde{m}_{ij}}{\tilde{s}_{ij}}\right)\right], \tag{2}$$

where w_{ij} is the regulatory matrix element (i.e. the strength and direction that exerts node j to node i), v_j is the value of node j, m_i and s_i the midpoint and slope of the target-specific sigmoid function, $\tilde{m}_{ij}$ and $\tilde{s}_{ij}$ the midpoint and slope of the regulator specific sigmoid function, n is number of regulating nodes, $basal_i$ is the basal expression parameter.

3 Parallel Simulation Framework

The simulations described here are of unprecedented scale and scope, with integrated models of the environment, population, organism, biological network and molecular species. This level of detail is necessary in order to model phenomena that transcend multiple scales, as in the case of Horizontal Gene Transfer. We had to develop efficient algorithms for HPC communication, balancing and process migration, as cell death and division creates unforeseen loads to the various computational cores. In addition, as organisms adapt and evolve, the complexity of their internal networks constantly increases, and with that the need for computational power. Cells with larger networks can be more efficient in nutrients metabolism and therefore grow and divide faster in real time. On the contrary, the computational time for cells with extended genomes is always larger, and scales with $O(N^2)$, where N is the number of nodes within the cellular network. This calls for a synchronization point at each timepoint during our simulations, which may lead to poor scalability due to load imbalance (Fig. 3).

Initially, cells were distributed to MPI processes with one cell per process per computational core; MPI processes were synchronized at the end of each time step. However, in this initial implementation the imbalance was a problem even for a small number of cells, and the code did not scale beyond 64 cores. The model was improved when a group of cells were assigned to each MPI process, because of averaging effects (i.e. the average computational load was similar among processes). Strong scaling results (Fig. 4a) showed that for our problem size, a load of 8 cells per core was ideal as the imbalance between processes was minimal.

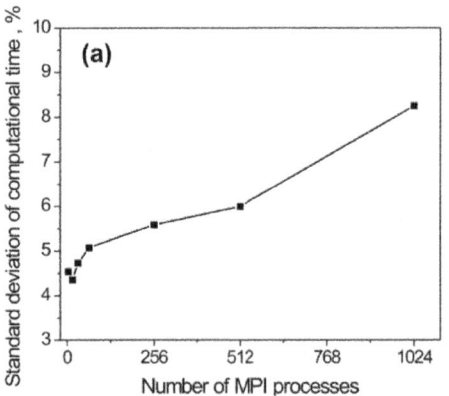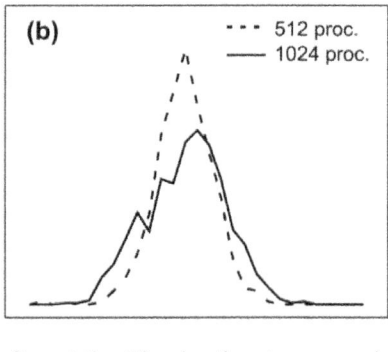

Fig. 3. Variation of the computational time per MPI process increases with the number processes. (a) Standard deviation of the computational time (per time step, per core) five epochs after cell distribution was balanced, at a constant load of 8 cells/process. (b) Distributions of computational time across MPI processes for jobs with 512 and 1024 processes (4096 and 8192 cells, respectively). In larger populations, the higher variance between cell size results in unbalanced computational loads and increased idling time at synchronization points.

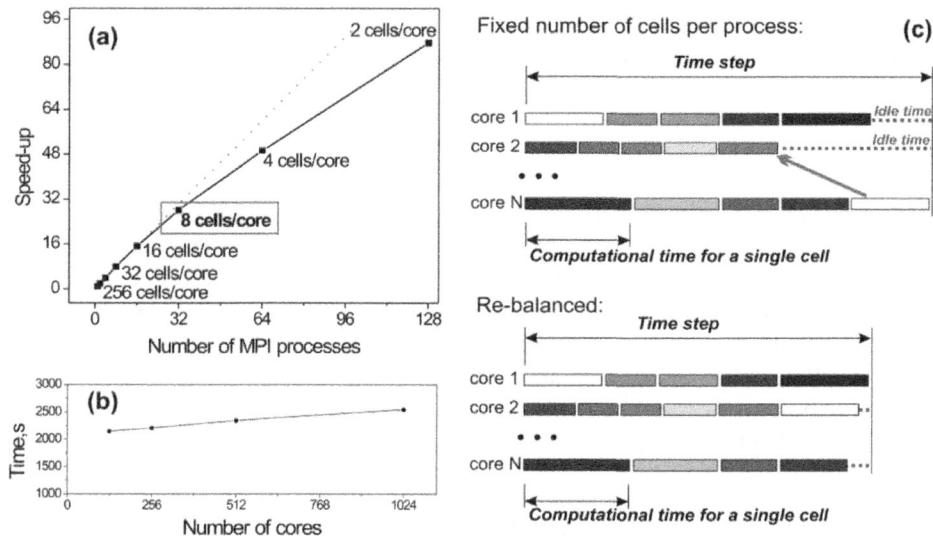

Fig. 4. (a) Strong scaling for a population 256 cells. Code scales well for loads of 8 cells/core or more. (b) Weak scaling up to 8192 cells for the hybrid MPI/OpenMP model. (c) Dynamic MPI load balancing. Top: in an evolving population, computational time for cells varies with the cell network size. This results in idle cores (dashed lines) with smaller (i.e. fast-to-compute) cells, as they synchronize at each time-point. Each bar segment depicts the computational time of a single cell, and multiple cells, of various complexities, are assigned to a single core. The maximum of these loads (here, the load of core N) defines the speed of the simulation. Bottom: with the addition of a dynamic load balancer, cells are redistributed to minimize idling time.

Next, we further extended our model by implementing a hybrid MPI/OpenMP solution: each MPI process is executed on a multi-core computational node; cells assigned to each MPI process are stored in the node's shared memory; computational cycles for each cell update in an MPI process are dynamically distributed between available cores. Dynamic cell distribution is carried out by creating a pool of cycles and cores, and aims to eliminate the idling time during communication. Weak and strong scaling shows scalability up to 8192 cells with near-linear speedup (Fig. 4b).

Finally, by adding dynamic MPI load balancing (Fig. 4c), computational time is monitored for each cell and is used to redistribute cells between MPI processes in order to have a more balanced load between cores. The cell growth rate is used to predict cell division/death events. Although the current load-balancing implementation is a distributed process, a scalable hierarchical implementation can further increase the performance of the simulator.

4   Application: Horizontal Gene Transfer

In our model, genes and its products are represented by triplets, and therefore HGT can be treated as inter-cellular transfer of one or more triplets (mobile genetic elements). For every HGT event a random fragment (i.e. subset of triplets) is copied from a donor cell and inserted into the regulatory network of a recipient cell. Fragment size is chosen using a probability density function (normalized sigmoid function):

$$P(n) = \frac{1 - tanh\left(\frac{n-m}{s}\right)}{m \cdot \left(2 + ln\left(e^{-\frac{2m}{s}} + 1\right)\right)}, \tag{3}$$

where n is the fragment size m and s are the middle point and slope of the probability density function, respectively. In most cases $s=m$ was used, and therefore 67% of all transferred fragments were not larger than m triplets (Fig. 5a). The original regulation of the response protein by the transferred sub-network is preserved (Fig. 5a, insert). Experimentally observed HGT rates between bacteria in natural environments vary between 10^{-7} and 10^{-11} per generation per cell [12-14], while in some cases the rate climbs between 10^{-3} to 10^{-1} [14, 15]. Default HGT rate used throughout the paper is $5 \cdot 10^{-6}$ per cell per time step.

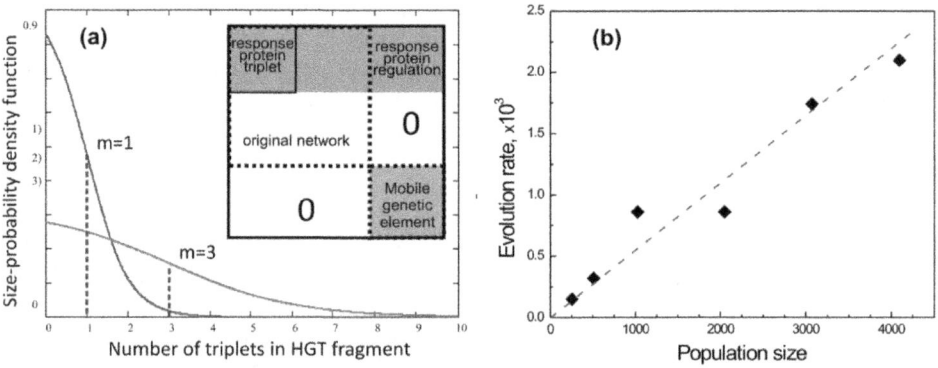

Fig. 5. (a) HGT transfer probability and genome integration: Probability density function profile used to select fragment sizes; (insert) incorporation of the transferred fragment into the regulatory matrix, where only the response pathway regulation is conserved. (b) Evolution rate is a linear function of the population size. Rate is calculated as an average slope of the maximum fitness increase for population sizes up to 4096 cells (16 replicates). Initial random populations evolved in the XOR environment until the maximum fitness is stabilized.

First, we looked at how the rate of evolution scales as a function of the population size. It is believed that evolution speed increases linearly with population size N for small populations, and with ln(N) for intermediate population sizes [16], while it approaches a saturation limit for large populations ($>10^{9}$) [17] . We observed a linear dependence of evolution rate to population size (Fig. 5b) in agreement with theoretical predictions. Although it is possible that HGT can eventually be beneficial at high population sizes, where simultaneous emergence of competing beneficial mutations may decrease the rate of evolution, we did not observe this effect in our simulations due to the smaller population size.

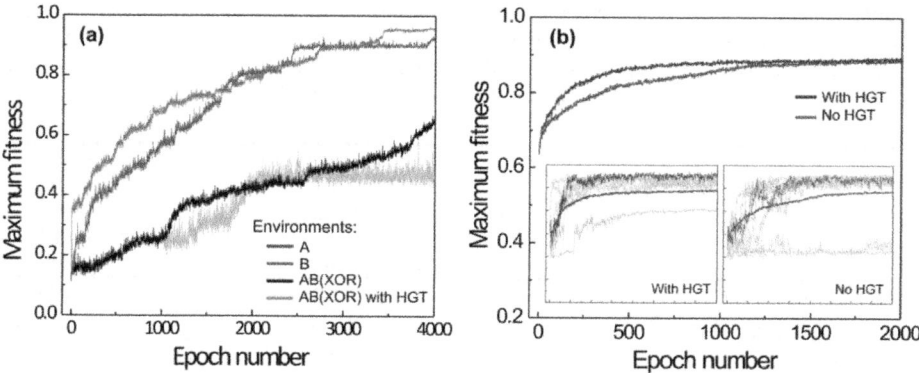

Fig. 6. (Color) Multi-step acceleration and HGT effect in "dual-step" evolution: (a) Evolution of random population of cells in *A*, *B* and *XOR* environments shown in red, blue, and gray respectively. Maximum fitness averaged over 64 simulations, (b) evolutionary trajectory under "dual-step" evolution, where population of evolved cells in A and B environments show remarkably fast adaptation to environment AB (64 simulations). HGT confers an additional acceleration of adaptation to new settings. Inset: Maximum fitness curves for 8 out of 64 individual simulations with (left inset) and without (right inset) HGT are shown in gray. One curve is highlighted with dark gray for clarity.

4.1 Evolution in Coupled Environments of Increasing Complexity

Recent theoretical predictions [18, 19] suggest that evolution generalizes to new environments through facilitated variation, a process in which genetic changes are channeled in useful phenotypic directions. Here we hypothesize that evolution can be accelerated by exposing evolving populations in similar, correlated environments of increasing complexity, and we assess whether HGT further accelerates evolution in such settings. When random populations are exposed directly to environment AB, more than 4,000 epochs are needed to evolve the delayed *XOR* function (Fig. 6, gray curve). In contrast, populations evolve faster in environments of lower complexity, such as the environments A and B (Fig. 6, red/blue lines). Remarkably, if we sample equal amounts of cells from A and B and expose the new population in the complex environment AB with all other parameters being equal (size of population, average nutrient concentration, etc.), XOR phenotypes of high fitness appear surprisingly fast (Fig. 6b). This effect is even more pronounced in the presence of HGT, where the fittest phenotype arises twice as fast as those without HGT present (Fig. 6b, insert). Analysis of individual simulation runs results in similar observations, with all experiments leading to phenotypes of increased fitness in the presence of HGT.

Detailed statistics of the evolution probability and speed is shown in Table 1. In "single-step" evolution (un-evolved → XOR) only 18 of 32 (56%) experiments were successful and terminated with an evolved *XOR* population (after 4,000 epochs). Success probability of the "dual-step" adaptation process was estimated as a product of "single-step" probabilities and equals 91% and 82% percent with and without HGT, respectively. HGT accelerates emergence of the combined phenotype in {*A, B*}

Table 1. Rate of adaptation a complex *XOR* environment in different experimental scenarios. The probability and the speed of phenotype emergence are shown for two fitness thresholds 0.75 (evolved organism) and 0.90 (refined evolved organism). Average speed is the average epoch number at which maximum fitness surpasses the threshold.

| | Emergence of the organism with fitness w | | | |
| | $w>0.75$ | | $w>0.90$ | |
	Success Rate	Average speed, *epochs*	Success Rate	Average speed, *epochs*
Un-evolved → *XOR*	18/32	2485	15/32	2489
Un-evolved → *OR*	29/32	1179	13/32	>4,000
OR → *XOR*	30/32	210	5/32	2093
Acceleration by stepwise adaptation		1.8		–
Un-evolved → *A*	30/32	1043	29/32	1067
Un-evolved → *B*	31/32	1217	31/32	1319
{*A* & *B*} → *XOR*	58/64	234	47/64	448
Acceleration by stepwise adaptation		1.7		1.4
{*A* & *B*} → *XOR* + **HGT**	64/64	138	48/64	406
Acceleration by HGT		1.7		1.1

mixed populations by a factor of 1.7. However the probability and the speed of the phenotypic refinement for fitness levels above 0.9 is less affected by HGT relative to the initial emergence of the phenotype above the 0.75 threshold (note that any phenotype with 0.75 Pearson correlation between metabolic pathway expression and nutrients exhibits the XOR I/O characteristic). This is to be expected, since subsequent fine-tuning is due to mutations, and not insertion of new functional fragments from other organisms. Evolution through a single environment of intermediate complexity (un-evolved → OR → XOR) accelerates the evolution of a XOR phenotype by a factor 1.8, but with a lower probability of highly fit cells to appear in the final population (only in 5 out of 32 experiments, cells with fitness higher than $w>0.90$ emerged).

4.2 Effect of Horizontal Gene Transfer on the Network Organization

The full gene regulatory and biochemical networks of evolved cells are usually too complex to analyze since many of the connections are not relevant to the observed phenotype. To address this, we employed a reduction algorithm described elsewhere [10] to extract the "minimal" network that encompasses only essential connections. As shown in Table 2, average fitness of reduced minimal networks is at least 95% of the full network's fitness, however the average number of regulatory edges is significantly reduced: from 338 to 14.1 and from 335 to 10.6 with and without HGT respectively. Presence of HGT events results in larger networks that are considerably more sparse (0.39 *vs.* 0.22), but with the same average sparsity and reduced network size difference when it comes to their minimal counterparts.

Table 2. Complete and minimal network statistics for populations evolved in a XOR phenotype with and without HGT

| | Full network | | | | Minimal network | | | |
	no HGT		HGT		no HGT		HGT	
Fitness *(St. Dev.)*	0.81	*(0.052)*	0.79	*(0.044)*	0.78	*(0.006)*	0.75	*(0.006)*
Triplets	8.8		13.8		5.5		6.7	
Links *(St. Dev.)*	335	*(157)*	338	*(136)*	10.6	*(0.03)*	14.1	*(0.03)*
Sparsity	0.39		0.22		0.11		0.10	
Modularity	3.8		10.1		3.3		3.1	

4.3 Distribution of Fitness Effect of Mutational and HGT Events

Mutations and HGT events differ in magnitude and direction when it comes to their fitness effect. Traditionally, models rely on theoretical or experimentally constructed distributions of fitness effect (DFE) when introducing mutations in a population. For mutations, these distributions have been measured experimentally for viruses and bacteria (e.g. [20-22]) and have also been obtained theoretically (e.g. [23] and references therein). In general, it is assumed that most mutations have a neutral or nearly neutral effect and the vast majority of mutations have a negative fitness effect [23]. In bacteriophage F1, 20% of single point mutations were found to be lethal, while the mean fitness decrease was around 11% [21]. In *E. coli*, the average effect of spontaneous deleterious mutations and random insertions is less than 1% and 3%, respectively [20, 22].

Table 3. Proprieties of Distribution of Fitness Effect for mutational and HGT events in evolved and un-evolved populations

	Mean fitness change, %	Fitness variance	Skewness	Kurtosis	Percent of lethal events
Mutations:					
Un-evolved populations	-6.6%	0.109	-0.014	3.10	33.0 %
Evolved populations	-4.5%	0.032	-1.028	6.39	3.2 %
Horizontal Gene Transfer:					
Un-evolved populations	-5.1%	0.063	-0.129	4.49	14.4 %
Evolved populations	-6.6%	0.048	-0.064	4.70	4.6 %

Here, we use our *in silico* simulation framework to investigate the shape and changes in the DFE for both mutations and HGT. Since each organism has its own regulatory network that results to a distinct phenotypic behavior, we are able to calculate fitness before and immediately after any HGT event by looking at the expression levels of the response pathway. This allows us to profile the shape of DFE along the evolutionary trajectory and to account for genetic drift, which can be a significant force in small populations. In both mutational and HGT DFEs, there is a profound decrease in the number of lethal events (i.e. fitness effect equal to -1) in evolved populations versus the non-evolved populations (Table 3), a clear indication of mutational robustness. Furthermore, we observe a decrease in variance and increase in kurtosis (sharpness) of the DFE as populations evolve, both for HGT and mutations, although the effect is more profound in the latter case. As population evolves, mutational DFE becomes more skewed towards negative fitness effects, which is to be expected as most mutations in an evolved organism result in a decreased fitness. Interestingly, the DFE of HGT events becomes more symmetric in evolved populations, as the probability for HGT to transfer a beneficial or disrupting fragment increases (the first because of the availability of beneficial sub-networks, the second because of the high ratio of already fine-tuned cells in the population which can be disrupted by a HGT event).

5   Discussion

To elucidate the effect of Horizontal Gene Transfer in bacterial evolution, we used *in silico* microbial organisms that compete and evolve under dynamic environments in the presence of HGT. The simulation framework presented here is the first that incorporates models of cellular and evolutionary processes, together with representations of the environment, population, organism, biological network and molecular species. This allows us to address questions that transcend many levels of biological organization and investigate the impact of phenomena, such as HGT and environmental perturbations in an unprecedented scale, albeit at the price of increased computational complexity.

Our results show that multi-step evolution accelerates the emergence of complex traits, especially in the presence of HGT, and illustrate its effect on adaptation and network organization. We showed that the distribution of fitness effects for HGT presents some notable differences from its mutational counterpart. There are many future directions to explore in order to increase the scope and biological realism of our simulator. The current framework will benefit from the addition of a spatial component, since in the current setting we assumed a well-mixed, homogeneous environment which clearly limits us on the number of hypotheses we can test. This will allow us to investigate individually the various HGT mechanisms, whose effect vary greatly with the spatial landscape of the environment and population structure. Furthermore, the biological realism of the underlying network can be improved by refining the models that capture cellular processes, as well as adding a metabolic layer and its corresponding models. Despite its limitations, this work is ground-breaking by creating an overarching model of biological phenomena, a synthetic environment, where hypotheses can be tested or automatically generated. This, in conjunction with advanced HPC techniques can prove to be transformative in predicting evolution and microbial behavior in general.

Acknowledgments. Authors would like to acknowledge Kitrick Sheets (NCSA) for assistance in parallel code development, Bob Miller and Dr. Kwan-Liu Ma (UC Davis) for the automated network visualization tool, and the members of the Tagkopoulos Lab for their comments. This research was supported by the NSF-OCI grant 0941360 and is part of the Blue Waters Project.

References

1. Ragan, M.A., Beiko, R.G.: Lateral genetic transfer: open issues. Philosophical Transactions of the Royal Society B-Biological Sciences 364, 2241–2251 (2009)
2. Boto, L.: Horizontal gene transfer in evolution: facts and challenges. Proc. Biol. Sci. 277, 819–827 (2010)

3. Koonin, E.V., Makarova, K.S., Aravind, L.: Horizontal gene transfer in prokaryotes: Quantification and classification. Annual Review of Microbiology 55, 709–742 (2001)

4. Gogarten, J.P., Doolittle, W.F., Lawrence, J.G.: Prokaryotic evolution in light of gene transfer. Molecular Biology and Evolution 19, 2226–2238 (2002)

5. Koslowski, T., Zehender, F.: Towards a quantitative understanding of horizontal gene transfer: A kinetic model. Journal of Theoretical Biology 237, 23–29 (2005)

6. Nielsen, K.M., Townsend, J.P.: Monitoring and modeling horizontal gene transfer. Nature Biotechnology 22, 1110–1114 (2004)

7. Novozhilov, A.S., Karev, G.P., Koonin, E.V.: Mathematical modeling of evolution of horizontally transferred genes. Molecular Biology and Evolution 22, 1721–1732 (2005)

8. Levin, B.R., Cornejo, O.E.: The Population and Evolutionary Dynamics of Homologous Gene Recombination in Bacteria. PLoS Genetics 5, Article No.: e1000601 (2009)

9. Philipsen, K.R., Christiansen, L.E., Hasman, H., Madsen, H.: Modelling conjugation with stochastic differential equations. Journal of Theoretical Biology 263, 134–142 (2010)

10. Tagkopoulos, I., Liu, Y.C., Tavazoie, S.: Predictive behavior within microbial genetic networks. Science 320, 1313–1317 (2008)

11. Duda, R.O., Hart, P.E.: Pattern classification and scene analysis. Wiley-Interscience, Hoboken (1973)

12. Ando, T., Itakura, S., Uchii, K., Sobue, R., Maeda, S.: Horizontal transfer of nonconjugative plasmid in colony biofilm of Escherichia coli on food-based media. World Journal of Microbiology & Biotechnology 25, 1865–1869 (2009)

13. Baur, B., Hanselmann, K., Schlimme, W., Jenni, B.: Genetic transformation in freshwater: Escherichia coli is able to develop natural competence. Appl. Environ. Microbiol. 62, 3673–3678 (1996)

14. Jiang, S.C., Paul, J.H.: Gene transfer by transduction in the marine environment. Applied and Environmental Microbiology 64, 2780–2787 (1998)

15. McDaniel, L., Young, E., Delaney, J., Ruhnau, F., Ritchie, K., Paul, J.: High Frequency of Horizontal Gene Transfer in the Oceans. Nature 330, 1 (2010)

16. Park, S.C., Simon, D., Krug, J.: The Speed of Evolution in Large Asexual Populations. Journal of Statistical Physics 138, 381–410 (2010)

17. Gerrish, P.J., Lenski, R.E.: The fate of competing beneficial mutations in an asexual population. Genetica 102-103, 127–144 (1998)

18. Kashtan, N., Noor, E., Alon, U.: Varying environments can speed up evolution. Proceedings of the National Academy of Sciences of the United States of America 104, 13711–13716 (2007)

19. Parter, M., Kashtan, N., Alon, U.: Facilitated Variation: How Evolution Learns from Past Environments To Generalize to New Environments. Plos Computational Biology 4 (2008)

20. Elena, S.F., Ekunwe, L., Hajela, N., Oden, S.A., Lenski, R.E.: Distribution of fitness effects caused by random insertion mutations in Escherichia coli. Genetica 102-103, 349–358 (1998)

21. Peris, J.B., Davis, P., Cuevas, J.M., Nebot, M.R., Sanjuan, R.: Distribution of Fitness Effects Caused by Single-Nucleotide Substitutions in Bacteriophage f1. Genetics 185, U308–U603 (2010)

22. Kibota, T.T., Lynch, M.: Estimate of the genomic mutation rate deleterious to overall fitness in E-coli. Nature 381, 694–696 (1996)

23. Eyre-Walker, A., Keightley, P.D.: The distribution of fitness effects of new mutations. Nature Reviews Genetics 8, 610–618 (2007)

Comparative Evaluation of Set-Level Techniques in Microarray Classification

Jiri Klema[1], Matej Holec[1], Filip Zelezny[1], and Jakub Tolar[2]

[1] Faculty of Electrical Engineering, Czech Technical University in Prague
[2] Department of Pediatrics, University of Minnesota, Minneapolis

Abstract. Analysis of gene expression data in terms of a priori-defined gene sets typically yields more compact and interpretable results than those produced by traditional methods that rely on individual genes. The set-level strategy can also be adopted in predictive classification tasks accomplished with machine learning algorithms. Here, sample features originally corresponding to genes are replaced by a much smaller number of features, each corresponding to a gene set and aggregating expressions of its members into a single real value. Classifiers learned from such transformed features promise better interpretability in that they derive class predictions from overall expressions of selected gene sets (e.g. corresponding to pathways) rather than expressions of specific genes. In a large collection of experiments we test how accurate such classifiers are compared to traditional classifiers based on genes. Furthermore, we translate some recently published gene set analysis techniques to the above proposed machine learning setting and assess their contributions to the classification accuracies.

Keywords: gene set, classifer, learning, predictive accuracy.

1 Introduction

Set-level techniques have recently attracted significant attention in the area of gene expression data analysis [20,9,13,18,14,23]. Whereas in traditional analysis approaches one typically seeks individual genes differentially expressed across sample classes (e.g. cancerous vs. control), the set-level approach aims to identify entire sets of genes that are significant e.g. in the sense that they contain an unexpectedly large number of differentially expressed genes. The gene sets considered for significance testing are defined prior to analysis, using appropriate biological background knowledge. The main advantage brought by set-level analysis is the improved interpretability of analysis results. Indeed, the long lists of differentially expressed genes characteristic of traditional expression analysis are replaced by shorter and more informative lists of actual biological processes.

Predictive classification [11] is a form of data analysis going beyond the mere identification of differentially expressed units. Here, units deemed significant for the discrimination between sample classes are assembled into formal models prescribing how to classify new samples whose class labels are not yet known. Predictive classification techniques are thus especially relevant to diagnostic tasks and

J. Chen, J. Wang, and A. Zelikovsky (Eds.): ISBRA 2011, LNBI 6674, pp. 274–285, 2011.

as such have been explored since very early studies on microarray data analysis
[10]. Predictive models are usually constructed by supervised machine learning
algorithms [11] that automatically discover patterns among samples whose labels are already available (so-called *training samples*). Learned classifiers may
take diverse forms ranging from geometrically conceived models such as *Support
Vector Machines* [24], which have been especially popular in the gene expression
domain, to symbolic models such as logical rules or decision trees that have also
been applied in this area [27,15].

The main motivation for extending the set-level framework to the machine
learning setting is again the interpretability of results. Informally, classifiers
learned using set-level features acquire forms such as "predict cancer if pathway P1 is active and pathway P2 is not" (where *activity* refers to aggregated
expressions of the member genes). In contrast, classifiers learned in the standard
setting derive predictions from expressions of individual genes; it is usually difficult to find relationships among the genes involved in such a classifier and to
interpret the latter in terms of biological processes.

The described feature transformation incurs a significant compression of the
training data since the number of considered gene sets is typically much smaller
than the number of interrogated genes. This raises the natural question whether
relevant information is lost in the transformation, and whether the augmented
interpretability will be traded off for decreased predictive accuracy. The main
objective of this study is to address this question experimentally.

A further important objective is to evaluate—from the machine learning
perspective—statistical techniques proposed recently in the research on set-level
gene expression analysis. These are namely the Gene Set Enrichment Analysis
(GSEA) method [20], the SAM-GS algorithm [7] and a technique known as the
Global test [9]. Informally, they rank a given collection of gene sets according to
their correlation with phenotype classes. The methods naturally translate into
the machine learning context in that they facilitate feature selection [17], i.e. they
are used to determine which gene sets should be provided as sample features to
the learning algorithm. We experimentally verify whether these methods work
reasonably in the classification setting, i.e. whether learning algorithms produce
better classifiers from gene sets ranked high by the mentioned methods than
from those ranking lower. We investigate classification conducted with a single
selected gene set as well as with a batch of high ranking sets.

To use a machine learning algorithm, a unique value for each feature of each
training sample must be established. Set-level features correspond to multiple
expressions and these must therefore be aggregated. We comparatively evaluate two aggregatation options. The first simply averages the expressions of the
involved genes, whereas the second relies on the more sophisticated method proposed by [23] and based on singular value decomposition.

Let us return to the initial experimental question concerned with how the final
predictive accuracy is influenced by the training data compression incurred by re-
formulating features to the gene set level. As follows from the above, two factors
contribute to this compression: selection (not every gene from the original sample

representation is a member of a gene set used in the set-level representation, i.e. some interrogated genes become ignored) and aggregation (for every gene set in the set-level representation, expressions of all its members are aggregated into a single value). We quantify the effects of these factors on predictive accuracy. Regarding selection, we experiment with set-level representations based on 10 best gene sets and 1 best gene set, respectively, and we do this for all three of the above-mentioned selection methods. We compare the obtained accuracies to the baseline case where all individual genes are provided as features to the learning algorithm. For each of the selection cases, we want to evaluate the contribution of the aggregation factor. This is done by comparing both of the above mentioned aggregation mechanisms to the control case where no aggregation is performed at all; in this case, individual genes combined from the selected gene groups act as features.

The contribution of the present study lies in the thorough experimental evaluation of a number of aspects and techniques of the gene set framework employed in the machine learning context. Our contribution is, however, also significant beyond the machine learning scope. In the general area of set-level expression analysis, it is undoubtedly important to establish a performance ranking of the various statistical techniques for the identification of significant gene sets in class-labeled expression data. This is made difficult by the lack of an unquestionable ranking criterion—there is in general no ground truth stipulating which gene sets should indeed be identified by the tested algorithms. The typical approach embraced by comparative studies (such as [7]) is thus to appeal to intuition (e.g. *the p53 pathway should be identified in p53-gene mutation data*). However legitimate such arguments are, evaluations based on them are obviously limited in generality and objectivity. We propose that the predictive classification setting supported by the cross-validation procedure for unbiased accuracy estimation, as adopted in this paper, represents exactly such a needed framework enabling objective comparative assessment of gene set selection techniques. In this framework, results of gene set selection are deemed good if the selected gene sets allow accurate classification of new samples. Through cross-validation, the accuracy can be estimated in an unbiased manner.

The rest of the paper is organized as follows. The next section describes the specific methods and data sets used in our experiments. In Section 3 we expose the experimental results. Section 4 summarizes the main conclusions and proposes directions for follow-up research.

2   Methods and Data

Here we first describe the methods adopted for gene set ranking, gene expression aggregation, and for classifier learning. Next we present the data sets used as benchmarks in the comparative experiments. Lastly, we describe the protocol followed by our experiments.

2.1   Gene Set Ranking

Three methods are considered for gene set selection. As inputs, all of the methods assume a set $G = \{g_1, g_2, \ldots g_n\}$ of interrogated genes, and a set S of m

expression samples where for each $s_i \in S$, $s_i = (e_{1,i}, e_{2,i}, \ldots e_{n,i}) \in \mathbb{R}^n$ where $e_{j,i}$ denotes the (normalized) expression of gene g_j in sample s_i. The sample set S is partitioned into phenotype classes $S = C_1 \cup C_2 \cup \ldots \cup C_o$ so that $C_i \cap C_j = \{\}$ for $i \neq j$. For simplicity in this paper we assume binary classification, i.e. $o = 2$. A further input is a collection of gene sets $\mathcal{G}$ such that for each $\Gamma \in \mathcal{G}$ it holds $\Gamma \subseteq G$. In the output, each of the methods ranks all gene sets in $\mathcal{G}$ by their estimated power to discriminate samples into the predefined classes.

Next we give a brief account of the three methods and refer to the original sources for a more detailed description. In experiments, we used the original implementations of the procedures as provided by the respective authors.

GSEA [20]. *Gene set enrichment analysis* tests a null hypothesis that gene rankings in a gene set Γ, according to an association measure with the phenotype, are randomly distributed over the rankings of all genes. It first sorts G by correlation with binary phenotype. Then it calculates an enrichment score (ES) for each $\Gamma \in \mathcal{G}$ by walking down the sorted gene list, increasing a running-sum statistic when encountering a gene $g_i \in \Gamma$ and decreasing it otherwise. The magnitude of the change depends on the correlation of g_i with the phenotype. The enrichment score is the maximum deviation from zero encountered in the random walk. The statistical significance of the ES is estimated by an empirical phenotype-based permutation test procedure that preserves the correlation structure of the gene expression data. GSEA was one of the first specialized gene-set analysis techniques. It has been reported to attribute statistical significance to gene sets that have no gene associated with the phenotype, and to have less power than other recent test statistics [7,9].

SAM-GS [7]. This method tests a null hypothesis that the mean vectors of the expressions of genes in a gene set do not differ by phenotype. Each sample s_i is viewed as a point in an n-dimensional Euclidean space. Each gene set $\Gamma \in \mathcal{G}$ defines its $|\Gamma|$-dimensional subspace in which projections s_i^Γ of samples s_i are given by coordinates corresponding to genes in Γ. The method judges a given Γ by how distinctly the clusters of points $\{s_i^\Gamma | s_i \in C_1\}$ and $\{s_j^\Gamma | s_j \in C_2\}$ are separated from each other in the subspace induced by Γ. SAM-GS measures the Euclidean distance between the centroids of the respective clusters and applies a permutation test to determine whether, and how significantly, this distance is larger than one obtained if samples were assigned to classes randomly.

Global Test [9]. The global test, analogically to SAM-GS, projects the expression samples into subspaces defined by gene sets $\Gamma \in \mathcal{G}$. In contrast to the Euclidean distance applied in SAM-GS, it proceeds instead by fitting a regression function in the subspace, such that the function value acts as the class indicator. The degree to which the two clusters are separated then corresponds to the magnitude of the coefficients of the regression function.

2.2   Expression Aggregation

Two methods are considered for assigning a value to a given gene set Γ for a given sample s_i by aggregation of expressions of genes in Γ.

Averaging. The first method simply produces the arithmetic average of the expressions of all Γ genes in sample s_i. The value assigned to the pair (s_i, Γ) is thus independent of samples s_j, $i \neq j$.

Singular Value Decomposition. A more sophisticated approach was employed by [23]. Here, the value assigned to (s_i, Γ) depends on other samples s_j. In particular, all samples in the sample set S are viewed as points in the $|\Gamma|$-dimensional Euclidean space induced by Γ the same way as explained in Section 2.1. Subsequently, the specific vector in the space is identified, along which the sample points exhibit maximum variance. Each point $s_k \in S$ is then projected onto this vector. Finally, the value assigned to (s_i, Γ) is the real-valued position of the projection of s_i on the maximum-variance vector in the space induced by Γ. We refer to the paper [23] for detailed explanation.

2.3   Machine Learning

We experimented with five diverse machine learning algorithms to avoid dependence of experimental results on a specific choice of a learning method, namely Support Vector Machine, 1-Nearest Neighbor, 3-Nearest Neighbors, Naive Bayes and Decision Tree. These algorithms are explained in depth for example by [11]. In experiments, we used the implementations available in the WEKA software due to [25], using the default settings. None of the methods above is in principle superior to the others, although the first one prevails in predictive modeling of gene expression data and is usually associated with high resistance to noise.

2.4   Expression and Gene Sets

We conducted our experiments using 20 public gene expression datasets, each containing samples pertaining to two classes. Table 1 shows for each dataset the number of samples in each class, the number of interrogated genes and the reference for further details. Some of the two-class datasets were derived from the three-class problems (Colitis and Crohn, Parkinson).

Besides expression datasets, we utilized a gene set database consisting of 1685 manually curated sets of genes obtained from the Molecular Signatures Database (MSigDB v2.0) [20]. These gene sets have been compiled from various online databases (e.g. KEGG, GenMAPP, BioCarta).

2.5   Experimental Protocol

Classifier learning in the set-level framework follows a simple workflow whose performance is influenced by several factors, each corresponding to a particular choice from a class of techniques (such as for gene set ranking). We evaluate the

Table 1. Number of genes interrogated and number of samples in each of the two classes of each dataset

Dataset	Genes	Class 1	Class 2	Reference
ALL/AML	10056	24	24	[1]
Brain/muscle	13380	41	20	[13]
Colitis and Crohn 1	14902	42	26	[4]
Colitis and Crohn 2	14902	42	59	[4]
Colitis and Crohn 3	14902	26	59	[4]
Diabetes	13380	17	17	[18]
Heme/stroma	13380	18	33	[13]
Gastric cancer	5664	8	22	[12]
Gender	15056	15	17	[20]
Gliomas	14902	26	59	[8]
Lung Cancer Boston	5217	31	31	[3]
Lung Cancer Michigan	5217	24	62	[2]
Melanoma	14902	18	45	[21]
p53	10101	33	17	[20]
Parkinson 1	14902	22	33	[19]
Parkinson 2	14902	22	50	[19]
Parkinson 3	14902	33	50	[19]
Pollution	37804	88	41	[16]
Sarcoma and hypoxia	14902	15	39	[26]
Smoking	5664	18	26	[5]

contribution of these factors to the predictive accuracy of the resulting classifiers through repeated executions of the learning workflow, varying the factors.

The learning workflow is shown in Fig. 1. Given a set of binary-labeled training samples from an expression dataset, the workflow starts by ranking the provided collection of a priori-defined gene sets according to their power to discriminate sample classes (see Sec. 2.1 for details). The resulting ranked list is subsequently used to select the gene sets used to form set-level sample features. Each such feature is then assigned a value for each training sample by aggregating the expressions in the gene set corresponding to the feature; an exception to this is the *none* alternative of the aggregation factor, where expressions are not aggregated, and features correspond to genes instead of gene sets. This alternative is considered for comparative purposes. Next, a machine learning algorithm produces a classifier from the reformulated training samples. Finally, the classifier's predictive accuracy is calculated as the proportion of samples correctly classifed on an independent testing sample fold. For compatibility with the learned classifier, the testing samples are also reformulated to the set level prior to testing, using the selected gene sets and aggregation as in the training phase.

Six factors along the workflow influence its result. The alternatives considered for each of them are summarized in Table 2. We want to assess the contributions of the first three factors (top in table). The remaining three auxiliary factors (bottom in table) are employed to diversify the experimental material and thus increase the robustness of the findings. Factor 6 (testing fold) is involved

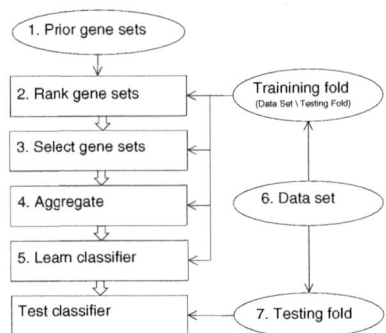

Fig. 1. The workflow of a set-level learning experiment conducted multiple times with varying alternatives in the numbered steps. For compatibility with the learned classifier, testing fold samples are also reformulated to the set level. This is done using gene sets selected in Step 3 and aggregation algorithm used in Step 4. The diagram abstracts from this operation.

automatically through the adoption of the 10-fold cross-validation procedure (see e.g. [11], chap. 7). We execute the workflow for each possible combination of factor alternatives, obtaining a factored sample of 198,000 predictive accuracy values.

While the measurements provided by the above protocol allow us to compare multiple variants of the set-level framework for predictive classification, we also want to compare these to the baseline gene-level alternative usually adopted in predictive classification of gene expression data. Here, each gene interrogated by a microarray represents a feature. This sample representation is passed directly to the learning algorithm without involving any of the pre-processing factors (1-3 in Table 2). The baseline results are also collected using the 5 different learning algorithms, the 20 benchmark datasets and the 10-fold crossvalidation procedure (i.e. factors 4-6 in Table 2 are employed). As a result, an additional sample of 1,000 predictive accuracy values are collected for the baseline variant.

Finally, to comply with the standard application of the cross-validation procedure, we averaged the accuracy values corresponding to the 10 cross-validation folds for each combination of the remaining factors. The subsequent statistical analysis thus deals with a sample of 19,800 and 100 measurements for the set-level and baseline experiments, described by the predictive accuracy value and the values of the relevant factors.

3   Results

All statistical tests in this section refer to the paired non-parametric Wilcoxon test (two-sided unless stated otherwise).[1] For pairing, we always related two

---

[1] Preliminary normality tests did not justify the application of the stronger t-test. Besides, the Wilcoxon test is argued [6] to be statistically safer than the t-test for comparing classification algorithms over multiple data sets.

Table 2. Alternatives considered for factors influencing the set-level learning workflow. The number left of each factor refers to the workflow step (Fig. 1) in which it acts.

Analyzed factors	Alternatives	#Alts
1. Ranking algo (Sec. 2.1)	{gsea, sam-gs, global}	3
2. Sets forming features*	{1, 2, . . . 10,	
	1676, 1677, . . . 1685,	
	1:10, 1676:1685}	22
3. Aggregation (Sec. 2.2)	{svd, avg, none}	3
Product		198

Auxiliary factors	Alternatives	#Alts
4. Learning algo (Sec. 2.3)	{svm, 1-nn, 3-nn, nb, dt}	5
5. Data set (Sec. 2.4)	{$d_1 \ldots d_{20}$}	20
6. Testing Fold	{$f_1 \ldots f_{10}$}	10
Product		1000

* identified by rank. 1685 corresponds to the lowest ranking set. $i{:}j$ denotes that all of gene sets ranking i to j are used to form features.

measurements equal in terms of all factors except for the one investigated. All significance results are at the 0.05 level.

Using the set-level experimental sample, we first verified whether gene sets ranked high by the established set-level analysis methods (GSEA, SAM-GS and Global test) indeed lead to construction of better classifiers by machine learning algorithms, i.e. we investigated how classification accuracies depend on *Factor 3* (see Table 2). In the top panel of Fig. 2, we plot the average accuracies for Factor 3 alternatives ranging 1 to 10, and 1676 to 1685. The trend line fitted by the least squares method shows a clear decay of accuracy as lower-ranking sets are used for learning. The bottom panel corresponds to Factor 3 values 1:10 (left) and 1676:1685 (right) corresponding to the situations where the 10 highest-ranking and the 10 lowest-ranking (respectively) gene sets are combined to produce a feature set for learning. Again, the dominance of the former in terms of accuracy is obvious.

Given the above, there is no apparent reason why low-ranking gene sets should be used in experiments. Therefore, to maintain relevance of the subsequent conclusions, we conducted further analyses only with measurements where Factor 2 (gene set rank) is either 1 or 1:10.

Firstly, we assessed the difference between the remaining alternatives 1 and 1:10 corresponding to more and less (respectively) compression of training data. Not surprisingly, the 1:10 variant, where sample features capture information from the ten best gene sets exhibits significantly ($p = 0.0007$) higher accuracies than the 1 variant using only the single best gene set to consitute features (a single feature if aggregation is employed).

We further compared the three gene-set ranking methods by splitting the set-level sample according to *Factor 1*. Since three comparisons are conducted in this case (one per pair), we used the Bonferroni-Dunn adjustment on the Wilcoxon test result. The Global test turned out to exhibit significantly higher accuracies

282   J. Klema et al.

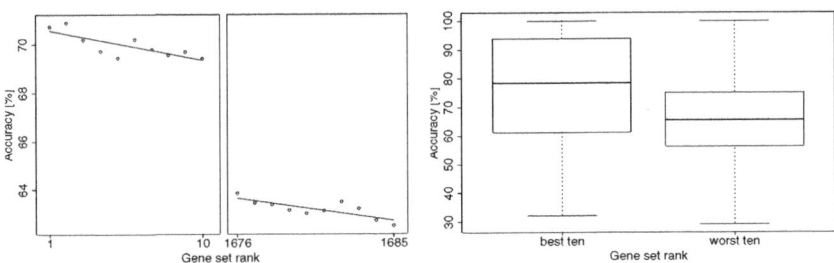

Fig. 2. Average predictive accuracy tends to fall as lower-ranking gene sets are used
to constitute features (see text for details). Each point in the left panels and each box
plot in the right panel follows from 16,000 learning experiments. The trend lines shown
in the left panels are the ones minimizing the residual least squares.

than either SAM-GS (p = 0.013) or GSEA (p = 0.027). The difference between
the latter two methods was not significant.

Concerning *Factor 3* (aggregation method), there are two questions of interest:
whether one aggregation method (svd, avg) outperforms the other, and whether
aggregation in general has a detrimental effect on performance. As for the first
question, no significant difference between the two methods was detected. The
answer to the second question turned out to depend on Factor 3 as follows.
In the more compressive (1) alternative, the answer is affirmative in that both
aggregation methods result in less accurate classifiers than those not incurring
aggregation (p = 0.015 for svd, p = 0.00052 for avg, both after Bonferroni-Dunn
adjustment). However, the detrimental effect of aggregation vanishes in the less
compressive (1:10) alternative of Factor 2, where none of the two comparisons
yield a significant difference.

The principle trends can also be well observed through the ranked list of
methodological combinations by median classification accuracy, again generated
from measurements not involving random or low-ranking gene sets. This is shown
in Table 3. Position 8 refers to the baseline method where sample features capture
expressions of all genes and prior gene set definitions are ignored (see Section 2.5
for details). In agreement with the statistical conclusions above, the ranked table
clearly indicates the superiority of the Global test for gene-set ranking, and of
using the 10 best gene sets (i.e., the 1:10 alternative) to establish features rather
than relying only on the single best gene set. It is noteworthy that all three
methods involving the combinations of the Global test and the 1:10 alternative
(i.e., ranks 1, 2, 4) outperform the baseline method. This is especially remarkable given that the two best of them (and two best overall) involve aggregation,
and the learning algorithm here receives training samples described by only 10
real-valued features. Thus, the gene-set framework allows for feature extraction
characterized by vast compression of data (from the original thousands of features corresponding to expressions of individual genes, to 10 features) and, at
the same time, by a boost in classification accuracy.

Table 3. Ranking of combinations of gene set methods by median predictive accuracy achieved on 20 datasets (Table 1, Section 2.4) with 5 machine learning algorithms (Section 2.3) estimated through 10-fold cross-validation (i.e. 1,000 experiments per row). The columns indicate, respectively, the resulting rank by median accuracy, the gene sets used to form features (1 – the highest ranking set, 1:10 – the ten highest ranking sets), the gene set selection method, the expression aggregation method (see Section 2 for details on the latter 3 factors), and the median, average, standard deviation and interquartile range of the accuracy.

Rank	*Methods*			*Accuracy*			
	Sets	*Rank. algo*	*Aggrgt*	*Median*	*Avg*	*σ*	*Iqr*
1	1:10	global	svd	86.5	79.8	17.3	32.0
2	1:10	global	avg	86.0	79.4	17.8	30.5
3	1:10	sam-gs	none	83.8	78.3	18.5	35.1
4	1:10	global	none	83.7	77.7	18.5	34.7
5	1:10	gsea	none	82.8	77.7	18.8	34.8
6	1	global	none	80.5	78.1	16.1	29.7
7	1:10	gsea	avg	79.7	76.3	17.1	28.0
8	*all genes used*			79.3	77.2	18.9	35.3
9	1	gsea	none	77.5	75.0	18.3	33.3
10	1	global	svd	77.5	74.8	15.0	25.6
11	1:10	gsea	svd	77.1	75.5	16.9	28.2
12	1:10	sam-gs	avg	74.2	75.1	16.8	28.5
13	1	sam-gs	none	73.9	74.1	15.1	26.3
14	1:10	sam-gs	svd	73.8	74.6	17.6	28.9
15	1	global	avg	72.8	72.2	14.0	22.2
16	1	gsea	avg	68.3	69.6	13.0	16.3
17	1	gsea	svd	67.4	68.5	13.2	14.4
18	1	sam-gs	avg	65.4	64.7	10.3	15.9
19	1	sam-gs	svd	64.2	65.0	12.7	13.0

4   Conclusions and Future Work

The set-level framework can be adopted in the machine learning setting without trading off classification accuracy. To identify the best a priori-defined gene sets for classification, the *Global test* [9] significantly outperforms the *GSEA* [20] and *SAM-GS* [7] methods. To aggregate expressions of genes contained in a gene set into a value assigned to that set acting as a feature, arithmetic average could not be differentiated from the method [23] based on singular value decomposition. Using only 10 features corresponding to genuine gene sets selected by the Global test, the learned set-level classifiers systematically outperform conventional gene-level classifiers learned with access to all measured gene expressions. Data compression and increased classification accuracy thus come as additional benefits to increased interpretability of set-level classifiers.

The above-mentioned effect of data shrinkage accompanied by increased predictive accuracy could, in principle, also be achieved by generic feature extraction methods (see e.g. [17]). The advantage of our approach is that our extracted features maintain direct interpretability since they correspond to gene sets that possess a biological meaning. In future work, it would be interesting to determine whether the generic feature extraction methods could outperform the present approach at least in terms of predictive accuracy achieved with a fixed target number of extracted features. By the same mail, the optimal number of set-level features employed will vary between data domains. For our experiments, we chose the ad hoc number of 10 features for all domains. In future experiments, the optimal domain-specific number may be estimated, e.g. through internal cross-validation [11].

We applied two previously suggested general methods enabling aggregation of multiple expression values into a single value assigned to a set-level feature. The downside of this generality is that substantial information available for specific kinds of gene sets is ignored. Of relevance to pathway-based gene sets, the recent study by [22] convincingly argues that the perturbation of a pathway depends on the expressions of its member genes in a non-uniform manner. It also proposes how to quantify the impact of each member gene on the perturbation, given the graphical structure of the pathway. It seems reasonable that a pathway-specific aggregation method should also weigh member genes by their estimated impact on the pathway. Such a method would likely result in more informative pathway-level features and could outperform the two aggregagation methods we have considered, potentially giving a futher boost to the good performance of predictive classification based on a small number of set-level features.

Acknowledgement

This research was supported by the Czech Science Foundation through project No. 201/09/1665 (FZ, MH) and the Czech Ministry of Education through research programme MSM 6840770012 (JK).

References

1. Armstrong, S.A., et al.: MLL translocations specify a distinct gene expression profile that distinguishes a unique leukemia. Nat. Genet. 30, 41–47 (2002)
2. Beer, D.G., et al.: Gene-expression profiles predict survival of patients with lung adenocarcinoma. Nat. Med. 8(8), 816–824 (2002)
3. Bhattacharjee, A., et al.: Classification of human lung carcinomas by mrna expression profiling reveals distinct adenocarcinoma subclasses. Proc. Natl. Acad. Sci. 98(24), 13790–13795 (2001)
4. Burczynski, M.E., et al.: Molecular classification of Crohn's disease and ulcerative colitis patients using transcriptional profiles in peripheral blood mononuclear cells. 8(1), 51–61 (2006)
5. Carolan, B.J., et al.: Up-regulation of expression of the ubiquitin carboxyl-terminal hydrolase L1 gene in human airway epithelium of cigarette smokers. Cancer Res. 66(22), 10729–10740 (2006)
6. Demšar, J.: Statistical comparisons of classifiers over multiple data sets. JMRL 7, 1–30 (2006)

7. Dinu, I.: Improving gene set analysis of microarray data by SAM-GS. BMC Bioinformatics 8(1), 242 (2007)
8. Freije, W.A., et al.: Gene expression profiling of gliomas strongly predicts survival. Cancer Res. 64(18), 6503–6510 (2004)
9. Goeman, J.J., Bühlmann, P.: Analyzing gene expression data in terms of gene sets: methodological issues. Bioinformatics 23(8), 980–987 (2007)
10. Golub, T.R., et al.: Molecular classification of cancer: Class discovery and class prediction by gene expression monitoring. Science 286(5439), 531–537 (1999)
11. Hastie, T., et al.: The Elements of Statistical Learning. Springer, Heidelberg (2001)
12. Hippo, Y., et al.: Global Gene Expression Analysis of Gastric Cancer by Oligonucleotide Microarrays. Cancer Res. 62(1), 233–240 (2002)
13. Holec, M., et al.: Integrating multiple-platform expression data through gene set features. In: Măndoiu, I., Narasimhan, G., Zhang, Y. (eds.) ISBRA 2009. LNCS, vol. 5542, Springer, Heidelberg (2009)
14. Huang, D.W., et al.: Bioinformatics enrichment tools: paths toward the comprehensive functional analysis of large gene lists. Nucleic Acids Res. (2008)
15. Huang, J., et al.: Decision forest for classification of gene expression data. Comput. Biol. Med. 40, 698–704 (2010)
16. Libalova, H., et al.: Gene expression profiling in blood of asthmatic children living in polluted region of the czech republic (project airgen). In: 10th International Conference on Environmental Mutagens (2010)
17. Liu, H., Motoda, H.: Feature Selection for Knowledge Discovery and Data Mining. Kluwer, Dordrecht (1998)
18. Mootha, V.K., et al.: Pgc-1-alpha-responsive genes involved in oxidative phosphorylation are coorinately down regulated in human diabetes. Nat. Genet. 34, 267–273 (2003)
19. Scherzer, C.R., et al.: Molecular markers of early Parkinson's disease based on gene expression in blood. Proc. Natl. Acad. Sci. 104(3), 955–960 (2007)
20. Subramanian, A., et al.: Gene set enrichment analysis: A knowledge-based approach for interpreting genome-wide expression profiles. Proc. Natl. Acad. Sci. 102(43), 15545–15550 (2005)
21. Talantov, D., et al.: Novel genes associated with malignant melanoma but not benign melanocytic lesions. Clin. Cancer Res. 11(20), 7234–7242 (2005)
22. Tarca, A.L., et al.: A novel signaling pathway impact analysis. Bioinformatics 25(1), 77–82 (2009)
23. Tomfohr, J., et al.: Pathway level analysis of gene expression using singular value decomposition. BMC Bioinformatics 6, 225 (2005)
24. Vapnik, V.N.: The Nature of Statistical Learning. Springer, Heidelberg (2000)
25. Witten, I.H., Frank, E.: Data Mining: Practical machine learning tools and techniques, 2nd edn. Morgan Kaufmann, San Francisco (2005)
26. Yoon, S.S., et al.: Angiogenic profile of soft tissue sarcomas based on analysis of circulating factors and microarray gene expression. J. Surg. Res. 135(2), 282–290 (2006)
27. Zintzaras, E., Kowald, A.: Forest classification trees and forest support vector machines algorithms: Demonstration using microarray data. Cell Cycle 40(5), 519–524 (2010)

Gene Network Modules-Based Liner Discriminant Analysis of Microarray Gene Expression Data

Pingzhao Hu[1], Shelley Bull[2], and Hui Jiang[1]

[1] Department of Computer Science and Engineering, York University,
4700 Keele Street, Toronto, Ontario M3J 1P3, Canada
[2] Samuel Lunenfeld Research Institute, Mount Sinai Hospital,
6000 University Avenue, Toronto, Ontario, M5G 1X5, Canada
phu@cse.yorku.ca

Abstract. Molecular predictor is a new tool for disease diagnosis, which uses gene expression to classify the diagnostic category of a patient. The statistical challenge for constructing such a predictor is that there are thousands of genes to predict for disease category, but only a small number of samples are available. Here we proposed a gene network modules-based linear discriminant analysis (MLDA) approach by integrating 'essential' correlation structure among genes into the predictor in order that the module or cluster structure of genes, which is related to diagnostic classes we look for, can have potential biological interpretation. We evaluated performance of the new method with other established classification methods using three real data sets. Our results show that the new approach has the advantage of computational simplicity and efficiency with lower classification error rates than the compared methods in most cases.

Keywords: Gene network modules, discriminant analysis, correlation-sharing, microarray.

1 Introduction

With the development of microarrays technology, more and more statistical methods have been developed and applied to the disease classification using microarray gene expression data. For example, Golub et al. developed a "weighted voting method" to classify two types of human acute leukemias [1]. Radmacher et al. constructed a 'compound covariate prediction' to predict the BRCA1 and BRCA2 mutation status of breast cancer [2]. The family of linear discriminant analysis (LDA) has been applied in such high-dimensional data [3-4]. LDA computes the optimal transformation, which minimizes the within-class distance and maximizes the between-class distance simultaneously, thus achieving maximum discrimination. Studies have shown that given the same set of selected genes, different classification methods often perform quite similarly and simple methods like diagonal linear discriminant analysis (DLDA) and k nearest neighbor (kNN) normally work remarkably well [3]. However, because the data points in microarray data sets are often from a very high-dimensional space and in general the sample size does not exceed this dimension, which presents unique challenges to feature selection and predictive modeling. Thus, finding the most

J. Chen, J. Wang, and A. Zelikovsky (Eds.): ISBRA 2011, LNBI 6674, pp. 286–296, 2011.
© Springer-Verlag Berlin Heidelberg 2011

informative genes is a crucial task in building predictive models from microarray gene expression data to handle the large p (number of genes) and small n (sample size) problem. To tackle this issue, different clustering-based classification approaches were proposed to reduce the data dimensions.

Li et al. developed cluster-Rasch models, in which a model-based clustering approach was first used to cluster genes and then the discretized gene expression values were input into a Rasch model to estimate a latent factor associated with disease classes for each gene cluster [5]. The estimated latent factors were finally used in a regression analysis for disease classification. They demonstrated that their results were comparable to those previously obtained, but the discretization of continuous gene expression levels usually results in a loss of information. Hastie et al. proposed a tree harvest procedure for find additive and interaction structure among gene clusters, in their relation to an outcome measure [6]. They found that the advantage of the method could not be demonstrated due to the lack of rich samples. Dettling et al. presented an algorithm to search for gene clusters in a supervised way. The average expression profile of each cluster was considered as a predictor for traditional supervised classification methods. However, using simple averages will discard information about the relative prediction strength of different genes in the same gene cluster [7]. Yu also compared different approaches to form gene clusters; the resulting information was used for providing sets of genes as predictors in regression [8]. However, clustering approaches are often subjective, and usually neglect the detailed relationship among genes.

Recently, gene co-expression networks have become a more and more active research area [9-12]. A gene co-expression network is essentially a graph where nodes in the graph correspond to genes, and edges between genes represent their co-expression relationship. The gene neighbor relations (such as topology) in the networks are usually neglected in traditional cluster analysis [11]. One of the major applications of gene co-expression network has been centered in identifying functional modules in an unsupervised way [9-10], which may be hard to distinguish members of different sample classes. Recent studies have shown that prognostic signature that could be used to classify the gene expression profiles from individual patients can be identified from network modules in a supervised way [12].

Based on these motivations, in this work we propose a new formulization of the traditional LDA. Specifically, we first use a seed based approach to identify gene network modules. Each of these modules includes a differentially expressed gene between sample classes, which is treated as seed, and a set of other genes highly co-expressed with the seed gene. Then we perform LDA in each module. The linear predictors in all the identified modules are then summed up. The new module-based classification approach returns signature components of tight co-expression with good predictive performance. The performance of this method is compared with other state-of-the-art classification methods. We demonstrate that the new approach has the advantage of computational simplicity and efficiency with lower classification error rates than the compared classification methods.

The remainder of this paper is organized as follows: Section 2 gives a detailed description of our new classification method and briefly discusses the methods to be compared as well as our evaluation strategy; Section 3 presents the results based on six classification methods in three real gene expression data sets; Section 4 summarizes our findings in the study.

2 Methods

2.1 MLDA Algorithm

Let assume there are A and B two sample groups (such as disease and normal groups), which have n_A and n_B samples, respectively. The data for each sample j consists of a gene expression profile $x_j = (x_{1j}, x_{2j}, ..., x_{pj})$, where x_{ij} be the log ratio expression measurement for gene $i = 1,2,...,p$ and sample $j = 1,2,...,n$, $n = n_A + n_B$ and y_j is equal to 1 if sample j belongs to group A, otherwise, y_j is equal to 0 if sample j belongs to group B. We assume that expression profiles x from group k ($k \in \{A,B\}$) are distributed as $N(\mu_k, \Sigma_k)$. The multivariate normal distribution has mean vector μ_k and covariance matrix Σ_k.

In a simplified way, we assume that $\Sigma = \Sigma_A = \Sigma_B = \{\sigma_{i,i'}\}_{i,i'=12,...,p}$, where $\sigma_{ii} = \sigma_i^2$, $\sigma_{ii'} = \sigma_{i'i}$ and $\sigma_{ii'}$ is the pooled covariance estimate of gene i and gene i' for sample groups A and B. Therefore, when $\hat{\Sigma}$ is a block-diagonal structure, we have

$$\hat{\Sigma} = \begin{bmatrix} \hat{\Sigma}_1 & 0 & 0 & \cdots & 0 \\ 0 & \hat{\Sigma}_2 & 0 & \cdots & 0 \\ 0 & 0 & \hat{\Sigma}_3 & \cdots & 0 \\ \vdots & \vdots & \vdots & \vdots & \vdots \\ 0 & 0 & 0 & \cdots & \hat{\Sigma}_C \end{bmatrix}_{p*p}$$

where C is the number of blocks (gene modules) and $\hat{\Sigma}_c$ is the estimated covariance matrix for block c $(c = 1,2,...,C)$.

The linear predictor (LP) with block-diagonal covariance structure is given by

$$LP = \sum_{c=1}^{C} \left[x_c - \tfrac{1}{2}(\mu_A^c + \mu_B^c) \right]^T \hat{\Sigma}_c^{-1} (\mu_A^c - \mu_B^c) \qquad (1)$$

where x_c^T is the expression measurements of the genes in module c for a new sample to be predicted and μ_k^c ($k \in \{A,B\}$ is the mean vector of the genes in module c. Obviously, linear discriminant analysis (LDA) and diagonal linear discriminant analysis (DLDA) [3] are the special cases of MLDA. That is, when $C = 1$,

$$LP = \left[x - \tfrac{1}{2}(\mu_A + \mu_B) \right]^T \hat{\Sigma}^{-1} (\mu_A - \mu_B),$$ where x^T is the expression measurements of p genes for a new sample to be predicted, so MLDA is simplified to LDA;

when $C = p$ (that is, each module has only one gene), $LP = \sum_{i=1}^{p} \left[x_i - \tfrac{1}{2}(\hat{\mu}_A^i + \hat{\mu}_B^i) \right]^T \{ (\hat{\mu}_A^i - \hat{\mu}_B^i) / \sigma_i^2 \}$, where x_i is the expression measurement of gene i for a new sample to be predicted, so MLDA is simplified to DLDA.

We estimate the mean vector μ_k^c of the genes in module c as $\bar{x}_k^c$ and use the pooled estimate of the common covariance matrix in each module c

$$\hat{\Sigma}_c = \frac{(n_A - 1)S_A^c + (n_B - 1)S_B^c}{n_A + n_B - 2} \tag{2}$$

where $S_k^c = \{\hat{\sigma}_{ii'}^c\}$, $i, i' = 1, 2, ..., p_c$ and p_c is the number of genes in the module c. $\hat{\sigma}_{ii'}^c$ is estimated as

$$\hat{\sigma}_{ii'}^c = \begin{cases} \hat{\sigma}_i^2 & for\ i = i' \\ \hat{\sigma}_i \hat{\sigma}_{i'} \hat{r}_c & for\ i \neq i' \end{cases} \tag{3}$$

where $\hat{r}_c = median\{\hat{r}_{ii'}\}$ $i, i' = 1, 2, ..., p_c$ and $i \neq i'$, $\hat{r}_{ii'}$ is the correlation estimate between gene i and gene i' in module c of sample group k.

Σ_c is inversible when $n \geq p_c$, that is,

$$\Sigma^{-1} = \begin{bmatrix} \Sigma_1^{-1} & 0 & 0 & \cdots & 0 \\ 0 & \Sigma_2^{-1} & 0 & \cdots & 0 \\ 0 & 0 & \Sigma_c^{-1} & \cdots & 0 \\ \vdots & \vdots & \vdots & \vdots & \vdots \\ 0 & 0 & 0 & \cdots & \Sigma_C^{-1} \end{bmatrix}$$

However, in some modules (say module c), it is possible that $n < p_c$. In this case, Σ_c is not inversible. We apply singular value decomposition (SVD) technology [13] to solve the problem. Assume Σ_c is a $p_c \times p_c$ covariance matrix, which can be discomposed uniquely as $\Sigma_c = UDV^T$, where U and V are orthogonal, and $D = diag(\sigma_1, \sigma_2, ..., \sigma_{p_c})$ with $\sigma_1 \geq \sigma_2 \geq, ..., \geq \sigma_{p_c} \geq 0$. If Σ_c is a $p_c \times p_c$ nonsingular matrix (iff $\sigma_i \neq 0$ for all $i (i = 1, 2, ..., p_c)$), then its inverse is given by $\Sigma_c^{-1} = VD^{-1}U^T$ where $D^{-1} = diag(1/\sigma_1, 1/\sigma_2, ..., 1/\sigma_{p_c})$.

The rule to assign a new sample j to group k is, thus, based on: $LP >= \log(\frac{n_B}{n_A})$, sample j is assigned to group A; otherwise, it is assigned to group B.

To identify gene modules used in Equation 1, we modify the correlation-sharing method developed by Tibshirani and Wasserman [14], which was originally proposed to detect differential gene expression. The revised approach works in the following steps:

1: Compute test statistic $T_i (i = 1,2,..., p)$ for each gene i using the standard t-statistics or a modified t-statistics, such as significance of microarrays (SAM) [15].

2: Rank the absolute test statistic values from the largest one to the smallest one and select the top m genes as seed genes.

3: Construct a gene co-expression network between the seed genes and all other genes. The pairwise Pearson correlation coefficient (r) is calculated between each of the selected seed genes and all other genes to generate the network.

4: Find the module membership s for each selected seed gene i^* in the co-expression network. The module assignments can be characterized by a many to one mapping. That is, one seeks a particular encoder $C_r(i^*)$ that maximizes

$$i_s^* = \max_{\{0 \le r \le 1\}} ave_{i \in C_r(i^*)} |T_i| \tag{4}$$

where $C_r(i^*) = \{s : abs(corr(x_{i^*}, x_s)) \ge r\}$. The set of genes s for each seed gene i^* is an adaptively chosen module, which maximizes the average (ave) differential expression signal around gene i^*. The set of identified genes s should have absolute (abs) correlation ($corr$) with i^* larger than or equal to r.

2.2 Comparisons of Different Classification Methods

We compared the prediction performances of MLDA with other established classification methods, which include diagonal quadratic discriminant analysis (DQDA), DLDA, one nearest neighbor method (1NN), support vector machines (SVM) with linear kernel and recursive partitioning and regression trees (Trees). We used the implementation of these methods in different R packages (http://cran.r-project.org/), which are *sma* for DQDA and DLDA, *class* for 1NN, *e1071* for SVM and *rpart* for Trees. Default parameters in *e1071* and *rpart* for SVM and Tree were used, respectively. For other methods (DQDA, DLDA, 1NN and MLDA), there are no tuning parameters to be selected. In the comparisons, seed genes were selected using t-test and SAM, respectively. We evaluated the performances of DQDA, DLDA, 1NN, SVM and Trees based on different number of the selected seed genes and that of MLDA based on different number of gene modules, which were built on the selected seed genes.

2.3 Cross-Validation

We performed 10-fold cross-validation to evaluate the performance of these classification methods. The basic principle is that we split all samples in a study into 10 subsets of (approximately) equal size, set aside one of the subsets from training and carried out seed gene selection, gene module construction and classifier fitting by the remaining 9 subsets. We then predicted the class label of the samples in the omitted subset based on the constructed classification rule. We repeated this process 10 times so that each sample is predicted exactly once. We determined the classification error rate as the proportion of the number of incorrectly predicted samples to the total number of samples in a given study. This 10-fold cross-validation procedure was repeated 10 times and the averaged error rate was reported.

3   Results

We applied the proposed algorithm and the established classification methods mentioned above to three real microarray data sets. The detailed description of these data sets is shown in **Table 1**. We got the preprocessed colon cancer microarray expression data from http://genomics-pubs.princeton.edu/oncology/. For prostate cancer and lung cancer microarray data sets, we downloaded their raw data from gene expression omnibus (http://www.ncbi.nlm.nih.gov/geo/) and preprocessed using robust multiarray average (RMA) algorithm [16].

Table 1. Descriptive characteristics of data sets used for classification

Disease	Response Type	No. Samples	No. Genes	Reference
Colon Cancer	Tumor/ Normal	40 / 22	2000	[17]
Prostate Cancer	Tumor/ Normal	50 / 38	12635	[18]
Lung Cancer	Tumor/ Normal	60 / 69	22215	[19]

Tables 2, 3 and **4** list the prediction performances of different classification methods applied to microarray gene expression data sets for colon, prostate and lung cancers, respectively. Here the different number of top seed genes (5, 10, 15, 20, 30, 40, 50) was selected by t-test. Since it is generally time-consuming to search for genes which are not only correlated with a given seed gene but maximize their averaged test statistic value (Equation 4), in order to save time, we only tested 10 cutoffs of correlation r from 0.5 to 0.95 with interval 0.05. We observed that the averaged correlation of genes in the identified modules is usually between 0.65 and 0.85 with the number of genes in the modules from 2 to 56, suggesting that the genes in the modules are highly co-expressed.

As we can see, the proposed MLDA has the best or comparable classification performances among all being compared classification methods in the three data sets. Other methods with better classification performances are DLDA and SVM. In general, all these methods except Tree works well for both colon and lung cancer data sets. The performances of these methods in prostate cancer data are slightly worse than those in colon and lung cancer data sets.

Table 2. Error rates of six classification methods applied to colon cancer data set

No. Genes	DQDA	DLDA	1NN	Tree	SVM	MLDA
5	0.113	0.113	0.210	0.226	0.113	**0.097**
10	0.177	0.177	0.161	0.290	**0.129**	**0.129**
15	**0.113**	0.129	0.129	0.242	0.145	**0.113**
20	0.145	**0.129**	0.161	0.258	**0.129**	**0.129**
30	0.145	0.129	0.161	0.194	0.145	**0.113**
40	0.145	**0.129**	0.145	0.210	0.145	**0.129**
50	0.145	0.145	0.194	0.226	0.145	**0.113**

Table 3. Error rates of six classification methods applied to prostate cancer data set

No. Genes	DQDA	DLDA	1NN	Tree	SVM	MLDA
5	0.227	0.239	0.261	0.227	0.216	**0.193**
10	0.205	0.193	0.284	0.318	**0.170**	0.182
15	0.250	**0.227**	0.261	0.295	0.261	**0.227**
20	0.216	0.227	0.250	0.273	**0.193**	0.205
30	**0.205**	0.216	0.239	0.295	0.216	**0.205**
40	0.261	0.250	0.295	0.318	0.250	**0.227**
50	0.227	0.227	0.341	0.330	0.216	**0.193**

Table 4. Error rates of six classification methods applied to lung cancer data set

No. Genes	DQDA	DLDA	1NN	Tree	SVM	MLDA
5	0.170	0.170	0.186	0.201	**0.162**	**0.162**
10	0.170	**0.147**	0.186	0.193	0.170	**0.147**
15	0.162	0.162	0.201	0.178	**0.132**	0.147
20	0.147	0.162	0.170	0.193	0.178	**0.132**
30	0.132	0.125	0.132	0.193	0.147	**0.116**
40	0.178	0.147	0.162	0.186	**0.132**	0.132
50	**0.125**	**0.125**	0.147	0.178	0.147	**0.125**

We also used SAM to select seed genes and evaluated their prediction performance using the same procedure as described above. Similar prediction results are observed as shown in **Tables 5, 6** and **7**. Overall, the MLDA has lower error rate than other being compared classification methods.

Table 5. Error rates of six classification methods applied to colon cancer data set

No. Genes	DQDA	DLDA	1NN	Tree	SVM	MLDA
5	0.129	0.129	0.177	0.242	**0.113**	**0.113**
10	0.161	0.161	0.161	0.226	**0.129**	0.145
15	0.129	**0.097**	0.129	0.226	0.129	0.113
20	0.145	0.145	0.145	0.177	0.145	**0.113**
30	0.145	0.129	0.194	0.290	0.145	**0.113**
40	**0.129**	**0.129**	0.210	0.258	0.145	**0.129**
50	0.145	0.145	0.210	0.290	0.145	**0.129**

Table 6. Error rates of six classification methods applied to prostate cancer data set

No. Genes	DQDA	DLDA	1NN	Tree	SVM	MLDA
5	0.125	0.136	**0.091**	**0.091**	0.114	**0.091**
10	0.114	0.136	0.148	**0.091**	0.114	0.102
15	**0.091**	0.102	0.182	0.136	**0.091**	**0.091**
20	0.136	0.170	0.148	0.136	**0.114**	**0.114**
30	0.114	0.114	**0.091**	0.159	0.114	0.114
40	0.125	0.125	**0.068**	0.170	0.102	**0.068**
50	0.136	0.148	**0.114**	0.170	0.125	0.125

Table 7. Error rates of six classification methods applied to lung cancer data set

No. Genes	DQDA	DLDA	1NN	Tree	SVM	MLDA
5	0.178	0.170	0.193	0.225	**0.162**	0.170
10	0.170	0.170	0.209	0.193	0.178	**0.147**
15	0.186	0.147	0.201	0.225	0.146	**0.116**
20	0.147	0.162	0.186	0.178	0.186	**0.132**
30	0.147	0.178	0.132	0.193	**0.101**	**0.101**
40	0.178	**0.132**	0.178	0.186	**0.132**	**0.132**
50	0.162	**0.132**	0.162	0.186	**0.132**	0.147

In many cases, we found that the simple method DLDA works well. Its performance is comparable with the advanced methods, such as SVM. We also observed that the performances of predictors with more genes are not necessarily better than those of the predictors with fewer genes. For example, when t-test was used to select the seed genes, the best performance was obtained with only 5 genes for MLDA predictor in colon cancer data set (**Table 2**), 10 genes for SVM predictor in prostate cancer data set (**Table 3**) and 30 genes for MLDA predictor in lung cancer data set (**Table 4**). When SAM was used to select the seed genes, the best performance was obtained with 15 genes for DLDA predictor in colon cancer data set (**Table 5**), 40 genes for MLDA and 1NN predictors in prostate cancer data set (**Table 6**) and 30 genes for SVM and MLDA predictors in lung data set (**Table 7**).

4 Discussions and Conclusions

In this study we developed a gene network modules-based linear discriminant analysis approach for tumor classification using microarray gene expression data. The core idea of the method is to incorporate 'essential' correlation structure among genes into a supervised classification procedure, which has been neglected or inefficiently applied in many benchmark classifiers. Our method takes into account the fact that genes act in networks and the modules identified from the networks act as the features in constructing a classifier. The rationale is that we usually expect tightly co-expressed genes to have a meaningful biological explanation. For example, if gene A and gene B has high correlation, which sometimes hints that the two genes belong to the same pathway or functional module. The advantage of this method over other methods has been demonstrated by three real data sets. Our results show that this algorithm works well for improving class prediction.

Our results are consistent with previous findings: the more advanced or complicated methods are not necessary to generate better classification results than simple methods [3]. This is very likely due to the fact that there are more parameters to be estimated in the advanced methods than in the simple methods, while our data sets usually have much smaller number of samples than features. We also tried to use more top genes (up to 100) in the classification models, but we did not find the classification results were improved (results were not shown). Although some previous results showed that better results can be obtained when the number of top genes (up to 200) used in the prediction models are much larger than the number of samples [20], the improved performance may be due to over fitting effect. Moreover, for clinical purpose, it is better to include fewer number of genes rather than larger number of genes in the prediction models.

Many other works have also extended the LDA framework for handle the large p (number of genes) and small n (sample size) problem [21-22]. The major difference between our method and those methods is that our framework is based on gene network. We built our classification models using module-specific features.

The MLDA framework can be extended in many ways. For example, we can first use principal component analysis (PCA) to extract the representing features (such as super-genes) from each of the identified modules, then any classification methods can be used to construct the prediction models based on the representing features. Also, it is possible to directly incorporate the module-specific features in other advanced discriminant learning approaches (such as SVM). In the future, we will explore these ideas in details.

Acknowledgments. The authors thank Dr. W He and S Colby for their helpful discussions and comments.

References

1. Golub, T.R., Slonim, D.K., Tamayo, P., Huard, C., Gaasenbeek, M., Mesirov, J.P., Coller, H., Loh, M.L., Downing, J.R., Caligiuri, M.A., Bloomfield, C.D., Lander, E.S.: Molecular classification of cancer: class discovery and class prediction by gene expression monitoring. Science 286, 531–536 (1999)
2. Radmacher, M.D., McShane, L.M., Simon, R.: A paradigm for class prediction using gene expression profiles. J. Comput. Biol. 9, 505–512 (2002)
3. Dudoit, S., Fridlyand, J., Speed, T.P.: Comparison of discrimination methods for the classification of tumors using gene expression data. Journal of the American Statistical Association 97, 77–87 (2002)
4. Guo, Y., Hastie, T., Tibshirani, R.: Regularized linear discriminant analysis and its application in microarrays. Biostatistics 8, 86–100 (2007)
5. Li, H., Hong, F.: Cluster-Rasch models for microarray gene expression data. Genome Biol. 2, 0031.1–0031.13 (2001)
6. Hastie, T., Tibshirani, R., Botstein, D., Brown, P.: Supervised harvesting of expression trees. Genome Biol. 2, 0003.1–0003.12 (2001)
7. Dettling, D., Bühlmann, P.: Supervised Clustering of Genes. Genome Biol. 3, 0069.1–0069.15 (2002)
8. Yu, X.: Regression methods for microarray data. Ph.D. thesis, Stanford University (2005)
9. Elo, L., Jarvenpaa, H., Oresic, M., Lahesmaa, R., Aittokallio, T.: Systematic construction of gene coexpression networks with applications to human T helper cell differentiation process. Bioinformatics 23, 2096–2103 (2007)
10. Presson, A., Sobel, E., Papp, J., Suarez, C., Whistler, T., Rajeevan, M., Vernon, S., Horvath, S.: Integrated weighted gene co-expression network analysis with an application to chronic fatigue syndrome. BMC Syst. Biol. 2, 95 (2008)
11. Horvath, S., Dong, J.: Geometric interpretation of gene coexpression network analysis. PLoS Comput. Biol. 4, e1000117 (2008)
12. Taylor, I.W., Linding, R., Warde-Farley, D., Liu, Y., Pesquita, C., Faria, D., Bull, S., Pawson, T., Morris, Q., Wrana, J.L.: Dynamic modularity in protein interaction networks predicts breast cancer outcome. Nat Biotechnol. 27, 199–204 (2009)
13. Jolliffe, I.T.: Principal component analysis. Springer, New York (2002)
14. Tibshirani, R., Wasserman, L.: Correlation-sharing for detection of differential gene expression. arXiv, math. ST, math/0608061 (2006)
15. Tusher, V., Tibshirani, R., Chu, G.: Significance analysis of microarrays applied to the ionizing radiation response. Proc. Natl. Acad. Sci. USA 98, 5116–5121 (2001)
16. Irizarry, R.A., Bolstad, B.M., Collin, F., Cope, L.M., Hobbs, B., Speed, T.P.: Summaries of Affymetrix GeneChip probe level data. Nucleic Acids Research 31, E15 (2003)
17. Alon, U., Barkai, N., Notterman, D.A., Gish, K., Ybarra, S., Mack, D., Levine, A.J.: Broad patterns of gene expression revealed by clustering analysis of tumor and normal colon tissues probed by oligonucleotide arrays. Proc. Natl. Acad. Sci. USA 96, 6745–6750 (1999)
18. Stuart, R.O., Wachsman, W., Berry, C.C., Wang-Rodriguez, J., Wasserman, L., Klacansky, I., Masys, D., Arden, K., Goodison, S., McClelland, M., Wang, Y., Sawyers, A., Kalcheva, I., Tarin, D., Mercola, D.: In silico dissection of cell-type-associated patterns of gene expression in prostate cancer. Proc. Natl. Acad. Sci. USA 101, 615–620 (2004)
19. Spira, A., Beane, J.E., Shah, V., Steiling, K., Liu, G., Schembri, F., Gilman, S., Dumas, Y.M., Calner, P., Sebastiani, P., Sridhar, S., Beamis, J., Lamb, C., Anderson, T., Gerry, N., Keane, J., Lenburg, M.E., Brody, J.S.: Airway epithelial gene expression in the diagnostic evaluation of smokers with suspect lung cancer. Nat. Med. 13, 361–366 (2007)

20. Antoniadis, A., Lambert-Lacroix, S., Leblanc, F.: Effective dimension reduction methods for tumor classification using gene expression data. Bioinformatics 19, 563–570 (2003)
21. Shen, R., Ghosh, D., Chinnaiyan, A.M., Meng, Z.: Eigengene based linear discriminant model for gene expression data analysis. Bioinformatics 22, 2635–2642 (2006)
22. Pang, H., Tong, T., Zhao, H.: Shrinkage-based diagonal discriminant analysis and its applications in high-dimensional data. Biometrics 65, 1021–1029 (2009)

A Polynomial Algebra Method for Computing Exemplar Breakpoint Distance

Bin Fu[1] and Louxin Zhang[2]

[1] Department of Computer Science,
University of Texas-Pan American,
Edinburg, TX 78539, USA
`binfu@cs.panam.edu`
[2] Department of Mathematics,
National University of Singapore,
Singapore 119076
`matzlx@nus.edu.sg`

Abstract. The exemplar breakpoint distance problem is NP-hard. Assume two genomes have at most n genes from m gene families. We develop an $O(2^m n^{O(1)})$ time algorithm to compute the exemplar breakpoint distance if one of them has no repetition. We develop an $O(2^m m! n^{O(1)})$ time algorithm to compute the exemplar breakpoint distance between two arbitrary genomes. If one of the given genomes has at most d repetitions for each gene, the computation time of the second algorithm is only $O((2d)^m n^{O(1)})$. Our algorithms are based on a polynomial algebra approach which uses a multilinear monomial to represent a solution for the exemplar breakpoint distance problem.

Keywords: Genome rearrangement, examplar breakpoint distance.

1 Introduction

As the full DNA sequence of an organism, the genome determines its shape and behavior. All the extant genomes have evolved from their common ancestor existed above four billion years ago through a series of evolutionary events including segment reversals, translocations, gene duplications, fissions and fusions (see [9] for example). As a result, closely related species often have highly similar but different genomes and hence two or more genomes are often compared to infer evolutionary renovations specific to a species lineage. As genes are the most important constitutes of a genome and they locate on either strand, a genome is often modeled as an ordered sequence of signed genes, where a negative sign before a gene indicates that the gene locates on the complement strand.

In the early study of genome rearrangement, a genome was modeled as a signed permutation over a set of genes; breakpoint distance and signed reversal distance were introduced in [19] and [16] respectively; and an elegant polynomial time algorithm for the signed reversal distance was discovered [10].

While the permutation model might be appropriate for studying small viruses and mitochondria genomes, it becomes problematic when applied to eukaryotic

J. Chen, J. Wang, and A. Zelikovsky (Eds.): ISBRA 2011, LNBI 6674, pp. 297–305, 2011.

genomes in which paralogous genes that have similar functions often form a large gene family. Therefore, a more generalized version of the genome rearrangement problem was formulated by Sankoff in order to taking multiple-gene families into account. In his seminal paper [17], a genome is modeled as a signed sequence over a set of genes in which a gene can appear several times. His approach is to delete all but one member of a gene family in each of the two genomes to minimize the breakpoint or reversal distance between the two reduced genomes, called exemplars. Under this model, one may infer gene duplications that are specific to each considered genome, occurring after their divergence by solving the exemplar breakpoint or reversal distance problem (see the section 2 for formal definition). The resulting exemplars indicates the ancestral genes in their common ancestor for the gene families being studied.

In the past decade, the exemplar breakpoint distance problem has been studied extensively. It is trivial to find out the breakpoint distance between two genomes modeled as signed permutations. However, the exemplar breakpoint distance problem is not only NP-hard [5], but also hard for approximation [6,14] even for special cases [2,3,4,11]. In addition, different efficient heuristic and exact algorithms have also been developed [1,4,15].

In this paper, we derive new efficient algorithms for the exemplar breakpoint distance problem by using monomial to encode a solution and transforming distance computation into polynomial manipulations. For a genome, its size is defined as the total number of genes it has. We develop an $O(2^m n^{O(1)})$ time algorithm to compute the breakpoint distance between two genomes A and B if one of them has no repetition, where m is the number of gene families contained in the genomes and n is the size of A or B whichever is larger. We also develop an $O(2^m m! n^{O(1)})$ time algorithm to compute the exemplar breakpoint distance. In the case that one of two genomes has at most d repetitions of each gene, the computation time is $O((2d)^m n^{O(1)})$ for finding the exemplar breakpoint distance.

2   Notations

A genome containing m gene families is considered as a signed sequence over an alphabet of m letters where each letter represents a gene family and one of its occurrences denotes a member of the corresponding gene family. In the theory of genome rearrangement, we do not distinguish the members of each gene family. For instance, the genome represented by ab-ac-bcd has four gene families; the 'a' family contains two gene members appearing on different strands.

If a genome does not have multiple-gene family, then, its representation is simply a signed permutation. For two such genomes $G = g_1 g_2 \cdots g_k$ and $H = h_1 h_2 \cdots h_k$, the breakpoint distance between them is defined to be the number of consecutive pairs $g_i g_{i+1}$ ($1 \leq i \leq k-1$) such that $g_i g_{i+1} \neq h_j h_{j+1}$ and $g_i g_{i+1} \neq$ $-h_{j+1} -h_j$ for any $j = 1, 2, \cdots, k-1$. We use $\mathrm{bd}(G, H)$ to denote the breakpoint distance between G and H. It is easy to see that $\mathrm{bd}(G, H) = \mathrm{bd}(H, G)$. For $G = 12$-34 and $H = 34$-2-1, there is a breakpoint in each of 2-3 and -34. Hence, $\mathrm{bd}(G, H) = 2$. For $G = g_1 g_2 \cdots g_k$, define $G^{-1} = -g_k \cdots -g_2 -g_1$.

For a genome represented by a signed sequence G, an exemplar genome of G is a genome G' obtained from G by deleting all but one occurrence of each letter. For example, $bcaadagef$ has two exemplar genomes: $bcadgef$ and $bcdagef$.

We use $ebd(G, H)$ to represent the exemplar breakpoint distance between the sequences G and H:

$$ebd(G, H) = \min\{bd(G', H') : G' \text{ and } H' \text{ are an exemplar of } G \text{ and } H \text{ resp.}\}.$$

The *exemplar breakpoint distance* problem to find out $ebd(G, H)$ on the input genomes G and H. Its decision version is as follows:

INSTANCE: For genomes G and H, each having the same gene families; and integer k.

QUESTION: $ebd(G, H) \le k$?

3   A Simple Algorithm

Let G and H be two genomes each consisting of k gene families $S = \{g_1, g_2, \cdots, g_m\}$. We consider G and H as two sequences over alphabet S. It is not hard to see that G and H have 0 exemplar breakpoint distance if and only if G and H have a common "permutation" subsequence. One could solve the zero exemplar breakpoint distance problem for G and H by using the following simple procedure:

To compute the exemplar breakpoint distance between G and H, we use the following search method:

Set $d = \infty$;
For each pair of permutations P, Q over the gene family set,
(i) check whether P and Q are a common subsequence of G and H respectivel or not;
(ii) if P is an exemplar of G and Q is an exmplar of H, do
(iii)   Compute $d = \min\{d, bd(P, Q)\}$;
Output d.

Let $G = a_1 a_2 \cdots a_n$ and $P = p_1 p_2 \cdots p_m$. Assume a_t is the first residue of G identical to p_1 then, P is a subsequence of G if and only if $p_2 p_3 \cdots p_m$ is a subsequence of $a_{t+1} a_{t+2} \cdots a_n$. This implies an $O(n)$ dynamic programming method for determining whether P is a subsequnce of G or not. Hence, Step (i) of the above procedure can be executed in $O(n)$ steps, where n is the sum of the lengths of G and H.

There are $(m!)^2$ pairs of permutations over the gene family set. For each pair of permutations, Step (i) and Step (ii) can be computed in $O(n)$ and $O(m)$ steps. Hence, the strategy described above has time complexity $O((m!)^2 n)$.

To make the above method more efficient, we have to consider how to enumerate the exemplars of the given genomes. Considering this problem leads to the following polynomial method algorithms that are much more efficient than the above methods.

4   Polynomial Algebra Method

We use the algebraic method that was originally developed by Koutis [12] and refined by Williams [18]. It was used in designing randomized fixed parameter tractable algorithms for the k-path problem in a directed graph. A theory for the connection between testing monomial and complexity theory are developed by Chen and Fu [7,8]. We use this method to derive deterministic algorithms for computing breakpoint distance between two genomes.

Definition 1. *Assume that $x_1, \cdots, x_k$ are variables.*

- *A monomial has format $x_1^{a_1} x_2^{a_2} \cdots x_k^{a_k}$, where a_i are nonnegative integers.*
- *A multilinear monomial is a monomial such that each variable has degree at most one. For examples, $x_3 x_5 x_6$ is a multilinear monomial, but $x_3 x_5^3 x_6^2$ is not.*
- *For a polynomial $p(x_1, \cdots, x_k)$, its sum of product expansion is $p(x_1, \cdots, x_k)$ $= \sum_j q_j(x_1, \cdots, x_k)$, where each $q_j(x_1, \cdots, x_k)$ is a monomial, which has a format $c_j x_1^{a_1} \cdots x_k^{a_k}$ with c_j as its coefficient.*

4.1   Breakpoint Distance for One Sequence without Repetition

We first consider how to compute the exemplar breakpoint distance between two genomes when one of them does not have gene repetition. This problem is still NP-hard [5].

Definition 2. *Let S_1 and S_2 be two input genomes. S_2 has no repetition with all of its genes. We say S_1 is directly convertible to S_2 if S_1 can become S_2 or S_2^{-1} by removing some genes. If S_1 is directly convertible to S_2 or S_2^{-1}, then we say S_1 is convertible to S_2.*

Lemma 1. *A checking if S_1 is convertible to S_2, which has no repetition with all its genes, can be done in $O(n)$ time.*

Proof. Let $n_2 = |S_2|$. Let $S_1[i]$ is the last character identical to $S_2[n_2]$. We have that S_1 is directly convertible to S_2 if and only if $S_1[1, i-1]$ is directly convertible to $S_2[1, n_2 - 1]$. This recursion brings an $O(n)$ time algorithm.

Lemma 2. *Let G_1 and G_2 be two genomes. G_2 has no repetition with all its genes. Then the breakpoint distance of G_1 and G_2 is u if and only if G_1 can be partitioned into $G_{1,1}, \ldots, G_{1,v+1}$, and G_2 can be partitioned into $G_{2,1}, \ldots, G_{2,v+1}$ such that each $G_{1,i}$ matches a unique G_{2,j_i}, and $G_{1,i}$ is convertible to G_{2,j_i} for $i = 1, \ldots, v+1$.*

Proof. It follows from the definition of breakpoint distance, and Definition 2.

Algorithm Construct-Distance-Polynomial(G_1, G_2, v)
Input: genome G_1, genome $G_2 = g_1 g_2 \ldots g_m$ without repetition over m gene families, and parameter v;
Output: a polynomial represented by a circuit.
Steps:

>For all $1 \leq s \leq t \leq m$, let $H_{s,t}$ to be $\prod_{i=s}^{t} x_i$;
>For each $1 \leq i \leq j \leq |G_1|$
>>let $L_{i,j} = \sum_{[s,t]:G_1[i,j]}$ is convertible to $G_2[s,t]$ $H_{s,t}$;
>
>For each $1 \leq i \leq |G_1|$, and $0 \leq u \leq v$,
>>Compute $F_{i,u}$ via a dynamic programming according to the following recursions
>>(i) $F_{i,0} = L_{1,i}$, and (ii) $F_{i,u+1} = \sum_{j<i} F_{j,u} L_{j+1,i}$;
>
>Output $F_{|G_1|,v}$;

End of Algorithm

Lemma 3. *There is a polynomial time algorithm such that given an integer parameter v, and two genomes G_1 and G_2 that G_2 has no repetition with its genes, it generates a $F_{|G_1|,v}$ that has a multilinear monomial containing every variable in its sum of product expansion if and only if the exemplar breakpoint distance of G_1 and G_2 is at most v.*

Proof. Let G_1 and G_2 be two input genomes. Sequence G_2 has no repetition with its genes. Without loss of generality, let $G_2 = g_1 \cdots g_m$. Let each gene g_i have a variable x_i to represent it. There is a polynomial time algorithm to check if $G_1[i, j]$ is convertible to $G_2[s, t]$.

If G_1 and G_2 have breakpoint distance equal to v, we can partition G_1 into $G_{1,1}, \ldots, G_{1,v+1}$, and partition G_2 into $G_{2,1}, \ldots, G_{2,v+1}$ such that each $G_{1,i}$ matches a unique G_{2,j_i}, and $G_{1,i}$ is convertible to G_{2,j_i} for $i = 1, \ldots, v + 1$. This induces a multilinear monomial $x_1 \ldots x_m$ in the sum of product expansion of $F_{n_1,v}$ via the recursions described in Construct-Distance-Polynomial(.).

A multilinear monomial M is a u-conversion monomial from G to G_2 if G can be partitioned into $G[y_1, z_1], \ldots, G[y_{u+1}, z_{u+1}]$ with $y_1 \leq z_1 < y_2 \leq z_2 < \ldots < y_{u+1} \leq z_{u+1}$, and there are u disjoint intervals $[s_1, t_1], \ldots, [s_{u+1}, t_{u+1}]$ for G_2 such that $G[y_i, z_i]$ is convertible to $G_2[s_i, t_i]$ for $i = 1, \ldots, u + 1$, and $M = \prod_{i=1}^{u+1} H_{s_i,t_i}$.
Claim A multilinear monomial M is u-conversion monomial from $G_1[1, i]$ to G_2 if and only if $F_{i,u}$ contains M in its sum of product expansion.

Proof. For the case $u = 0$, it follows from the definition of $L_{i,j}$ and $F_{i,0}$, which is assigned to $L_{1,i}$ in the recursion (i) of the algorithm Construct-Distance-Polynomial(.). Assume the claim is true for $u - 1$. Consider $F_{i,u}$.

Let $G_1[y_1, z_1], \ldots, G_1[y_{u+1}, z_{u+1}]$ be a partition for $G_1[1, i]$ with $y_1 \leq z_1 < y_2 \leq z_2 < \ldots < y_{u+1} \leq z_{u+1}$. There are disjoint intervals $[s_1, t_1], \ldots, [s_{u+1}, t_{u+1}]$ for G_2 such that $G_1[y_i, z_i]$ is convertible to $G_2[s_i, t_i]$ for $i = 1, \ldots, u + 1$. By our assumption for the case u, we have that $M' = \prod_{i=1}^{u} H_{s_i,t_i}$ is a monomial in sum of product expansion of $F_{y_{u+1}-1,u-1}$. By the definition of $L_{y_{u+1},z_{u+1}}$, we have $H_{s_{u+1},t_{u+1}}$ is one of the monomials in $L_{y_{u+1},z_{u+1}}$ since $G_1[y_{u+1}, z_{u+1}]$ is convertible to $G_2[s_{u+1}, t_{u+1}]$. Thus, $M = \prod_{i=1}^{u+1} H_{s_i,t_i}$ is in the sum of product expansion of $F_{i,u}$ by the recursion (ii) in the algorithm Construct-Distance-Polynomial(.).

We have that the G_1 and G_2 have breakpoint distance at most v, if and only if $F_{n_1,v}$ has a multi-linear monomial $x_1 \cdots x_m$ in its sum of product expansion. The polynomial $F_{n_1,v}$ can be expressed a polynomial size circuit since we can have a polynomial time dynamic programming method to compute the entire recursion.

Lemma 4. *There is a $2^m n^{O(1)}$ time algorithm to find out all multilinear monomials in the sum of product expansion of the polynomial generated by Construct-Distance-Polynomial(G_1, G_2, v) with $v < |G_2|$, where G_1 is a general genome, and G_2 is a genome without repetition with all of its genes.*

Proof. Since G_2 has no gene repetition, $|G_1| \geq |G_2|$. Let $n = |G_1|$. Use a bottom up approach to evaluate a polynomial $F_{|G_1|,v}$ generated by Construct-Distance-Polynomial(G_1, G_2, v). The total number of multilinear monomials is at most 2^m since each gene g_i is represented by a unique variable x_i. For each multiplication $F_{j,u} L_{j+1,i}$ involved in case (ii) of the recursion, $L_{j+1,i}$ has at most a polynomial number of monomials in terms of n. Therefore, each multiplication can be computed in $2^m n^{O(1)}$ time. The circuit size for $F_{|G_1|,v}$ is bounded by $n^{O(1)}$. We can output all the multilinear monomials in the sum of product expansion of $F_{|G_1|,v}$ in $O(2^m n^{O(1)})$ time.

> **Algorithm Compute-Breakpoint-Distance(G_1, G_2)**
> Input: genome G_1, genome $G_2 = g_1 g_2 \ldots g_m$ without repetition;
> Output: the exemplar breakpoint distance between G_1 and G_2.
> Steps:
>   For each $0 \leq v \leq |G_1|$
>   Let $F_{|G_1|,v}$ =Construct-Distance-Polynomial(G_1, G_2, v);
>   Find all of the multilinear monomials in the sum of product of
> $F_{|G_1|,v}$ (see Lemma 4);
>   If there is a monomial $x_1 \ldots x_m$
>   then output v and exit;
> **End of Algorithm**

Theorem 1. *There is an $O(2^m n^{O(1)})$ time algorithm to compute the exemplar breakpoint distance between two given genomes if one of them does not contain gene repetition, where m is the number of genes and n is the maximal length of genomes.*

Proof. Let G_1 and G_2 be two input genomes. Let $n = \max(|G_1|, |G_2|)$. Sequence G_2 has no repetition with its genes. Without loss of generality, let $G_2 = g_1 \cdots g_m$.

By Lemma 3, using $n^{O(1)}$ time, we can obtain a polynomial $F_{|G_1|,v}$ such that G_1 and G_2 have breakpoint distance at most v if and only if $F_{|G_1|,v}$ has the multilinear monomial $x_1 \ldots x_m$ in its sum of product expansion.

By Lemma 4, the existence of such a multilinear monomial $x_1 \ldots x_m$ can be checked in $O(2^m n^{O(1)})$ time for each $0 \le v \le |G_1|$. Therefore, the total time is $O(2^m n^{O(1)})$.

4.2   Exemplar Breakpoint Distance between Arbitrary Genomes

In this section, we develop an algorithm for computing the exemplar breakpoint distance between two general sequences. Its computational complexity is higher than those in the last two sections, but it is much more efficient than the brute force algorithm.

Algorithm Compute-General-Breakpoint-Distance(G_1, G_2)
Input: genome G_1, and genome G_2;
Output: the breakpoint distance between G_1 and G_2.
Steps:
    Let $x = \infty$;
    For each permutation P of all genes
        Check if P is a subsequence of G_2 via a dynamic programming method;
        If yes, let $x = \min(x, \text{Compute-General-Breakpoint-Distance}(G_1, P))$;
    Output x.
End of Algorithm

Theorem 2. *Compute-General-Breakpoint-Distance(.) is an $O(2^m m! n^{O(1)})$ time algorithm to compute the exemplar breakpoint distance between two genomes of length at most n and at most m genes.*

Proof. We convert the problem into the problem that one of genomes has no repetition. For each permutation $g_{i_1} \ldots g_{i_m}$ of m genes, using the dynamic programming method, we can check if such a permutation can be derived via deleting some gene in a genome G. After a permutation G_2' of genes is derived from G_2, the problem is converted into computing the exemplar breakpoint distance between G_1 and G_2' that has no repetition. By Theorem 1, it takes $O(2^m n^{O(1)})$ time to compute the distance between G_1 and G_2'. Since there are $m!$ cases for G_2', the total time is $O(2^m m! n^{O(1)})$.

For the case that one genome has a bounded number of repetitions for each gene, we have the following improved algorithm stated in Theorem 3.

Algorithm Compute-General-Breakpoint-Distance2(G_1, G_2)
Input: genome G_1, and genome G_2 with at most d repetitions for each gene;
Output: the breakpoint distance between G_1 and G_2.
Steps:
    Let $x = \infty$;
    For each permutation P of all genes by selecting one copy of each gene in G_2
        let $x = \min(x, \text{Compute-General-Breakpoint-Distance}(G_1, P))$;
    Output x.
End of Algorithm

Theorem 3. *Compute-General-Breakpoint-Distance2(.) is a $O(2^m d^m n^{O(1)})$ time algorithm to compute the exemplar breakpoint distance if one of them two input sequences has at most d repetitions for each gene.*

Proof. Assume that G_2 has at most d repetitions for each gene. There are at most d^m ways to select a permutation of all genes. The problem is converted into d^m problems with one genome without repetition that takes $O(2^m n^{O(1)})$ time each by Theorem 1. Therefore, the total time is $O(2^m d^m n^{O(1)})$.

5   Conclusions

We have developed a novel polynomial algebra approach for computing the exemplar breakpoint problem between two genomes, which is NP-hard even for checking zero breakpoint distance and has no any factor polynomial time approximation unless P=NP. An interesting open problem is if the exemplar breakpoint distance between two genomes can be computed in $O(2^{O(m)} n^{O(1)})$ time if the genomes contain at most n genes from m gene families.

Acknowledgements

Bin Fu is supported in part by National Science Foundation Early Career Award 0845376. Louxin Zhang is partially financially supported by AcRF R146-000-109-112.

References

1. Angibaud, S., Fertin, G., Rusu, I., Thévenin, A., Vialette, S.: A pseudo-Boolean programming approach for computing the breakpoint distance between two genomes with duplicate genes. In: Tesler, G., Durand, D. (eds.) RECMOB-CG 2007. LNCS (LNBI), vol. 4751, pp. 16–29. Springer, Heidelberg (2007)
2. Angibaud, S., Fertin, G., Rusu, I., Thévenin, A., Vialette, S.: On the approximability of comparing genomes with duplicates. J. Graph Algor. Appli. (accepted)
3. Blin, G., Chauve, C., Fertin, G., Rizzi, R., Vialette, S.: Comparing genomes with duplications: A computational complexity point of view. IEEE/ACM IEEE Trans. Comput. Biol. Bioinform. 4, 523–534 (2007)
4. Blin, G., Fertin, G., Sikora, F., Vialette, S.: The exemplar breakpoint distance for non-trivial genomes cannot be approximated. In: Das, S., Uehara, R. (eds.) WALCOM 2009. LNCS, vol. 5431, pp. 357–368. Springer, Heidelberg (2009)
5. Bryant, D.: The complexity of calculating exemplar distance. In: Sankoff, D., Nadeau, J. (eds.) Comparative Genomics: Empirical and Analytical Approaches to Gene Order Dynamics, Map Alignment, and the Evolution of Gene Families, pp. 207–212 (2000)
6. Chen, Z., Fu, B., Zhu, B.: Approximations for the exemplar breakpoint distance problem. In: Cheng, S.-W., Poon, C.K. (eds.) AAIM 2006. LNCS, vol. 4041, pp. 291–302. Springer, Heidelberg (2006)

7. Chen, Z., Fu, B.: Approximating Multilinear Monomial Coefficients and Maximum Multilinear Monomials in Multivariate Polynomials, Electronic Colloquium on Computational Complexity (ECCC-TR10-124)

8. Chen, Z., Fu, B.: The Complexity of Testing Monomials in Multivariate Polynomials, Electronic Colloquium on Computational Complexity (ECCC-TR10-114)

9. Eichler, E.E., Sankoff, D.: Structural dynamics of eukaryotic chromosome evolution. Science 301, 793–797 (2003)

10. Hannenhalli, S., Pevzner, P.: Transforming cabbage into turnip: polynomial algorithm for sorting signed permutations by reversals. J. Assoc. Comput. Mach. 46, 1–27 (1999)

11. Jiang, M.: The zero exemplar distance problem. In: Tannier, E. (ed.) RECOMB-CG 2010. LNCS, vol. 6398, pp. 74–82. Springer, Heidelberg (2010)

12. Koutis, I.: Faster algebraic algorithms for path and packing problems. In: Aceto, L., Damgård, I., Goldberg, L.A., Halldórsson, M.M., Ingólfsdóttir, A., Walukiewicz, I. (eds.) ICALP 2008, Part I. LNCS, vol. 5125, pp. 575–586. Springer, Heidelberg (2008)

13. Nadeau, J.H., Taylor, B.A.: Lengths of chromosomal segments conserved since divergence of man and mouse. Proc. Natl. Acad. Sci. USA 81, 814–818 (1984)

14. Nguyen, C.T.: Algorithms for calculating exemplar distances. Honors Thesis, Department of Computer Science, National University of Singapore (2005)

15. Nguyen, C.T., Tay, Y.C., Zhang, L.X.: Divide-and-conquer approach for the exemplar breakpoint problem. Bioinformatics 21, 2171–2176 (2005)

16. Sankoff, D.: Mechanisms of genome evolution: models and inference. Bull. Int. Stat. Institut. 47, 461–475 (1989)

17. Sankoff, D.: Genome rearrangement with gene families. Bioinformatics 15, 909–917 (2009)

18. Williams, R.: Finding paths of length k in $o^*(2^k)$ time. Information Processing Letters 109, 315–318 (2009)

19. Watterson, G.A., Ewens, W.J., Hall, T.E., Morgan, A.: The chromosome inversion problem. J. Theor. Biol. 99, 1–7 (1982)

The Maximum Clique Enumeration Problem: Algorithms, Applications and Implementations

John D. Eblen, Charles A. Phillips, Gary L. Rogers, and Michael A. Langston

Department of Electrical Engineering and Computer Science
University of Tennessee, Knoxville TN 37996-3450, USA

Abstract. Algorithms are designed, analyzed and implemented for the maximum clique enumeration (MCE) problem, which asks that we identify all maximum cliques in a finite, simple graph. MCE is closely related to two other well-known and widely-studied problems: the maximum clique optimization problem, which asks us to determine the size of a largest clique, and the maximal clique enumeration problem, which asks that we compile a listing of all maximal cliques. Naturally, these three problems are $\mathcal{NP}$-hard, given that they subsume the classic version of the $\mathcal{NP}$-complete clique decision problem.

MCE can be solved in principle with standard enumeration methods due to Bron, Kerbosch, Kose and others. Unfortunately, these techniques are ill-suited to graphs encountered in our applications. We must solve MCE on instances deeply seeded in data mining and computational biology, where high-throughput data capture often creates graphs of extreme size and density. MCE can also be solved in principle using more modern algorithms based in part on vertex cover and the theory of fixed-parameter tractability (FPT). While FPT is an improvement, these algorithms too can fail to scale sufficiently well as the sizes and densities of our datasets grow.

An extensive testbed of benchmark MCE instances is devised, based on applications in transcriptomic data analysis. Empirical testing reveals crucial but latent features of such high-throughput biological data. In turn, it is shown that these features distinguish real data from random data intended to reproduce salient topological features. In particular, with real data there tends to be an unusually high degree of maximum clique overlap. Armed with this knowledge, novel decomposition strategies are tuned to the data and coupled with the best FPT MCE implementations. It is demonstrated that the resultant run times are frequently reduced by several orders of magnitude, and that instances once prohibitively time-consuming to solve are now often brought into the domain of realistic feasibility.

Keywords: maximum clique enumeration, maximal clique, gene expression analysis, software tools and applications.

1 Introduction

Clique is one of the best known and most widely studied combinatorial problems. Although classically formulated as an $\mathcal{NP}$-complete decision problem [17], the

J. Chen, J. Wang, and A. Zelikovsky (Eds.): ISBRA 2011, LNBI 6674, pp. 306–319, 2011.

search and optimization formulations are probably most often encountered in practice. In computational biology, one needs to look no farther than PubMed to gauge clique's utility in a variety of applications. A notable example is the search for putative molecular response networks in high-throughput biological data. Popular clique-centric tools include clique community algorithms for clustering [24] and paraclique-based methods for QTL analysis and noise abatement [9,10].

A clique is *maximal* if it cannot be augmented by adding additional vertices. A clique is *maximum* if it is of largest size. A maximum clique is particularly useful in our work on graphs derived from biological datasets. It provides a dense core that can be extended to produce plausible biological networks [13]. Other biological applications include the thresholding of normalized microarray data [6,25], searching for common *cis*-regulatory elements [3], and solving the compatibility problem in phylogeny [16]. See [5] for a survey of additional applications of maximum clique.

Any algorithm that relies on maximum clique, however, has the potential for inconsistency. This is because graphs often have more than just one clique of largest size. Thus it is that algorithmic idiosyncrasies, not scientific reason, are apt to lead to an arbitrary choice of cliques. This motivates us to find an efficient mechanism to enumerate all maximum cliques in a graph. These can then be examined using a variety of relevant criteria, such as the average weight of correlations driven by strain or stimulus [2].

We therefore seek to solve the *Maximum Clique Enumeration (MCE)* problem. Unlike maximal clique enumeration, for which a substantial body of literature exists, very little seems to be known about MCE. The only exception we have found is a game-theoretic approach for locating a predetermined number of largest cliques [8]. We begin by creating a testbed of graphs derived from gene expression data on which to test MCE performance. We concentrate on transcriptomic data, given its abundance, and eschew synthetic data, having learned long ago that effective algorithms for one have little bearing on the other. (The pathological matchings noted in [15] for vertex cover can be extended to clique, but likewise they too are of course hugely irrelevant to real data.) We then implement and compare both standard enumeration algorithms and more advanced codes based on the theory of fixed-parameter tractability (FPT) [1,11]. In an effort to improve performance, we scrutinize the structure of transcriptomic graphs and explore the notion of maximum clique covers and essential vertex sets. Indeed, we find that with the right preprocessing we are able to tailor algorithms to the sorts of data we routinely encounter, and that we can now solve instances previously considered unassailable.

2   Implementation Environment

To test the performance of different algorithmic approaches, we created a testbed of 75 graphs, 25 from each of three different transcriptomic datasets. The datasets were selected for diversity within the domain of mRNA microarray expression experiments. Since experimental conditions drive the correlations used to create the graphs, we selected datasets with three different types of conditions: strain, time, and stimulus. Two of the datasets were from experiments on *M. musculus* (mouse); the third was from an experiment on *S. cerevisiae* (yeast). In all three, mRNA microarrays were used to measure the intensity of mRNA expression.

Our first dataset used microarrays containing 45127 probes and consisted of expression data collected for adult mouse specimens from 41 different BXD strains. The data was segregated by sex, so we constructed 13 graphs from female data and 12 from male data. Our second dataset used microarrays containing 46632 probes and measured expression on mice on successive days during prenatal and postnatal development on two different strains, C57BL/6 and DBA/2J, at 12 and 13 time points respectively. Our third dataset used microarrays containing 6214 probes and measured expression on yeast at 16 different oxygen levels and 15 glucose concentrations.

To analyze expression data, we first constructed weighted graphs in which vertices represented probes and edge weights were Pearson correlation coefficients computed across experimental conditions. We then converted the weighted graphs into unweighted graphs by retaining only those edges whose weights were at or above some chosen threshold, t. By employing incremental values for t between 0.7 and 0.94, a range typical of correlation values used to analyze microarray data, we obtained a testbed of graphs of various sizes and densities. All size/density values were within the spectrum typically seen in our work with biological datasets. The smallest graph had 5,300 vertices and 292,829 edges; the largest had 30,033 vertices and 1,818,945 edges.

The number of maximum cliques for the graphs in our testbed ranged from 5 to 47496, with no discernible pattern based on graph size or density. One might ask why there is such wide, unpredictable variability. It turns out that the number of maximum cliques can be extremely sensitive to small changes in the graph. Even the modification of a single edge can have a huge effect. Consider, for example, a graph with a unique maximum clique of size k, along with a host of disjoint cliques of size $k - 1$. The removal of just one edge from what was the largest clique may now result in many maximum cliques of size $k - 1$. Edge addition can of course have similar effects. See Figure 1 for an illustrative example.

3   Fundamental Approaches to MCE

While very little prior work seems to have been done on MCE, the problem of maximal clique enumeration has been studied extensively. Since any algorithm that enumerates all maximal cliques also enumerates all maximum cliques, it is reasonable to approach MCE by attempting first to adapt existing maximal clique enumeration algorithms. An implementation of an existing maximal clique enumeration algorithm also provides a useful runtime benchmark that should be improved upon by any new approach. For completeness, we also consider other enumeration algorithms. One possibility is to compute the maximum clique size and then test all possible combinations of vertices of that size for complete connectivity. While this approach may be reasonable for very small clique sizes, as the maximum clique size increases the runtime of this approach quickly becomes prohibitive.

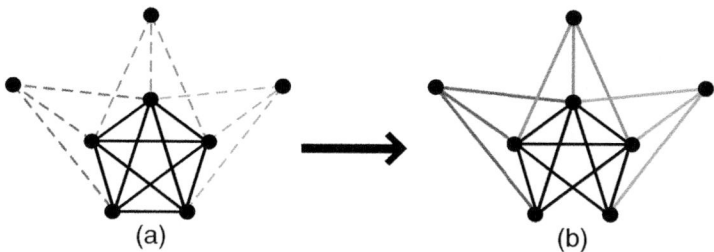

Fig. 1. The number of maximum cliques in a graph can be highly subject to perturbations due, for example, to noise. For example, a graph may contain a single maximum clique C representing a putative network of size k, along with any number of vertices connected to $k - 2$ vertices in C. In (a), there is a single maximum clique of size $k = 5$, with "many" other vertices (only three are shown) connected to $k - 2 = 3$ of its nodes. In (b), noise results in the removal of a single edge, creating many maximum cliques now of size $k - 1 = 4$.

Current maximal clique enumeration algorithms can be classified into two general types: iterative enumeration (breadth-first traversal of a search tree) and backtracking (depth-first traversal of a search tree). Iterative enumeration algorithms, such as the method suggested by Kose *et al* [19], enumerate all cliques of size k at each stage, test each one for maximality, then use the remaining cliques of size k to build cliques of size $k + 1$. The process is typically initialized for $k = 3$ by enumerating all vertex subsets of size 3 and testing for connectivity. In practice, such an approach can have staggering memory requirements, because all cliques of a given size must be retained at each step. In [29], this approach is improved by using efficient bitwise operations to prune the number of cliques that must be saved. Nevertheless, storage needs can be excessive, since all maximal cliques of one size must still be made available before moving on to the next larger size. Figure 2 shows the number of maximal cliques of each size in a graph near the mean size in our testbed. This graphic illustrates the enormous lower bounds on memory that can be encountered with iterative enumeration algorithms.

Many variations of backtracking algorithms for maximal clique enumeration have been published in the literature. To the best of our knowledge, all can be traced back to the algorithms of Bron and Kerbosch first presented in [7]. Some subsequent modifications tweak the data structures used. Others change the order in which vertices are traversed. See [18] for a performance comparison between several variations of backtracking algorithms. As a basis for improvement, however, we implemented the original, highly efficient algorithm of [7]. We made this choice for three reasons. First, an enormous proportion of the time consumed by enumeration algorithms is spent in outputting the maximal cliques that are generated. This output time is a practical limitation on any such approach. Second, a graph can theoretically contain as many as $3^{(n/3)}$ maximal cliques [23]. It was shown in [28] that the algorithm in [7] achieves this bound in the worst case. No algorithm with a theoretically lower asymptotic runtime can thus exist. Third, and most importantly, the improvements we introduce do not

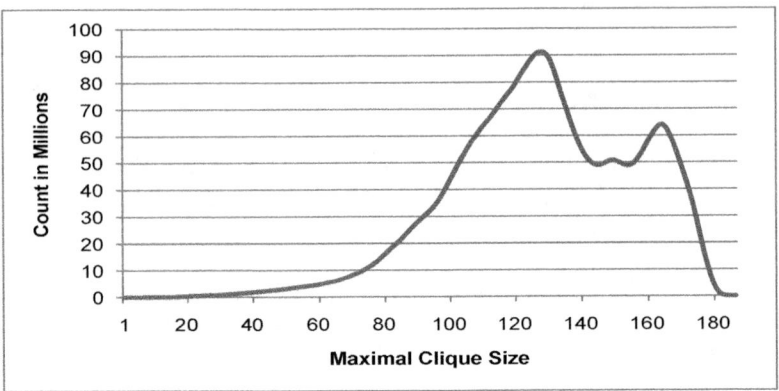

Fig. 2. The maximal clique profile of graph 70 in our testbed. MCE algorithms that are based on a breadth-first traversal of the search tree will retain at each step all maximal cliques of a given size. This can lead to titanic memory requirements. This graph, for example, contains more than 90 million maximal cliques of size 131. These sorts of memory demands tend to render non-backtracking methods impractical.

depend on the particulars of any one backtracking algorithm; they can be used in conjunction with any and all of them.

In the following sections, we test multiple approaches to solving MCE. We first test *Basic Backtracking*, the algorithm of [7]. We next modify this approach to take advantage of the fact that we only want to find maximum cliques. We call this approach *Intelligent Backtracking*. We then modify our existing tool for finding a single maximum clique to enumerate all maximum cliques. We call this approach *Parameterized Maximum Clique*, or *Parameterized MC*. This can be seen as another backtracking approach that goes even further to exploit the fact that we only want to find maximum cliques. Finally, based on observations about the properties of biological graphs, we introduce the concepts *maximum clique covers* and *essential vertex sets*, and apply them to improve the runtime of backtracking algorithms.

3.1   Basic Backtracking

The seminal maximal clique publication [7] describes two algorithms. A detailed presentation of the second, which is an improved version of the first, is provided. It is this second, more efficient, method that we implement and test. We shall refer to it here as Basic Backtracking. All maximal cliques are enumerated with a depth-first search tree traversal. The primary data structures employed are three global sets of vertices: COMPSUB, CANDIDATES and NOT. COMPSUB contains the vertices in the current clique, and is initially empty. CANDIDATES contains unexplored vertices that can extend the current clique, and initially contains all vertices in the graph. NOT contains explored vertices that cannot extend the current clique, and is initially empty. Each recursive call performs the following steps:

- Select a vertex v in CANDIDATES and move it to COMPSUB.
- Save CANDIDATES and NOT lists.
- Remove all vertices not adjacent to v from both CANDIDATES and NOT. At this point, if both CANDIDATES and NOT are empty, then COMPSUB is a maximal clique. If so, output COMPSUB as a maximal clique and continue the next step. If not a maximal clique, then make recursive call.
- Restore CANDIDATE and NOT lists.
- Move v from COMPSUB to NOT. Make recursive call.

Note that NOT is used to keep from generating duplicate maximal cliques. The search tree can be pruned by terminating a branch early if some vertex of NOT is connected to all vertices of CANDIDATES. Vertices are selected in a way that causes this pruning to occur as soon as possible, We omit the details since they are not pertinent to our modifications of the algorithm.

The storage requirements of Basic Backtracking are relatively modest. No information about previous maximal cliques needs to be retained. In the improvements we will test, we focus on speed but also improve memory usage. Thus, such limitations are in no case prohibitive for any of our tested methods. Nevertheless, in some environments, memory utilization can be extreme. We refer the interested reader to [29].

Our implementation of Basic Backtracking solved 29 of the 75 graphs in our testbed within 24 hours. See Figure 4. We therefore have now an initial benchmark upon which we can now try to improve.

3.2 Backtracking with Knowledge of Maximum Clique Size

Given the relative effectiveness with which we can find a single maximum clique, it seems logical to consider whether knowledge of that clique's size can be helpful in enumerating all maximum cliques. We first describe how our software implementation does its job. We then discuss subtree prunings that work to improve the backtracking approach. Finally we implement code changes and test the resultant algorithm on our testbed.

We use the term Maximum Clique Finder (MCF) to denote the software we have implemented and refined for finding a single clique of largest size [12]. MCF employs a suite of preprocessing rules along with a branching strategy that mirrors the well-known FPT approach to vertex cover [1,26]. It first invokes a simple greedy heuristic to find a reasonably large clique rapidly. This clique is then used for preprocessing, since it puts a lower bound on the maximum clique size. The heuristic works by choosing the highest degree vertex, v, then choosing the highest degree neighbor of v. These two vertices form an initial clique C, which is then iteratively extended by choosing the highest degree vertex adjacent to all of C. On each iteration, any vertex not adjacent to all of C is removed. The process continues until no more vertices exist outside C. Since $|C|$ is a lower bound on the maximum clique size, all vertices with degree less than $|C - 1|$ can be permanently removed from the original graph. Next, all vertices with degree $n - 1$ are temporarily removed from the graph, but retained in a list since they must be part of any maximum clique. MCF exploits a novel form of color preprocessing [12], used previously in [27] to guide branching. This form of preprocessing attempts to reduce the graph as follows. Given a known lower bound k on the size of the maximum clique, for each vertex v we apply fast greedy coloring to v and its neighbors. If these vertices can be colored with fewer than k colors, then v cannot be part of a maximum clique and is removed from the graph. Once the graph is thus reduced, MCF uses standard recursive branching on vertices, where each branch assumes that the vertex either is or is not in the maximum clique.

Intelligent Backtracking. Given that MCF rapidly finds a maximum clique, we now modify Basic Backtracking to make use of this information. First we compute the maximum clique size k using MCF and apply color preprocessing as previously described to reduce the graph. Only a slight modification to the internals of Basic Backtracking is necessary to make use of k to prune the search tree. Specifically, at each node in the search tree we check if there are fewer than k vertices in the union of COMPSUB and CANDIDATES. If so, that branch cannot lead to a clique of size k, and so we return.

While the modification may seem slight, the resultant pruning of the search tree can lead to a substantial reduction in the search space. As seen in Figure 4, Intelligent Backtracking results in a significant runtime improvement. We are now able to solve 58 of the 75 graphs in our testbed, twice the number solvable by Basic Backtracking.

Parameterized Enumeration. We now modify MCF to find not just one, but all maximum cliques. The modification is straightforward. We maintain a global list of all cliques of maximum size found thus far. Whenever a larger maximum clique is found, the list is flushed and refreshed to contain only the new maximum clique. When the search space has been exhausted, the list of maximum cliques is output.

We must take special care, however, to note that certain preprocessing rules used during interleaving are no longer valid. Consider, for example, the removal of a leaf vertex. The clique analogue is to find a vertex with degree $n-2$ and remove its lone non-neighbor. This rule patently assumes that only a single maximum clique is desired, because it ignores any clique depending on the discarded vertex.

After implementing these modifications, we are now able to solve 63 of the 75 graphs in our testbed. See Figure 4. While this is encouraging, our algorithms are still inadequate to handle readily all our test graphs. We will therefore explore two new approaches that are tailored to exploit properties of the sort of graphs we need to solve.

4   Maximum Clique Covers

We view MCF as a subroutine that can be called repeatedly. This provides us with a simple greedy algorithm for computing a maximal set of disjoint maximum cliques. We merely compute a maximum clique, remove it from the graph, and iterate until the size of a maximum clique decreases. To explore the advantages of computing such a set, we introduce the following notion:

Definition 1. *A* maximum clique cover *of* $G = (V, E)$ *is a set* $V' \subseteq V$ *with the property that each maximum clique of* G *contains some vertex in the cover.*

The union of all vertices contained in a maximal set of disjoint maximum cliques is of course a maximum clique cover (henceforth MCC), because all maximum cliques must overlap with such a set. This leads to a useful reduction algorithm. Any vertex not adjacent to at least one member of an MCC cannot be in a maximum clique, and can thus be removed.

In practice, we find that applying MCC before the earlier algorithms yields only marginal improvement. When coupled with Intelligent Backtracking, we can solve 59 of our 75 graphs. Even when coupled with Parameterized MC, we are still only able to solve 65 of our 75 graphs. Nevertheless, the concept of MCC leads us to a more useful approach based on individual vertices.

5   Essential Vertex Sets

Our investigation of the MCC algorithm showed us that it typically does not reduce the size of the graph more than the preprocessing rules already incorporated into our clique codes. For example, our clique codes already quickly find a lower bound on the maximum clique size and removes any vertex with degree lower than this bound. Upon closer examination, however, we found that for 74 of our 75 graphs, only one clique was needed in an MCC. Moreover, the lone outlier was sparse and easy to solve. In fact this coincides closely with our experience, in which we typically see high overlap among large cliques in the transcriptomic graphs we encounter on a regular basis. Thus, based on this observation, we shall now refine the concept of MCC. Rather than covering maximum cliques with cliques, we will cover maximum cliques with individual vertices.

We define an *essential vertex* as one that is contained in every maximum clique. Of course no such vertex may exist. On the contrary, however, based on computations we suspect that there are often numerous essential vertices. But even just one may suffice. An essential vertex has the potential to be extremely helpful, because it allows us to remove all its non-neighbors. We employ the following observation. For any graph G, $\omega(G) > \omega(G/v)$ if and only if v covers all maximum cliques, where $\omega(G)$ is the maximum clique size of G.

We define an *essential set* to be the set of all essential vertices, and propose the Essential Set (ES) Algorithm as described in Figure 3. The goal of this procedure is to find all essential vertices in hopes that we can compress the graph as much as possible.

When we run the ES algorithm before Intelligent Backtracking, we are able to solve 74 of 75 graphs. When we run it before Parameterized MC, we are able to solve the entire testbed (see Figure 4). Furthermore, the runtime improvement is great enough to suggest that the algorithm will scale to considerably larger graphs.

```
The Essential Set (ES) Algorithm for Graph Reduction
input: a simple graph G
output: a reduced graph G'

begin
    M = MCF(G) (M is a single maximum clique)
    For each vertex v in M
        G' = G/v
        M' =MCF(G')
        if (|M'| < |M|) then G = G/vertices not adjacent to v (v
covers all maximum cliques)
    end
    output G'
end
```

Fig. 3. Pseudocode for the Essential Set Algorithm

6   Analysis and Discussion

We timed the performance of Basic Backtracking, Intelligent Backtracking, and
Parameterized MC on graphs built from biological data. Basic Backtracking
was found to be non-competitive. We then reduced the graphs using reduction by MCC and ES and retested with Intelligent Backtracking and Parameterized MC. Run times include both the reduction step and Intelligent
Backtracking or Parameterized MC. ES is shown to perform much faster on
such instances. The effect is more pronounced at lower correlations, which correspond to larger graphs, though we do not observe strict monotonicity; such
a non-monotonic runtime progression is not unusual for $\mathcal{NP}$-complete
problems.

ES serves as a practical example of an innovative algorithm tailored to handle
a difficult combinatorial problem by exploiting knowledge of the input space. It
succeeds by exploiting properties of the graphs of interest, in this case the overlapping nature of maximum cliques. More broadly, these experiments underscore
the importance of considering graph types when testing algorithms.

It may be useful to examine graph size after applying MCC and ES, and
compare to both the size of the original graph and the amount of reduction
achieved by color preprocessing alone. Figure 5 depicts original and reduced
graph sizes for a selected subgroup of graphs.

While MCC seems as if it should produce better results, in practice we find it
not to be the case for two reasons. First, the vertices in an MCC may collectively
be connected to a large portion of the rest of the graph, and so very little
reduction in graph size takes place. And second, any reduction in graph size
may be redundant with FPT-style preprocessing rules already in place.

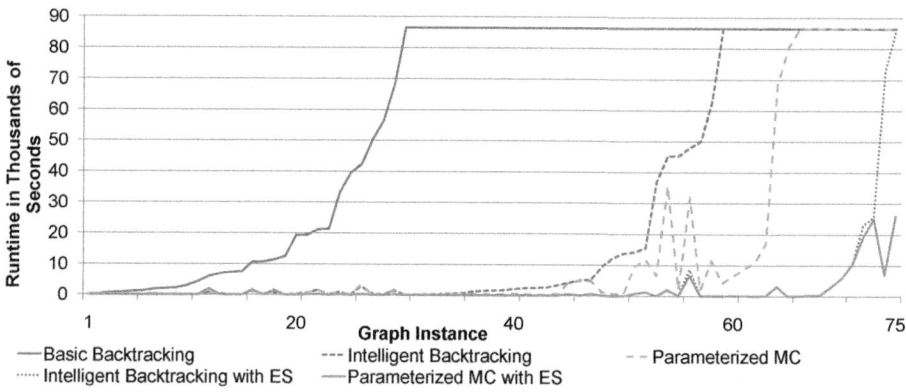

Fig. 4. Timings on various approaches to MCE on the testbed of 75 biological graphs. The tests were conducted under the Debian 3.1 Linux operating system on dedicated machines with Intel Xeon processors running at 3.20 GHz and 4 GB of main memory. Timings include all preprocessing, as well as the time to find the maximum clique size, where applicable. Runs were halted after 24 hours and deemed to have not been solved, as represented by those shown to take 86400 seconds. The graph instances are sorted first in order of runtimes for Basic Backtracking, then in order of runtimes for Intelligent Backtracking. This is a reasonable way to visualize the timings, though not perfect, since graphs that are difficult for one method may not be as difficult for another, hence the subsequent timings are not monotonic.

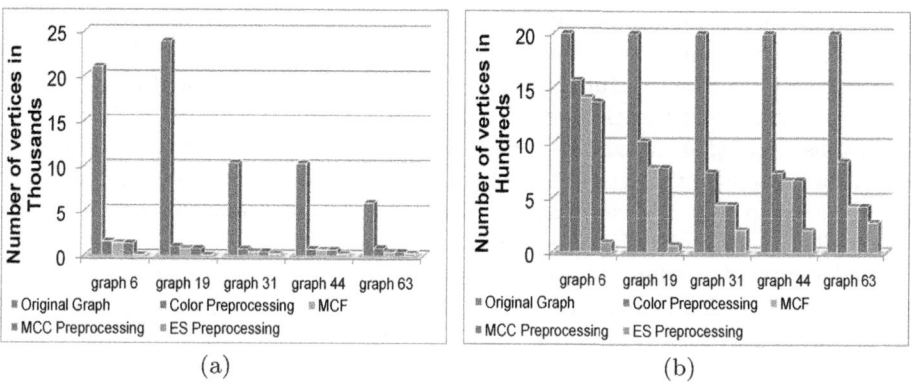

Fig. 5. Reduction in graph size thanks to preprocessing. (a) Five representative graphs are chosen from our testbed. The resulting number of vertices from post-processed graphs are plotted. (b) A closer view of graph sizes.

7   Contrast to Random Graphs

It would have probably been fruitless to test and design our algorithms around random graphs. (Yet practitioners do just that with some regularity.) In fact it has long been observed that the topology of graphs derived from real

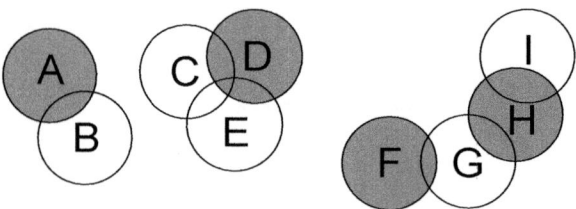

Fig. 6. The Subset Cover Problem. The decision version asks if there are k or fewer subsets that cover all other subsets. A satisfying solution for $k = 4$ is highlighted.

relationships differs drastically from the Erdös-Rényi random graph model introduced in [14]. Attempts to characterize the properties of real data graphs have been made, such as the notion of scale-free graphs, in which the degrees of the vertices follow a power-law distribution [4]. While work to develop the scale-free model into a formal mathematical framework continues [21], there remains no generally accepted formal definition. More importantly, the scale-free model is an inadequate description of real data graphs. We have observed that constructing a graph so the vertices follow a power law (scale-free) degree distribution, but where edges are placed randomly otherwise using the vertex degrees as relative probabilities for edge placement, still results in graphs with numerous small disjoint maximum cliques. For instance, constructing graphs with the same degree distribution as each of the 75 biological graphs in our testbed resulted in maximum clique sizes no greater than 5 for even the highest density graphs. Compare this to maximum clique sizes that ranged into hundreds of vertices in the corresponding biological graphs. Other metrics have been introduced to attempt to define important properties, such as cluster coefficient and diameter. Collectively, however, such metrics remain inadequate to model fully the types of graphs derived from actual biological data. The notions of maximum clique cover and essential vertices stem from the observation that transcriptomic data graphs tend to have one very large highly-connected region, and most (very often all) of the maximum cliques lie in that space. Furthermore, there tends to be a great amount of overlap between maximum cliques, perhaps as a natural result of gene pleiotropism. Such overlap is key to the runtime improvement achieved by the ES algorithm.

8   Future Research Directions

Our efforts with MCE suggest a number of areas with potential for further investigation. A formal definition of the class of graphs for which ES achieves runtime improvements may lead to new theoretical complexity results, perhaps based upon parameterizing by the amount of maximum clique overlap. Furthermore, such a formal definition may form the basis of a new model for real data graphs. We have noted that the number of disjoint maximum cliques that can be extracted provides an upper bound on the size of an MCC. If we parameterize by the maximum clique size and the number of maximum cliques, does an FPT algorithm exist? In addition, formal mathematical results may be achieved on the sensitivity of the number of maximum cliques to small changes in the graph.

Note that any MCC forms a hitting set over the set of maximum cliques, though not necessarily a minimum one. Also, a set D of disjoint maximum cliques, to which no additional disjoint maximum clique can be added, forms a *subset cover* over the set of all maximum cliques. That is, any maximum clique $C \notin D$ contains at least one vertex $v \in$ some maximum clique $D' \in D$. See Figure 6. To the best of our knowledge, this problem has not previously been studied. All we have found in the literature is one citation that erroneously reported it to be one of Karp's original $\mathcal{NP}$-complete problems [22].

For the subset cover problem, we have noted that it is $\mathcal{NP}$-hard by a simple reduction from hitting set. But in the context of MCE we have subsets all of the same size. It may be that this alters the complexity of the problem, or that one can achieve tighter complexity bounds when parameterizing by the subset size. Alternately, consider the problem of finding the minimum subset cover given a known minimum hitting set. The complexity of this tangential problem is not at all clear, although we conjecture it to be $\mathcal{NP}$-complete in and of itself. Lastly, as a practical matter, exploring whether an algorithm that addresses the memory issues of the subset enumeration algorithm presented in [19] and improved in [29] may also prove fruitful. As we have found here, it may well depend at least in part on the data.

Acknowledgments

This research was supported in part by the National Institutes of Health under grants R01-MH-074460, U01-AA-016662 and R01-AA-018776, and by the U.S. Department of Energy under the EPSCoR Laboratory Partnership Program. Mouse data were provided by the Goldowitz Lab at the Centre for Molecular Medicine and Therapeutics, University of British Columbia, Canada. Yeast data were obtained from the experimental work described in [20].

References

1. Abu-Khzam, F.N., Langston, M.A., Shanbhag, P., Symons, C.T.: Scalable parallel algorithms for FPT problems. Algorithmica 45, 269–284 (2006)
2. Baldwin, N.E., Chesler, E.J., Kirov, S., Langston, M.A., Snoddy, J.R., Williams, R.W., Zhang, B.: Computational, integrative, and comparative methods for the elucidation of genetic coexpression networks. J. Biomed. Biotechnol. 2(2), 172–180 (2005)
3. Baldwin, N.E., Collins, R.L., Langston, M.A., Leuze, M.R., Symons, C.T., Voy, B.H.: High performance computational tools for motif discovery. In: Proceedings of 18th International Parallel and Distributed Processing Symposium (2004)
4. Barabási, A.-L., Albert, R.: Emergence of scaling in random networks. Science 286, 509–512 (1999)
5. Bomze, I., Budinich, M., Pardalos, P., Pelillo, M.: The maximum clique problem. Handbook of Combinatorial Optimization 4 (1999)

6. Borate, B.R., Chesler, E.J., Langston, M.A., Saxton, A.M., Voy, B.H.: Comparison of thresholding approaches for microarray gene co-expression matrices. BMC Research Notes 2 (2009)
7. Bron, C., Kerbosch, J.: Algorithm 457: finding all cliques of an undirected graph. Commun. ACM 16(9), 575–577 (1973)
8. Bul, S.R., Torsello, A., Pelillo, M.: A game-theoretic approach to partial clique enumeration. Image and Vision Computing 27(7), 911–922 (2009); 7th IAPR-TC15 Workshop on Graph-based Representations (GbR 2007)
9. Chesler, E.J., Langston, M.A.: Combinatorial genetic regulatory network analysis tools for high throughput transcriptomic data. In: RECOMB Satellite Workshop on Systems Biology and Regulatory Genomics (2005)
10. Chesler, E.J., Lu, L., Shou, S., Qu, Y., Gu, J., Wang, J., Hsu, H.C., Mountz, J.D., Baldwin, N.E., Langston, M.A., Hogenesch, J.B., Threadgill, D.W., Manly, K.F., Williams, R.W.: Complex trait analysis of gene expression uncovers polygenic and pleiotropic networks that modulate nervous system function. Nature Genetics 37, 233–242 (2005)
11. Downey, R.G., Fellows, M.R.: Parameterized Complexity. Springer, New York (1999)
12. Eblen, J.D.: The Maximum Clique Problem: Algorithms, Applications, and Implementations. PhD thesis, University of Tennessee (2010), http://trace.tennessee.edu/utk_graddiss/793/
13. Eblen, J.D., Gerling, I.C., Saxton, A.M., Wu, J., Snoddy, J.R., Langston, M.A.: Graph algorithms for integrated biological analysis, with applications to type 1 diabetes data. In: Clustering Challenges in Biological Networks, pp. 207–222. World Scientific, Singapore (2008)
14. Erdős, P., Rényi, A.: Random graphs, pp. 17–61. Publication of the Mathematical Institute of the Hungarian Academy of Science (1960)
15. Fernau, H.: On parameterized enumeration. In: Proceedings of the 8th Annual International Conference on Computing and Combinatorics (2002)
16. Fernndez-Baca, D.: The perfect phylogeny problem. In: Cheng, X., Du, D.-Z. (eds.) Steiner Trees in Industry (2002)
17. Garey, M.R., Johnson, D.S.: Computers and Intractability: A Guide to the Theory of NP-Completeness. WH Freeman & Co., New York (1979)
18. Harley, E.R.: Comparison of clique-listing algorithms. In: Proceedings of the International Conference on Modeling, Simulation and Visualization Methods, pp. 433–438 (2004)
19. Kose, F., Weckwerth, W., Linke, T., Fiehn, O.: Visualizing plant metabolomic correlation networks using clique-metabolite matrices. Bioinformatics 17, 1198–1208 (2001)
20. Lai, L.C., Kosorukoff, A.L., Burke, P.V., Kwast, K.E.: Metabolic-state-dependent remodeling of the transcriptome in response to anoxia and subsequent reoxygenation in saccharomyces cerevisiae. Eukaryotic Cell 5(9), 1468–1489 (2006)
21. Li, L., Alderson, D., Doyle, J.C., Willinger, W.: Towards a theory of scale-free graphs: Definition, properties, and implications (extended version). Internet Mathematics (2005)
22. Malouf, R.: Maximal consistent subsets. Computational Linguistics 33, 153–160 (2007)
23. Moon, J.W., Moser, L.: On cliques in graphs. Israel Journal of Mathematics 3, 23–28 (1965)
24. Palla, G., Derényi, I., Farkas, I., Vicsek, T.: Uncovering the overlapping community structure of complex networks in nature and society. Nature 435, 814–818 (2005)

25. Perkins, A.D., Langston, M.A.: Threshold selection in gene co-expression networks using spectral graph theory techniques. BMC Bioinformatics 10 (2009)
26. Rogers, G.L., Perkins, A.D., Phillips, C.A., Eblen, J.D., Abu-Khzam, F.N., Langston, M.A.: Using out-of-core techniques to produce exact solutions to the maximum clique problem on extremely large graphs. In: ACS/IEEE International Conference on Computer Systems and Applications (AICCSA 2009), IEEE Computer Society, Los Alamitos (2009)
27. Tomita, E., Kameda, T.: An efficient branch-and-bound algorithm for finding a maximum clique with computational experiments. Journal of Global Optimization 37, 95–111 (2007)
28. Tomitaa, E., Tanakaa, A., Takahashia, H.: The worst-case time complexity for generating all maximal cliques and computational experiments. Theoretical Computer Science 363(1), 28–42 (2006)
29. Zhang, Y., Abu-Khzam, F.N., Baldwin, N.E., Chesler, E.J., Langston, M.A., Samatova, N.F.: Genome-scale computational approaches to memory-intensive applications in systems biology. In: Supercomputing (2005)

Query-Adaptive Ranking with Support Vector Machines for Protein Homology Prediction*

Yan Fu[1], Rong Pan[2], Qiang Yang[3], and Wen Gao[4]

[1] Institute of Computing Technology and Key Lab of Intelligent Information
Processing, Chinese Academy of Sciences, Beijing 100190, China
[2] School of Information Science and Technology,
Sun Yat-sen University, Guangzhou 510275, China
[3] Department of Computer Science and Engineering,
Hong Kong University of Science and Technology, Hong Kong, China
[4] Institute of Digital Media, Peking University, Beijing 100871, China
yfu@ict.ac.cn, panr@mail.sysu.edu.cn, qyang@cse.ust.hk, wgao@pku.edu.cn

Abstract. Protein homology prediction is a crucial step in template-based protein structure prediction. The functions that rank the proteins in a database according to their homologies to a query protein is the key to the success of protein structure prediction. In terms of information retrieval, such functions are called ranking functions, and are often constructed by machine learning approaches. Different from traditional machine learning problems, the feature vectors in the ranking-function learning problem are not identically and independently distributed, since they are calculated with regard to queries and may vary greatly in statistical characteristics from query to query. At present, few existing algorithms make use of the query-dependence to improve ranking performance. This paper proposes a query-adaptive ranking-function learning algorithm for protein homology prediction. Experiments with the support vector machine (SVM) used as the benchmark learner demonstrate that the proposed algorithm can significantly improve the ranking performance of SVMs in the protein homology prediction task.

Keywords: Protein homology prediction, information retrieval, ranking function, machine learning, support vector machine.

1 Introduction

A good ranking function is crucial for a successful information retrieval system [1]. A ranking function is based on the measurement of the relevance of database items to a query. Usually, there are multiple ways to measure the relevance of

* This work was supported by the Research Initiation Funds for President Scholarship Winners of Chinese Academy of Sciences (CAS), the National Natural Science Foundation of China (30900262, 61003140 and 61033010), the CAS Knowledge Innovation Program (KGGX1-YW-13), and the Fundamental Research Funds for the Central Universities (09lgpy62).

J. Chen, J. Wang, and A. Zelikovsky (Eds.): ISBRA 2011, LNBI 6674, pp. 320–331, 2011.

database items. An important issue is how to automatically and intelligently combine these relevance measures into a powerful single function using machine learning technologies [2,3,4,5].

Protein structures play an important role in biological functions of proteins. Experimental approach to protein structure determination is both slow and expensive. Since homologous proteins (evolved from the same ancestor) usually share similar structures, predicting protein structures based protein homologies has been one of the most important problems in bioinformatics [6,7,8,9]. Protein homology prediction is a key step of protein structure prediction and is a typical ranking problem[10]. In this problem, the database items are protein sequences with known three-dimensional structures, and the query is a protein sequence with unknown structure. The objective is to find those proteins in the database that are homologous to the query protein so that the homologous proteins can be used as structural templates.

The homology between two proteins can be captured from multiple views, such as sequence alignment, sequence profile and threading [11]. In this paper, we will not focus on these homology measures or features, but on the machine learning algorithms that integrate these features into a single score, i.e., a ranking function, in order to rank the proteins in a database. Since the proposed algorithm is, in principle, applicable to general ranking-function learning tasks, we will discuss the problem and describe the algorithm in a somewhat general manner. For example, when we say a 'query' or 'database item', it corresponds to a 'protein' in our protein homology prediction problem, and 'relevant'/'relevance' means 'homologous'/'homology'.

In ranking-function learning, the items (proteins in our case) in a database are represented as vectors of query-dependent features, and the objective is to learn out a function that can rank the database items in order of their relevances to the query. Each query-dependent feature vector corresponds to a query-item pair. Training data also consist of relevance (homology in our case) judgments for query-item pairs, which can be either absolute (e.g., item A is relevant, item B is not, while item C is moderate, etc.) or relative (e.g., item A is more relevant than item B). The relevance judgments can be acquired from the manual annotation of domain experts.

Algorithms for ranking-function learning mainly differ in the form of training data (e.g., absolute or relative relevance judgments), the type of ranking function (e.g., linear or nonlinear), and the way to optimize coefficients. In early years, various regression models were used to infer probability of relevance from binary judgments, e.g., the polynomial regression [3] and the logistic regression [12,13]. Information retrieval can also be viewed as a binary classification problem: given a query, classify all database items into two classes - relevant or irrelevant[2,14]. An advantage of viewing retrieval as a binary classification problem is that powerful discriminative models in machine learning, e.g., SVM, can be directly applied and the resultant ranking function is discriminative. Between regression and classification is the ordinal regression. Ordinal regression differs from conventional regression in that the targets are not continuous but finite and differs from classification in that the finite targets are not nominal but ordered [15,16]. Methods were also proposed to directly learn to rank things instead of learning the concept of relevance. For example, Joachims addressed the ranking-function learning problem in the framework of large margin criterion, resulting in the Ranking SVM algorithm [5]. Learning to rank has drawn more and more attention from the machine learning field in recent years (e.g., [17,18]).

A major characteristic of ranking-function learning is that the feature vector of each database item is computed with regard to a query. Therefore, all feature vectors are partitioned into groups by queries (each group of data associated with a query is called a *block* in this paper). Unlike traditional learning tasks, e.g., classification and regression, in which data are assumed to be independently and identically distributed, the ranking data belonging to the same block are correlated via the same query. We have observed that the data distributions may vary greatly from block to block [19]. The same value of a feature may indicate relevance in one block but irrelevance in another block. In the pure ranking algorithms that take preference judgments as input and do not aim to estimate relevance, e.g., Ranking SVM [5], training is performed so that rankings are only consistent within training queries. In this case, the difference between queries does not pose an obstacle to learning a pure ranking function. However, few efforts have so far been devoted to explicitly making use of the difference between queries to *improve* the generalization performance of learned ranking functions. Since queries in practice differ in various ways, no single ranking function performs well for all queries. A possible way to improve ranking performance is to use different ranking functions for different queries [20].

In this paper, we describe a query-adaptive ranking-function learning algorithm for the protein homology prediction task. Our approach, called K-Nearest-Block Ensemble Ranking, is motivated by the intuitive idea of learning a ranking function for a query using its similar queries in training data instead of using all available training data. Note that by similar we do not mean sequence similarity, but rather similarity in distributions of query-dependent data. To avoid online training, we employ an ensemble method. On each data block (corresponding to a query protein) in training data, an individual ranking model is trained offline in advance. Given the data block derived from a new query, the k ranking models trained on the k most similar blocks to the given block are applied separately to the given block and the k groups of ranks are aggregated into a single rank. In this way, incremental learning is also supported. As support vector machines (SVMs) [21] have been extensively studied in recently years for ranking-function learning and have been shown to have excellent performance (see, e.g., [5,15,14,17]), we use the SVM as the benchmark learner in this work. Experiments on a public dataset of protein homology prediction show that the proposed algorithm performs excellently in terms of both ranking accuracy and training speed, significantly improving the ranking performance of SVMs.

2   Algorithm

In this section, we describe our K-Nearest-Block (KNB) approach to ranking. Below is the terminology used in this paper.

Query-Dependent Feature Vector. Given a query (a protein sequence here), a database item (a protein sequence with known structure) is represented as a vector of (query-dependent) features that measure the relevance (homology) of the database item to the query. Each query-dependent feature vector corresponds to a query-item pair.

Block. A block B is a group of instances of query-dependent feature vectors associated with the same query. Each block corresponds to a query. A block usually includes several hundreds or thousands of feature vectors, which are computed from the most homologous proteins in a database according to some coarse scoring function.

Training Block. A training block is a block with relevance judgements for all of the feature vectors in it.

Test Block. A test block is a block in which the relevance judgments are unavailable and are to be made.

Block Distance. A block distance $D(B_i, B_j)$ is a mapping from two blocks B_i and B_j to a real value that measures the dissimilarity between the two blocks.

k Nearest Blocks. Just as the name implies, the k nearest blocks to a block are the k training blocks that are most similar to this blocks according to a block distance definition.

The block structure of data is a unique feature of the ranking problem. We believe that the differences among blocks, if appropriately used, can be very valuable information for training more accurate ranking models. One straightforward idea is that given the test block corresponding to a new query, all n training blocks should not be used to learn a ranking model, but only the $k(\ll n)$ most similar training blocks to the test block should be used. This is the block-level version of the traditional K-Nearest Neighbors method for classification or regression and thus can be called the K-Nearest Block (KNB) approach to ranking.

Three important sub-problems in the KNB approach are:

1. How to find the k nearest training blocks?
2. How to learn a ranking model using the k nearest blocks?
3. How to choose the value of k?

Different resolutions to the above three sub-problems lead to different implementations of the KNB method for ranking. High speed is a most crucial factor for a real-world retrieval system. Generating a new ranking model for each new query seems to apparently conflict with the above criterion. Therefore, the second sub-problem is especially important and needs to be carefully addressed.

2.1 K Nearest Blocks (KNB)

To find the k nearest blocks to a given block, a distance between blocks must be defined in advance. In general, the block distance can be defined in various ways, either domain-dependent or independent.

The most intuitive way is to represent each block as a vector of block features. Block features are variables that characterize a block from some views. For example, block features can be the data distribution statistics of a block. Given a vector representation of blocks, any vector-based distance can serve as a block distance, such as Euclidean distance or Mahalanobis' distance.

For simplicity and generality, we employ a domain-independent vector presentation of blocks and use Euclidean distance in this paper. Each block is represented by the statistics of each feature in the block, that is,

$$\Phi(B_i) = \langle \mu_{i1}, \sigma_{i1}, \mu_{i2}, \sigma_{i2}, \cdots, \mu_{id}, \sigma_{id} \rangle, \tag{1}$$

where μ_{ik} and σ_{ik} are the mean and the standard deviation of the k-th feature in block B_i, respectively. The distance between two blocks B_i and B_j then is

$$D(B_i, B_j) = \sqrt{\| \Phi(B_i) - \Phi(B_j) \|^2}. \tag{2}$$

2.2 KNB Ensemble Ranking

Given the k nearest blocks, the next step is to learn a ranking model from the selected training data. The most simple resolution is to train a **global model** on the union of all selected blocks. However, as we pointed out previously, a fatal drawback of doing this is that a training process has to be conducted online for each new query while the training time to be needed is totally unknown. This is unacceptable for a practical retrieval system.

To overcome the difficulty of online training, we propose to use an **ensemble model**. First, on each training block, a ranking model (called a local model) is trained offline. All local models are saved for future use. When a new query comes, a test block is generated and is compared to all training blocks. The k nearest training blocks to the test block are identified and the k corresponding local models are separately applied to the test block. Then, the k groups of relevance predictions are aggregated together to generate the final single rank for the query. Figure 1 gives the flowchart of the KNB Ensemble Ranking method.

In principle, any ranking model can serve as the local model in the KNB Ensemble Ranking method. Since SVMs have recently been extensively explored for ranking, we choose the classification SVM as the base learner and compare the resulted KNB ensemble SVM with other SVM-based ranking methods. Another reason for using SVMs is that previous best results on the data set used in this paper (see next section for detail) were mostly obtained with SVMs [19,22,23,24].

For aggregation, we simply sum over all the predictions made by selected local models with block distances as weights; that is,

$$Relevance(B^*) = \sum_{i \in I^*} \frac{1}{D(B^*, B_i)} Model_i(B^*), \tag{3}$$

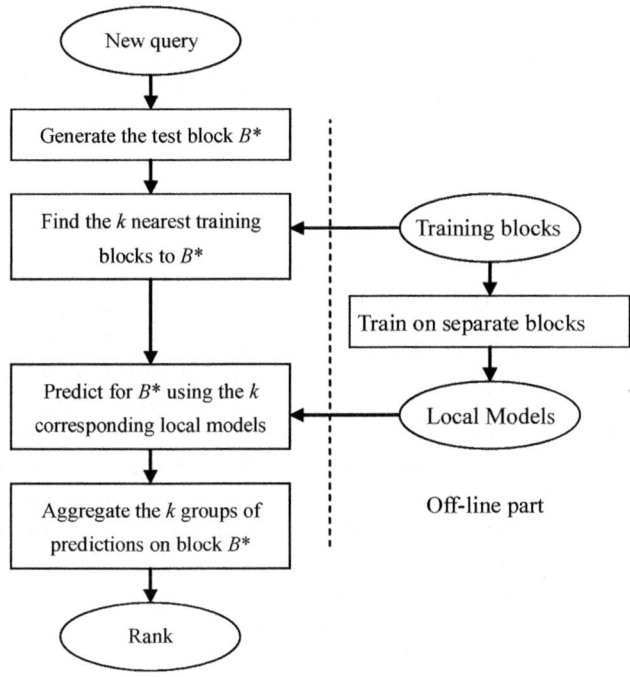

Fig. 1. Flowchart of the KNB Ensemble Ranking method

where I^* is the index of the k nearest training blocks to the test block B^* and $Model_i(B^*)$ denotes the predictions made by the local model trained on block B_i.

Compared to the global model, the ensemble model used in the KNB ranking method has several advantages:

- Firstly, training is fast. The time needed for training a learner (e.g. SVM) often increases nonlinearly with the number of training examples. Training on separate blocks is a divide-and-conquer strategy and thus is faster.
- Secondly, test is fast. KNB-based global model must be trained online for each new query, while local models can be trained offline in advance.
- Thirdly, incremental learning is supported. When a new training block comes, a new local model can be trained on it and be added into the repository of local models.
- Fourthly, a training block can be easily weighted according to its distance from the test block.
- Finally, an ensemble model often outperforms a global model in complex learning problems.

3   Experiments

In this section, we apply the KNB ensemble ranking algorithm to the protein homology prediction problem and demonstrate that the algorithm is superior to other SVM-based ranking methods in both ranking accuracy and training speed.

Table 1. Statistics of the protein homology data set

	#Queries	Block size	#Examples	#Features
Training data	153	~ 1000	145,751	74
Test data	150		139,658	

3.1 Data Set

Protein homology search is a routine task in current biology and bioinformatics researches. The task of protein homology prediction is to rank/predict the homologies of database proteins to a query protein. In this paper, we use the KDDCUP2004 data set of protein homology prediction [25]. In this data set, each database protein is characterized by 74 features measuring its homology to the query protein. The data are generated by the program LOOPP (Learning Observing and Outputting Protein Patterns), a protein fold recognition program [26]. The homology features include length of alignment, percentage of sequence identity, z-score for global sequence alignment, etc. [11]. On this data set, we have obtained the Tied for First Place Overall Award in the KDDCUP2004 competition. In the original winning solution, we successfully developed and used the intra-block data normalization and support-vector data sampling technologies for ranking-function learning [19].

The statistics of the data are summarized in Table 1. They include 153 training queries and 150 training queries. For each query, the examples (candidate homologous proteins) were obtained from a preliminary scoring/ranking function. The labels (homology judgments) are in binary form (0 for homology and 1 for non-homology) and are available for training data. Labels for test data are not published. The predictions for test data can be evaluated online at the competition web site. This provides a relatively fair manner for researchers to test and compare their methods.

For computing the block distance, we globally normalize the query-dependent features so that the mean is zero and the variance is one in the whole training data. For training local SVMs, we locally normalize the features so that the mean is zero and the variance is one within each block. We found that this kind of intra-block normalization resulted in improved prediction performance compared to the global normalization [19].

3.2 Performance Evaluation

Four metrics are used to evaluate the performance of a ranking method. They are TOP1, RKL (average rank of the last relevant item), APR (mean average precision), and RMS (root mean squared error). Each of the four metrics is first computed on individual blocks, and then averaged over all blocks.

TOP1. (maximize) TOP1 is defined as the fraction of blocks with a relevant item ranked highest. It measures how frequently a search engine returns a relevant item to the user at the top position.

RKL. (minimize) RKL is defined as the average rank of the last relevant item. It measures how many returned items have to be examined sequentially in average so that all relevant items can be found. If the purpose is to find out all relevant items, then RKL is a more suitable metric than TOP1.

APR. (maximize) APR is defined as the average of a kind of average ranking precision on each block. The average precision on single block is quite similar to the AUC (Area Under presision/recall Curve) metric. APR provides an overall measurement of the ranking quality.

RMS. (minimize) RMS is the average root mean square error. It evaluates how accurate the prediction values are if they are used as estimates of relevance (1 for absolute relevance and 0 for absolute irrelevance).

The first three metrics exclusively depend on the relative ranking of the items in a block while RMS needs relevance estimates. Since the target values are binary, we found that to a large extent RMS seems to rely on a good normalization of the prediction values more than on a good ranking or classification. Therefore, we place emphasis on the first three metrics, although we have obtained the best result of RMS on test data.

To evaluate a learning algorithm and perform model selection, cross validation is the most widely used strategy. Since the performance measures for ranking are calculated based on blocks, it becomes natural to divide the training data by blocks for cross validation. We extend the traditional cross validation method Leave-One-Out (LOO) to a block-level version which we call Leave-One-Block-Out (LOBO). Given a training data set, the LOBO cross validation puts one block aside as the validation set at a time and uses other blocks for training. After all blocks have their turns as the validation set, an averaged performance measure is computed at the end. The LOBO cross validation is different from both the traditional LOO and the n-fold cross validation. It is a graceful combination of these two common cross validation strategies, taking advantage of the block structure of ranking data.

3.3   Results

Four algorithms are compared, including the standard classification SVM, the Ranking SVM, the KNB global SVM, and the KNB ensemble SVM. For Ranking SVM, relative relevance judgments are derived from binary labels. In all experiments, the SVM^{light} package [27] is used. In most cases, the linear kernel is used for SVM training, based on the following considerations:

- High speed is crucial for a real-word retrieval system and linear ranking function are more efficient than nonlinear ones. Especially, a linear SVM can be represented by a weight vector while a nonlinear SVM has to be represented by a group of support vectors, the number of which is in general uncontrollable.
- Nonlinear kernels introduce additional parameters, thus increasing the difficulty of model selection. In our case, experiments have shown that training with nonlinear kernels, e.g., RBF kernel, on the entire data set we used is extremely slow and does not show better results.

Table 2. Cross-validation performance on training data set (the results of Standard SVM and Ranking SVM were obtained after intra-block data normalization, a method we previously proposed [19]; otherwise they would perform much worse. It is the same with Table 3).

Method	TOP1 (maximize)	RKL (minimize)	APR (maximize)	RMS (minimize)
Standard SVM	0.8758	51.94	0.8305	0.0367
Ranking SVM	0.8562	**36.83**	0.8257	N/A
KNB global SVM	**0.8889**	45.18	**0.8560**	**0.0354**
KNB ensemble SVM	**0.8889**	39.50	0.8538	0.0357

Table 3. Performance on test data set

Method	TOP1 (maximize)	RKL (minimize)	APR (maximize)	RMS (minimize)
Standard SVM	**0.9133**	59.21	0.8338	**0.0357**
Ranking SVM	0.9000	45.80	0.8369	N/A
KNB global SVM	0.9067	45.90	0.8475	0.0379
KNB ensemble SVM	0.9067	**40.50**	**0.8476**	0.0364

- In the KNB SVM ensemble approach, the linear local SVMs can be easily combined into a single linear function before prediction, thus decreasing the online computational burden. Moreover, it has been shown that in ensemble machine learning, a good generalization ability is often achieved using weak (e.g., linear) base learners rather than strong (e.g., nonlinear) ones.
- The use of linear kernel is fair for all the SVM-based methods compared.

The only parameter in the linear SVM is the parameter C, the tradeoff between training error and learner complexity. Another parameter in the KNB-based methods is k, the number of selected local models. To do model selection and performance evaluation on the training data, we use the LOBO cross-validation as described above. Table 2 gives the best results obtained using various methods. It shows that on the TOP1, APR and RMS metrics, KNB-based methods are superior to the other two methods. On the RKL metric, the Ranking SVM obtains the best result. On all metrics, the KNB ensemble SVM is comparable or superior to the KNB global SVM.

On the test data, predictions, made by models trained with the parameter values optimized for each metric, were evaluated on online. Table 3 gives the test results. The KNB ensemble SVM obtains the best results on the RKL and APR metrics among the four methods. In fact, they are the **best** known results on these two metrics (up to the time that this paper is submitted). On the other two metrics, KNB ensemble SVM does not perform best. However, the differences are very small and once again we argue that the RMS metric is very sensitive to the normalization of prediction values. On average, the solution of KNB ensemble SVM is the best result among all the original and subsequent submissions to the competition (http://kodiak.cs.cornell.edu/cgi-bin/newtable.pl?prob=bio).

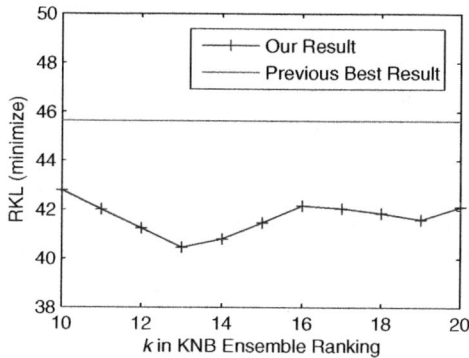

Fig. 2. RKL performance on test data set vs. k in KNB ensemble SVM in comparison with previous best result

Table 4. Training speed comparison

Method	Training mode	Training time (seconds)
Standard SVM	Offline	95
Ranking SVM	Offline	32255
KNB global SVM	Online	dependent on k
KNB ensemble SVM	Offline	**9**

It is also found that the KNB ensemble SVM is not sensitive to k, the number of selected nearest blocks. Figure 2 shows the RKL result of KNB ensemble SVM on the test data with the k as a variable. It can be seen that between a large range of the value of k (from 10 to 20), the RKL is rather stable and is considerably better than the previous best result obtained by Foussette et al. [22].

Besides the ranking accuracy, the training speed is another important factor. Table 4 lists the training mode and training time of the four methods. Standard classification SVM and Ranking SVM are trained on all available blocks. KNB global SVM is trained *online* on selected k nearest blocks, and the training time is dependent on k and is unpredictable in practice. For KNB ensemble SVM, local SVMs are trained *offline* on separate blocks. We can see that the offline training of KNB ensemble SVM only costs 9 seconds, 3600 times faster than ranking SVM. These experiments were performed on a Solaris/SPARC Server with 8 Sun Microsystems Ultra-SPARC III 900Mhz CPUs and 8GB RAM.

4 Conclusion and Future Work

In this paper, we have proposed a K-Nearest-Blocks (KNB) ensemble ranking algorithm with SVMs used as the base learners, and applied it to the protein homology prediction problem. Experiments show that compared to several other SVM-based ranking algorithms, the proposed one is significantly better on most performance evaluation metrics and meanwhile is extremely fast in training speed. Here, we have used a public data set. It is possible to develop more measures of protein homology to improve the accuracy of protein homology prediction. On the other hand, since the method is domain-independent, it is in principle applicable to general ranking problems. A potential problem with the KNB approach is that when the number of training blocks becomes very large, finding the k nearest blocks may be computationally expensive. However, all methods for expediting the traditional K-Nearest Neighbor method can also be used for the KNB method. In addition, we used a very simply definition of block distance in this paper. In fact, it can be improved in various ways, for example, refinement of block features, feature selection, distance learning, etc. We will try to address some of these aspects in our future work.

References

1. Baeza-Yates, R., Ribeiro-Neto, B.: Modern Information Retrieval. Addison-Wesley-Longman, Harlow (1999)
2. Robertson, S.E., Sparck Jones, K.: Relevance weighting of search terms. Journal of American Society for Information Sciences 27, 129–146 (1976)
3. Fuhr, N.: Optimal polynomial retrieval functions based on the probability ranking principle. ACM Transactions on Information Systems 7, 183–204 (1989)
4. Cohen, W., Shapire, R., Singer, Y.: Learning to order things. Journal of Artificial Intelligence Research 10, 243–270 (1999)
5. Joachims, T.: Optimizing Search Engines Using Clickthrough Data. In: 8th ACM Conference on Knowledge Discovery and Data Mining, pp. 133–142. ACM Press, New York (2002)
6. Baker, D., Sali, A.: Protein structure prediction and structural genomics. Science 294, 93–96 (2001)
7. Zhang, Y., Skolnick, J.: The protein structure prediction problem could be solved using the current PDB library. Proc. Natl. Acad. Sci. USA 102, 1029–1034 (2005)
8. Ginalski, K.: Comparative modeling for protein structure prediction. Current Opinion in Structural Biology 16, 172–177 (2006)
9. Zhang, Y.: Progress and challenges in protein structure prediction. Current Opinion in Structural Biology 18, 342–348 (2008)
10. Soding, J.: Protein homology detection by HMMCHMM comparison. Bioinformatics 2, 951–960 (2005)
11. Teodorescu, O., Galor, T., Pillardy, J., Elber, R.: Enriching the sequence substitution matrix by structural information. Proteins: Structure, Function and Bioinformatics 54, 41–48 (2004)
12. Cooper, W., Gey, F., Chen, A.: Information retrieval from the TIPSTER collection: an application of staged logistic regression. In: 1st NIST Text Retrieval Conference, pp. 73–88. National Institute for Standards and Technology, Washington, DC (1993)
13. Gey, F.: Inferring Probability of Relevance Using the Method of Logistic Regression. In: 17th Annual International ACM Conference on Research and Development in Information Retrieval, Dublin, Ireland, pp. 222–231 (1994)
14. Nallapati, R.: Discriminative Models for Information Retrieval. In: 27th Annual International ACM Conference on Research and Development in Information Retrieval, pp. 64–71. ACM Press, New York (2004)

15. Herbrich, R., Obermayer, K., Graepel, T.: Large margin rank boundaries for ordinal regression. In: Smola, A.J., Bartlett, P., Schölkopf, B., Schuurmans, C. (eds.) Advances in Large Margin Classifiers, pp. 115–132. MIT Press, Cambridge (2000)
16. Crammer, K., Singer, Y.: Pranking with ranking. In: Advances in Neural Information Processing Systems, vol. 14, pp. 641–647. MIT Press, Cambridge (2002)
17. Chapelle, O., Keerthi, S.S.: Efficient algorithms for ranking with SVMs. Information Retrieval Journal 13, 201–215 (2010)
18. McFee, B., Lanckriet, G.: Metric Learning to Rank. In: 27th International Conference on Machine Learning, Haifa, Israel (2010)
19. Fu, Y., Sun, R., Yang, Q., He, S., Wang, C., Wang, H., Shan, S., Liu, J., Gao, W.: A Block-Based Support Vector Machine Approach to the Protein Homology Prediction Task in KDD Cup 2004. SIGKDD Explorations 6, 120–124 (2004)
20. Fu, Y.: Machine Learning Based Bioinformation Retrieval. Ph.D. Thesis, Institute of Computing Technology, Chinese Academy of Sciences (2007)
21. Vapnik, V.N.: The Nature of Statistical Learning Theory. Springer, New York (1995)
22. Foussette, C., Hakenjos, D., Scholz, M.: KDD-Cup 2004 - Protein Homology Task. SIGKDD Explorations 6, 128–131 (2004)
23. Pfahringer, B.: The Weka Solution to the 2004 KDD Cup. SIGKDD Explorations 6, 117–119 (2004)
24. Tang, Y., Jin, B., Zhang, Y.: Granular Support Vector Machines with Association Rules Mining for Protein Homology Prediction. Special Issue on Computational Intelligence Techniques in Bioinformatics, Artificial Intelligence in Medicine 35, 121–134 (2005)
25. Caruana, R., Joachims, T., Backstrom, L.: KDD Cup 2004: Results and Analysis. SIGKDD Explorations 6, 95–108 (2004)
26. Tobi, D., Elber, R.: Distance dependent, pair potential for protein folding: Results from linear optimization. Proteins, Structure Function and Genetics 41, 16–40 (2000)
27. Joachims, T.: Making large-Scale SVM Learning Practical. In: Schölkopf, B., Burges, C., Smola, A. (eds.) Advances in Kernel Methods - Support Vector Learning, pp. 115–132. MIT Press, Cambridge (1999)

A Novel Core-Attachment Based Greedy Search Method for Mining Functional Modules in Protein Interaction Networks

Chaojun Li[1,2], Jieyue He[1,2,*], Baoliu Ye[2], and Wei Zhong[3]

[1] School of Computer Science and Engineering, Southeast University,
Nanjing, 210018, China
[2] State Key Laboratory for Novel Software Technology, Nanjing University,
Nanjing, 210093, China
[3] Division of Mathematics and Computer Science, University of South Carolina Upstate
800 University Way, Spartanburg, SC 29303, USA
chaojunli@126.com, jieyuehe@seu.edu.cn, yebl@nju.edu.cn,
wzhong@uscupstate.edu

Abstract. As advances in the technologies of predicting protein interactions, huge data sets portrayed as networks have been available. Therefore, computational methods are required to analyze the interaction data in order to effectively detect functional modules from such networks. However, these analysis mainly focus on detecting highly connected subgraphs in PPI networks as protein complexes but ignore their inherent organization. A greedy search method (GSM) based on core-attachment structure is proposed in this paper, which detects densely connected regions in large protein-protein interaction networks based on the edge weight and two criteria for determining core nodes and attachment nodes. The proposed algorithm is applied to the protein interaction network of S.cerevisiae and many significant functional modules are detected, most of which match the known complexes. The comparison results show that our algorithm outperforms several other competing algorithms.

Keywords: protein interaction network, functional module, clustering.

1 Introduction

Most real networks typically contain parts in which the nodes are more highly connected to each other than to the rest of the network. The sets of such nodes are usually called clusters, communities, or modules [1,2,3,4]. The presence of biologically relevant functional modules in Protein-Protein Interaction (PPI) graphs has been confirmed by many researchers [4,5]. Identification of functional modules is crucial to the understanding of the structural and functional properties of networks [6,7]. There is a major distinction between two biological concepts, namely, protein complexes and functional modules [7]. A protein complex is a physical aggregation of several proteins (and possibly other molecules) via molecular interaction (binding) with each other at the same location and time. A functional module also consists of a number of proteins (and

---

* Corresponding author.

J. Chen, J. Wang, and A. Zelikovsky (Eds.): ISBRA 2011, LNBI 6674, pp. 332–343, 2011.
© Springer-Verlag Berlin Heidelberg 2011

other molecules) that interact with each other to control or perform a particular cellular function. Unlike protein complexes, proteins in a functional module do not necessarily interact at the same time and location. In this paper, we do not distinguish protein complexes from functional modules because the protein interaction data used for detecting protein complex in this work do not provide temporal and spatial information.

Many graph clustering approaches have been used for mining functional modules [8,9]. These studies are mainly based on the observation that densely connected regions in the PPI networks often correspond to actual protein functional modules. In short, they detect densely connected regions of a graph that are separated by sparse regions. Some graph clustering approaches using PPI networks as the dataset for mining functional modules are introduced in the following.

Bader and Hogue [10] proposed the Molecular COmplex Detection (MCODE) algorithm that utilizes connectivity values in protein interaction graphs to mine for protein complexes. The algorithm first computes the vertex weight value (vertex weighting step) from its neighbor density and then traverses outward from a seed protein with a high weighting value (complex prediction step) to recursively include neighboring vertices whose weights are above a given threshold. However, since the highly weighted vertices may not be highly connected to each other, the algorithm does not guarantee that the discovered regions are dense.

Amin et al. [11] proposed a cluster periphery tracking algorithm (DPClus) to detect protein complexes by keeping track of the periphery of a detected cluster. DPClus first weighs each edge based on the common neighbors between its two proteins and further weighs nodes by their weighted degree. To form a protein complex, DPClus first selects the seed node, which has the highest weight as the initial cluster and then iteratively augments this cluster by including vertices one by one, which are out of but closely related with the current cluster.

Adamcsek et al. [12] provided a software called CFinder to find functional modules in PPI networks. CFinder detects the k-clique percolation clusters as functional modules using a Clique Percolation Method [13]. In particular, a k-clique is a clique with k nodes and two k-cliques are adjacent if they share (k − 1) common nodes. A k-clique percolation cluster is then constructed by linking all the adjacent k-cliques as a bigger subgraph.

Above computational studies mainly focus on detecting highly connected subgraphs in PPI networks as protein complexes but ignore their inherent organization. However, recent analysis indicates that experimentally detected protein complexes generally contain Core/attachment structures. Protein complexes often include cores in which proteins are highly co-expressed and share high functional similarity. And core proteins are usually more highly connected to each other and may have higher essential characteristics and lower evolutionary rates than those of peripheral proteins [19]. A protein complex core is often surrounded by some attachments, which assist the core to perform subordinate functions. Gavin et al.'s work [21] also demonstrates the similar architecture and modularity for protein complexes. Therefore, protein complexes have their inherent organization [19, 20,22] of core-attachment.

To provide insights into the inherent organization of protein complexes, some methods [14,19,22] are proposed to detect protein complexes in two stages. In the first stage, protein complex cores, as the heart of the protein complexes, are first detected. In the second stage, protein complexes are generated by including attachments into the protein complex cores. Wu et al. [14] presented a COre-AttaCHment based method (COACH) and Leung et al. also developed an approach called CoreMethod to detect protein complexes in PPI networks by identifying their cores and attachments separately [22]. To detect cores, COACH performs local search within vertex's neighborhood graphs while the CoreMethod [22] computes the p-values between all the proteins in the whole PPI networks.

In this paper, a greedy search algorithm called GSM is introduced. Comparing with the other methods of core-attachment, the new edge weight calculation method and evaluation criterion for judging a node as a core node or an attachment node are proposed in our GSM algorithm. GSM uses a pure greedy procedure to move a node between two different sets. The detected clusters are also core-attachment structures. In particular, GSM firstly defines seed edges of the core from the neighborhood graphs based on the highest weight and then detects protein-complex cores as the hearts of protein complexes. Finally, GSM includes attachments into these cores to form biologically meaningful structures. The new algorithm is applied to the protein interaction network of S.cerevisiae and the identified modules are mapped to the MIPS [15] benchmark complexes and validated by GO [16] annotations. The experimental results show that the identified modules are statistically significant. In terms of prediction accuracy, our GSM method outperforms several other competing algorithms. Moreover, most of the previous methods cannot detect the overlapping functional modules by generating separate subgraphs. But our algorithm can not only generate non-overlapping clusters, but also overlapping clusters.

Briefly then, the outline of this paper is as follows. In Section 2 we describe in detail the implementation of our methods. In Section 3, we apply our algorithm to the protein interaction network of *S.cerevisiae* yeast and analyze the results. In Section 4, we give our conclusions.

2 Methods

2.1 Definitions

Protein interaction network can be represented as an undirected graph $G = (V, E)$, where V is the set of vertices and $E = \{(u, v) \mid u, v \in V\}$ is the set of edges between the vertices. For a node $v \in V$, the set of $v's$ direct neighbors is denoted as N_v where $N_v = \{u \mid u \in V, (u, v) \in E\}$. Before introducing details of the algorithm, some terminologies used in this paper are defined.

The closeness cn_{nk} of any node n with respect to some node k in cluster c is defined by (1).

$$cn_{nk} = \frac{\mid NC_n \cap NC_k \mid}{\mid NC_k \mid} \tag{1}$$

Here, NC_n is the set of $n's$ direct neighbors in cluster c, and NC_k is the set of $k's$ direct neighbors in cluster c.

The DPClus algorithm defines the weight w_{uv} of an edge $(u,v) \in E$ as the number of the common neighbors of the nodes u and v. It is likely that two nodes that belong to the same cluster have more common neighbors than two nodes that do not. For two edges having the same number of common neighbors, the one that has more interactions between the common neighbors is more likely to belong to the same cluster.

Therefore, the definition of w_{uv} is modified in the paper by (2)

$$w_{uv} = |N_{uv}| + \alpha * |E_{uv}| \tag{2}$$

Here $N_{uv} = N_u \cap N_v$, $E_{uv} = \{(v_j, v_k) \mid (v_j, v_k) \in E, v_j, v_k \in N_{uv}\}$ and α is the interaction factor to indicate how important the interactions are. α's default value is set as 1.

The number of common neighbors between any two nodes is actually equal to the number of paths of length 2 between them. This definition of weight is used to cluster the graphs that have densely connected regions separated by sparse regions. In relatively sparse graphs, the nodes on the paths of length 3 or length 4 between the two nodes of one edge can be considered.

The highest edge weight of a node n is defined as $hw_n = \max(w_{nu})$ for all u such that $(n,u) \in E$. The highest weight edge (n,v) of node n is the edge satisfying the condition that $w_{nv} = hw_n$.

2.2 Greedy Search Method (GSM)

Because core and peripheral proteins may have different roles and properties due to their different topological characteristics, a Greedy Search Method (GSM) is proposed based on the definition of the edge weight and two evaluation criterion for judging a node as a core node or an attachment node. GSM uses a greedy procedure to get the suitable set of clusters. It first generates the core of a cluster, and then selects reliable attachments cooperating with the core to form the final cluster. The algorithm is divided into six steps: 1) Input & initialization; 2) Termination check; 3) Seed selection; 4) Core formation; 5) attachments selection; 6) Output & update. The functional modules are determined by final clusters. The whole description of algorithm GSM is shown in the following.

2.2.1 Input and Initialization
The input to the algorithm is an undirected simple graph and hence the associated matrix of the graph is read first. The user need decide the minimum value for closeness in cluster formation. The minimum value will be referred to as cn_{in}. Each edge's weight is computed. It is computed just once and will not be recalculated in the following steps.

2.2.2 Termination Check

Once a cluster is generated, it is removed from the graph. The next cluster is then formed in the remaining graph and the process goes on until no seed edge whose weight is above one can be found in the remaining graph.

2.2.3 Seed Selection

Each cluster starts at a deterministic edge called the seed edge. The highest weight edge of the remaining graph is considered as the seed edge.

2.2.4 Core Formation

A protein complex core is a small group of proteins which show a high co-expression patterns and share high degree of functional similarity. It is the key functional unit of the complex and largely determines the cellular role and essentiality of the complex [14,19,20,21]. For example, a protein in a core often has many interacting partners and protein complex cores often correspond to small, dense and reliable subgraphs in PPI networks [21].

The core starts from a single edge and then grows gradually by adding nodes one by one from the neighbors. The neighbors of a core are the nodes connected to any node of the core but not part of the core. The core is referred to as C. For a neighbor u of C, if u's neighbor v linked by u's highest weight edge (u,v) is in C, we consider to adding it to the core. Before adding u to C, we check the condition: $cn_{uv} >= cn_{in}$. We add the neighbor whose highest edge weight is biggest in all satisfying the condition. This process goes on until no such neighbor can be found, and then the core of one cluster is generated.

2.2.5 Attachments Selection

After the core of one cluster has been detected, we will extract the peripheral information of each core and select reliable attachments cooperating with it to form the final cluster. For each neighbor u of the core C, if u's neighbor v linked by u's highest weight edge (u,v) is in C, $\dfrac{|V_{uv}|}{|N_{uv}|}$ is computed. V_{uv} is the common neighbors of u and v in the core C. N_{uv} is the common neighbors of u and v in graph G. If $\dfrac{|V_{uv}|}{|N_{uv}|} > 0.5$, u will be selected as an attachment. After all neighbors of the core are checked, the final cluster is generated.

2.2.6 Output and Update

Once a cluster is generated, graph G is updated by removing the present cluster, i.e. The nodes belonging to the present cluster and the incident edges on these nodes are marked as clustered and not considered in the following. Then in the remaining graph, each node's highest edge weight is updated by not considering the edges that have been marked. The pseudocode of GSM algorithm is shown in Table 1.

Table 1. Algorithm of GSM

```
Algorithm GSM
input: a graph G = (V,E), parameters cn_in;
output: identified modules;
```

(1) **for** each edge $e(u, v) \in E$ **do**

   compute its weight;

   end for

(2) select the edge $e(u,v)$ with highest weight in G;

   if $e(u,v)$ does not exist **then** break;

   initial core $C=\{u,v\}$;

   while neighbor i whose neighbor j linked by $i's$ highest weight edge is in C and $cn_{ij}>=cn_{in}$ can be found **do**

   //among them, neighbor $n's$ highest edge weight is biggest

   $C.add(n)$;

   end while

(3) **for each** neighbor u of the core C **do**

   if $u's$ neighbor v linked by $u's$ highest weight edge is in C, and $\dfrac{|V_{uv}|}{|N_{uv}|} > 0.5$ **then**

   u is selected as attachment of C

   end if

   end for each

(4) output C and mark C as clustered, update each vertex's highest edge weight in the remaining graph

   goto(2)

2.3 Generation of Overlapping Clusters

In the above algorithm, once a cluster is generated it is marked as clustered and not considered in the following, and the next cluster is generated in the remaining graph. Therefore, non-overlapping clusters are generated. In order to generate overlapping clusters, we extend the existing non-overlapping clusters by adding nodes to them from their neighbors in the original graph (considering the marked nodes and edges). Then in the original graph excluding the edges between the nodes that have been marked as clustered, each node's highest edge weight is updated.

3 Experiments and Results

In order to evaluate effectiveness of the new system, our algorithm is applied to the full DIP (the Database of Interacting Proteins) [17] yeast dataset, which consists of 17201 interactions among 4930 proteins [14]. The full dataset is more complex and more difficult to identify the modules than the core dataset. The performance of our method is compared with several competing algorithms including MCODE, CFinder, DPClus, and COACH. The values of the parameters in each algorithm are selected from those recommended by the author. For comprehensive comparisons, several evaluation measures, including f-measure and p-value are employed.

The experimental results using a reference dataset of known yeast protein complexes retrieved from the MIPS are evaluated [15]. While it is probably one of the most comprehensive public datasets of yeast complexes available, it is by no means a complete dataset—there are still many yeast complexes that need to be discovered. After filtering the predicted protein complexes and complexes composed of a single protein from the dataset, a final set of 214 yeast complexes as our benchmark for evaluation are obtained.

The overlapping score [10] between a predicted complex and a real complex in the benchmark, $OS(p,b) = i^2/(p*b)$, is used to determine whether they match with each other, where i is the size of the intersection set of a predicted complex with a known complex, p is the size of the predicted complex and b is the size of the known complex. If $OS(p,b) \geq \omega$, they are considered to be matching (ω is set as 0.20 which is adopted in the MCODE paper [10]). We assume that P is the sets of complexes predicted by a computational method and B is the sets of target complexes in the benchmark respectively. The set of true positives (TP) is defined as $TP = \{p \mid p \in P, \exists b \in B, OS(p,b) \geq \omega\}$, while the set of false negatives (FN) is defined as $FN = \{b \mid p \in P, b \in B, \forall p(OS(p,b) < \omega)\}$. The set of false positives (FP) is $FP = P - TP$, while the set of known benchmark complexes matched by predicted complexes (TB) is $TB = B - FN$. The sensitivity and specificity [10] are defined as:

$$sensitivity = |TP|/(|TP|+|FN|) \qquad (3)$$

$$specificity = |TP|/(|TP|+|FP|) \qquad (4)$$

s-measure, as the harmonic mean of sensitivity and specificity, can be used to evaluate the overall performance of the different techniques.

$$s-measure = 2*sen*spe/(sen+spe) \qquad (5)$$

Table 2. Results of various algorithms compared with MIPS complexes using DIP data

Algorithms	MCODE	CFinder	DPClus	COACH	GSM
#predicted complexes	59	245	1143	745	**353**
ITPI	18	52	133	155	**105**
ITBI	19	61	144	106	**119**
s-measure	0.132	0.231	0.198	0.307	**0.380**

Table 2 shows the results of several methods compared with MIPS benchmark complexes. In Table 2, for MCODE, CFinder, DPClus and GSM, the number of correctly predicted complexes is less than the number of benchmark complexes matched by predicted complexes. But COACH is opposite. Because COACH detects the clusters from each node, the overlapping rate is high. Although the redundancy-filtering procedure is used, some predicted complexes are still similar and match the same benchmark complex. From Table 2, we notice that the s-measure of COACH (0.307) is highest among the methods of MCODE, CFinder, DPClus, and the s-measure of GSM (0.380) is significantly higher than that of COACH. In addition, the overall performance of COACH is much better than CoreMethod[14] which is another approach based on core-attachment structure.

To evaluate the biological significance of our predicted complexes, we calculate their p-values, which represent the probability of co-occurrence of proteins with common functions. Low p-value of a predicted complex generally indicates that the collective occurrence of these proteins in the module does not happen merely by chance and thus the module has high statistical significance. In our experiments, the p-values of complexes are calculated by the tool, SGD's *Go::TermFinder* [16], using all the three types of ontology *Biological Process(BP), Molecular Function(MF) and Cellular Component(CC)*. The cutoff of p-value is set as 0.01. The average $-\log(p-value)$ of all modules is calculated by mapping each module to the annotation with the lowest p value.

Let the total number of proteins be N with a total of M proteins sharing a particular annotation. The p-value of observing m or more proteins that share the same annotation in a cluster of n proteins, using the Hyper-geometric Distribution is defined as (6):

$$p-value = \sum_{i=m}^{n} \frac{\binom{M}{i}\binom{N-M}{n-i}}{\binom{N}{n}} \tag{6}$$

The average f-measure is used to evaluate the overall significance of each algorithm. f-measure of an identified module is defined as a harmonic mean of its recall and precision [18].

$$f-measure = \frac{2 * recall * precision}{recall + precision} \tag{7}$$

$$recall = \frac{|M \cap F_i|}{|F_i|} \tag{8}$$

$$precision = \frac{|M \cap F_i|}{|M|} \tag{9}$$

Where F_i is a functional category mapped to module M. The proteins in functional category F_i are considered as true predictions, the proteins in module M are considered as positive predictions, and the common proteins of F_i and M are considered as true positive predictions. Recall is the fraction of the true-positive predictions out of all the true predictions, and precision is the fraction of the true-positive predictions out of all the positive predictions [18]. The average f-measure value of all modules is calculated by mapping each module to the function with the highest f-measure value.

Table 3. Comparison of the results before and after adding attachments

	Average Size	f-measure of BP	-log(p-value)
before	5.29	0.356	7.2
after	7.37	0.362	8.6

Table 4. Statistical significance of functional modules predicted by various methods

Algorithms	No. of Modules size>=3	No. of Significant Modules	Average Size	Maximum Size	f-measure of BP	-log(p-value)	Parameters
MCODE	59	54	83.8	549	0.296	10.87	fluff=0.1; VWP=0.2
CFinder	245	157	10.2	1409	0.246	4.49	K=3
DPClus	217	187	5.23	25	0.335	6.78	Density=0.7; CP_{in} =0.5
COACH	746	608	8.54	44	0.272	6.96	Null
GSM	**187**	**168**	**7.37**	**79**	**0.362**	**8.60**	**CN_{in}=0.5**

Table 5. Comparison of f-measure based the three types of GO of GSM and other algorithms

Algorithms	f-measure		
	BP	MF	CC
MCODE	0.296	0.245	0.374
CFinder	0.246	0.145	0.266
DPClus	0.335	0.231	0.402
COACH	0.272	0.180	0.315
GSM	**0.362**	**0.241**	**0.453**

Comparison of the results before and after adding attachments is shown in Table 3. The comparison shows that after adding attachments the average size of modules grows. f-measure and $-\log(p-value)$ also increase.

Comparisons of biological significance of modules predicted by several algorithms are shown in Table 4. Because MCODE just generates a little number of modules, we don't consider it. From Table 4, we can see that the proportion of significant modules predicted by GSM is highest, and $-\log(p-value)$ are also higher than the other algorithms. Moreover, in all of the other methods, the average f-measure of DPClus is highest (0.335), however, the average f-measure of GSM is 0.362, which is higher than that of DPClus. The detailed comparison of f-measure using the three types of ontology is shown in Table 5. From Table 5, we can see the average f-measure of *Cellular Component* GSM is also highest (0.453).

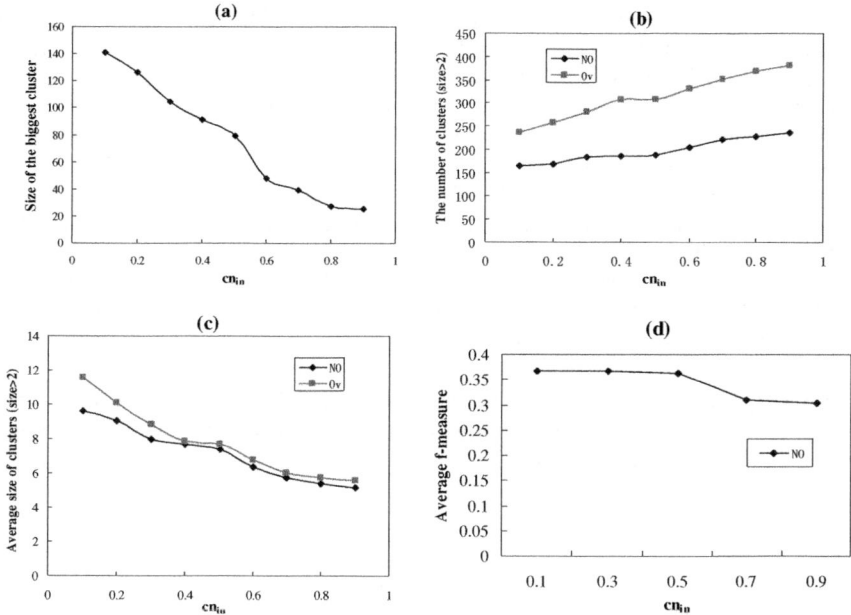

Fig. 1. The effects of cn_{in} on clustering. (a) the size of the biggest cluster, (b) the total number of the clusters of size>2, (c) the average size of the clusters of size>2,(d) the average f-measure

We also study the generated clusters by changing the parameter cn_{in}. The effects of the parameter cn_{in} are shown in Figure 1. When cn_{in} changes from 0.1 to 0.9, the size of the biggest cluster and the average size of clusters decrease but the number of cluster increases. The sizes of the biggest overlapping clusters are same as that of the non-overlapping clusters, so Figure 1(a) just draws one line. In Figure 1(b), the total number of the overlapping clusters is more than that of the non-overlapping clusters. In Figure 1(c), the average size of the overlapping clusters is bigger than that of the non-overlapping clusters. The effect of cn_{in} on f-measure is shown in Figure 1(d). From the figure, we can see when $cn_{in} > 0.5$, the f-measure is relatively lower. Because when cn_{in} is close to 1, the core of cluster is almost clique. It may be too strict to match well with the known annotations. When $cn_{in} <= 0.5$, the f-measure is basically stable. So we set cn_{in} as 0.5.

4 Conclusions

Identification of functional modules is crucial to the understanding of the structural and functional properties of protein interaction networks. The increasing amount of protein interaction data has enabled us to detect protein functional modules. In this paper, a greedy search clustering algorithm called GSM is proposed to mine functional modules from the protein interaction networks. Because core and peripheral proteins may have different roles and properties due to their different topological characteristics, GSM defines edge weight and two criterion for determining core nodes and attachment nodes. It first generates the core of a module, and then forms the module by including attachments into the core. GSM is applied to the typical PPI networks of *S.cerevisiae*. The MIPS benchmark and the GO annotation is used to validate the identified modules and compare the performances of our algorithm and several other algorithms including MCODE, CFinder, DPClus, COACH. The evaluation and analysis show that most of the functional modules predicted by our algorithm have high functional similarity and match well with the benchmark. The quantitative comparisons reveal that our algorithm outperforms the other competing algorithms. But there are some sparse modules in actual PPI network. In the future, we hope to detect the sparse modules and also apply the algorithm to the weighted graph.

Acknowledgments. This research work is supported by State Key Laboratory for Novel Software Technology of Nanjing University (KFKT2010B03) and Open Research Foundation of Key Laboratory for Computer Network and Information Integration, Southeast University (K93-9-2010-19).

References

1. Everitt, B.S.: Cluster Analysis, 3rd edn. Edward Arnold, London (1993)
2. Newman, M.E.J.: Detecting community structure in networks. Eur. Phys. J. B 38, 321–330 (2004)
3. Watts, D.J., Dodds, P.S., Newman, M.E.J.: Identity and search in social networks. Science 296, 1302–1305 (2002)
4. Girvan, M., Newman, M.E.: Community structure in social and biological networks. Proc. Natl. Acad. Sci. 99, 7821–7826 (2002)
5. Brun, C., Herrmann, C., Guenoche, A.: Clustering proteins from interaction networks for the prediction of cellular functions. BMC Bioinformatics 5(95) (July 2004)
6. Wu, L.F., Hughes, T.R., Davierwala, A.P., Robinson, M.D., Stoughton, R., Altschuler, S.J.: Large-scale prediction of saccharomyces cerevisiae gene function using overlapping transcriptional clusters. Nature Genetics 31, 255–265 (2002)
7. Spirin, V., Mirny, L.A.: Protein complexes and functional modules in molecular networks. Proc. Natl Acad. Sci. USA 100, 12123–12128 (2003)
8. Gao, L., Sun, P.G.: Clustering Algorithms for detecting functional modules in protein interaction networks. Journal of Bioinformatics and Computational Biology 7, 1–26 (2009)
9. Li, X., Wu, M., Kwoh, C.-K., Ng, S.-K.: Computational approaches for detecting protein complexes from protein interaction networks: a survey. BMC Genomics 11(Suppl 1), S3 (2010)
10. Bader, G.D., Hogue, C.W.: An Automated Method for Finding Molecular Complexes in Large Protein Interaction Networks. BMC Bioinformatics 4, 2 (2003)
11. Altaf-Ul-Amin, M., Shinbo, Y., Mihara, K., Kurokawa, K., Kanaya, S.: Development and implementation of an algorithm for detection of protein complexes in large interaction networks. BMC Bioinformatics 7, 207 (2006)
12. Adamcsek, B., Palla, G., Farkas, I.J., Derényi, I., Vicsek, T.: CFinder: locating cliques and overlapping modules in biological networks. Bioinformatics 22(8), 1021–1023 (2006)

13. Palla, G., Dernyi, I., Farkas, I., et al.: Uncoverring the overlapping community structure of complex networks in nature and society. Nature 435(7043), 814–818 (2005)
14. Wu, M., Li, X.L., Kwoh, C.K., Ng, S.K.: A Core-Attachment based Method to Detect Protein Complexes in PPI Networks. BMC Bioinformatics 10, 169 (2009)
15. Mewes, H.W., et al.: MIPS: analysis and annotation of proteins from whole genomes. Nucleic Acids Res. 32(Database issue), D41–D44 (2004)
16. Dwight, S.S., et al.: Saccharomyces Genome Database provides secondary gene annotation using the Gene Ontology. Nucleic Acids Research 30(1), 69–72 (2002)
17. Xenarios, I., et al.: DIP: the Database of Interaction Proteins: a research tool for studying cellular networks of protien interactions. Nucleic Acids Res. 30, 303–305 (2002)
18. Cho, Y.R., Hwang, W., Ramanmathan, M., Zhang, A.D.: Semantic integration to identify overlapping functional modules in protein interaction networks. BMC Bioinformatics 8, 265 (2007)
19. Luo, F., Li, B., Wan, X.-F., Scheuermann, R.H.: Core and periphery structures in protein interaction networks. BMC Bioinformatics 10, S8 (2009)
20. Dezso, Z., Oltvai, Z.D., Barabasi, A.L.: Bioinformatics Analysis of Experimentally Determined Protein Complexes in the Yeast Saccharomyces cerevisiae. Genome Res. 13, 2450–2454 (2003)
21. Gavin, A., Aloy, P., Grandi, P., Krause, R., Boesche, M., Marzioch, M., Rau, C., Jensen, L.J., Bastuck, S., Dumpelfeld, B., et al.: Proteome survey reveals modularity of the yeast cell machinery. Nature 440(7084), 631–636 (2006)
22. Leung, H., Xiang, Q., Yiu, S., Chin, F.: Predicting protein complexes from ppi data: A core-attachment approach. Journal of Computational Biology 16(2), 133–144 (2009)

ProPhyC: A Probabilistic Phylogenetic Model for Refining Regulatory Networks

Xiuwei Zhang and Bernard M.E. Moret

Laboratory for Computational Biology and Bioinformatics
EPFL (Ecole Polytechnique Fédérale de Lausanne), Switzerland
and Swiss Institute of Bioinformatics
{xiuwei.zhang,bernard.moret}@epfl.ch

Abstract. The experimental determination of transcriptional regulatory networks in the laboratory remains difficult and time-consuming, while computational methods to infer these networks provide only modest accuracy. The latter can be attributed in part to the limitations of a single-organism approach. Computational biology has long used comparative and, more generally, evolutionary approaches to extend the reach and accuracy of its analyses. We therefore use an evolutionary approach to the inference of regulatory networks, which enables us to study evolutionary models for these networks as well as to improve the accuracy of inferred networks.

We describe *ProPhyC*, a probabilistic phylogenetic model and associated inference algorithms, designed to improve the inference of regulatory networks for a family of organisms by using known evolutionary relationships among these organisms. *ProPhyC* can be used with various network evolutionary models and any existing inference method. We demonstrate its applicability with two different network evolutionary models: one that considers only the gains and losses of regulatory connections during evolution, and one that also takes into account the duplications and losses of genes. Extensive experimental results on both biological and synthetic data confirm that our model (through its associated refinement algorithms) yields substantial improvement in the quality of inferred networks over all current methods.

1 Introduction

Transcriptional regulatory networks are models of the cellular regulatory system that governs transcription. Because establishing the topology of the network from bench experiments is very difficult and time-consuming, regulatory networks are commonly inferred from gene-expression data. Various computational models, such as Boolean networks [1], Bayesian networks [9], dynamic Bayesian networks (DBNs) [15], and differential equations [6], have been proposed for this purpose. Results, however, have proved mixed: the high noise level in the data, the paucity of well studied networks, and the many simplifications in the models all combine to make inference difficult, in terms of both accuracy and computation.

Bioinformatics has long used evolutionary approaches to improve the accuracy of computational analyses. Recent work on the evolution of regulatory networks has

J. Chen, J. Wang, and A. Zelikovsky (Eds.): ISBRA 2011, LNBI 6674, pp. 344–357, 2011.

demonstrated the applicability of such approaches to regulatory networks. Although regulatory networks produced from bench experiments are available for only a few model organisms, other types of data have been used to assist in the comparative study of regulatory mechanisms across organisms. For example, gene-expression data [22], sequence data such as transcription factor binding site (TFBS) [7,21], and *cis*-regulatory elements [22] have all been used in this context. Moreover, a broad range of model organisms have been studied, including bacteria [3], yeast [7,22], and fly [21]. These studies have identified a number of evolutionary events, such as adding or removing network edges, and the duplication and loss of genes [3,20,23]. Results have also appeared on the evolution of metabolic networks and protein interaction networks [4,17].

Phylogenetic relationships are well established for many groups of organisms; as the regulatory networks evolved along the same lineages, the phylogenetic relationships informed this evolution and so can be used to improve the inference of regulatory networks. Indeed, Bourque and Sankoff [5] developed an integrated algorithm to infer regulatory networks across a group of species whose phylogenetic relationships are known, under a simple parsimony criterion. In previous work [25,26], we presented *refinement* algorithms, based on phylogenetic information and using a likelihood framework, that boost the performance of any chosen network inference method, hereafter called a *base* method. These refinement algorithms, *RefineFast* and *RefineML*, are two-step iterative algorithms. The networks to be refined are placed at the corresponding leaves of the known phylogeny. In the first step, ancestral networks for the phylogeny (strings labelling internal nodes) are inferred; in the second step, these ancestral networks are used to refine the leaf networks. These two steps are then repeated as needed. On both simulated and biological data, the *receiver-operator characteristic (ROC)* curves for our algorithms consistently dominated those of the base methods used alone.

We present *ProPhyC*, a probabilistic phylogenetic model and associated algorithms, designed to refine regulatory networks for a family of organisms. *ProPhyC* can accommodate a large variety of evolutionary models of regulatory networks with only slight modifications, as we demonstrate in the results section. Given that the evolution of regulatory networks is not yet well understood and given the several different models for regulatory network evolution [7,23,5], such flexibility is highly desirable. We present algorithms and experimental results in this refinement model for two network evolutionary models: a basic model that includes only gains and losses of regulatory interactions, and an extended model that also accounts for duplications and losses of genes. We also show how to take advantage of position-specific confidence values, if any, assigned to the input networks by the base inference method. Our probabilistic phylogenetic model confirms the usefulness of phylogenetic information in obtaining better inference of regulatory networks. Extensive experiments show that *ProPhyC* model not only brings significant improvement to base network inference algorithms, but also dominates the performance of existing refinement algorithms.

2 Background

Our approach posits that the evolution of regulatory networks correlates strongly with the evolution of the respective organisms, so that independent network inference errors can be corrected by using the phylogenetic relationships between the networks.

2.1 Base Network Inference Methods

We chose dynamic Bayesian inference (*DBI*), the method devised for DBNs, as the base inference method in our experiments. When DBNs are used to model regulatory networks, an associated structure-learning algorithm is used to infer the networks from gene-expression data [15,16]; so as to avoid overly complex networks, a penalty on graph structure complexity is usually added to the ML score, thereby reducing the number of false positive edges. In [25] we used a coefficient k_p to adjust the weight of this penalty and studied different tradeoffs between sensitivity and specificity, yielding the optimization criterion $\log Pr(D|G, \hat{\Theta}_G) - k_p \# G \log N$, where D denotes the dataset used in learning, G is the (structure of the) network, $\hat{\Theta}_G$ is the ML estimate of parameters for G, $\# G$ is the number of free parameters of G, and N is the number of samples in D.

2.2 Reconciliation of Species Tree and Gene Trees

To recover the gene contents of ancestral networks under the extended model, we need a full history of gene duplications and losses. We reconstruct this history by reconciling the gene trees and the species tree, that is, by using the differences between these trees to infer past duplication and loss events. While reconciliation is a hard computational problem, algorithms have been devised for it in a Bayesian framework [2] or using a simple parsimony criterion, as in the software Notung [8].

3 Models and Methods

We begin by presenting two network evolutionary models, then describe the *ProPhyC* refinement framework, and finally give associated refinement algorithms, one for each network evolutionary model.

3.1 Network Evolutionary Models

We present a basic model and an extended model. In both models, the networks are represented by binary adjacency matrices. For the basic model, the evolutionary operations are: *edge gain*, in which an edge between two genes is generated with probability p_{01}, and *edge loss*, an existing edge is deleted with probability p_{10}. The model parameters are thus (i) the base frequencies of 0 and 1 entries in the given networks $\Pi = (\pi_0 \ \pi_1)$, and (ii) the substitution matrix of 0s and 1s, $P = (p_{ij})$. The extended model has two additional evolutionary operations, *gene duplication* and *gene loss*, with corresponding additional model parameters p_d and p_l. In gene duplication, a gene is duplicated with probability p_d; after duplication, edges for the newly generated copy are assigned according to (i) *neutral initialization*, where the new copy gets connected to other genes randomly according to the proportion π_1 of edges in the background network; or (ii) *inheritance initialization*, where the new copy inherits the connections of the original, then loses or gains connections at some fixed rate, following reports of strong correlations between the connections of the new copy and those of the original copy [3,20,23]. In gene loss, a gene is deleted along with all its connections with probability p_l.

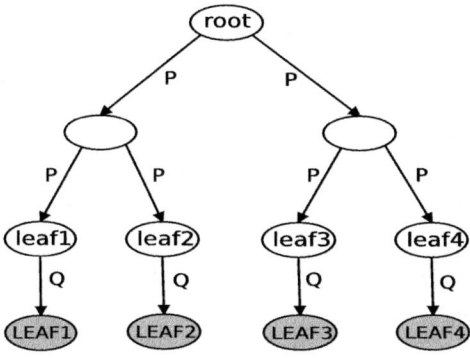

Fig. 1. *The* ProPhyC *model*

3.2 The *ProPhyC* Framework

ProPhyC is a probabilistic phylogenetic model designed to refine the inferred (and error-prone) regulatory networks for a family of organisms by making use of known phylogenetic information for the family. ProPhyC is also a graphical model: the phylogeny of this family is the main information to determine its structure as illustrated in Fig. 1. The shaded nodes labeled in upper case represent the input noisy networks, while the nodes labeled in lower case represent the correct networks for these organisms that we want to infer. In turn, the correct networks are the leaves of the rooted phylogenetic tree of these organisms, while internal nodes correspond to ancestral regulatory networks. The edges in this graph fall into two categories: (i) edges in the phylogenetic tree, representing the evolution from a parent network to a child network, and (ii) edges from correct leaf networks to noisy ones, representing the error-prone process of inferring networks from latent correct networks. The parameters for this model are thus the substitution matrices P and Q, where P represents the transition parameters from an ancestral network to its child network—subject to the network evolutionary model—and Q represents the difference between the "true" networks and the inferred (observed, from the point of view of the *ProPhyC* model) noisy networks—associated with one's confidence in the base network inference method.

The input information is thus the evolutionary model, the phylogenetic tree, and the noisy leaf networks. With a dynamic programming algorithm to maximize the likelihood of the whole graph, we can infer the ancestral networks and the "true" leaf networks. These "true" leaf networks inferred are the refined networks for these organisms and the output of the refinement algorithm. The framework can easily be generalized to fit different network evolutionary models. Some base inference methods can predict regulatory networks with different confidence on different edges or non-edges of the networks, so in this case Q can vary for different entries of different leaf networks. Our model can incorporate these position-specific confidence values to get better refinements. We name this version of the refinement algorithm *ProPhyCC*.

3.3 *ProPhyC* **Under the Basic Model**

Under the basic model, all networks have the same size and gene contents. Each network is represented by its binary adjacency matrix, so the character set is $S = \{0,1\}$. The parameters to calculate the likelihood are those from the evolutionary model, Π and P, and the error parameter for the base inference method, $Q = (q_{ij})$. We assume independence between the network entries, so that we can process separately each entry in the adjacency matrices. Let i, j, k denote nodes in the tree and $a, b, c \in S$ denote possible values of a character. For each character a at each node i, we maintain two variables:

- $L_i(a)$: the likelihood of the best reconstruction of the subtree with root i, given that the parent of i is assigned character a.
- $C_i(a)$: the optimal character for i, given that its parent is assigned character a.

When the phylogenetic tree is binary, our inference algorithm works as follows:

1. For each leaf node i, if its corresponding noisy network has character b, then for each $a \in S$, set $L_i(a) = \max_{c \in S} p_{ac} \cdot q_{cb}$ and $C_i(a) = \arg\max_{c \in S} p_{ac} \cdot q_{cb}$.
2. If i is an internal node and not the root, its children are j and k, and it has not yet been processed, then for each $a \in S$, set $L_i(a) = \max_{c \in S} p_{ac} \cdot L_j(c) \cdot L_k(c)$ and $C_i(a) = \arg\max_{c \in S} p_{ac} \cdot L_j(c) \cdot L_k(c)$.
3. If there remain unvisited nonroot nodes, return to Step 2.
4. If i is the root node, with children j and k, assign it the value $a \in S$ that maximizes $\pi_a \cdot L_j(a) \cdot L_k(a)$.
5. Traverse the tree from the root, assigning to each node its character by $C_i(a)$.

3.4 *ProPhyC* **Under the Extended Model**

The extended model includes gene duplications and losses, so that the gene content may vary across networks. While the gene content of the leaf networks is known, we need to reconstruct the gene content for ancestral networks, that is, to reconstruct the history of gene duplications and losses. This part can be solved by using an algorithm to reconcile the gene trees and species tree [2,8,19] or by the algorithms that we presented in earlier work under the *duplication-only* or *loss-only* model [27].

Under the basic model, we assumed independence among the entries of the adjacency matrices and so greatly simplified the computation. To enable us to do the same under the extended model, we embed each network into a larger one that includes every gene that appears in any network. We then represent a network with a ternary adjacency matrix, where the rows and columns of the missing genes are filled with a special character x. All networks are thus represented with adjacency matrices of the same size. Since the gene contents of ancestral networks are known thanks to reconciliation, the entries with x are already identified in their matrices; the other entries are reconstructed by the refinement algorithm using the new character set $S' = \{0,1,x\}$. The substitution matrix P' for S' can be derived from the model parameters, without introducing new parameters. Assuming that at most one gene duplication and one gene loss can happen at each evolutionary step, we have:

$$P' = \begin{pmatrix} p'_{00} & p'_{01} & p'_{0x} \\ p'_{10} & p'_{11} & p'_{1x} \\ p'_{x0} & p'_{x1} & p'_{xx} \end{pmatrix} = \begin{pmatrix} (1-p_l) \cdot p_{00} & (1-p_l) \cdot p_{01} & p_l \\ (1-p_l) \cdot p_{10} & (1-p_l) \cdot p_{11} & p_l \\ p_d \cdot \pi_0 & p_d \cdot \pi_1 & 1-p_d \end{pmatrix} .$$

We also extend the parameter Q to be Q' to fit the new character set S':

$$Q' = \begin{pmatrix} q'_{00} & q'_{01} & q'_{0x} \\ q'_{10} & q'_{11} & q'_{1x} \\ q'_{x0} & q'_{x1} & q'_{xx} \end{pmatrix} = \begin{pmatrix} q_{00} & q_{01} & 0 \\ q_{10} & q_{11} & 0 \\ 0 & 0 & 1 \end{pmatrix} .$$

The transition probabilities in Q' remain the same as in Q, since the gene contents of the "true" and corresponding noisy network are the same. For each character a at each tree node i, we calculate $L_i(a)$ and $C_i(a)$ for each site with the following procedure:

1. For each leaf node i, if its corresponding noisy network has character b, then for each $a \in S'$, set $L_i(a) = \max_{c \in S'} p'_{ac} \cdot q'_{cb}$ and $C_i(a) = \arg\max_{c \in S'} p'_{ac} \cdot q'_{cb}$.
2. If i is an internal node and not the root, its children are j and k, and it has not yet been processed, then
   - if i has character x, for each $a \in S'$, set $L_i(a) = p'_{ax} \cdot L_j(x) \cdot L_k(x)$ and $C_i(a) = x$;
   - otherwise, for each $a \in S'$, set $L_i(a) = \max_{c \in S} p'_{ac} \cdot L_j(c) \cdot L_k(c)$ and $C_i(a) = \arg\max_{c \in S} p'_{ac} \cdot L_j(c) \cdot L_k(c)$.
3. If there remain unvisited nonroot nodes, return to Step 2.
4. If i is the root node, with children j and k, assign it the value $a \in S$ that maximizes $\pi_a \cdot L_j(a) \cdot L_k(a)$, if the character of i is not already identified as x.
5. Traverse the tree from the root, assigning to each node its character by $C_i(a)$.

3.5 Refinement Algorithm *ProPhyCC* Using Confidence Values

Parameter Q (or Q') models the errors introduced in the base inference process; its values are obtained from one's confidence in that method and in the source data. The *ProPhyC* algorithm uses the same matrix for all entries in all leaf networks. When sufficient information is available to produce different confidence values for different entries in different networks, we can take advantage of the extra information through the *ProPhyCC* algorithm.

If the noisy networks are predicted from gene-expression data by DBN models, to obtain the confidence values, we first estimate the conditional probability tables (CPTs) of the *DBI* inferred networks from the gene-expression data on the inferred structure [11], and then calculate the confidence values from the CPTs. Following [16], we use binary gene-expression levels in our experiments, where 1 and 0 indicate that the gene is, respectively, *on* and *off*. For each gene g_i, if m_i nodes have arcs directed to g_i in the network, let the expression levels of these nodes be denoted by the vector $y = y_1 y_2 \cdots y_{m_i}$ and the confidence values of their arcs by the vector $c = c_1 c_2 \cdots c_{m_i}$. We use signed weights to represent the strength of these arcs, denoted by $w = w_1 w_2 \cdots w_{m_i}$. Considering that if an arc is predicted with high weight, then this arc is very likely to be true, we assign high confidence values to the arcs predicted with high absolute weight values. Let k be a coefficient value to normalize probabilities, we have $k \cdot w \cdot y = Pr(g_i \text{ is } on|y)$. Since there are 2^{m_i} configurations of y, there are 2^{m_i} such equations. The value of $Pr(g_i \text{ is } on|y)$ can be directly taken from the CPTs. So w can be obtained by solving these equations, and c derived directly from w.

4 Experimental Design

We designed a comprehensive collection of experiments to assess our model and its associated algorithms. The accuracy of the output is calculated by comparing the output with the "true" networks for the chosen family of organisms, where the "true" networks are either obtained through simulation or collected from biological datasets. We compare the accuracies of the networks produced by the base method *DBI* and of the networks after refinement, to get *absolute* assessments. We also use our previous refinement algorithms [25,26,27] to refine the same networks and compare the outcome with that of our new refinement model, to get *relative* assessments.

Since regulatory networks are usually reconstructed from gene-expression data, we follow the same path in our assessment. With networks inferred from gene-expression data as input for *ProPhyC*, *ProPhyCC*, *RefineFast* and *RefineML*, we run experiments with different combinations of networks evolutionary models and types of datasets. Under each setting, we show both absolute and relative assessments.

4.1 Biological Data Collection

Transcription factor binding site (TFBS) data is used to study regulatory networks, assuming that the regulatory interactions determined by transcription factor binding share many properties with the real interactions [7,10,21]. Given this close relationship between regulatory networks and TFBSs and given the large amount of available data on TFBSs, we chose to use TFBS data to derive regulatory networks for the organisms as their "true" networks. The TFBS data is drawn from the work of Kim *et al.* [14], where the TFBSs are annotated for the *Drosophila* family (whose phylogeny is well studied) with 12 species. They reported TFBS annotations for 7 transcription factors on 51 cis-regulatory modules (CRMs) for all 12 species. Since each CRM corresponds to a target gene, we get a regulatory network with 58 nodes for each organism as the "true" network for this organism. We add noise into these "true" networks to obtain noisy networks as input to our refinement algorithm.

4.2 Data Simulation

In simulation experiments, we generate gene-expression data from simulated leaf networks. This step helps in decoupling the generation and the reconstruction phases. The data simulation procedure consists of two main steps: (i) generate the "true" leaf networks according to the evolutionary model and (ii) generate the gene-expression data. The whole process starts from three pieces of information: the phylogenetic tree, the network at its root, and the evolutionary model. Since we need quantitative relationships in the networks in order to generate gene-expression data from each network, in the network generation process, we use adjacency matrices with signed weights.

We take specific precautions against systematic bias during data simulation and result analysis. We use a wide variety of phylogenetic trees from the literature (of modest sizes: between 20 and 60 taxa) and several choices of root networks, the latter variations on part of the yeast network from the KEGG database [13]. The root network has between 14 and 17 genes, a relatively easy case for inference algorithms and thus a more challenging case for a refinement algorithm. We explore a wide range of evolutionary rates, including rates of gene duplication and loss and rates of edge gain and loss.

Simulating networks. Denote the weighted adjacency matrix of the root network as A_p. Under the basic model, we obtain the adjacency matrix for its child A_c by mutating A_p according to the substitution matrix. By repeating this process as we traverse down the tree we obtain weighted adjacency matrices at the leaves. In other words, we evolve the weighted networks down the tree according to the model parameters, following standard practice in the study of phylogenetic reconstruction [12,18]. Under the extended model, to get the adjacency matrix for the child network of A_p, we follow two steps: evolve the gene contents and evolve the regulatory connections. First, genes are duplicated or lost by p_d and p_l. If a duplication happens, a row and column for this new copy will be added to A_p, the values initialized either according to the *neutral initialization* model or the *inheritance initialization* model. Call this intermediate adjacency matrix A'_c. Now edges in A'_c are mutated according to p_{01} and p_{10} to get A_c. Again we repeat this process as we traverse down the tree to obtain weighted adjacency matrices at the leaves.

Generating gene-expression data. From the "true" networks, we use *DBNSim* [25], based on the DBN model, to generate time-series gene-expression data. Note that, while *DBNSim* and *DBI* are both based on Bayesian networks, which might artificially improve the performance of *DBI*, this bias can only make it more difficult for *ProPhyC* to achieve significant improvements.

For all experiments on simulated gene-expression data, we run the generation process 10 times for each choice of tree structure and parameters to compute a mean and a standard deviation. Under the basic model, for each leaf network, we generate 200 time points for its gene-expression matrix. Under the extended model, we generate $13 \times n$ time points for a leaf network with n genes, since larger networks generally need more samples to gain inference accuracy comparable to smaller ones.

4.3   Measurements

We want to examine the predicted networks at different levels of sensitivity and specificity. With *DBI*, we use a penalty coefficient on structure complexity so as to obtain different tradeoffs between sensitivity and specificity. On each dataset, we apply different penalty coefficients to predict regulatory networks, from 0 to 0.5, with an interval of 0.05, which results in 11 discrete coefficients. For each penalty coefficient, we apply our approach (and any method chosen for comparison) on the predicted networks, measure specificity and sensitivity, and plot the values into ROC curves. (In these ROC plots, the closer the curves are to the top left corner of the coordinate space, the better the results.)

5   Results and Analysis

5.1   Performance Under the Basic Model on Simulated Data

Absolute results. We show experimental results on two representative trees: one has 37 nodes on 7 levels and the other has 41 nodes on 6 levels. We only plot part of the curves within the 11 penalty coefficients to give a more detailed view of the comparison. Fig. 2 shows the results of *ProPhyC* and *ProPhyCC* on the networks predicted by *DBI*. We can see that *ProPhyC* and *ProPhyCC* significantly improve both sensitivity and specificity over the base inference algorithm *DBI*. The improvement remains similar on different

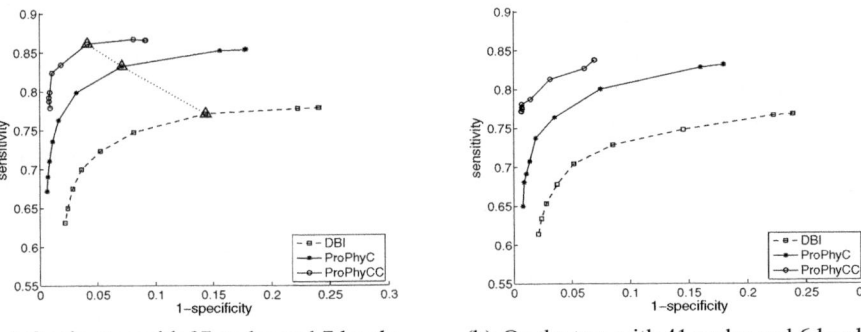

(a) On the tree with 37 nodes and 7 levels   (b) On the tree with 41 nodes and 6 levels

Fig. 2. Comparison of *ProPhyC* and *ProPhyCC* with base inference algorithm *DBI* under the basic model. In part (a), the dotted lines join data points for the same model penalty coefficient.

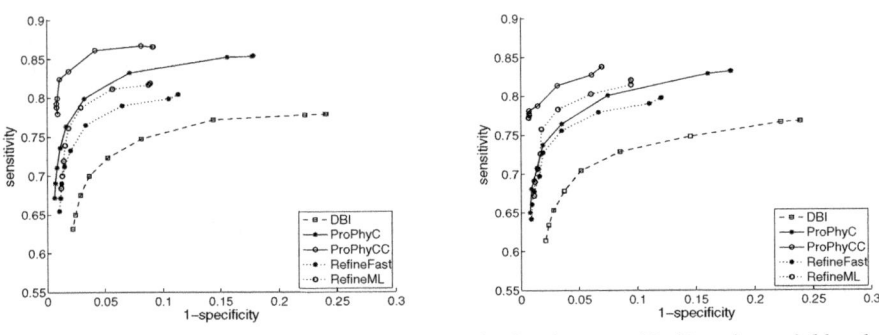

(a) On the tree with 37 nodes and 7 levels   (b) On the tree with 41 nodes and 6 levels

Fig. 3. Comparison of *ProPhyC* and *ProPhyCC* with *RefineFast* and *RefineML* under the basic model

tree structures. *ProPhyCC* further improves *ProPhyC*, which shows the advantage of using position-specific confidence values. For example, the dots in Fig. 2(a) marked by triangles correspond to the same penalty coefficient on the three curves, showing that, in going from *DBI* to *ProPhyCC*, the sensitivity increases from 77% to 86% while the specificity increases from 86% to 96%. Similar improvements can be observed with other trees, other evolutionary rates, and other base methods.

Relative results. Fig. 3 shows the same experiments as in Fig. 2, but adds curves for *RefineFast* and *RefineML* to provide a comparison between different refinement approaches. Among the four refinement algorithms, *ProPhyCC* and *RefineML* take advantage of the position-specific confidence values, which gives them better performance than *ProPhyC* and *RefineFast*. *ProPhyCC* is obviously the best among all refinement algorithms, while *ProPhyC* outperforms *RefineFast*. From Figs. 2 and 3, we conclude that refinement algorithms under our new model outperform not only base inference algorithms, but also our previous refinement algorithms on simulated data.

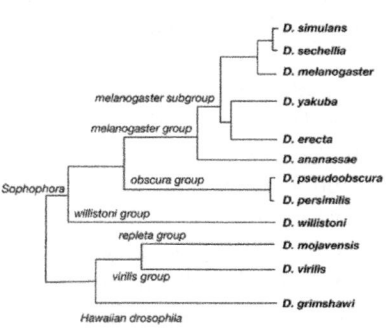

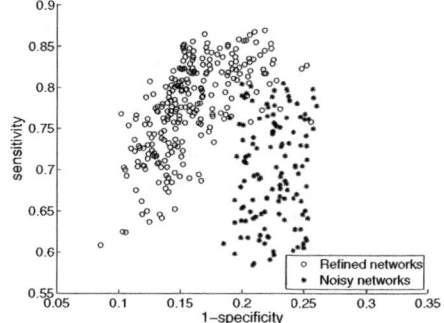

Fig. 4. The phylogeny connecting the 12 Drosophila species [24]

Fig. 5. Results of *ProPhyC* under the basic model on biological datasets

Table 1. Performance of ProPhyC under different tradeoffs

	sensitivity (noisy → refined)	specificity (noisy → refined)
improve both	59.9% → 66.3%	80.0% → 86.5%
focus on sensitivity	59.5% → 69.2%	69.3% → 72.7%
focus on specificity	57.7% → 58.5%	70.1% → 80.0%

5.2 Performance Under the Basic Model on Biological Data

Here we show our results on the datasets for 12 species of *Drosophila*, whose phylogenetic tree is illustrated in Fig. 4. We use different noise rates to get noisy networks with different false positives and false negatives. Then for each set of noisy networks we use *ProPhyC* to obtain refined networks with different parameter settings. Fig. 5 shows the accuracies of these networks plotted as points. The cloud of points for *ProPhyC* clearly dominates that of the noisy networks, and the two clouds are well separated; the average improvement brought by *ProPhyC* is roughly 7% in each of sensitivity and specificity.

By adjusting the penalty parameter, we can choose whether to emphasize sensitivity over specificity or the reverse, i.e., we can choose in which part of the ROC curve to operate; Table 1 gives some examples.

5.3 Performance Under the Extended Model on Simulated Data

In evaluating performance under the extended model, we must first consider the effect of the first phase, in which the history of gene duplications and losses is reconstructed. In [27] we analyzed various duplication-loss history models and their effect on the performance of *RefineFast* and *RefineML*. Our experiments showed that accurate history information with reliable orthology assignments helps the refinement algorithms to get good performance. Here we test *ProPhyC* and *ProPhyCC* with two representative histories. One is the "true" history which is available in the framework of simulation experiments; with this history we can exclude the error introduced by the history inference step, and test purely the performance of the refinement algorithms. The other is the history inferred by gene tree and species tree reconciliation algorithms without any prior information. As the rates of gene duplication and loss during evolution can also affect the performance of refinement algorithms, we conduct simulation experiments with different rates of duplication and loss.

We run refinement algorithms with the two gene duplication and loss histories: the true history and the history reconstructed by Notung [8]. In the following we show results on one representative phylogenetic tree with 35 nodes on 7 levels, and a root network of 15 genes. Since the results of using the neutral initialization model or the inheritance initialization model in data generation are very similar, we only show results with the former. For each experiment we show two plots: the left plot has relatively low rates of gene loss (resulting in 19 duplications and 15 losses along the tree on average), while the right one has significantly higher rates(with 20 duplications and 23 losses).

Absolute results, with true history. Fig. 6 shows the comparison of *ProPhyC, ProPhyCC* and the base inference algorithm *DBI*, using the true history of duplications and losses. Given the size of the tree and the root network, the rates of gene duplication and loss are quite high, yet the improvement gained by our refinement algorithms remains significant in both plots – almost as much as the improvement gained under the basic model shown in Fig. 2. *ProPhyCC* further dominates *ProPhyC* in both sensitivity and specificity, thanks to the appropriate use of the position-specific confidence values. Once again, we obtain similar improvements with other trees, other evolutionary rates, and other base methods.

Relative results, with true history. Fig. 7 shows the results of the same experiments as in Fig. 6, with *RefineFast* and *RefineML* added. Although *RefineFast* and *RefineML* still clearly improve on *DBI*, the improvement is less pronounced than with the basic model (Fig. 3). Gene duplications and losses give rise to a large overall gene population, yet many of them exist only in a few leaf networks; for these underrepresented genes, phylogenetic information is much reduced and so the refinement is less successful. *RefineFast* and *RefineML* are affected by this shortage, however, *ProPhyC* and *ProPhyCC* are more robust and easily outperform *RefineFast* and *RefineML*.

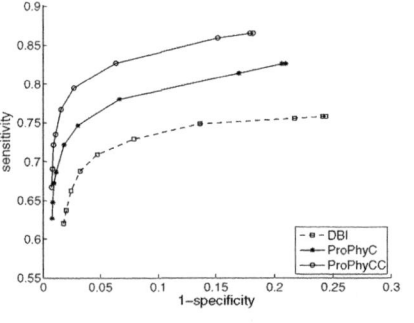

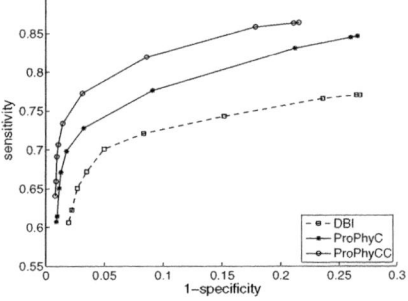

(a) with lower duplication and loss rates   (b) with higher duplication and loss rates

Fig. 6. Results of refinement algorithms with extended network evolutionary model, comparison of *ProPhyC* and *ProPhyCC* with *DBI*, with true gene duplication and loss history

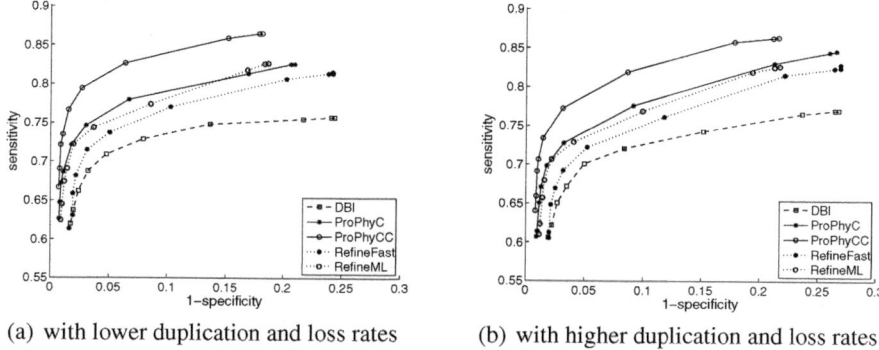

(a) with lower duplication and loss rates   (b) with higher duplication and loss rates

Fig. 7. Results of refinement algorithms with extended model, comparison of *ProPhyC* and *Pro-PhyCC* with *RefineFast* and *RefineML*, with true gene duplication and loss history

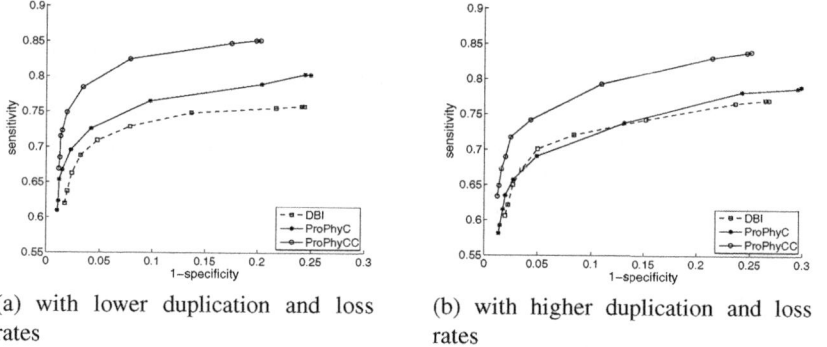

(a) with lower duplication and loss rates   (b) with higher duplication and loss rates

Fig. 8. Results of refinement algorithms with extended network evolutionary model, comparison of *ProPhyC* and *ProPhyCC* with *DBI*, with inferred gene duplication and loss history by Notung

Absolute results, with inferred history. Here we use Notung to reconstruct the history of duplications and losses without any orthology input. In these experiments, with reliable gene tree input, Notung correctly predicts duplication events (modulo changes in the networks), but usually misses recent loss events (it shows those events as happening earlier on the lineages). Furthermore, Notung not only infers the gene contents for ancestral networks, but also alters the gene contents of the leaves, which causes some difficulty for the refinement procedure. Fig. 8 shows the results of *ProPhyC* and *ProPhyCC* with Notung reconstructed gene contents for the ancestral networks. We see that in Fig. 8(a), the two ends of the *ProPhyC* curve have lost a little specificity while gaining sensitivity or vice versa, a tradeoff rather than an outright gain. However, *Pro-PhyC* dominates *DBI* through the useful range of specificity and sensitivity. In Fig. 8(b), *ProPhyC* barely improves *DBI*, because the high rate of gene loss reduces the performance of refinement algorithms in two ways: first a high rate affects the performance of Notung (which does a poor job at inferring losses); secondly it increases the total population of genes and decreases the frequency of occurrence of an ortholog in the leaf networks, thus limiting the phylogenetic information. However, *ProPhyCC* still improves *DBI* significantly in both plots. Our probabilistic framework can incorporate the prior information in an appropriate way, so as to gain good performance even when the phylogenetic information, including the history of gene duplication and loss, is noisy and incomplete.

6   Conclusions

We described *ProPhyC*, a probabilistic phylogenetic model designed to improve the inference of regulatory networks for a family of organisms by using the phylogenetic relationships among these organisms. This model and its associated refinement algorithms can easily be adapted to work with different network evolutionary models. We conducted experiments on both simulated and biological data to test the performance of the refinement algorithms. With both the basic and extended network evolutionary models, the corresponding versions of *ProPhyC* and *ProPhyCC* outperformed those of our previous algorithms *RefineFast* and *RefineML*, and all four refinement algorithms outperformed the base inference algorithm. The improvement of *ProPhyC* and *ProPhyCC* over *RefineFast* and *RefineML* was more significant under the extended model, where the performance of *RefineFast* and *RefineML* was affected by the decrease of the phylogenetic information for each ortholog, while *ProPhyC* and *ProPhyCC* were hardly influenced. Our probabilistic phylogenetic model is thus quite robust against changes in these network evolutionary models. Our probabilistic phylogenetic model can easily be extended into a probabilistic graphical model to incorporate the evolution of both regulatory networks and binding sites.

References

1. Akutsu, T., Miyano, S., Kuhara, S.: Identification of genetic networks from a small number of gene expression patterns under the Boolean network model. In: Proc. 4th Pacific Symp. Biocomp. (PSB 1999), pp. 17–28. World Scientific, Singapore (1999)
2. Arvestad, L., Berglund, A.-C., Lagergren, J., et al.: Gene tree reconstruction and orthology analysis based on an integrated model for duplications and sequence evolution. In: Proc. 8th Conf. Comput. Mol. Bio. (RECOMB 2004), pp. 326–335. ACM Press, New York (2004)
3. Babu, M.M., Teichmann, S.A., Aravind, L.: Evolutionary dynamics of prokaryotic transcriptional regulatory networks. J. Mol. Bio. 358(2), 614–633 (2006)
4. Berg, J., Lassig, M., Wagner, A.: Structure and evolution of protein interaction networks: a statistical model for link dynamics and gene duplications. BMC Evol. Bio. 4(1), 51 (2004)
5. Bourque, G., Sankoff, D.: Improving gene network inference by comparing expression time-series across species, developmental stages or tissues. J. Bioinform. Comput. Bio. 2(4), 765–783 (2004)
6. Chen, T., He, H.L., Church, G.M.: Modeling gene expression with differential equations. In: Proc. 4th Pacific Symp. Biocomp. (PSB 1999), pp. 29–40. World Scientific, Singapore (1999)
7. Crombach, A., Hogeweg, P.: Evolution of evolvability in gene regulatory networks. PLoS Comput. Biol. 4(7), e1000112 (2008)
8. Durand, D., Halldórsson, B.V., Vernot, B.: A hybrid micro-macroevolutionary approach to gene tree reconstruction. J. Comput. Bio. 13(2), 320–335 (2006)
9. Friedman, N., Linial, M., Nachman, I., Pe'er, D.: Using Bayesian networks to analyze expression data. J. Comput. Bio. 7(3-4), 601–620 (2000)

10. Harbison, C.T., Gordon, D.B., Lee, T.I., et al.: Transcriptional regulatory code of a eukaryotic genome. Nature 431, 99–104 (2004)
11. Heckerman, D.: Learning in graphical models. In: A Tutorial on Learning with Bayesian Networks, pp. 301–354. MIT Press, Cambridge (1999)
12. Hillis, D.M.: Approaches for assessing phylogenetic accuracy. Sys. Bio. 44, 3–16 (1995)
13. Kanehisa, M., Goto, S., Hattori, M., et al.: From genomics to chemical genomics: new developments in KEGG. Nucleic Acids Res. 34, D354–D357 (2006)
14. Kim, J., He, X., Sinha, S.: Evolution of regulatory sequences in 12 *Drosophila* species. PLoS Genet. 5(1), e1000330 (2009)
15. Kim, S.Y., Imoto, S., Miyano, S.: Inferring gene networks from time series microarray data using dynamic Bayesian networks. Briefings in Bioinf. 4(3), 228–235 (2003)
16. Liang, S., Fuhrman, S., Somogyi, R.: REVEAL, a general reverse engineering algorithm for inference of genetic network architectures. In: Proc. 3rd Pacific Symp. Biocomp. (PSB 1998), pp. 18–29. World Scientific, Singapore (1998)
17. Mithani, A., Preston, G.M., Hein, J.: A Bayesian approach to the evolution of metabolic networks on a phylogeny. PLoS Comput. Bio. 6(8), e1000868 (2010)
18. Moret, B.M.E., Warnow, T.: Reconstructing optimal phylogenetic trees: A challenge in experimental algorithmics. In: Fleischer, R., Moret, B.M.E., Schmidt, E.M. (eds.) Experimental Algorithmics. LNCS, vol. 2547, pp. 163–180. Springer, Heidelberg (2002)
19. Page, R.D.M., Charleston, M.A.: From gene to organismal phylogeny: Reconciled trees and the gene tree/species tree problem. Mol. Phyl. Evol. 7(2), 231–240 (1997)
20. Roth, C., Rastogi, S., Arvestad, L., et al.: Evolution after gene duplication: models, mechanisms, sequences, systems, and organisms. J. Exper. Zoology Part B: Mol. Devel. Evol. 308B(1), 58–73 (2007)
21. Stark, A., Kheradpour, P., Roy, S., Kellis, M.: Reliable prediction of regulator targets using 12 Drosophila genomes. Genome Res. 17, 1919–1931 (2007)
22. Tanay, A., Regev, A., Shamir, R.: Conservation and evolvability in regulatory networks: The evolution of ribosomal regulation in yeast. Proc. Nat'l Acad. Sci. 102(20), 7203–7208 (2005)
23. Teichmann, S.A., Babu, M.M.: Gene regulatory network growth by duplication. Nature Genetics 36(5), 492–496 (2004)
24. Tweedie, S., Ashburner, M., Falls, K., et al.: Flybase: enhancing Drosophila Gene Ontology annotations. Nucleic Acids Res. 37, D555–D559 (2009)
25. Zhang, X., Moret, B.M.E.: Boosting the performance of inference algorithms for transcriptional regulatory networks using a phylogenetic approach. In: Crandall, K.A., Lagergren, J. (eds.) WABI 2008. LNCS (LNBI), vol. 5251, pp. 245–258. Springer, Heidelberg (2008)
26. Zhang, X., Moret, B.M.E.: Improving inference of transcriptional regulatory networks based on network evolutionary models. In: Salzberg, S.L., Warnow, T. (eds.) WABI 2009. LNCS, vol. 5724, pp. 415–428. Springer, Heidelberg (2009)
27. Zhang, X., Moret, B.M.E.: Refining transcriptional regulatory networks using network evolutionary models and gene histories. BMC Algs. for Mol. Bio. 5(1), 1 (2010)

Prediction of DNA-Binding Propensity of Proteins by the Ball-Histogram Method

Andrea Szabóová[1], Ondřej Kuželka[1], Sergio Morales E.[2],
Filip Železný[1], and Jakub Tolar[3]

[1] Czech Technical University, Prague, Czech Republic
[2] Instituto Tecnológico de Costa Rica ITCR
[3] University of Minnesota, Minneapolis, USA
szaboand@fel.cvut.cz

Abstract. We contribute a novel, *ball-histogram* approach to DNA-binding propensity prediction of proteins. Unlike state-of-the-art methods based on constructing an ad-hoc set of features describing the charged patches of the proteins, the ball-histogram technique enables a systematic, Monte-Carlo exploration of the spatial distribution of charged amino acids, capturing joint probabilities of specified amino acids occurring in certain distances from each other. This exploration yields a model for the prediction of DNA binding propensity. We validate our method in prediction experiments, achieving favorable accuracies. Moreover, our method also provides interpretable features involving spatial distributions of selected amino acids.

1 Introduction

The process of protein-DNA interaction has been an important subject of recent bioinformatics research, however, it has not been completely understood yet. DNA-binding proteins have a vital role in the biological processing of genetic information like DNA transcription, replication, maintenance and the regulation of gene expression. Several computational approaches have recently been proposed for the prediction of DNA-binding function from protein structure.

In the early 80's, when the first three-dimensional structures of protein-DNA complexes were studied, Ohlendorf and Matthew noticed that the formation of protein-DNA complexes energetically driven by the electrostatic interaction of asymmetrically distributed charges on the surface of the proteins complement the charges on DNA [1]. Large regions of positive electrostatic potentials on protein surfaces has been suggested to be a good indication of DNA-binding sites.

Stawiski et al. proposed a methodology for predicting Nucleic Acid-binding function based on the quantitative analysis of structural, sequence and evolutionary properties of positively charged electrostatic surfaces. After defining the electrostatic patches they found the following features for discriminating the DNA-binding proteins from other proteins: secondary structure content, surface area, hydrogen-bonding potential, surface concavity, amino acid frequency and

J. Chen, J. Wang, and A. Zelikovsky (Eds.): ISBRA 2011, LNBI 6674, pp. 358–367, 2011.
© Springer-Verlag Berlin Heidelberg 2011

composition and sequence conservation. They used 12 parameters to train a neural network to predict the DNA-binding propensity of proteins [2].

Jones et al. analysed residue patches on the surface of DNA-binding proteins and developed a method of predicting DNA-binding sites using a single feature of these surface patches. Surface patches and the DNA-binding sites were analysed for accessibility, electrostatic potential, residue propensity, hydrophobicity and residue conservation. They observed that the DNA-binding sites were amongst the top 10% of patches with the largest positive electrostatic scores [3].

Tsuchiya et al. analysed protein-DNA complexes by focusing on the shape of the molecular surface of the protein and DNA, along with the electrostatic potential on the surface, and constructed a statistical evaluation function to make predictions of DNA interaction sites on protein molecular surfaces [4].

Ahmad and Sarai trained a neural network based on the net charge and the electric dipole and quadrupole moments of the protein. It was found that the magnitudes of the moments of electric charge distribution in DNA-binding protein chains differ significantly from those of a non-binding control data set. It became apparent that the positively charged residues are often clustered near the DNA and that the negatively charged residues either form negatively charged clusters away from the DNA or get scattered throughout the rest of the protein. The entire protein has a net dipole moment, because of the topological distribution of charges. The resulting electrostatic force may steer proteins into an orientation favorable for binding by ensuring that correct side of the protein is facing DNA [5].

Bhardwaj et al. examined the sizes of positively charged patches on the surface of DNA-binding proteins. They trained a support vector machine classifier using positive potential surface patches, the protein's overall charge and its overall and surface amino acid composition [6]. In case of overall composition, noticeable differences were observed (in binding and non-binding cases) with respect to the frequency of Lys and Arg. These are positively charged amino acids, so their over-representation in DNA-binding proteins is evident.

Szilágyi and Skolnick created a logistic regression classifier based on ten variables to predict whether a protein is DNA-binding from its sequence and low-resolution structure. To find features that discriminate between DNA-binding and non-DNA-binding proteins, they tested a number of properties. The best combination of parameters resulted in the amino acid composition, the asymmetry of the spatial distribution of specific residues and the dipole moment of the protein. When ranking these parameters by relative importance, they found out that the Arginine content was the strongest predictor of DNA-binding, followed by the Glycine and Lysine. The dipole moment was the fourth most important variable [7].

Here we contribute a novel approach to DNA-binding propensity prediction, called the *ball-histogram* method, which improves on the state-of-the-art approaches in the following way. Rather than constructing an ad-hoc set of features describing the charged patches of the proteins, we base our approach on a systematic, Monte-Carlo-style exploration of the spatial distribution of charged amino acids (under normal circumstances, Arg and Lys are positively charged, whereas Glu and Asp are charged negatively). For this purpose we employ so-called ball histograms, which are capable of capturing joint probabilities of specified amino acids occurring in certain distances from each other. Another positive aspect of our method is that it provides us with interpretable features involving spatial distributions of selected amino acids.

The rest of the paper is organized as follows. Section 2 describes the protein data sets we use in the study. In Section 3 we explain the ball histogram method. Section 4 exposes the results of applying the method on the protein data. In Section 5 we conclude the paper.

2   Data

DNA-binding proteins are proteins that are composed of DNA-binding domains. A DNA-binding domain is an independently folded protein domain that contains at least one motif that recognizes double- or single-stranded DNA. We decided to investigate structural relations within these proteins following the spatial distributions of certain amino acids in available DNA-protein complexes.

We decided to work with a positive data set (PD138) of 138 DNA-binding protein sequences in complex with DNA. It was created using the Nucleic Acid Database by [7] - it contains a set of DNA-binding proteins in complex with DNA strands with a maximum pairwise sequence identity of 35% between any two sequences. Proteins have $\leq 3.0\text{Å}$ resolution. An example DNA-binding protein in complex with DNA is shown in Fig.1.

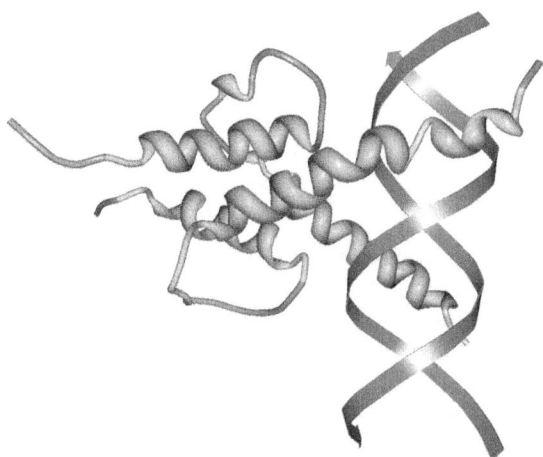

Fig. 1. Exemplary DNA-binding protein in complex with DNA shown using the protein viewer software [8]. Secondary structure motifs are shown in green (α-helices), light blue (turns) and pink (coils); the two DNA strands are shown in blue.

Rost and Sander constructed a dataset (RS126) for secondary structure pre-diction. Ahmad & Sarai [5] removed the proteins related to DNA binding from it, thus getting a final dataset of non-DNA-binding proteins. We used this set of non-DNA-binding proteins as our negative dataset (NB110).

From the structural description of each protein we extracted the list of all contained amino acids with information on their type and spatial structure.

3   Method

In this section we describe our novel method for predictive classification of pro-teins using so-called *ball histograms*. The classification method consists of three main steps. First, *ball histograms* are constructed for all proteins in a training set and then a transformation method is used to convert these histograms to a form usable by standard machine learning algorithms. Finally, a random forest classifier [9] is learned on this transformed data and then it is used for classifi-cation. Random forest classifier is known to be able to cope with large numbers of attributes such as in our case of *ball histograms* [10].

3.1   Ball Histograms

In this section we describe *protein ball histograms*. We start by defining several auxiliary terms. A *template* is a list of amino acid types or amino acid properties. An example of a template is (Arg, Lys) or $(Positive, Negative, Neutral)$. We say that an amino acid *complies with a property* f_1 if it has the corresponding property. For example, if A is an Arginine then it complies both with property *Arg* and with property *Positive*. A *bounding sphere* of a protein is a sphere with center located in the geometric center of the protein and with radius equal to the distance from the center to the farthest amino acid of the protein plus the diameter of *sampling ball* which is a parameter of the method. We say that an amino acid *falls* within a sampling ball if the alpha-carbon of that amino acid is contained in the sampling ball in geometric sense.

Given a protein, a template $\tau = (f_1, \ldots, f_k)$, a sampling-ball radius R and a bounding sphere S, a *ball histogram* is defined as:

$$H_\tau(t_1, \ldots, t_k) = \frac{\int \int \int_{(x,y,z) \in S} I_{T,R}(x, y, z, t_1, \ldots, t_k) dx dy dz}{\sum_{(t'_1, \ldots, t'_k)} \int \int \int_{(x,y,z) \in S} I_{T,R}(x, y, z, t'_1, \ldots, t'_k) dx dy dz} \quad (1)$$

where $I_{T,R}(x, y, z, t_1, \ldots, t_k)$ is an indicator function which we will define in turn. The expression $\sum_{(t'_1, \ldots, t'_k)} \int \int \int_{(x,y,z) \in S} I_{T,R}(x, y, z, t'_1, \ldots, t'_k) dx dy dz$ is meant as a normalization factor - it ensures that $\sum_{(t_1, \ldots, t_k)} H_T(t_1, \ldots, t_k) = 1$. In order to define the indicator function $I_{T,R}$ we first need to define an auxiliary indicator function $I'_{T,R}(x, y, z, t_1, \ldots, t_k)$

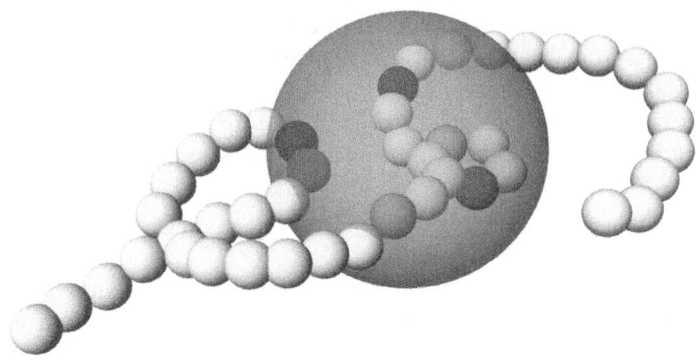

Fig. 2. Illustration of the Ball Histogram Method - Amino acids are shown as small balls in sequence forming an amino acid chain. A *sampling ball* is shown in violet. Some of the amino acids which comply with properties of an example template are highlighted inside the *sampling ball* area. They have different colors according to their type.

$$
I'_{T,R}(x,y,z,t_1,\ldots,t_k) =
\begin{cases}
1 & \text{if there are exactly } t_1 \text{amino acids complying} \\
& \text{with } f_1, t_2 \text{ amino acids complying with} f_2 \text{ etc.} \\
& \text{in a sampling ball with center } x,y,z \text{ and radius } R \\
0 & \text{otherwise}
\end{cases}
$$

Notice that $I'_{T,R}(x,y,z,0,\ldots,0)$ does not make any distinction between a sampling ball that contains no amino acid at all and a sampling ball that contains some amino acids none of them complying with the parameters in the template T. Therefore if we used $I'_{T,R}$ in place of $I_{T,R}$ the histograms would be affected by the amount of empty space in the bounding spheres. Thus, for example, there might be a big difference between histograms of otherwise similar proteins where one would be oblong and the other one would be more curved. In order to get rid of this unwanted dependence of the indicator function $I_{T,R}$ on proportion of empty space in sampling spheres we define $I_{T,R}$ in such a way that it ignores the empty space. For $(t_1,\ldots,t_k) \neq 0$ we set

$$
I_{T,R}(x,y,z,t_1,\ldots,t_k) = I'_{T,R}(x,y,z,t_1,\ldots,t_k).
$$

In the cases when $(t_1,\ldots,t_k) = 0$ we set $I_{T,R}(x,y,z,t_1,\ldots,t_k) = 1$ if and only if $I'_{T,R}(x,y,z,t_1,\ldots,t_k) = 1$ and if the sampling ball with radius R at (x,y,z) contains at least one amino acid.

Ball histograms capture the joint probability that a randomly picked *sampling ball* (See Fig. 2) containing at least one amino acid will contain exactly t_1 amino acids complying with f_1, t_2 amino acids complying with f_2 etc. They are invariant to rotation and translation of proteins which is an important property for classification. Also note that the histograms would not change if we increased the size of the bounding sphere.

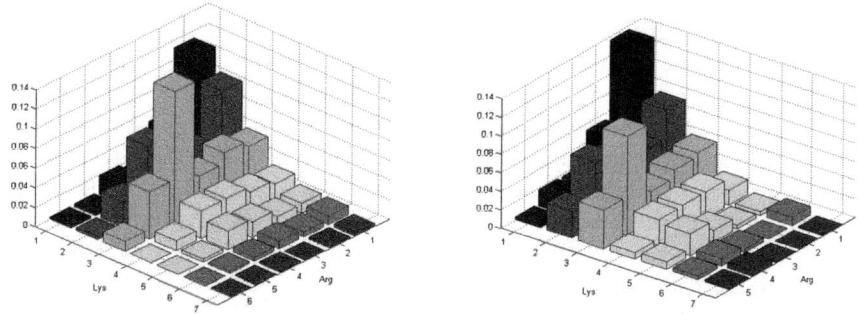

Fig. 3. Example ball histograms with template (Arg, Lys) and sampling-ball radius $R = 12\overset{\circ}{A}$ constructed for proteins 1A31 and 1A3Q from PD138

The indicator function $I_{T,R}$ makes crisp distinction between *amino acid falls within a sampling ball* and *amino acid does not fall within a sampling ball*. It could be changed to capture a more complex case by replacing 1 by the fraction of the amino acid that falls within the sampling ball, however, for simplicity we will not consider this case in this extended abstract.

3.2   Ball-Histogram Construction

Computing the integral in Eq. 1 precisely is infeasible therefore we decided to use a Monte-Carlo method. The method starts by finding the bounding sphere. First, the geometric center C of all amino acids of a given protein P is computed (each amino acid is represented by coordinates of its alpha-carbon). The radius R_S of the sampling sphere for the protein P is then computed as

$$R_S = \max_{Res \in P} \left(distance(Res, C)\right) + R$$

where R is a given sampling-ball radius. After that the method collects a predefined number of samples from the bounding sphere. For each sampling ball the algorithm counts the number of amino acids in it, which comply with the particular properties contained in a given template and increments a corresponding bin in the histogram. In the end, the histogram is normalized.

Example 1. Let us illustrate the process of histogram construction by a small example. Let us have a template (Arg, Lys). We assume that we already have a bounding sphere. The algorithm starts by placing a sampling ball randomly inside the bounding sphere. Let us assume that the first such sampling ball contained the following amino acids: *2 Arginins and 1 Leucine* therefore we increment a counter in the histogram associated with vector $(2, 0)$. Then in the second sampling ball, we get *1 Histidine and 1 Aspartic acid* so we increment a counter associated with vector $(0, 0)$. We continue in this process until we have gathered a sufficient number of samples. In the end we normalize the histogram. Examples of such histograms are shown in Fig. 3.

3.3   Predictive Classification Using Ball Histograms

In the preceding sections we have explained how to construct ball-histograms but we have not explained how we can use them for predictive classification. One possible approach would be to define a metric on the space of normalized histograms and then use either a nearest neighbour classifier or a nearest-centroid classifier. Since our preliminary experiments with these classifiers did not give us satisfying predictive accuracies, we decided to follow a different approach inspired by a method from relational learning known as *propositionalization* [11] which is a method for transferring complicated relational descriptions to attribute-value representations.

The transformation method is quite straightforward. It looks at all histograms generated from the proteins in a training set and creates a numerical *attribute* for each vector of property occurrences which is non-zero at least in one of the histograms. After that an attribute vector is created for each training example using the collected attributes. The values of the entries of the attribute-vectors then correspond to heights of the bins in the respective histograms. After this transformation a random forest classifier is learned on the attribute-value representation. This random forest classifier is then used for the predictive classification.

In practice, there is a need to select an optimal sampling-ball radius. This can be done by creating several sets of histograms and their respective attribute-value representations corresponding to different radiuses and then selecting the optimal parameters using an internal cross-validation procedure[1].

4   Results

In this section we present experiments performed on real-life data described in Section 2. We performed two types of experiments. First, we decided to study distribution of charged amino acids (represented by *ball histograms*). We constructed histograms with template (Arg, Lys, Glu, Asp) and three different sampling-ball radiuses: 6, 8 and $10\mathring{A}$. We trained random forest classifiers selecting optimal sampling-ball radius and an optimal number of trees for each fold by internal cross-validation. The estimated accuracy is shown in Table 1. As we can see, the accuracy of our method is comparable with the accuracy obtained by the method used in [7]. They used properties of the following amino acids: Arg, Lys, Gly, Asp, Asn, Ser and Ala, whereas we used only the distribution of the charged amino acids. A natural question is whether our results could be improved by taking into account also this set of amino acids. Therefore, we performed the second set of experiments. In this case, the accuracy obtained by our method exceeded the accuracy of [7].

---

[1] When evaluating the classifiers' performance in Section 4 using 10-fold cross-validation, we optimize the sampling-ball radius parameter always on the nine training folds and then use it for the remaining testing fold, which is a standard way to obtain an unbiased estimate of the predictive performance of a classifier with tunable parameters.

Table 1. Accuracies estimated by 10-fold cross-validation on PD138/NB110

Method	Accuracy [%]
Szilágyi et al.	81.4
Ball Histogram using Charged Amino Acids	80.2
Ball Histogram using the Second Set of Amino Acids	84.7

Table 2. The three most informative features according to the χ^2 criterion using the distribution of the charged amino acids

	Arg	Lys	Glu	Asp
1st feature	1	1	0	0
2nd feature	2	0	0	0
3rd feature	1	0	0	0

Table 3. The three most informative features according to the χ^2 criterion using the distribution of the selected set of amino acids

	Arg	Lys	Gly	Asp	Asn	Ser	Ala
1st feature	1	1	0	0	0	0	0
2nd feature	1	0	0	0	0	0	0
3rd feature	2	0	0	0	0	0	0

In addition to improved accuracy, our method provides us with interpretable features involving spatial distributions of selected amino acids. We show the three most informative *features* according to the χ^2 criterion for the first set of experiments in Table 2 and for the second set of experiments in Table 3. Given a protein each feature captures the fraction of sampling balls, which contain the specified numbers of amino acids of given types. For example: the first feature from Table 2 denotes the fraction of sampling balls, which contain exactly one Arginine, one Lysine, no Glutamic acid and no Aspartic acid.

A quick look at the most informative features in the presented Tables 2 and 3 suggests that the major role is played by the amino acids: Arginine and Lysine. These amino acids are known to often interact with the negatively charged backbone as well as with the bases [12,13,14]. In order to get a global view on the differences between spatial distributions of these amino acids in DNA-binding and non-DNA-binding proteins, we computed the average ball histograms for these two classes of proteins. They are shown in Fig. 4. The histogram obtained by subtracting the average ball histogram for DNA-binding proteins from the average ball histogram for non-DNA-binding proteins is shown in Fig. 5. We can notice a remarkable difference of spatial distribution of Arginine and Lysine between DNA-binding and non-DNA-binding proteins.

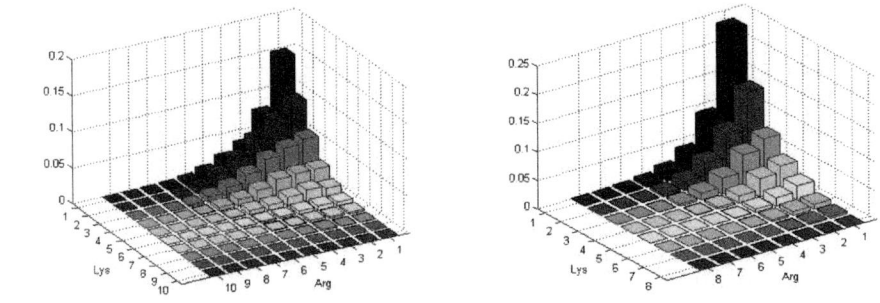

Fig. 4. Ball histograms with template (Arg, Lys) and sampling-ball radius $R = 12\text{Å}$ averaged for all proteins from PD138 (left panel) and all proteins from NB110 (right panel)

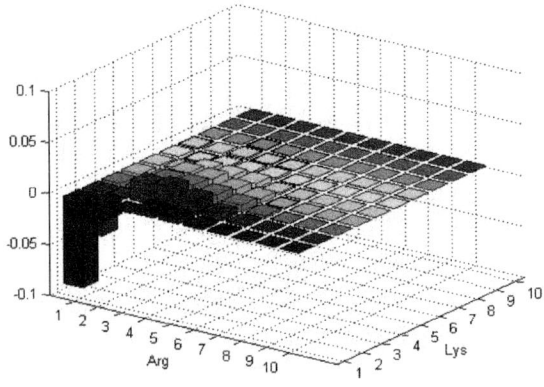

Fig. 5. Difference between histograms from Fig. 4

5   Conclusion

We contributed a novel, *ball-histogram* approach to the prediction of DNA-binding propensity of proteins. We validated the method in prediction experiments with favorable results. Observing only the distribution of charged amino acids we achieved accuracies comparable to the state of the art [7]. Importantly though, the results reported by [7] had been achieved using the results of a prior systematic search for the best combination of parameters. In particular, the following amino acids were identified as critical: Arg, Lys, Gly, Asp, Asn, Ser and Ala. Using this set of amino acids instead of the original set of all charged amino acids, our accuracies in fact exceeded those of [7]. Moreover, our method provides us with interpretable features involving spatial distributions of selected amino acids.

Acknowledgement. Andrea Szabóová and Filip Železný were supported by external project ME10047 granted by the Czech Ministry of Education. Andrea Szabóová was further supported by the Czech Technical University internal grant #10-801940. Ondřej Kuželka was supported by the Czech Technical University internal grant OHK3-053/11. Sergio Morales was supported by Costa Rica Council for Scientific and Technological Research.

References

1. Ohlendorf, D.H., Matthew, J.B.: Electrostatics and flexibility in protein-DNA interactions. Advances in Biophysics 20, 137–151 (1985)
2. Stawiski, E.W., Gregoret, L.M., Mandel-Gutfreund, Y.: Annotating nucleic acid-binding function based on protein structure. J. Mol. Biol. (2003)
3. Jones, S., Shanahan, H.P., Berman, H.M., Thornton, J.M.: Using electrostatic potentials to predict DNA-binding sites on DNA-binding proteins. Nucleic Acid Research 31(24), 7189–7198 (2003)
4. Tsuchiya, Y., Kinoshita, K., Nakamura, H.: Structure-based prediction of DNA-binding sites on proteins using the empirical preference of electrostatic potential and the shape of molecular surfaces. Proteins: Structure, Function, and Bioinformatics 55(4), 885–894 (2004)
5. Ahmad, S., Sarai, A.: Moment-based prediction of DNA-binding proteins. Journal of Molecular Biology 341(1), 65–71 (2004)
6. Bhardwaj, et al.: Kernel-based machine learning protocol for predicting DNA-binding proteins. Nuc. Acids Res. (2005)
7. Szilágyi, A., Skolnick, J.: Efficient Prediction of Nucleic Acid Binding Function from Low-resolution Protein Structures. Journal of Molecular Biology 358, 922–933 (2006)
8. Moreland, J.L., Gramada, A., Buzko, O.V., Zhang, Q., Bourne, P.E.: The Molecular Biology Toolkit (MBT): A Modular Platform for Developing Molecular Visualization Applications. BMC Bioinformatics (2005)
9. Breiman, L.: Random Forests. Machine Learning 45, 5–32 (2001)
10. Caruana, R., Karampatziakis, N., Yessenalina, A.: An empirical evaluation of supervised learning in high dimensions. In: International Conference on Machine Learning (ICML), pp. 96–103 (2008)
11. Lavrač, N., Flach, P.: An Extended Transformation Approach to Inductive Logic Programming. ACM Transactions on Computational Logic 2, 458–494 (2001)
12. Pabo, C.O., Sauer, R.T.: Transcription factors: structural families and principles of DNA recognition. Annual Review of Biochemistry 20, 137–151 (1992)
13. Mandel-Gutfreund, Y., Schueler, O., Margalit, H.: Comprehensive analysis of hydrogen bonds in regulatory protein DNA-complexes: in search of common principles. Journal of Molecular Biology 253, 370–382 (1995)
14. Jones, S., van Heyningen, P., Berman, H.M., Thornton, J.M.: Protein-DNA interactions: a structural analysis. Journal of Molecular Biology 287, 877–896 (1999)

Multi-label Correlated Semi-supervised Learning for Protein Function Prediction

Jonathan Q. Jiang

Department of Computer Science
City University of Hong Kong
Tat Chee Avenue, Kowloon, Hong Kong
qiajiang@cityu.edu.hk

Abstract. The advent of large volume of molecular interactions has led to the emergence of a considerable number of computational approaches for studying protein function in the context of network. These algorithms, however, treat each functional class independently and thereby suffer from a difficulty of assigning multiple functions to a protein simultaneously. We propose here a new semi-supervised algorithm, called MCSL, by considering the correlations among functional categories which improves the performance significantly. The guiding intuition is that a protein can receive label information not only from its neighbors annotated with the same category in *functional-linkage* network, but also from its partners labeled with other classes in *category* network if their respective neighborhood topologies are a good match. We encode this intuition as a two-dimensional version of network-based learning with local and global consistency. Experiments on a *Saccharomyces cerevisiae* protein-protein interaction network show that our algorithm can achieve superior performance compared with four state-of-the-art methods by 5-fold cross validation with 66 second-level and 77 informative MIPS functional categories respectively. Furthermore, we make predictions for the 204 uncharacterized proteins and most of these assignments could be directly found in or indirectly inferred from SGD database.

1 Introduction

Although the high-throughput techniques have made it possible to monitor hundreds or even thousands of molecular simultaneously, most of their functions still remain mysteries for us. For example, even for a well-investigated species, such as *Saccharomyces cerevisiae*, there are 944 uncharacterized ORFs ($\sim 14\%$ out of all 6607 genes) and 811 dubious ORFs ($\sim 12\ \%$) in the MIPS database[1] until January 20, 2011. Therefore, function annotation of proteins is one of the fundamental issues in the post-genomic era.

Traditional computational approaches for protein annotation mainly relied on collecting a set of features from each protein and then applying machine learning algorithms (particularly, the support vector machine) to get the final decisions [12].

---

[1] http://mips.helmholtz-muenchen.de/genre/proj/yeast/

J. Chen, J. Wang, and A. Zelikovsky (Eds.): ISBRA 2011, LNBI 6674, pp. 368–379, 2011.

Thanks to the availability of large volume of molecular interactions, either physical or genetic, we can possibly adopt a top-down point of view and investigate protein function in the context of a network. This idea has led to the emergence of a considerable number of network-based studies [8,10,14,17]. Although these implementations are seemingly quite different, their common start point is exploiting the same underlying assumption, i.e., a concept often referred to as *guilt-by-association* [14]. That is, proteins are firstly large-scale linked in case they are physically interacted [6,14,9], co-expressed [4] or co-regulated [7] and even have similar phylogenetic profiles [13]. After constructing the comprehensive functional-linkage map, one can infer the protein function in the following two types of approaches: direct annotation schemes and module-assisted schemes. See recent review [15] for more details. We focus here only on the first scenario.

The straightforward algorithm is the *neighborhood counting* that determines the function of a protein based on the known function of proteins lying in its immediate neighborhood [14,8,3]. Although it is simple and effective, we are not clear what the appropriate radius is and whether neighborhoods of different sizes should be selected for different nodes or for different functions. In contrast to the local methods, graph algorithms are global and take the full structure of the network into consideration. The recent study [17] utilized cut-based methodology so as to minimize the number of times that different annotations are associated with neighboring proteins. Such idea has been applied in [10] by a local search procedure and in [11] by integer linear programming (ILP). In addition, Nabieva et al [11] also introduce an algorithm by simulating the spread over time of "functional flow" through the network. Thus, this method considers both local and global effects.

The main drawback of these existing methods is that they treat each class independently. Thus, all of them ignored, more or less, the correlations among different classes, which often could be an important hint for deciding the class memberships. In fact, the majorities of 12 known functions of protein YAL041W are indeed correlated. For example, the two functional categories 18.02.03: "guanyl-nucleotide exchange factor (GEF)" and 18.02.05:"regulator of G-protein signaling" share a common parent 18.02 "regulation of protein activity"[2]. Therefore, we expect that incorporating the inherent correlations among multiple functional classes into semi-supervised learning strategy could improve predictive performance which has been confirmed in other fields, such as text categorization and image annotation [18,2].

In this paper, we propose a new semi-supervised algorithm for protein function annotation by considering the correlations among functional categories. This method can be encoded as a two-dimensional version of graph-based learning with local and global consistency [19]. We use a *Saccharomyces cerevisiae* protein-protein interaction network consisting of 2988 proteins and 15806 interactions derived form the BioGRID database (Release 2.0.58)[3]. 5-fold cross

---

[2] http://mips.helmholtz-muenchen.de/genre/proj/yeast/singleGeneReport.
   html?entry=YAL041w
[3] http://thebiogrid.org/download.php

validation experiments show that our algorithm can achieve superior performance compared with four state-of-the-art approaches (see Section 4.4). Furthermore, we give the predictions for 204 uncharacterized proteins, most of which are found in or supported by SGD database.

2   Multi-label Correlated Semi-supervised Learning

For a multiple function prediction problem where we have K functional categories, given a protein set $\mathcal{P} = \{p_i\}_{i=1,\ldots,n}$ where the first l proteins are labeled as $\{\mathbf{y}_1, \ldots, \mathbf{y}_l\}$ with $y_{ik} = 1$ in case the i-th protein is associated with the k-th function, our goal is to predict the labels $\{\mathbf{y}_{l+1}, \ldots, \mathbf{y}_n\}$ for the remaining unlabeled proteins $\{p_{l+1}, \ldots, p_n\}$. The functional-linkage network of these proteins can be represented as a graph $G = (\mathcal{V}, E, W)$, with nodes set $\mathcal{V} = \mathcal{L} \cup \mathcal{U}$ where $\mathcal{L}$ corresponds to labeled proteins and $\mathcal{U}$ corresponds to uncharacterized proteins. The element w_{uv} of the affinity matrix $W \in \Re^{n \times n}$ indicates the strength of functional linkage between protein u and v.

We aim at developing a method to predict the labels of the unannotated proteins from both labeled and unlabeled ones. Such a learning problem is often called semi-supervised or transductive. The key to this strategy is the *consistent assumption* [19], which means: (1) nearby points are likely to share the same label; and (2) points on the same structure (typically referred to as a cluster or a manifold) are likely to have the same label. This can be expressed in a regularization framework where the first term is a loss function to penalize the deviation from the given labels, and the second term is a regularizer to prefer the label smoothness. We follow this pipeline and give its two-dimensional version. Our work is inspired by these prior works [19,18,2], additionally by the recent study [16].

To make use of the correlations among multiple functional classes, we build a *category network* $G' = (V', E', W')$ where each node represent one functional class. The symmetric weight matrix $W' \in \Re^{K \times K}$ captures the functional similarity[4] between each class. It is worthwhile to point out that $w'_{ii} = \sqrt{\sum_{j \neq i} w'_{ij}}, \forall i$. This allows that different proteins nearby can be associated with the same function class and is rather distinct with other existing one-dimensional approaches. Let $\mathbf{Y} \in \Re^{n \times K}$ be the known label matrix, and $\mathbf{F} := [\mathbf{f}_1, \ldots, \mathbf{f}_n]^T$ where f_{ik} is the confidence that p_i can be annotated with label k. Following the framework [19], our objective function consists of two components: a loss function $\mathcal{H}_l(\mathbf{F})$ and a *smoothness* regularizer $\mathcal{H}_f(\mathbf{F})$. Specifically, $\mathcal{H}_l(\mathbf{F})$ corresponds to the first property to penalize the deviation from the given multi-label assignments, and $\mathcal{H}_f(\mathbf{F})$ address the multi-label smoothness on the whole functional-linkage network. Then, the proposed framework can be formulated to minimize

$$\mathcal{H}(\mathbf{F}) = \mathcal{H}_f(\mathbf{F}) + \mu \mathcal{H}_l(\mathbf{F}) \tag{1}$$

---

[4] We postpone the discussions of similarity between different functional class until Section 4.3.

where $\mu > 0$ is the regularization parameter. The terms $\mathcal{H}_f(\mathbf{F})$ and $\mathcal{H}_l(\mathbf{F})$ can be specified in a way similar to that adopted in existing graph-based methods [19,2,18]. Specifically, we define them as

$$\mathcal{H}_f(\mathbf{F}) = \sum_{u,v=1}^{n} \sum_{i,j}^{K} w_{uv} w'_{ij} \left\| \frac{f_{ui}}{\sqrt{D_{uu} D'_{ii}}} - \frac{f_{vj}}{\sqrt{D_{vv} D'_{jj}}} \right\|^2 \tag{2}$$

$$\mathcal{H}_l(\mathbf{F}) = \|\mathbf{F} - \mathbf{Y}\|_F^2 \tag{3}$$

where $\|\cdot\|_F$ denote the Frobenius norm of matrix and $D = \operatorname{diag}(\sum_{j \neq i} w_{ij})$, $D' = \operatorname{diag}(\sum_{j \neq i} w'_{ij})$ respectively. We can understand the smoothness term as the sum of local variations whose justification will be given in Section 3.2.

Differentiating $\mathcal{H}(\mathbf{F})$ with respect to $\mathbf{F}$ and denoting $S = D^{-1/2} W D^{-1/2}$, $S' = D'^{-1/2} W' D'^{-1/2}$, we have

$$\frac{\partial \mathcal{H}}{\partial \mathbf{F}}|_{\mathbf{F}=\mathbf{F}^*} = \operatorname{Vec}(\mathbf{F}^*) - S'^T \otimes S^T \operatorname{Vec}(\mathbf{F}^*) + \mu(\mathbf{F}^* - \mathbf{Y}) = 0 \tag{4}$$

where $\operatorname{Vec}(\cdot)$ denote the vectorization of a matrix[5] and $\otimes$ the Kronecker product. Let $\alpha = \frac{1}{\mu+1}$ and follow a similar procedure as [19], then we can get

$$\operatorname{Vec}(\mathbf{F}^*) = (I - \alpha S'^T \otimes S^T)^{-1} \operatorname{Vec}(\mathbf{Y}) \tag{5}$$

3   Algorithm, Justification and Extension

In this section, we propose an iterative algorithm to solve the minimization problem (1) similar as that presented in [19]. Then, there follows the justification why we design the regularizer $\mathcal{H}_f$ as Eq.(2) and some natural extensions.

3.1   Iterative Algorithm

The algorithm is given in Algorithm 1. We can see clearly that in this algorithm each protein receives the functional category information from its neighbors (the first term) and also retains its initial label information (the second term) during each iteration of the second step. The parameter α specifies the relative amount of these two information. Different with previous uncorrelated learning [19,18,2], the first term depicts the strength of match between one specific protein's respective neighborhood topologies of the functional-linkage network and functional category network (more details, see 3.2). It is worth mentioning that *self-reinforcement* is removed in functional-linkage network as suggested by [19,2,18] but remains in functional category network since proteins nearby can be assigned the same functional class which is rather distinct with the network alignment problem [16].

---

[5] http://en.wikipedia.org/wiki/Vectorization_(mathematics)

Algorithm 1. MCSL: Iterative algorithm for multi-label correlated semi-supervised learning

Input

- $W = (w_{uv}) \in \Re^{n \times n}$: weighted matrix of functional-linkage network
- $W' = (w'_{ij}) \in \Re^{K \times K}$: weighted matrix of functional category network
- $\mathbf{Y} = (y_{ui}) \in \Re^{n \times K}$: the known initial label matrix

Output

- $\mathbf{F} = (f_{ui}) \in \Re^{n \times K}$: the final label matrix

Steps

1. Construct the matrices $S = D^{-1/2}WD^{-1/2}$ and $S' = D'^{-1/2}W'D'^{-1/2}$;
2. Iterate $\text{Vec}(\mathbf{F}(t+1)) = \alpha S'^T \otimes S^T \text{Vec}(\mathbf{F}(t)) + (1-\alpha)\text{Vec}(\mathbf{Y}(t))$ until convergence;
3. let $\mathbf{F}^*$ denote the limit of the sequence $\{\mathbf{F}(t)\}$. Assignment each functional class to proteins by a threshold selection method proposed in [5] and return $\mathbf{F}$;

3.2   Justification and Extensions

As mentioned in Section 2, the term $\mathcal{H}_f$ is the *smoothness* regularizer that addressed the multi-label smoothness on the whole functional-linkage network. Intuitively, we can divide this term into two components: (1) the *intra-class* term that describes the sum of the local variations of functional assignment for one specific functional class; and (2) the *inter-class* term that depicts the smoothness of functional assignment on proteins nearby for different classes. The first component is just the regularizer of traditional graph-based semi-supervised learning. That has been introduced in [19] and utilized in following works[2,18]. We extend it to multiple classes as

$$\mathcal{H}_f^{\text{intra}} := \sum_{c=1}^{K} \mathcal{H}_f^{\text{intra}(c)} = \sum_{c=1}^{K} \sum_{u,v=1}^{n} w_{uv} \left\| \frac{f_{uc}}{\sqrt{D_{uu}}} - \frac{f_{vc}}{\sqrt{D_{vv}}} \right\|^2 \tag{6}$$

For the second component, we pursue the intuition [16]: one protein should be assigned to one specific function class in case their respective neighbors are a good match with each other. See the illustrative example given in Fig.1 where we want to associate an uncharacterized protein p_u with the function c_i "Transcription". Besides receiving label information from its four neighbors, the protein can also receive information with two of these four partners annotated with functions closely related with c_i, for example, c_j "Metabolism". Consider the pair of functional assignment (p_u, c_i) and (p_v, c_j), and its local variation can be reformulated as

$$\mathcal{H}_f^{\text{inter}(u,i)(v,j)} := w_{uv}w'_{ij} \left\| \frac{f_{ui}}{\sqrt{D_{uu}D'_{ii}}} - \frac{f_{vj}}{\sqrt{D_{vv}D'_{jj}}} \right\|^2 \tag{7}$$

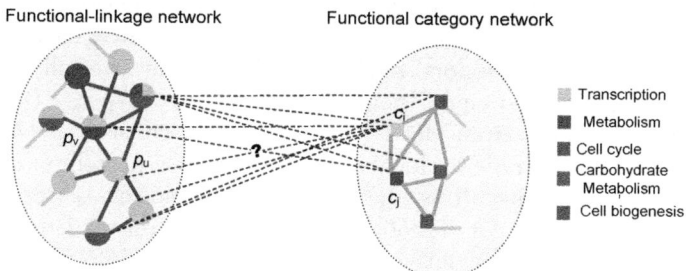

Fig. 1. An illustrative example for our algorithm. We here want to associate an uncharacterized protein p_u with the function c_i: "Transcription" (red dashed line). Clearly, p_u can not only receive label information from its four neighbors for function c_i, but also receive information from its two partners annotated with functions closely correlated with c_i, for example, c_j. The respective neighborhood topologies of p_u and c_i form a good match.

Therefore, summing all the potential pairs, we get

$$\mathcal{H}_f^{\text{inter}} = \sum_{u,v=1}^{n} \sum_{i,j=1,i\neq j}^{K} w_{uv} w'_{ij} \left\| \frac{f_{ui}}{\sqrt{D_{uu}D'_{ii}}} - \frac{f_{vj}}{\sqrt{D_{vv}D'_{jj}}} \right\|^2 \tag{8}$$

Obviously, the smoothness regularizer Eq.(2) is the sum of Eq.(6) and Eq.(8). Further, we can naturally generalize this term as a linear combination of these two components, i.e., $\mathcal{H}_f = \mathcal{H}_f^{\text{intra}} + \beta \mathcal{H}_f^{\text{inter}}$, where $0 \leq \beta \leq 1$ is a trade-off between these two competing constraints, which can be understood as comparative reliability of inter-class information. Note that the formulation Eq.(7) is slightly different with that used in [16]. If we use that term for our framework, Eq.(5) can be illustrated as a random walk on the assembled bi-correlation graph of proteins and functional categories. The matrix $(I - \alpha S'^T \otimes S^T)^{-1}$ in fact is the diffusion kernel which propagates the label information through the two-dimensional assembled network. We omit the detail here due to out of the scope of this paper.

4   Experiment

4.1   Setup

We construct the functional-linkage network using the protein interaction dataset complied from BioGRID (release 2.0.58)[1]. In order to reduce the false positive, we used only those interactions that were confirmed by at least two publications. The largest connected component of such network consists of 2988 proteins with 15806 interactions. It is well known that the edges' weight has a very important influence on the results, even if networks are based on the same underlying topology [11,10,3]. We introduce two types of schemes for applying our algorithm.

The first variant attempts to capture only qualitative functional links between proteins by PPI. In the second scheme, we simply weighted each edge by the number of its experiment support. In the rest of this paper, we call these variants as "PPI-only" and "PPI-weight" network, respectively. We use the functional annotation scheme taken from MIPS Funcat-2.0[6] that consists of 473 functional classes (FCs) arranged in hierarchical order. Note that a protein annotated with a FC is also annotated with all its super-classes. To avoid this biases, we consider two different studies, one for the 66 second-level FCs (referred to as "2-level") and another for the 75 informative FCs (referred to as "infor"). We define an informative FC as the one having (1) at least 30 proteins annotated with it and (2) no subclass meeting the requirement [11].

4.2   Evaluation Metrics

In traditional classification problems, the standard evaluation criteria is the accuracy. By contrast, we can not so simply to determine whether a prediction is right or wrong because of the *partially correct* phenomenon in multi-label learning [5]. We adopt here the widely-used performance metric, Mean Average Precision (MAP), as suggested by TRECVID[7]. In addition, we also choose the F-score to evaluate both the precision and recall together. The F-score for kth category and the macro-average F-score are defined as

$$F_k = \frac{2 \sum_{u=1}^{n} \widehat{y}_{uk} y_{uk}}{\sum_{u=1}^{n} \widehat{y}_{uk} + \sum_{u=1}^{n} y_{uk}} \quad \text{macro} F = \frac{1}{K} \sum_{k=1}^{K} F_k$$

where $\widehat{y}_{uk}$ is the predicted label.

4.3   Construct Functional Category Network

The key for building functional category network is how to measure the similarity between different FCs. Intuitively, we can use the data-driven manner via cosine similarity

$$w'_{ij} = \cos(\mathbf{z}_i, \mathbf{z}_j) = \frac{\langle \mathbf{z}_i, \mathbf{z}_j \rangle}{\|\mathbf{z}_i\| \|\mathbf{z}_j\|} \tag{9}$$

where $\mathbf{Z} = [\mathbf{z}_1, \ldots, \mathbf{z}_K] = \mathbf{Y}$. Note that the hierarchical architecture of MIPS Funcat is a tree. So we design the following biology-driven manner: (1) the similarity between FCs in different branches equals to zero; (2) the similarity between FCs in the same branch is defined as

$$w'_{ij} = 1/\log C(i, j) \tag{10}$$

where $C(i, j)$ denotes the number of proteins annotated with the lowest common super-class of FC c_i and c_j. We refer to our methods for these two different category networks as MCSL-d and MCSL-b hereafter.

---

[6] `ftp://ftpmips.gsf.de/catalogue/`
[7] `http://www-nlpir.nist.gov/projects/trecvid/`

Table 1. 5-fold cross validation for five algorithms

		2level		infor	
		macroF	MAP	macroF	MAP
PPI-only	Majority	0.3024 ± 0.0026	0.3052 ± 0.0055	0.4960 ± 0.0025	0.5035 ± 0.0078
	$\chi^2 - 1$[1]	0.3028 ± 0.0018	0.3155 ± 0.0061	0.5092 ± 0.0021	0.5164 ± 0.0069
	$\chi^2 - 2$	0.1747 ± 0.0013	0.1424 ± 0.0019	0.3791 ± 0.0019	0.3323 ± 0.0023
	$\chi^2 - 3$	0.1704 ± 0.0003	0.1690 ± 0.0014	0.2918 ± 0.0016	0.2885 ± 0.0050
	GMC	0.2961 ± 0.0020	0.3141 ± 0.0039	0.5042 ± 0.0025	0.5135 ± 0.0066
	FCflow	0.2954 ± 0.0022	0.3135 ± 0.0036	0.5033 ± 0.0024	0.5127 ± 0.0075
	MCSL-b	0.3055 ± 0.0021	**0.3242 ± 0.0090**	**0.5163 ± 0.0029**	**0.5319 ± 0.0075**
	MCSL-d	**0.3167 ± 0.0040**	0.3154 ± 0.0048	0.5017 ± 0.0027	0.4949 ± 0.0062
PPI-weight	Majority	0.3023 ± 0.0025	0.3245 ± 0.0033	0.5186 ± 0.0021	0.5125 ± 0.0049
	GMC	0.3070 ± 0.0014	0.3225 ± 0.0051	0.5338 ± 0.0025	0.5275 ± 0.0032
	FCflow	0.3071 ± 0.0013	0.3234 ± 0.0061	0.5323 ± 0.0020	0.5246 ± 0.0024
	MCSL-b	0.3222 ± 0.0029	0.3354 ± 0.0051	**0.5441 ± 0.0025**	**0.5426 ± 0.0020**
	MCSL-d	**0.3358 ± 0.0041**	**0.3478 ± 0.0061**	0.5333 ± 0.0024	0.5264 ± 0.0023

[1] $\chi^2 - k$ denotes the χ^2-like score method with radius k.
[2] $\pm$ means standard variation.

4.4 5-Fold Cross Validation

We test the performance using 5-fold cross-validation, i.e, these yeast proteins is randomly divided into 5 groups, and each group, in turn, is separated from original dataset and used for testing. In our implementation, we fixed the parameter $\alpha = 0.99$ as suggested by [19]. The trade-off parameter β is selected through the validation process. We compare our algorithm to four state-of-the-art methods: (1) Majority approach [14], (2) χ^2-like score [8], (3) GenMultiCut (GMC) [17,10] and (4) FunctionFlow (FCflow) [11]. Note that the χ^2-like score can only perform on the "PPI-only" network. The results are summarized in Table 1.

From these results, we have the following observations:

1. Our algorithm MCSL consistently, sometimes significantly, outperforms the other four approaches. In particular, macroF improves about 5.86%, 2.61%, 9.93% and 3% on the 2-level "PPI-only", infor "PPI-only", 2-level "PPI-weight" and infor "PPI-weight" experiment, respectively. Similarly, as expected, MAP improves about 3.89%, 3.98%, 6.54% and 4.04%, respectively.
2. Generally speaking, the metric values of infor experiment are much higher than its counterparts of 2-level for all the approaches. The reason is that there are fewer annotated proteins in some 2-level FCs, i.e., data sparsity problem occurs. Typical examples are 02.10 "tricarboxylic-acid pathway", 20.09 " transport routes" and 42.03 " cytoplasm" (see Fig.2)
3. Consistent with previous works[10,11,3], edge weights of the functional-linkage network have an important influence on the prediction results. In our study, the macroF and MAP are improved 6.03%, 5.38% and 7.28%, 2.01% on the 2-level and infor experiment, respectively, even if the weighted strategy is so simple.

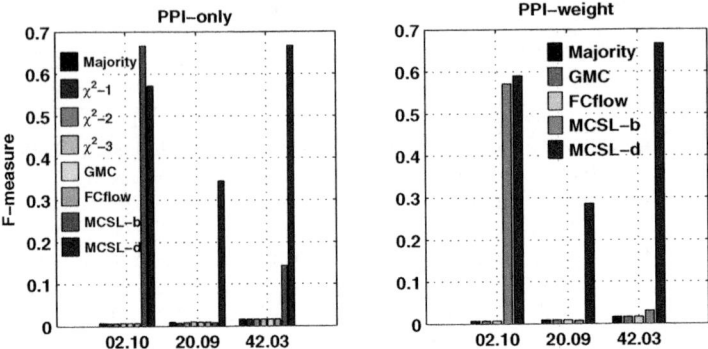

Fig. 2. F-measure of three functional classes in 2-level where the sparsity problem occurs. Clearly, the performance of our method MCSL is significantly improved compared with other approaches. It is because proteins can receive label information from other functional classes that are closely correlated with the category of interest.

4. For the FC similarity measure, the data-driven manner Eq.(9) achieves the best performance on the 2-level experiment. By contrast, the biology-driven manner Eq.(10) outperforms on the infor experiment. This phenomenon can be explain as follows. The 2-level experiment only considers the FCs belonging to the second-level of the MIPS Funcat hierarchy. Therefore, each FC has sparse correlation with others if we use biology-driven manner since they are under different branches; meanwhile, they are more closely correlated if the data-driven manner utilized. Similar explanations can made for the infor experiment.

We further check the F-measure for each functional category on both experiments. Fig.2 gives the F-measures of three functional classes on 2-level experiment where the sparsity problem occurs. We can see clearly that the four previous methods almost can not successfully associate proteins with the three categories which there are only 9,4 and 1 proteins annotated with in the functional-linkage network. As the benefit of receiving label information from other classes that are closely correlated with the three categories, our algorithm, as expected, significantly improved the predictive performances. In particular, the F-measure of MCSL-d and MCSL-b respectively improved 40.41 and 35.75 times for FC 42.03 "cytoplasm" even though there is only one protein labeled with this category.

4.5   Predict Function for Uncharacterized Proteins

There are still 204 "unclassified proteins" in our functional-linkage network. We further apply our algorithm MCSL-b to the "PPI-weight" network for associated these uncharacterized proteins with "infor" categories. Only the first ten predictions are listed in Table 2 due to the page limits of this paper. Most of our predictions could be found in or supported by SGD database[8]. For example,

---

[8] http://www.yeastgenome.org/

Table 2. Function predictions for the first ten uncharacterized proteins in our study. Most of them can be found in or supported by SGD database.

ORF	Predicted function	SGD annotation
YAL027W	32.01.09 10.01.05.01	GO:0000736, GO:0006974
YAL053W	16.19.03 34.01.01.01	GO:0015230, GO:0015883
YBL046W	11.02.03.01.04	GO:0019888,GO:0005634
YBR025C	10.01.02 10.01.03.05 10.01.05.03.01 10.01.09.05 16.19.03	GO:0000166 GO:0016887
YBR187W	01.04 14.07.03	GO:0000324
YBR273C	14.13.01.01 16.19.03	GO:0006511 GO:0005783
YBR280C	14.01 16.19.03 01.04 10.03.02	GO:0004842, GO:0031146,GO:0019005
YDL139C	10.03.01.01.11 10.03.04.05	GO:0000086 GO:0007059
YDL204W	34.11.03.13	GO:0032541,GO:0005789,GO:0005635
YDR084C	20.09.13	GO:0016192,GO:0030173

the first protein YAL027W is associated with two functional categories 32.01.09 "DNA damage response" and 10.01.05.01 "DNA repair". This prediction can be directly found in SGD database where this protein is annotated with GO:0000736 "double-strand break repair via single-strand annealing" and GO:0006974 "response to DNA damage stimulus". Several predictions cannot be directly found but supported by an indirect inference. We predict the protein YBR187W belonging to two classes 01.04 "phosphate metabolism" and 14.07.03 "modification by phosphorylation, dephosphorylation, autophosphorylation". In fact, such protein has been labeled with GO:0000324 "fungal-type vacuole". As we known, the fungal vacuole is a large, membrane-bounded organelle that functions as a reservoir for the storage of small molecules (including polyphosphate, amino acids, several divalent cations (e.g. calcium), other ions, and other small molecules) as well as being the primary compartment for degradation. All these results illustrate that our predictions are accuracy and effective.

5   Concluding Remarks

We propose a new semi-supervised learning framework for associating proteins with multiple functional categories simultaneously. It can effectively overcome the sparsity of label instance problem that previous approaches[10] often suffer from (see Fig.2), as a benefit of taking the correlation between different labels into account. In our implementations, during each iteration, the proteins receives the label information not only from their neighbors annotated with the same class in the functional-linkage network, but also from partners annotated with other closely related classes. Therefore, our framework could be understood as a two-dimensional version of the traditional graph-based semi-supervised learning with local and global consistency [19]. 5-fold cross validations on a yeast protein-protein interaction network derived from the BioGRID database show that our algorithm can achieve superior performance compared with four state-of-the-art approaches. Further, we assign functions to 204 uncharacterized proteins, most of which can be directly found in or indirectly support by the SGD database.

As we known, the weight of edge has a very important influence on the prediction results. The performances of our algorithm are significantly improved even if a rather simple weighted strategy is introduced in our study. We expect that the performance can further improved once we adopt the more comprehensive weighted scheme, for example, the terms used in [3,11]. We fix here the parameter $\alpha = 0.99$ as suggested by [19] and select the parameter β through the cross validation process. In fact, choosing these two parameters is a combinatorial optimization problem and should be pay more attentions on it. We left all above-mentioned issues for our further investigation.

Acknowledgments. The author would like to thank Maoying Wu and Lisa J. McQuay for their useful discussions and comments, and the anonymous reviewers for their constructive suggestions to improve the work.

References

1. Breitkreutz, B.J., Stark, C., Reguly, T., et al.: The BioGRID Interaction Database: 2008 update. Nucleic Acids Res. 36(Database issue), D637–D640 (2008)
2. Chen, G., Song, Y., Wang, F., Zhang, C.: Semi-supervised Multi-label Learning by Solving a Sylvester Equation. In: SIAM International Conference on Data Mining (2008)
3. Chua, H.N., Sung, W.K., Wong, L.: Exploiting indirect neighbours and topological weight to predict protein function from protein-protein interactions. Bioinformatics 22, 1623–1630 (2006)
4. Edgar, R., Domrachev, M., Lash, A.E.: Gene Expression Omnibus: NCBI gene expression and hybridization array data repository. Nucleic Acids Res. 30, 207–210 (2002)
5. Fan, R.-E., Lin, C.-J.: A Study on Threshold Selection for Multi-label Classification. Technical Report, National Taiwan University (2007)
6. Gavin, A.C., Bosche, M., Krause, R., Grandi, P., Marzioch, M., Bauer, A., Schultz, J., Rick, J.M., Michon, A.M., Cruciat, C.M., et al.: Functional organization of the yeast proteome by systematic analysis of protein complexes. Nature 415, 141–147 (2002)
7. Harbison, C.T., Gordon, D.B., Lee, T.I., Rinaldi, N.J., Macisaac, K.D., Danford, T.W., Hannett, N.M., Tagne, J.-B., Reynolds, D.B., Yoo, J., et al.: Transcriptional regulatory code of a eukaryotic genome. Nature 431, 99–104 (2004)

8. Hishigaki, H., Nakai, K., Ono, T., Tanigami, A., Takagi, T.: Assessment of prediction accuracy of protein function from proteinCprotein interaction data. Yeast 18, 523–531 (2001)
9. Ito, T., Tashiro, K., Muta, S., Ozawa, R., Chiba, T., Nishizawa, M., Yamamoto, K., Kuhara, S., Sakaki, Y.: Toward a protein-protein interaction map of the budding yeast: a comprehensive system to examine two-hybrid interactions in all possible combinations between the yeast proteins. Proc. Natl Acad. Sci. USA 97, 1143–1147 (2000)
10. Karaoz, U., Murali, T.M., Letovsky, S., Zheng, Y., Ding, C., Cantor, C.R., Kasif, S.: Whole-genome annotation by using evidence integration in functional-linkage networks. Proc. Natl. Acad. Sci. USA 101, 2888–2893 (2004)
11. Nabieva, E., Jim, K., Agarwal, A., Chazelle, B., Singh, M.: Whole-proteome prediction of protein function via graph-theoretic analysis of interaction maps. Bioinformatics 21(Suppl 1), i302–i310 (2005)
12. Pavlidis, P., Weston, J., Cai, J., Grundy, W.N.: Gene functional classification from heterogeneous data. In: Proceedings of the Fifth Annual International Conference on Computational Biology. ACM Press, Montreal (2001)
13. Pellegrini, M., Marcotte, E.M., Thompson, M.J., Eisenberg, D., Yeates, T.O.: Assigning protein functions by comparative genome analysis: protein phylogenetic profiles. Proc. Natl. Acad. Sci. USA 96, 4285–4288 (1999)
14. Schwikowski, B., Uetz, P., Fields, S.: A network of proteinCprotein interactions in yeast. Nat. Biotechnol. 18, 1257–1261 (2000)
15. Sharan, R., Ulitsky, I., Shamir, R.: Network-based prediction of protein function. Molecular Systems Biology 3, 88 (2007)
16. Singh, R., Xu, J., Berger, B.: Global alignment of multiple protein interaction networks with application to functional orthology detection. Proc. Natl. Acad. Sci. USA 105, 12763–12768 (2008)
17. Vazquez, A., Flammini, A., Maritan, A., Vespignani, A.: Global protein function prediction from proteinCprotein interaction networks. Nat. Biotechnol. 21, 697–700 (2003)
18. Zha, Z., Mei, T., Wang, J., Wang, Z., Hua, X.: Graph-based semi-supervised learning with multi-label. In: IEEE International Conference on Multiamedia and Expo (2008)
19. Zhou, D., Bousquet, O., Lal, T.N., Weston, J., Scholkopf, B.: Learning with local and global consistency. In: Advances in Neural Information Processing Systems (NIPS), vol. 16, pp. 321–328. MIT Press, Cambridge (2004)

Regene: Automatic Construction of a Multiple Component Dirichlet Mixture Priors Covariance Model to Identify Non-coding RNA

Felipe Lessa[1], Daniele Martins Neto[2], Kátia Guimarães[3], Marcelo Brigido[4], and Maria Emilia Walter[1]

[1] Department of Computer Science, University of Brasilia
[2] Department of Mathematics, University of Brasilia
[3] Center of Informatics, Federal University of Pernambuco
[4] Institute of Biology, University of Brasilia

Abstract. Non-coding RNA (ncRNA) molecules do not code for proteins, but play important regulatory roles in cellular machinery. Recently, different computational methods have been proposed to identify and classify ncRNAs. In this work, we propose a covariance model with multiple Dirichlet mixture priors to identify ncRNAs. We introduce a tool, named Regene, to derive these priors automatically from known ncRNAs families included in Rfam. Results from experiments with 14 families improved sensitivity and specificity with respect to single component priors.

1 Introduction

The classic central dogma of molecular biology says that genetic information flows from DNA to proteins with different types of RNAs as intermediates, the DNA storing information of the genotype, and the proteins producing phenotypes. This orthodox view of the central dogma suggests that RNA is an auxiliary molecule involved in all stages of protein synthesis and gene expression. But recent research [7] have shown that some types of RNA may indeed regulate other genes and control gene expression and phenotype by themselves. Many other functions of RNAs are already known, and new functions are continuously being discovered. Roughly speaking, RNAs can be divided into two classes, mRNAs – which are translated into proteins, and non-coding RNAs (ncRNAs) – which play several important roles besides protein coding in the cellular machinery.

Supporting the biologists findings to distinguish mRNAs from ncRNAs, recently, many computational methods based on different theories and models have been proposed. It is remarkable that methods that successfully identified mRNAs, such as BLAST, in general fail when used to identify ncRNAs, although they work in some cases [13]. Examples of these models use thermodynamics [19,10], Support Vector Machine (SVM) like CONC [12], CPC [11] and PORTRAIT [2], or Self-Organizing Maps [17].

J. Chen, J. Wang, and A. Zelikovsky (Eds.): ISBRA 2011, LNBI 6674, pp. 380–391, 2011.

Particularly, for the problem of RNA similarity search, Eddy et al. [6] proposed a covariance model (CM) and implemented it in the tool Infernal (from *INFErence of RNA aLignment*) [15]. Infernal is used by the Rfam [8] database, which is continously growing, but contains 1,372 RNA families in its current version 9.1. Nawrocki and Eddy [14] obtained better sensitivity and specificity results including informative Dirichlet priors through the use of Bayesian inference. The principle of this theory is to combine prior information with information obtained from the sample (likelihood) to construct the posterior distribution, which synthesizes all the information. So, this method adds, in a transparent way, prior knowledge to the data. Priors play a key role in this case, since they allow to include subjective information and knowledge about the analyzed problem. Informative prior expresses specific information about a variable, and non-informative prior refers to vague and general information about this variable.

The objectives of this work are to create a new tool to automatically derive priors and to evaluate the use of Dirichlet mixture priors with more than one component in CMs. The main justification is to make more precise the identification of this CM transition distribution within empirical data. The tool *Regene* (from the character *Regene* Regetta of the japanese anime Gundam 00 and *gene*) to derive the Dirichlet mixture priors from Rfam is built, as well as a module to automatically draw the CM and the parse tree diagrams.

In Section 2, we describe CMs, particularly the CM proposed by Eddy and Durbin [6] that was used in the tests. In Section 3, we first show how the Dirichlet mixture priors are estimated using the Expectation-Maximization (EM) method, and then describe the Conjugate Gradient Descent method, used to minimize the objective function. In Section 4, we present our method and show some implementation details. In Section 5, we discuss experiments with 14 Rfam families and compare the obtained results with those of Nawrocki and Eddy [14]. Finally, in Section 6, we conclude and suggest future work.

2   Covariance Models

The covariance model (CM) proposed by Eddy and Durbin [6] represents a specific family of non-coding RNAs. These CM s are a subset of the stochastic context-free grammars (SCFGs), called a "profile SCFG" [5]. The main types of production used in these CMs are $P \rightarrow aXb$ for base pairs in a stem, $L \rightarrow aX$ and $R \rightarrow Xb$ for single stranded bases, and $B \rightarrow SS$ for bifurcations that are used to separate a loop with multiple stems. Each non-terminal is called a "state", and terminals (i.e. bases of the sequences) are called "emissions", noting that this terminology has been borrowed from hidden Markov models (HMMs) [4].

These grammars are "profiles" because states are mechanically constructed from the secondary structure consensus. A "guide tree" is then built, where the nodes represent single stranded bases, base pairs, and loops with multiple stems.

For example, in Figure 1 we present the consensus secondary structure of a fictitious family of tRNAs and the generated data structures. In this figure, MATP nodes represent both sides of a stem, and two BIF nodes are used to represent the internal loop. See Eddy [5] for a detailed explanation about the CM building process. We note that the guide tree diagrams shown in this figure were automatically created by our Regene tool (`regene-diagrams` module).

We now explain how the model is used. Each state may emit zero, one or two bases and indicate a set of allowed transitions. We generate RNA sequences from the model by "walking" through the states, collecting its emissions. Each "walk" is called a *parse tree*.

Parse trees may be used to align a new sequence to an existing CM. Although there are possibly many parse trees that emit a given sequence, each parse tree has an associated probability, which allows us to find the tree presenting the highest probability. However, in this paper we are more interested in a particular type of a parse tree, named *fake parse tree* [1]. It is constructed by walking over consensus-representing states, not including probability values, and going to insertion or deletion states only when that column of the sequence is not in the consensus. In Figure 2 we show the diagrams of fake parse trees constructed from the CM shown in Figure 1.

In order to calculate the probabilities of a CM, a fake parse tree is created for each one of the sequences in the multiple alignment. The transitions and emissions generated in each state of all parse trees are counted, and these counts are then converted into probabilites using the posterior Dirichlet mixture probabilites as described in Sections 3 and 4.

Fake parse trees are used to obtain the Dirichlet mixture priors in a first step. For every family of the corresponding set used for training, a CM and its fake parse trees are constructed. Instead of observing transition counts of individual states, transitions of a particular type are counted [14]. Each type defines the guide tree node, the origin state of a transition, and its destination node. For example, transitions of type "MATL/D → MATR" may go to IL, MR or D state type. Each family count is then used as a training vector for the corresponding Dirichlet mixture.

3   Dirichlet Mixture Priors

To identify ncRNAs using CMs, it is essential to estimate the transition parameters. In Infernal [15], transition parameters are mean posterior estimates, combining observed counts from an input that is a RNA multiple alignment with an informative Dirichlet prior. Nawrocki and Eddy [14] use this prior through a Dirichlet mixture with a single component, based on the fact that they accurately predict target subsequence lengths, mainly when there are few query alignments sequences. The use of informative priors is more efficient for a small number of observations. Considering this, and with the purpose of facilitating the identification of transitions within empirical data, we use as informative prior a mixture of Dirichlet densities with more than one component. Sjölander et al. [18] has used such a tool for estimating amino acids distributions. In this work, we estimate the transition probabilities of a CM in the process of identifying ncRNAs.

Following we present the model and the structure used to obtain the transition estimates of a CM, and after that the CG_DESCENT method to minimize the objective function.

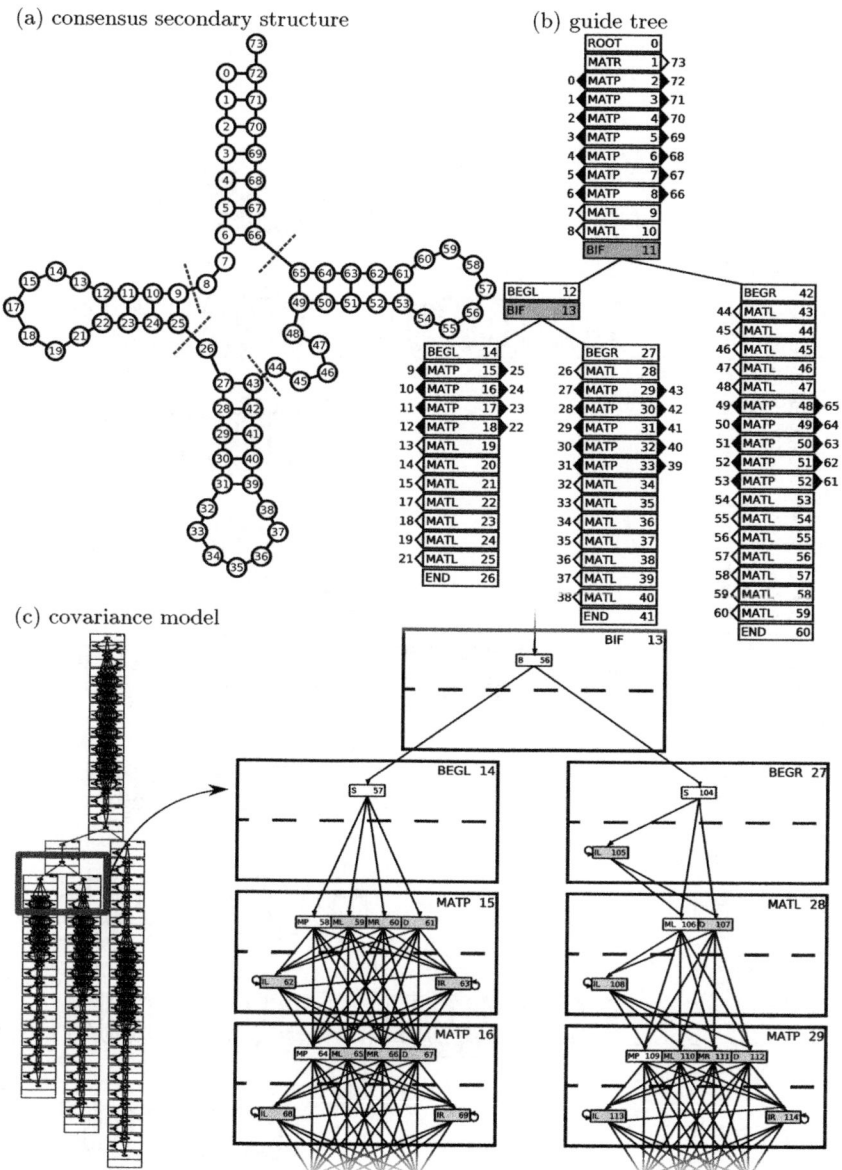

Fig. 1. Diagrams of a fictitious tRNA family extracted from the `trna.5.sto` file of Infernal's tutorial, noting that (b) and (c) were automatically drawn by Regene. (a) The consensus secondary structure. Dashed lines show bifurcations created by BIF nodes to separate stems. (b) The corresponding guide tree. (c) The CM created from this guide tree.

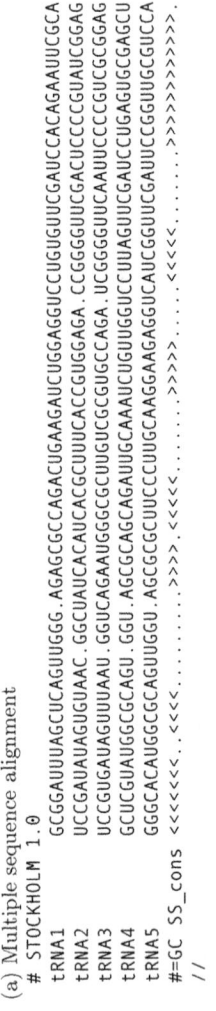
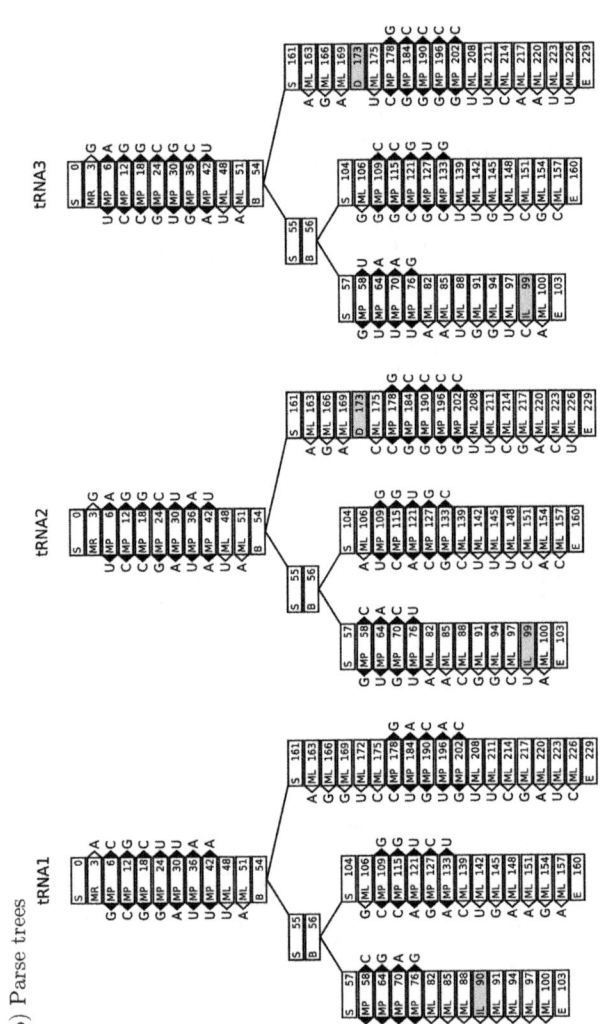

Fig. 2. (a) Multiple sequence alignment of the CM shown in Figure 1. (b) Diagrams of the fake parse trees for the three sequences, tRNA1, tRNA2 and tRNA3, created by Regene (regene-diagrams module).

3.1   Obtaining the Mixture Prior

Let $\mathbf{N} = (N_1, \ldots, N_m)$ be a random vector of transition counts, where N_i is a random number of times that each ith transition occurs. Assume that $\mathbf{N}$ is multinomial distributed with a parameter $\mathbf{p}$. In this case, $\mathbf{p} = (p_1, \ldots, p_m)$ is the vector of transition probabilities to be estimated, where p_i is the ith transition probability, and m is the fixed number of transitions.

The Bayesian approach is used, with some prior knowledge of $\mathbf{p}$, to estimate the transition parameters $p_1, \ldots, p_m$. In this paper, we consider that the prior density of $\mathbf{p}$, denoted by ρ, is the following Dirichlet mixture density:

$$\rho = q_1 \rho_1 + \cdots + q_l \rho_l \, , \tag{1}$$

where each ρ_j is a single Dirichlet density with parameter $\boldsymbol{\alpha}_j = (\alpha_{j1}, \ldots, \alpha_{jm})$, with $\alpha_{ji} > 0$, $i = 1, \ldots, m$, $q_j > 0$, and $\sum_{j=1}^{l} q_j = 1$. The ρ_j densities are called *mixture components*, and the q_j values are called *mixture coefficients*. The entire set of parameters defining a prior in the case of a mixture, which are obtained from the Rfam database, is $\Theta = (\boldsymbol{\alpha}_1, \ldots, \boldsymbol{\alpha}_l, q_1, \ldots, q_l)$.

The $\widehat{p}_i$ estimated probability of the ith transition is the mean posterior estimate, given by:

$$\widehat{p}_i = E(p_i \mid \mathbf{N} = \mathbf{n}) = \sum_{j=1}^{l} P(\boldsymbol{\alpha}_j \mid \mathbf{N} = \mathbf{n}) \, \frac{\alpha_{ji} + n_i}{\alpha_{j0} + n_0} \, , \tag{2}$$

where $n_0 = \sum_{i=1}^{m} n_i$, $\alpha_{j0} = \sum_{i=1}^{m} \alpha_{ji}$, and $\mathbf{n} = (n_1, \ldots, n_m)$ is the data observed from variable $\mathbf{N}$. Probability $P(\boldsymbol{\alpha}_j \mid \mathbf{N} = \mathbf{n})$, calculated from the Bayes rule, with $P(\mathbf{N} = \mathbf{n} \mid \Theta, n_0) = \sum_{k=1}^{l} q_k P(\mathbf{N} = \mathbf{n} \mid \boldsymbol{\alpha}_k, n_0)$, is:

$$P(\boldsymbol{\alpha}_j \mid \mathbf{N} = \mathbf{n}) = \frac{q_j P(\mathbf{N} = \mathbf{n} \mid \boldsymbol{\alpha}_j, n_0)}{P(\mathbf{N} = \mathbf{n} \mid \Theta, n_0)} \, . \tag{3}$$

From the model, and taking $\Gamma(.)$ as the Gamma function, we have:

$$P(\mathbf{N} = \mathbf{n} \mid \boldsymbol{\alpha}_j, n_0) = \frac{\Gamma(n_0 + 1)\Gamma(\alpha_{j0})}{\Gamma(n_0 + \alpha_{j0})} \prod_{i=1}^{m} \frac{\Gamma(n_i + \alpha_{ji})}{\Gamma(n_i + 1)\Gamma(\alpha_{ji})} \, . \tag{4}$$

Note that, in the case of a single-component density ($l = 1$), we have $\Theta = \boldsymbol{\alpha} = (\alpha_1, \ldots, \alpha_m)$ and the estimate for the ith transition probability is:

$$\widehat{p}_i = E(p_i \mid \mathbf{N} = \mathbf{n}) = \frac{\alpha_i + n_i}{\alpha_0 + n_0} \, . \tag{5}$$

In this context, to introduce prior knowledge about $\mathbf{p}$ through the density ρ, it is necessary to inform about Θ. Therefore, we have used the maximum likelihood estimator for Θ, observing the count vectors of the ncRNA families of the Rfam database.

Given a set of r fake parse trees, which are constructed from r Rfam families, we get a count vector of transitions for each family, based on the multiple alignment obtained for each family. The result is a set of count vectors $\mathbf{n_1} = (n_{11}, \ldots, n_{1m}), \ldots, \mathbf{n_r} = (n_{r1}, \ldots, n_{rm})$, where n_{ji} is the number of times that the ith transition occurs in the jth fake parse tree.

Our objective is to estimate, using maximum likelihood, the parameters of mixture priors obtained from the set of count vectors. Therefore, we want to find Θ that maximizes $\prod_{t=1}^{r} P(\mathbf{N} = \mathbf{n}_t \mid \Theta, n_{t0})$, where $n_{t0} = \sum_{i=1}^{m} n_{ti}$, or equivalently, Θ that minimizes the following *objective function*:

$$f(\Theta) = -\sum_{t=1}^{r} \log P(\mathbf{N} = \mathbf{n}_t \mid \Theta, n_{t0}) . \tag{6}$$

Next, we present the procedure to estimate the parameters of the mixture priors, using the EM (expectation-maximization) algorithm, which has been shown to be efficient to find the maximum likelihood estimate in such cases (see Duda and Hart [3]).

In a mixture density, there are two sets of parameters, $\boldsymbol{\alpha}_j$ and q_j, $j = 1, \ldots, l$, that are jointly estimated in an iterative two steps process. First one is estimated, keeping the other fixed, and then the process is reverted. Next, we present two procedures to separately estimate each set of parameters. For more details, see Sjölander et al. [18].

Estimating the α Parameters. The α_{ji} parameters are strictly positive and we use the CG_DESCENT method (see Section 3.2) to estimate them via EM. In this case, a reparametrization is needed, which takes $\alpha_{ji} = e^{w_{ji}}$, where w_{ji} is a real number with no restrictions. Thus, the partial derivate of the objective function with respect to w_{ji}, taking $\Psi(.) = \frac{\Gamma'(.)}{\Gamma(.)}$ as the digamma function, is:

$$\frac{\partial f(\Theta)}{\partial w_{ji}} = -\sum_{t=1}^{r} \frac{\partial \log P(\mathbf{N} = \mathbf{n}_t \mid \Theta, n_{t0})}{\partial \alpha_{ji}} \alpha_{ji} \tag{7}$$

$$= -\sum_{t=1}^{r} \alpha_{ji} [\Psi(\alpha_{j0}) - \Psi(n_{t0} + \alpha_{j0}) + \Psi(n_{ti} + \alpha_{ji}) - \Psi(\alpha_{ji})] . \tag{8}$$

Estimating the q Parameters. To estimate the mixture coefficients q_j, $j = 1, \ldots, l$, which must be non-negative and sum to 1, a reparametrization is also needed. Let $q_j = \frac{Q_j}{Q_0}$, where Q_j is strictly positive and $Q_0 = \sum_j Q_j$. The partial derivate of the objective funtion with respect to Q_j is:

$$\frac{\partial f(\Theta)}{\partial Q_j} = -\sum_{t=1}^{r} \frac{\partial \log P(\mathbf{N} = \mathbf{n}_t \mid \Theta, n_{t0})}{\partial Q_j} = \frac{m}{Q_0} - \frac{\sum_{t=1}^{r} P(\boldsymbol{\alpha}_j \mid \mathbf{N} = \mathbf{n}_t)}{Q_j} . \tag{9}$$

It follows that the maximum likelihood estimate of Q_j is:

$$\widehat{Q}_j = \frac{Q_0}{r} \sum_{t=1}^{r} P\left(\boldsymbol{\alpha}_j \mid \mathbf{N} = \mathbf{n}_t\right), \tag{10}$$

and hence the estimate for q_j is:

$$\widehat{q}_j = \frac{Q_j}{Q_0} = \frac{1}{r} \sum_{t=1}^{r} P\left(\boldsymbol{\alpha}_j \mid \mathbf{N} = \mathbf{n}_t\right). \tag{11}$$

3.2   Conjugate Gradient Method

EM method requires minimization of the objective function. Since the objective function is not simple, analytic methods are not good choices to compute its minimum. Instead, we compute its gradient and iteratively optimize it.

There are many different ways of optimizing a nonlinear function. Sjölander et al. [18] suggested a gradient descent method, while Nawrocki used a nonlinear conjugate gradient method with the Polak-Ribière formula [14]. We used the nonlinear conjugate gradient method proposed by Hager and Zhang [9], called CG_DESCENT. This relatively new method was chosen for two reasons. First, it performs well on a set of different nonlinear functions. The other advantage is that the CG_DESCENT 3.0 library, written by the authors, is robust, fast and well-tested.

4   Multiple Component Dirichlet Mixture Priors

We first present our method and then important implementation details.

4.1   Regene Method

We created the new Regene tool that combines the techniques described in the previous sections into a complete program that maps a set of ncRNA families into a Dirichlet mixture. Particularly, we used the EM method for Dirichlet mixtures and the CG_DESCENT method (Section 3), and counted the transitions of CMs built from those ncRNA families (Section 2). Note that we did not estimate the emission priors, but the transition priors. We used the same emission priors from Nawrocki and Eddy [14].

We first inform the objective function and its gradient as a black box to the CG_DESCENT method, that implements the EM method. EM method was used with our countings in a similar way. Since each counting has a meaning, we used these countings as training sequences for the EM method. We did not need to code any knowledge about CMs in our EM method, nor change our counting method to estimate the Dirichlet mixtures. In a nutshell, we separately implemented each module, and combined them through a pipeline (Figure 3).

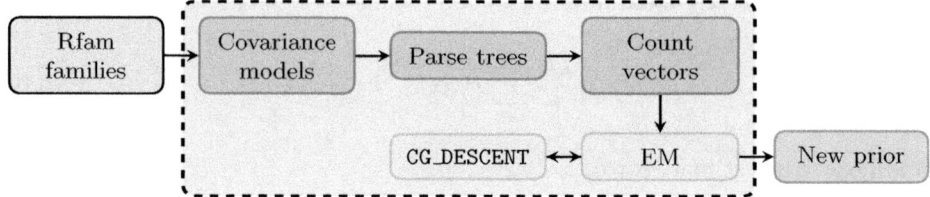

Fig. 3. The Regene tool pipeline to derive a multiple component Dirichlet mixture prior. The new prior is used to construct CMs.

4.2   Regene Implementation Details

The pipeline of Figure 3 was developed in 5,900 source lines of code in the Haskell language. Bindings were created to use the CG_DESCENT 3.0 library with Haskell functions, which were packed with the name nonlinear-optimization.

The EM method was implemented into a generic library that may be used for any kind of Dirichlet density or mixture. It is now hardcoded to use the CG_DESCENT method, although another method could be easily used as well. The library is packed into the statistics-dirichlet library.

Counting the transitions is somewhat more involved, as there are many details to be considered. Infernal has a non documented option for printing the counts of a given family, as clarified by personal communication with E. Nawrocki. We reimplemented whatever needed into the regene library, from parsers of the Stockholm file format (used to store the ncRNA families' data) to CM data structures. Using the library, we created the regene-priori program that glues everything together.

As a nice result of our reimplementation of the data structures, we also created the regene-diagrams program. It may be used to create diagrams of guide trees, CMs (in a flat or in a graph-like view) and fake parse trees. Besides having an easy-to-use graphical user interface, it also has a console interface that may be used to automatically create diagrams from other programs, such as a website. Our code is available in the Hackage's public repository under the free GNU Public License (GPL) [16].

Unfortunately, the objective function did not behave very well in all cases. Sometimes, CG_DESCENT would find solutions with high α value, especially when dealing with a high number of components. High α values imply that data from the specific family would not be used. Our implementation allows to discard values above a given threshold. In our experiments, we used threshold 10^4.

5   Results and Discussion

To measure the sensitivity and specificity of our priors, we used the same cmark-1 benchmark used by Nawrocki and Eddy [14], which is included in Infernal. From the seed alignments of Rfam 7.0, they selected 51 families, from which we used 14 families in our tests. These families were chosen due to their different sequence

Table 1. Minimum error rate (MER) calculated from our tests. Column "#QSs" represent the number of query sequences, column "N" is the MER using Nawrocki's prior. Columns "A1", "A2", "A3", "A4", "A5" and "A10" show the MER using our priors constructed from Rfam 6.1 data (same data as Nawrocki's) with 1, 2, 3, 4, 5 and 10 components, respectively. Columns "B1" to "B10" show the MER using priors constructed from Rfam 9.1 data with 1 to 10 components.

Rfam 7.0 family				Using Rfam 6.1						Using Rfam 9.1					
ID	Name	#QSs	N	A1	A2	A3	A4	A5	A10	B1	B2	B3	B4	B5	B10
RF00004	U2	76	0	0	0	0	0	0	0	0	0	0	0	0	0
RF00009	RNaseP_nuc	26	19	19	19	19	19	19	19	19	19	19	19	19	19
RF00011	RNaseP_bact_b	30	0	0	0	0	0	0	0	0	0	0	0	0	0
RF00017	SRP_euk_arch	28	7	6	6	6	6	6	6	9	9	8	9	9	9
RF00023	tmRNA	19	12	11	13	11	11	11	11	11	11	11	11	11	11
RF00029	Intron_gpII	7	2	1	1	1	1	1	1	2	1	2	2	2	1
RF00030	RNase_MRP	18	3	3	3	3	3	3	3	3	3	3	3	3	3
RF00031	SECIS	11	16	17	17	17	16	17	17	19	19	19	19	19	19
RF00037	IRE	36	1	1	1	1	1	1	1	1	1	1	1	1	1
RF00168	Lysine	33	0	0	0	0	0	0	0	0	0	0	0	0	0
RF00174	Cobalamin	87	0	0	0	0	0	0	0	0	0	0	0	0	0
RF00177	SSU_rRNA_5	145	3	4	4	4	4	4	4	4	4	4	4	0	4
RF00234	glmS	8	0	0	0	0	0	0	0	0	0	0	0	0	0
RF00448	IRES_EBNA	7	1	1	1	1	1	1	1	1	1	1	1	1	1
Summed across all families			64	63	65	63	62	63	63	69	68	68	69	65	68
Summary MER statistics			78	76	79	78	76	74	76	79	78	78	79	77	79

lengths and distinct biological functions. They also used some sequences to create CMs for the benchmark, while others were used as test sequences. Those test sequences were inserted into a 1 Mb pseudo-genome with independent and identically distributed background distribution of bases. The test consists in the use of the CMs to find those test sequences. Results were reported as minimum error rate (MER), the sum of the number of false positives and false negatives for the best found threshold. We did exactly the same steps to realize our tests.

But while Nawrocki and Eddy [14] used Infernal v0.72 bit scores, we used Infernal v1.0 E-values, and then we had to "calibrate" the CMs. Therefore, our results were slightly different, although using the same data. E-values were chosen since they present the best criterion regarding to statistical significance [1]. Besides, biologists commonly use E-values when analyzing the Infernal results.

The results for Nawrocki's prior and twelve priors created by `regene-priori` are presented in Table 1. In the line "Summary MER statistics", note that, among the priors constructed with Rfam 6.1 data, we reach the optimal point with 5 components (column "A5", presenting the lowest "Summary MER Statistics"), which is the best prior data that we tested. Priors constructed using Rfam 9.1 data did not behave very well in the tests, with the best one having 5 components (column "B5"). Perhaps that could be explained by the fact that the tests used the Rfam 7.0 families.

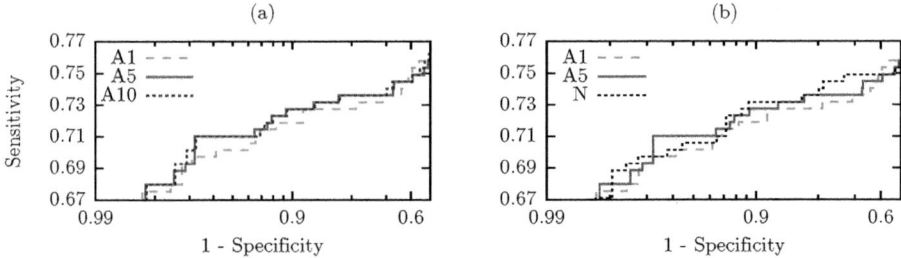

Fig. 4. ROC curves for representatives priors. (a) Comparisons of our priors A1, A5 and A10. (b) Comparisons of our A1 and A5 priors with Nawrocki's (N).

Figure 4 shows two ROC curves for the most representative priors, as indicated in the "Summary MER Statistics" line of Table 1. ROC curves show how the sensitivity increases while the specificity decreases. The right part of the graphics show that high specificity leads to low sensitivity, while the left part shows that most sequences can be found with low specificity. Figure 4(a) shows that multiple component mixtures improve sensitivity and specificity relative to the single component mixture. Note that the mixture with 10 components (A10) does not improve the mixture with 5 components (A5), so using more than 5 components is not necessary. Besides, our 5 component prior slightly improved Nawrocki's single component prior (N). However, Figure 4(b) shows that our single component prior (A1) is worse when compared to the one obtained by Nawrocki, although both priors were constructed from the same Rfam 6.1 data. This result could be explained by the fact that the estimation methods are different. We believe that using more than one component in the method used by Nawrocki and Eddy [14] would improve their results.

6   Conclusions and Future Work

In this work we created a new tool to automatically derive priors and evaluated the use of Dirichlet mixture priors with more than one component in CMs. We built a tool named *Regene* to automatically derive the priors from Rfam data, as well as a module to automatically draw guide trees, Eddy's CMs (in a flat or in a graph-like view) and fake parse trees. Experiments were done with 14 Rfam families, comparing 13 different priors – with 1, 2, 3, 4, 5 and 10 components using Rfam 6.1 and Rfam 9.1 data, and with Nawrocki's prior with 1 component. The use of multiple components improved sensitivity and specificity.

We plan to run experiments using Rfam 9.1 instead of Rfam 7.0, and also to extend the experiments to more families. We will also look into other estimation methods such as Monte-Carlo EM, which could improve the results.

Acknowledgements

FL, DMN and MEW thank Brazilian sponsoring agency FINEP, and KG, MB and MEW thank Brazilian sponsoring agency CNPq for financial support.

References

1. The Infernal's user guide, http://infernal.janelia.org/
2. Arrial, R., Togawa, R., Brigido, M.: Screening non-coding RNAs in transcriptomes from neglected species using PORTRAIT: case study of the pathogenic fungus Paracoccidioides brasiliensis. BMC Bioinformatics 10, 239 (2009)
3. Duda, R.O., Hart, P.E.: Pattern Classification and Scene Analysis. Wiley, Chichester (1973)
4. Eddy, S.R.: Profile hidden Markov models. Bioinformatics 14(9), 755–763 (1998)
5. Eddy, S.R.: A memory-efficient dynamic programming algorithm for optimal alignment of a sequence to an RNA secondary structure. BMC Bioinformatics 3, 18 (2002)
6. Eddy, S.R., Durbin, R.: RNA sequence analysis using covariance models. Nucleic Acids Research 22(11), 2079–2088 (1994)
7. Griffiths-Jones, S.: Annotating Noncoding RNA Genes. Annu. Rev. Genomics Hum. Genet. 8, 279–298 (2007)
8. Griffiths-Jones, S., Moxon, S., Marshall, M., Khanna, A., Eddy, S.R., Bateman, A.: Rfam: annotating non-coding RNAs in complete genomes. Nucleic Acids Research 33, D121–D124 (2005), http://www.sanger.ac.uk/Software/Rfam/
9. Hager, W.W., Zhang, H.: A new conjugate gradient method with guaranteed descent and an efficient line search. SIAM J. on Optimization 16(1), 170–192 (2005)
10. Hofacker, I.L., Fekete, M., Stadler, P.F.: Secondary Structure Prediction for Aligned RNA Sequences. Journal of Molecular Biology 319(5), 1059–1066 (2002)
11. Kong, L., Zhang, Y., Ye, Z.-Q., Liu, X.-O., Zhao, S.-O., Wei, L., Gao, G.: CPC: assess the protein-coding potential of transcripts using sequence features and support vector machine. Nucleic Acids Res. 35, 345–349 (2007)
12. Liu, J., Gough, J., Rost, B.: Distinguishing protein-coding from non-coding RNAs through Support Vector Machines. PLoS Genet. 2(4), e29–e36 (2006)
13. Mount, S.M., Gotea, V., Lin, C.F., Hernandez, K., Makalowski, W.: Spliceosomal Small Nuclear RNA Genes in Eleven Insect Genomes. RNA 13, 5–14 (2007)
14. Nawrocki, E.P., Eddy, S.R.: Query-Dependent Banding (QDB) for Faster RNA Similarity Searches. PLoS Computational Biology 3(3), e56 (2007)
15. Nawrocki, E.P., Kolbe, D.L., Eddy, S.R.: Infernal 1.0: Inference of RNA alignments. Bioinformatics 25, 1335–1337 (2009)
16. Regene, http://regene.exatas.unb.br
17. Silva, T.C., et al.: SOM-PORTRAIT: Identifying Non-coding RNAs Using Self-Organizing Maps. In: Guimarães, K.S., Panchenko, A., Przytycka, T.M. (eds.) BSB 2009. LNCS, vol. 5676, pp. 73–85. Springer, Heidelberg (2009)
18. Sjölander, K., et al.: Dirichlet mixtures: a method for improved detection of weak but significant protein sequence homology. Computer Applications in the Biosciences 12(4), 327–345 (1996)
19. Zucker, M., Matthews, D.H., Turner, D.H.: Algorithms and thermodynamics for RNA secondary structure prediction: A practical guide. In: RNA Biochemistry and Biotechnology. NATO ASI Series, pp. 11–43. Kluwer Academic, Dordrecht (1999)

Accurate Estimation of Gene Expression Levels from DGE Sequencing Data

Marius Nicolae and Ion Măndoiu

Computer Science & Engineering Department, University of Connecticut
371 Fairfield Way, Storrs, CT 06269
{man09004,ion}@engr.uconn.edu

Abstract. Two main transcriptome sequencing protocols have been proposed in the literature: the most commonly used shotgun sequencing of full length mRNAs (RNA-Seq) and 3'-tag digital gene expression (DGE). In this paper we present a novel expectation-maximization algorithm, called DGE-EM, for inference of gene-specific expression levels from DGE tags. Unlike previous methods, our algorithm takes into account alternative splicing isoforms and tags that map at multiple locations in the genome, and corrects for incomplete digestion and sequencing errors. The open source Java/Scala implementation of the DGE-EM algorithm is freely available at http://dna.engr.uconn.edu/software/DGE-EM/.

Experimental results on real DGE data generated from reference RNA samples show that our algorithm outperforms commonly used estimation methods based on unique tag counting. Furthermore, the accuracy of DGE-EM estimates is comparable to that obtained by state-of-the-art estimation algorithms from RNA-Seq data for the same samples. Results of a comprehensive simulation study assessing the effect of various experimental parameters suggest that further improvements in estimation accuracy could be achieved by optimizing DGE protocol parameters such as the anchoring enzymes and digestion time.

1 Introduction

Massively parallel transcriptome sequencing is quickly replacing microarrays as the technology of choice for performing gene expression profiling due to its wider dynamic range and digital quantitation capabilities. However, accurate estimation of expression levels from sequencing data remains challenging due to the short read length delivered by current sequencing technologies and still poorly understood protocol- and technology-specific biases. To date, two main transcriptome sequencing protocols have been proposed in the literature. The most commonly used one, referred to as RNA-Seq, generates short (single or paired) sequencing tags from the ends of randomly generated cDNA fragments. An alternative protocol, referred to as 3'-tag Digital Gene Expression (DGE), or high-throughput sequencing based Serial Analysis of Gene Expression (SAGE-Seq), generates single cDNA tags using an assay including as main steps transcript capture and cDNA synthesis using oligo(dT) beads, cDNA cleavage with an anchoring restriction enzyme, and release of cDNA tags using a tagging restriction

J. Chen, J. Wang, and A. Zelikovsky (Eds.): ISBRA 2011, LNBI 6674, pp. 392–403, 2011.

enzyme whose recognition site is ligated upstream of the recognition site of the anchoring enzyme.

While computational methods for accurate inference of gene (and isoform) specific expression levels from RNA-Seq data have attracted much attention recently (see, e.g., [4,6,8]), analysis of DGE data still relies on direct estimates obtained from counts of uniquely mapped DGE tags [1,10]. In part this is due to salient features of the DGE protocol, which, unlike RNA-Seq, guarantees that each mRNA molecule in the sample generates at most one tag and obviates the need for length normalization. Nevertheless, ignoring ambiguous DGE tags (which, due to the severely restricted tag length, can represent a sizeable fraction of the total) is at best discarding useful information, and at worst may result in systematic inference biases. In this paper we seek to address this shortcoming of existing methods for DGE data analysis. Our main contribution is a rigorous statistical model of DGE data and a novel expectation-maximization algorithm for inference of gene and isoform expression levels from DGE tags. Unlike previous methods, our algorithm, referred to as DGE-EM, takes into account alternative splicing isoforms and tags that map at multiple locations in the genome, and corrects for incomplete digestion and sequencing errors. Experimental results show that DGE-EM outperforms methods based on unique tag counting on a multi-library DGE dataset consisting of 20bp tags generated from two commercially available reference RNA samples that have been well-characterized by quantitative real time PCR as part of the MicroArray Quality Control Consortium (MAQC).

We also take advantage of the availability of RNA-Seq data generated from the same MAQC samples to directly compare estimation performance of the two transcriptome sequencing protocols. While RNA-Seq is clearly more powerful than DGE at detecting alternative splicing and novel transcripts such as fused genes, previous studies have suggested that for gene expression profiling DGE may yield accuracy comparable to that of RNA-Seq at a fraction of the cost [7]. We find that the two protocols achieve similar cost-normalized accuracy on the MAQC samples when using state-of-the-art estimation methods. However, the current protocol versions are unlikely to be optimal. Indeed, the results of a comprehensive simulation study assessing the effect of various experimental parameters suggest that further improvements in DGE accuracy could be achieved by using anchoring enzymes with degenerate recognition sites and using partial digest of cDNA with the anchoring enzyme during library preparation.

2   DGE Protocol

The DGE protocol generates short cDNA tags from a mRNA population in several steps (Figure 1). First, PolyA+ mRNA is captured from total RNA using oligo-dT magnetic beads and used as template for cDNA synthesis. The double stranded cDNA is then digested with a first restriction enzyme, called *Anchoring Enzyme* (AE), with known sequence specificity (e.g., the NlaIII enzyme cleaves cDNA at sites at which the four nucleotide motif CATG appears). We refer to

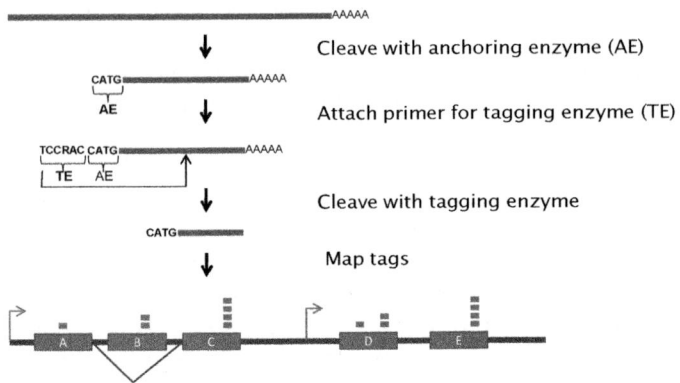

Fig. 1. Schematic representation of the DGE protocol

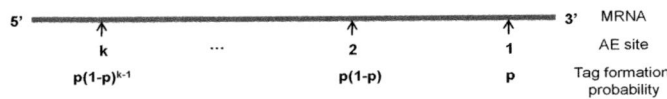

Fig. 2. Tag formation probability: p for the rightmost AE site, geometrically decreasing for subsequent sites

the cDNA sites cleaved by the anchoring enzyme as *AE sites*. The recognition site of a second restriction enzyme, called *Tagging Enzyme* (TE) is ligated to the fragments of cDNA that remain attached to the beads after cleavage with the AE, immediately upstream of the AE site. The cDNA fragments are then digested with TE, which cleaves several bases away from its recognition site. This results in very short cDNA tags (10 to 26 bases long, depending on the TE used), which are then sequenced using any of the available high-throughout technologies.

Since the recognition site of AE is only 4 bases long, most transcripts contain multiple AE sites. Under perfect experimental conditions, full digest by AE would ensure that DGE tags are generated only from the most 3' AE site of each transcript. In practice some mRNA molecules release tags from other AE sites, or no tag at all. As in [10], we assume that the cleavage probability of the AE, denoted by p, is the same for all AE sites of all transcripts. Since only the most 3' cleaved AE site of a transcript releases a DGE tag, the probability of generating a tag from site $i = 1, \ldots, k$ follows a geometric distribution with ratio $1 - p$ as shown in Figure 2, where sites are numbered starting from the 3' end. Note that splicing isoforms of a gene are likely to share many AE sites. However, the probability of generating a tag from a site is *isoform specific* since it depends on the number downstream AE sites on each isoform. Thus, although the primary motivation for this work is inference of gene expression levels from DGE tags, the algorithm presented in next section must take into account alternative splicing isoforms to properly allocate ambiguous tags among AE sites.

3   DGE-EM Algorithm

Previous studies have either discarded ambiguous DGE tags (e.g. [1,10]) or used simple heuristic redistribution schemes for rescuing some of them. For example, in [9] the rightmost site in each transcript is identified as a "best" site. If a tag matches several locations, but only one of them is a best site, then the tag is assigned to that site. If a tag matches multiple locations, none of which is a best site, the tag is equally split between these locations. In this section we detail an Expectation Maximization algorithm, referred to as DGE-EM, that probabilistically assigns DGE tags to candidate AE sites in different genes, different isoforms of the same gene, as well as different sites within the same isoform.

In a pre-processing step, a weight is assigned to each (DGE tag, AE site) pair, reflecting the conditional probability of the tag given the site that releases it. This probability is computed from base quality scores assuming that sequencing errors at different tag positions arise independently of one another. Formally, the weight for the alignment of tag t with the jth rightmost AE site in isoform i is $w_{t,i,j} \propto \prod_{k=1}^{|t|}[(1-\varepsilon_k)M_{t_k} + \frac{\varepsilon_k}{3}(1-M_{t_k})]$, where $M_{t,k}$ is 1 if position k of tag t matches the corresponding position at site j in the transcript, 0 otherwise, while ε_k denotes the error probability of the k-th base of t, derived from the corresponding Phred quality score reported by the sequencing machine. In practice we only compute these weights for sites at which a tag can be mapped with a small (user selected) number of mismatches, and assume that remaining weights are 0. To each tag t we associate a "tag class" y_t which consists of the set of triples (i, j, w) where i is an isoform, j is an AE site in isoform i, and $w > 0$ is the weight associated as above to tag t and site j in isoform i. The collection of tag classes, $y = (y_t)_t$, represents the observed DGE data.

Let m be the number of isoforms. The parameters of the model are the relative frequencies of each isoform, $\theta = (f_i)_{i=1,\dots,m}$. Let $n_{i,j}$ denote the (unknown) number of tags generated from AE site j of isoform i. Thus, $x = (n_{i,j})_{i,j}$ represents the complete data. Denoting by k_i the number of AE sites in isoform i, by $N_i = \sum_{j=1}^{k_i} n_{i,j}$ the total number of tags from isoform i, and by $N = \sum_{i=1}^{m} N_i$ the total number of tags overall, we can write the complete data likelihood as

$$g(x|\theta) \propto \prod_{i=1}^{m}\prod_{j=1}^{k_i}\left[\frac{f_i(1-p)^{j-1}p}{S}\right]^{n_{i,j}} \tag{1}$$

where $S = \sum_{i=1}^{m}\sum_{j=1}^{k_i} f_i(1-p)^{j-1}p = \sum_{i=1}^{m} f_i\left(1 - (1-p)^{k_i}\right)$. Put into words, the probability of observing a tag from site j in isoform i is the frequency of that isoform (f_i) times the probability of not cutting at any of the first $j - 1$ sites and cutting at the jth $[(1 - p)^{j-1}p]$. Notice that the algorithm effectively downweights the matching AE sites far from the 3′ end based on the site probabilities shown in Figure 2. Since for each transcript there is a probability that no tag is actually generated, for the above formula to use proper probabilities we have to normalize by the sum S over all observable AE sites.

Taking logarithms in (1) gives the complete data log-likelihood:

$$\log g(x|\theta) = \sum_{i=1}^{m} \sum_{j=1}^{k_i} n_{i,j} \left[\log f_i + (j-1) \log (1-p) + \log p - \log S \right] + \text{constant}$$

$$= \sum_{i=1}^{m} \sum_{j=1}^{k_i} n_{i,j} \left[\log f_i + (j-1) \log(1-p) \right]$$

$$+ N \log p - N \log \left(\sum_{i=1}^{m} f_i \left(1 - (1-p)^{k_i} \right) \right) + \text{constant}$$

3.1 E-Step

Let $c_{i,j} = \{y_t | \exists w \text{ s.t. } (i,j,w) \in y_t\}$ be the collection of all tag classes that are compatible with AE site j in isoform i. The expected number of tags from each cleavage site of each isoform, given the observed data and the current parameter estimates $\theta^{(r)}$, can be computed as

$$n_{i,j}^{(r)} := E(n_{i,j}|y, \theta^{(r)}) = \sum_{y_t \in c_{i,j}, (i,j,w) \in y_t} \frac{f_i(1-p)^{j-1}pw}{\sum_{(l,q,z) \in y_t} f_l(1-p)^{q-1}pz} \qquad (2)$$

This means that each tag class is fractionally assigned to the compatible isoform AE sites based on the frequency of the isoform, the probability of cutting at the cleavage sites where the tag matches, and the confidence that the tag comes from each location.

3.2 M-Step

In this step we want to select θ that maximizes the Q function,

$$Q(\theta|\theta^{(r)}) = E\left[\log g(x|\theta)|y, \theta^{(r)} \right] = \sum_{i=1}^{m} \sum_{j=1}^{k_i} n_{i,j}^{(r)} \left[\log f_i + (j-1) \log(1-p) \right]$$

$$+ N \log p - N \log \left(\sum_{i=1}^{m} f_i \left(1 - (1-p)^{k_i} \right) \right) + \text{constant}$$

Partial derivatives of the Q function are:

$$\frac{\delta Q(\theta|\theta^{(r)})}{\delta f_i} = \frac{1}{f_i} \sum_{j=1}^{k_i} n_{i,j}^{(r)} + N \frac{1 - (1-p)^{k_i}}{\sum_{l=1}^{m} f_l \left(1 - (1-p)^{k_l} \right)}$$

Letting $C = N/(\sum_{l=1}^{m} f_l \left(1 - (1-p)^{k_l} \right))$ and equating partial derivatives to 0 gives

$$\frac{N_i^{(r)}}{f_i} + C \left(1 - (1-p)^{k_i} \right) = 0 \implies f_i = -\frac{N_i^{(r)}}{C \left(1 - (1-p)^{k_i} \right)}$$

Since $\sum_{i=1}^{m} f_i = 1$ it follows that

$$f_i = \frac{N_i^{(r)}}{1 - (1-p)^{k_i}} \left(\sum_{l=1}^{m} \frac{N_l^{(r)}}{1 - (1-p)^{k_l}} \right)^{-1} \tag{3}$$

3.3   Inferring p

In the above calculations we assumed that p is known, which may not be the case in practice. Assuming the geometric distribution of tags to sites, the observed tags of each isoform provide an independent estimate of p [10]. However, the presence of ambiguous tags complicates the estimation of p on an isoform-by-isoform basis. In order to globally capture the value of p we incorporate it in the DGE-EM algorithm as a hidden variable and iteratively re-estimate it as the distribution of tags to isoforms changes from iteration to iteration.

We estimate the value of p as N^1/D, where D denotes the total number of RNA molecules with at least one AE site, and $N^1 = \sum_{i=1}^{m} n_{i1}$ denotes the total number of tags coming from first AE sites. The total number of RNA molecules representing an isoform is computed as the number of tags coming from that isoform divided by the probability that the isoform is cut. This gives $D = \sum_{i=1}^{m} N_i/(1 - (1-p)^{k_i})$, which happens to be the normalization term used in the M step of the algorithm.

3.4   Implementation

For an efficient implementation, we pre-process AE sites in all the known isoform sequences. All tags that can be generated from these sites, assuming no errors, are stored in a trie data structure together with information about their original locations. Searching for a tag is performed by traversing the trie, permitting for as many jumps to neighboring branches as the maximum number of mismatches allowed. The Expectation Maximization part of DGE-EM, which follows after mapping, is given in Algorithm 1 (for simplicity, the re-estimation of p is omitted).

In practice, for performance reasons, tags with the same matching sites and weights are collapsed into one, keeping track of their multiplicity. Then the EM algorithm can process them all at once by factoring in their multiplicity when increasing the n(iso, site) counter. This greatly reduces the running time and memory footprint.

4   Results

4.1   Experimental Setup

We conducted experiments on both real and simulated DGE and RNA-Seq datasets. In addition to estimates obtained by DGE-EM, for DGE data we also computed direct estimates from uniquely mapped tags; we refer to this method

---

Algorithm 1. DGE-EM algorithm

---
assign random values to all $f(i)$
while not converged **do**
  initialize all $\mathbf{n(iso, site)}$ to 0
  for each tag class t **do**
    sum $= \sum_{(\text{iso,site},w) \in t} w \times f(\text{iso}) \times (1-p)^{\text{site}-1}$
    for (iso, site, w) $\in t$ **do**
      n(iso, site)$+ = w \times f(\text{iso}) \times (1-p)^{\text{site}-1}/$sum
    end for
  end for
  for each isoform i **do**
    $N_i = \sum_{j=1}^{\text{sites}(i)} n(i,j)$
    $f(i) = N_i/(1-(1-p)^{\text{sites}(i)})$
  end for
end while

---

as "Uniq". RNA-Seq data was analyzed using both our IsoEM algorithm [6], which was shown to outperform existing methods of isoform and gene expression level estimation, and the well-known Cufflinks algorithm [8]. As in previous works [4,6], estimation accuracy was assessed using the *median percent error (MPE)*, which gives the median value of the relative errors (in percentage) over all genes.

Real DGE datasets included nine libraries kindly provided to us (in fastq format) by the authors of [1]. These libraries were independently prepared and sequenced at multiple sites using 6 flow cells on Illumina Genome Analyzer (GA) I and II platforms, for a total of 35 lanes. The first eight libraries were prepared from the Ambion Human Brain Reference RNA, (Catalog #6050), henceforth referred to as HBRR and the ninth was prepared from the Stratagene Universal Human Reference RNA (Catalog #740000) henceforth referred to as UHRR. *DpnII*, with recognition site GATC, was used as anchoring enzyme and *MmeI* as tagging enzyme, resulting in approximately 238 million tags of length 20 across the 9 libraries. Unless otherwise indicated, Uniq estimates are based on uniquely mapped tags with 0 mismatches (63% of all tags) while for DGE-EM we used all tags mapped with at most 1 mismatch (83% of all tags) since preliminary experiments (Section 4.2) showed that these are the optimal settings for each algorithm.

For comparison, we downloaded from the SRA repository two RNA-Seq datasets for the HBRR sample and six RNA-Seq datasets for the UHRR sample (SRA study SRP001847 [2]). Each RNA-Seq dataset contains between 47 and 92 million reads of length 35. We mapped RNA-Seq reads onto Ensembl known isoforms (version 59) using bowtie [3] after adding a polyA tail of 200 bases to each transcript. Allowing for up to two mismatches, we were able to map between 65% and 72% of the reads. We then ran IsoEM and Cufflinks assuming a mean fragment length of 200 bases with standard deviation 50.

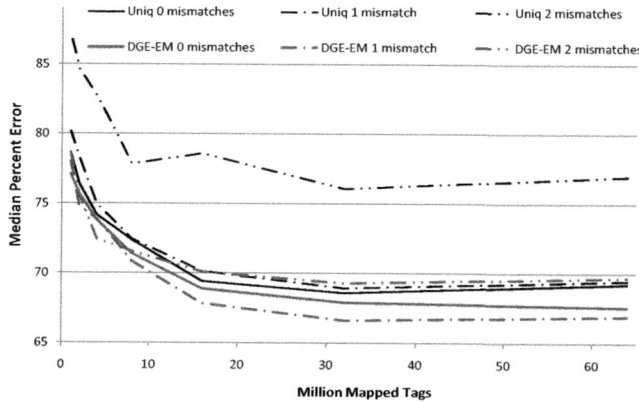

Fig. 3. Median Percent Error of DGE-EM and Uniq estimates for varying number of allowed mismatches and DGE tags generated from the HBRR library 4

To assess accuracy, gene expression levels estimated from real DGE and RNA-Seq datasets were compared against TaqMan qPCR measurements (GEO accession GPL4097) collected by the MicroArray Quality Control Consortium (MAQC). As described in [5], each TaqMan Assay was run in four replicates for each measured gene. POLR2A (ENSEMBL id ENSG00000181222) was chosen as the reference gene and each replicate CT was subtracted from the average POLR2A CT to give the log2 difference (delta CT). For delta CT calculations, a CT value of 35 was used for any replicate that had $CT > 35$. Normalized expression values are reported: $2^{(\text{CT of POLR2A})-(\text{CT of the tested gene})}$. We used the average of the qPCR expression values in the four replicates as the ground truth. After mapping gene names to Ensembl gene IDs using the HUGO Gene Nomenclature Committee (HGNC) database, we got TaqMan qPCR expression levels for 832 Ensembl genes. Expression levels inferred from DGE and RNA-Seq data were similarly divided by the expression level inferred for POLR2A prior to computing accuracy.

Synthetic error-free DGE and RNA-Seq data was generated using an approach similar to that described in [6]. Briefly, the human genome sequence (hg19, NCBI build 37) was downloaded from UCSC and used as reference. We used isoforms in the UCSC KnownGenes table ($n = 77,614$), and defined genes as clusters of known isoforms in the GNFAtlas2 table ($n = 19,625$). We conducted simulations based on gene expression levels for five different tissues in GNFAtlas2. The simulated frequency of isoforms within gene clusters followed a geometric distribution with ratio 0.5. For DGE we simulated data for all restriction enzymes with 4-base long recognition sites from the Restriction Enzyme Database (REBASE), assuming either complete digestion ($p = 1$) or partial digestion with $p = 0.5$. For RNA-Seq we simulated fragments of mean length 250 and standard deviation 25 and simulated polyA tails with uniform length of 250bp. For all simulated data mapping was done without allowing mismatches.

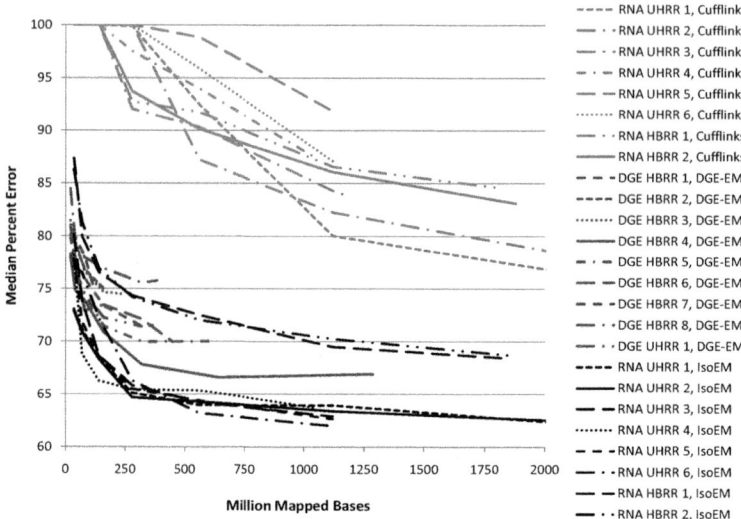

Fig. 4. Median Percent Error of DGE-EM, IsoEM, and Cufflinks estimates from varying amounts of DGE/RNA-Seq data generated from the HBRR MAQC sample

4.2   DGE-EM Outperforms Uniq

The algorithm referred to as Uniq quantifies gene expression based on the number of tags that match one or more cleavage sites in isoforms belonging to the same gene. These tags are unique with respect to the source gene. Figure 3 compares the accuracy of Uniq and DGE-EM on library 4 from the HBRR sample, with the number of allowed mismatches varying between 0 and 2. As expected, counting only perfectly mapped tags gives the best accuracy for Uniq, since with the number of mismatches we increase the ambiguity of the tags, and thus reduce the number of unique ones. When run with 0 mismatches, DGE-EM already outperforms Uniq, but the accuracy improvement is limited by the fact that it cannot tolerate any sequencing errors (tags including errors are either ignored, or, worse, mapped at an incorrect location). Allowing 1 mismatch per tag gives the best accuracy of all compared methods, but further increasing the number of mismatches to 2 leads to accuracy below that achieved when using exact matches only, likely due to the introduction of excessive tag ambiguity for data for which the error rate is well below 10%.

4.3   Comparison of DGE and RNA-Seq Protocols

Figure 4 shows the gene expression estimation accuracy for 9 DGE and 8 RNA-Seq libraries generated from the HBRR and UHRR MAQC sample. All DGE estimates were obtained using the DGE-EM algorithm, while for RNA-Seq data we used both IsoEM [6] and the well-known Cufflinks algorithm [8]. The cutting

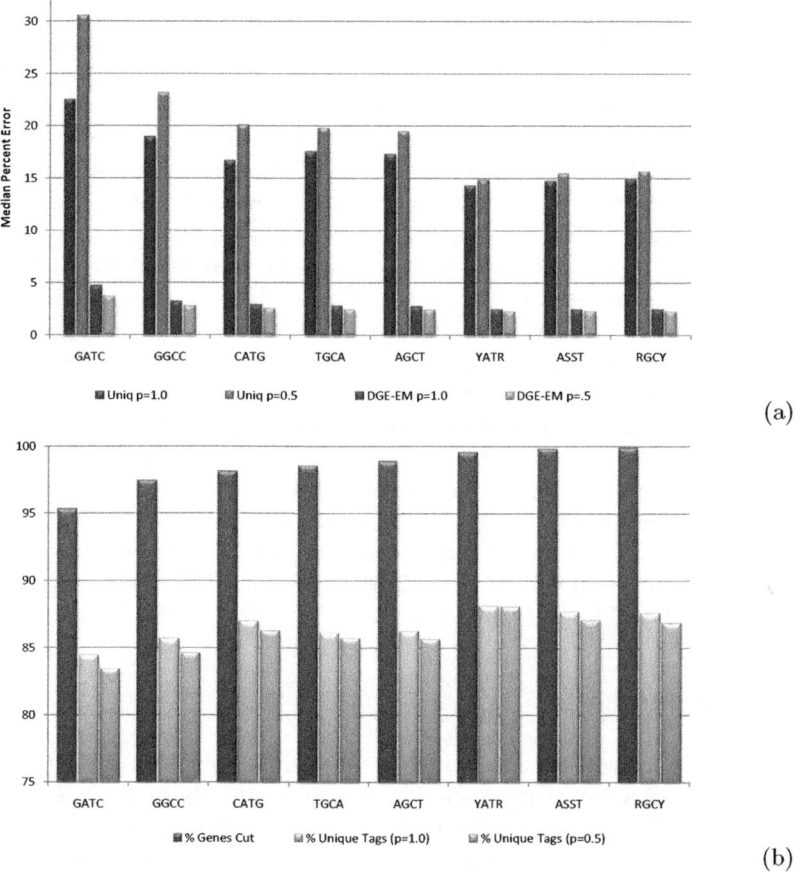

Fig. 5. (a) Median Percent Error of Unique and DGE-EM estimates obtained from 30 million 21bp DGE tags simulated for anchoring enzymes with different restriction sites (averages over 5 GNF-Atlas tissues) (b) Percentage of genes cut and uniquely mapped tags for each anchoring enzyme.

probability inferred by DGE-EM is almost the same for all libraries, with a mean of 0.8837 and standard deviation 0.0049. This is slightly higher than the estimated value of $70 - 80\%$ suggested in the original study [1], possibly due to their discarding of non-uniq or non-perfectly matched tags. Normalized for sequencing cost, DGE performance is comparable to that of RNA-Seq estimates obtained by IsoEM, with accuracy differences between libraries produced using different protocols within the range of library-to-library variability within each of the two protocols. The MPE of estimates generated from RNA-Seq data by Cufflinks is significantly higher than that of IsoEM and DGE-EM estimates, suggesting that accurate analysis methods are at least as important as the sequencing protocol.

4.4   Possible DGE Assay Optimizations

To assess accuracy of DGE estimates under various protocol parameters, we conducted an extensive simulation study where we varied the anchoring enzyme used, the number of tags, the tag length and the cutting probability. We tested all restriction enzymes with 4-base long recognition sites from REBASE. Figure 5(a) gives MPE values obtained by the Unique and DGE-EM algorithms for a subset of these enzymes on synthetic datasets with 30 million tags of length 21, simulated assuming either complete or $p = .5$ partial digest. Figure 5(b) gives the percentage of genes cut and the percentage of uniquely mapped DGE tags for each of these enzymes. These results suggest that using enzymes with high percentage of genes cut leads to improvements in accuracy. In particular, enzymes like NlaIII (previously used in [9]) with recognition site CATG and CviJI with degenerate recognition site RGCY (R=G or A, Y=C or T) cut more genes than the DpnII (GATC) enzyme used to generate the MAQC DGE libraries, and yield better accuracy for both Uniq and DGE-EM estimates. Furthermore, for every anchoring enzyme, partial digestion with $p = .5$ yields an improved DGE-EM accuracy compared to complete digestion. Interestingly, Unique estimates are less accurate for partial digest due to the smaller percentage of uniquely mapped reads. For comparison, IsoEM estimates based on 30 million RNA-Seq tags of length 21 yield an MPE of 8.3.

5   Conclusions

In this paper we introduce a novel expectation-maximization algorithm, called DGE-EM, for inference of gene-specific expression levels from DGE tags. Our algorithm takes into account alternative splicing isoforms and tags that map at multiple locations in the genome within a unified statistical model, and can further correct for incomplete digestion and sequencing errors. Experimental results on both real and simulated data show that DGE-EM outperforms commonly used estimation methods based on unique tag counting. DGE-EM has cost-normalized accuracy comparable to that achieved by state-of-the-art RNA-Seq estimation algorithms on the tested real datasets, and outperforms them on error-free synthetic data. Simulation results suggest that further accuracy improvements can be achieved by tuning DGE protocol parameters such as the degeneracy of the anchoring enzyme and cutting probability. It would be interesting to experimentally test this hypothesis.

Acknowledgment

This work has been supported in part by NSF awards IIS-0546457 and IIS-0916948. The authors wish to thank Yan Asmann for kindly providing us with the DGE data from [1].

References

1. Asmann, Y., Klee, E.W., Thompson, E.A., Perez, E., Middha, S., Oberg, A., Therneau, T., Smith, D., Poland, G., Wieben, E., Kocher, J.-P.: 3' tag digital gene expression profiling of human brain and universal reference RNA using Illumina Genome Analyzer. BMC Genomics 10(1), 531 (2009)
2. Bullard, J., Purdom, E., Hansen, K., Dudoit, S.: Evaluation of statistical methods for normalization and differential expression in mRNA-Seq experiments. BMC Bioinformatics 11(1), 94 (2010)
3. Langmead, B., Trapnell, C., Pop, M., Salzberg, S.: Ultrafast and memory-efficient alignment of short DNA sequences to the human genome. Genome Biology 10(3), R25 (2009)
4. Li, B., Ruotti, V., Stewart, R.M., Thomson, J.A., Dewey, C.N.: RNA-Seq gene expression estimation with read mapping uncertainty. Bioinformatics 26(4), 493–500 (2010)
5. MAQC Consortium: The Microarray Quality Control (MAQC) project shows inter- and intraplatform reproducibility of gene expression measurements. Nature Biotechnology 24(9), 1151–1161 (2006)
6. Nicolae, M., Mangul, S., Măndoiu, I., Zelikovsky, A.: Estimation of Alternative Splicing isoform Frequencies from RNA-Seq Data. In: Moulton, V., Singh, M. (eds.) WABI 2010. LNCS, vol. 6293, pp. 202–214. Springer, Heidelberg (2010)
7. 't Hoen, P.A., Ariyurek, Y., Thygesen, H.H., Vreugdenhil, E., Vossen, R.H., de Menezes, R.X., Boer, J.M., van Ommen, G.-J.J., den Dunnen, J.T.: Deep sequencing-based expression analysis shows major advances in robustness, resolution and inter-lab portability over five microarray platforms. Nucleic Acids Research 36(21), e141 (2008)
8. Trapnell, C., Williams, B.A., Pertea, G., Mortazavi, A., Kwan, G., van Baren, M.J., Salzberg, S.L., Wold, B.J., Pachter, L.: Transcript assembly and quantification by RNA-Seq reveals unannotated transcripts and isoform switching during cell differentiation. Nature Biotechnology 28(5), 511–515 (2010)
9. Wu, Z.J., Meyer, C.A., Choudhury, S., Shipitsin, M., Maruyama, R., Bessarabova, M., Nikolskaya, T., Sukumar, S., Schwartzman, A., Liu, J.S., Polyak, K., Liu, X.S.: Gene expression profiling of human breast tissue samples using SAGE-Seq. Genome Research 20(12), 1730–1739 (2010)
10. Zaretzki, R., Gilchrist, M., Briggs, W., Armagan, A.: Bias correction and Bayesian analysis of aggregate counts in SAGE libraries. BMC Bioinformatics 11(1), 72 (2010)

An Integrative Approach for Genomic Island Prediction in Prokaryotic Genomes

Han Wang[1], John Fazekas[1], Matthew Booth[1], Qi Liu[2,*], and Dongsheng Che[1,*]

[1] Department of Computer Science, East Stroudsburg University,
East Stroudsburg, PA 18301, USA
dche@po-box.esu.edu
[2] College of Life Science and Biotechnology, Tongji University,
Shanghai, 200092, P.R. China
qiliu@tongji.edu.cn

Abstract. A *genomic island* (GI) is a segment of genomic sequence that is horizontally transferred from other genomes. The detection of genomic islands is extremely important to the medical research. Most of current computational approaches that use sequence composition to predict genomic islands have the problem of low prediction accuracy. In this paper, we report, for the first time, that gene information and inter-genic distance are different between genomic islands and non-genomic islands. Using these two sources and sequence information, we have trained the genomic island datasets from 113 genomes, and developed a decision-tree based bagging model for genomic island prediction. In order to test the performance our approach, we have applied it on three genomes: *Salmonella typhimurium* LT2, *Streptococcus pyogenes* MGAS315, and *Escherichia coli* O157:H7 str. Sakai. The performance metrics have shown that our approach is better than other sequence composition based approaches. We conclude that the incorporation of gene information and intergenic distance could improve genomic island prediction accuracy. Our prediction software, Genomic Island Hunter (GIHunter), is available at http://www.esu.edu/cpsc/che_lab/software/GIHunter.

Keywords: Genomic islands, gene information, intergenic distance, sequence composition.

1 Introduction

Genomic islands (GIs) are chromosomal sequences in some bacterial genomes, whose origins can be either from viruses or other bacteria [1]. The studies of GIs are very important to biomedical research, due to the fact that such knowledge can be used to explain why some strains of bacteria within the same species are pathogenic while others are not [2], or the phenomena that some strains of bacteria can adapt to extreme environments while others cannot. Therefore, it is urgent to develop computational tools to detect GIs in bacterial genomes.

---

* Corresponding authors.

J. Chen, J. Wang, and A. Zelikovsky (Eds.): ISBRA 2011, LNBI 6674, pp. 404–415, 2011.
© Springer-Verlag Berlin Heidelberg 2011

In general, GIs do have sequence compositional biases such as G + C content difference [3], codon usage bias [4,5], oligo-nucleotide frequency bias [6], when compared with the remaining sequences in the host genome. Some GIs also contain insertion sequences [7], flanking tRNA genes [1] and mobile genes such as integrase and transposes [8]. Despite of strutural features, GIs are difficult to be characterized as not all GIs have such features, and the sizes of GIs can range from 5 kb to 500 kb.

Current practices of detecting GIs include comparative genomic analysis and sequence composition approaches [9]. A typical procedure of using comparative genome analysis consists of collecting the genome sequences of phylogenetically closely related species, aligning these genome sequences, and then considering those genome segments present in a query genome but not present in others to be GIs [10]. The tools using comprative genome analysis include MobilomeFinder [11] and IslandPick [10]. While this type of approach is very reliable, it is limited to those genomes which have closely related genomes as reference genomes, and sometimes it needs manual involvement [12].

Sequence composition based approach, on the other hand, does not require reference genomes and manual adjustment. It is generally believed that each genome has an unique genomic sequence signature. Thus, GIs that are integrated into the host genome can be detected by analyzing the sequence composition. The sequence compositional biases can be evaluated by analyzing G + C content, k-mer frequency, codon adaption index (CAI). Such sequence composition based tools include AlienHunter [13], Centroid [14], IslandPath [15], PAI-IDA [16] and SIGI-HMM [17].

While sequence composition based approaches can be used to predict GIs for any prokaryotic genome, the prediction accuracy is typically low. This is due to the amelioration of the genome [4]. After several generations that GIs are integrated into the host genome, GIs are subject to the same mutational process as those of the remaining sequences in the host genome, and thus making it difficult to differentiate GIs and host genome sequences. On the other hand, some parts of host genome sequences, such as those involved in ribosomal activity [6], are different from the rest of the host genomes in terms of sequence composition. Therefore, it may not be sufficient to use the sequence composition bias to determine a sequence is a GI or not, and it is necessary to incorporate more GI-associated features for accurate GI detection.

In this paper, we report two features, gene information and intergenic distance that are useful for accurate GI prediction. We show that the gene information and intergenic distance are different between GIs and non-GIs, and they can be combined with sequence information for more accurate GI prediction. We present our computational framework that integrates three sources of data, and build an ensemble of decision tree models for genome scale GI prediction. We show our decision tree based bagging model performs better than other sequence composition based approaches in terms of prediction accuracy and other performance metrics. The remainder of the paper is organized as follows: Section 2 describes the Materials and Methods. Section 3 analyzes the features, as well as prediction results of our approach. The paper is concluded in Section 4.

2   Materials and Methods

2.1   Dataset

The GIs and non-GIs datasets that we used in this study were obtained from [10]. The dataset contains 713 GIs and 3,517 non-GIs, covering 113 genomes in total. The whole list of these 113 genomes can be found in Supplementary Data (`http://www.esu.edu/cpsc/che_lab/software/GIHunter/SupplementaryData.pdf`).

The sequence lengths of original GIs and non-GIs vary from 8 kb to 31 kb, making it less meaningful to compare the associated feature values among GIs (or non-GIs) with different lengths, and thus making the prediction of genomic island difficult. In order to simplify this problem, we split each whole genomic island (or non-genomic island) into the same sizes of segments. We chose the segment of 8 kb in this study because this is smallest genomic island in the dataset. This splitting process led to 2,232 GI segments and 8,525 non-GI segments.

In order to obtain the feature values for GIs and non-GIs, we also downloaded the corresponding complete genome sequences and the annotations from the National Center for Biotechnology Information (NCBI) FTP server (`fpt::://ftp.ncbi.nih.gov/genomes/Bacteria`).

2.2   Computational Framework

Our computational framework for genomic island prediction consists of the following steps:

1. Extract genomic island-related features. The feature values corresponding to GI segments and non-GI segments are calculated (See Section 2.3).
2. Build the genomic island model. The feature values obtained in Step 1 are used to train a general genomic island model (See Section 2.4).
3. Predict genomic islands using the trained model. The model is used to scan the whole genome, and the genomic island regions in the whole genome can be predicted (See Section 2.5).

We describe the details for each of these steps as follows.

2.3   Feature Extraction

Sequence composition. The sequence composition feature values for each GI segment (or non-GI segment) are evaluated by AlienHunter [13]. AlienHunter evaluates compositional biases of genomic sequences by using Interpolated Variable Order Motifs (IVOMs), and it has been reported to capture the local composition of a sequence accurately [10]. AlienHunter accepts the whole genome sequence as an input, and calculates the IVOM scores for each sequence segment. Generally speaking, the higher the IVOM score, the more GI segment the genomic sequence looks like.

Gene information. Karlin *et al.* [5] discovered that the codon frequencies of highly expressed genes deviated significantly from those of average genes of the genome. Thus, gene information such as highly expressed genes, when combined with sequence information, might be useful to differentiate a genome segment is GI or non-GI. To test this feature, we enumerate the number of highly expressed genes within the segments of GIs and non-GIs. In this study, we consider ribosomal protein (RP) genes, translation and transcriptional processing factor (TF) genes (e.g., RNA helicase, RNA polymerase, tRNA synthetase), chaperone-degradation (CH) genes, and genes involved in energy metabolism (e.g., NADH) to be highly expressed genes [18].

Intergenic distance. The *intergenic distance* between two adjacent genes is the number of base pairs between them. The intergenic distance reflects the transcriptional or functional relatedness between them to some degree. For example, the genes under the same transcription unit tend to stay together, and thus the intergenic distance between them is usually short. As genomic islands involve the integration process of alien genes into the recipient genome, we hypothesize that the distribution of intergenic distances of genomic islands is different from that of non-genomic islands.

Let N be the number of all genes within the genomic segment, and $id(g_i, g_{i+1})$ be the intergenic distance between adjacent genes g_i and g_{i+1}, then the average intergenic distance within a GI segment (or non-GI segment), $avgID$, can be calculated as follows,

$$avgID = \frac{1}{N-1} \sum_{i=1}^{N-1} id(g_i, g_{i+1}) \tag{1}$$

2.4   Model Construction

The genomic island model is a bootstrap aggregating (bagging) [19] of base classifiers, in which each base classifier is a decision tree in this study. We chose this model since it has proven to perform better than other machine learning methods for GI classification in our previous study [20]. The decision trees are constructed based on training examples. The training examples are a set of tuples <x, c>, where c is the class label, and x is the set of features. In this study, the training examples include 2,232 GI segments and 8,525 non-GI segments, x is the set of features: sequence composition, gene information and intergenic distance, and c is the label of GI or non-GI. The architecture of decision tree bagging model for genomic island is sketched in Fig. 1. The construction of decision trees, as well as bagging, is described as follows.

Decision tree. The decision trees built are based on the ID3 algorithm [21], which implements a top-down greedy search schema to search through all possible tree spaces. It starts with all training set (S) and chooses the best feature as the root node. It then splits the set based on the possible values of the selected best feature. If the all instances in a subset have the same classification, then the

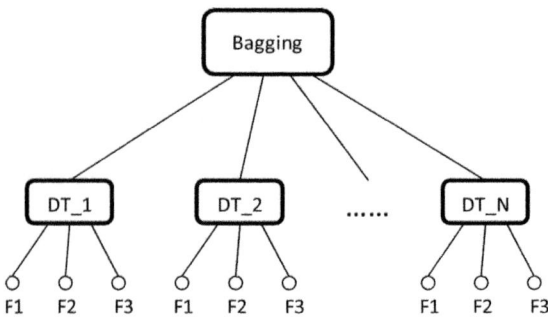

Fig. 1. Architecture of decision-tree based bagging. F1, F2, and F3 are the features used for constructing decision-tree classifiers. The features used in this study are sequence composition, gene information and intergenic distance. DT_1, DT_2, and DT_N are decision tree classifiers.

algorithm assigns the node to a leaf node, with the label of this classification. Otherwise, and the next best feature is chosen again based on the subset of the training examples. The whole process will be repeated until there are no further distinguishing features.

The best feature is the one that has the highest *information gain (IG), i.e.,*

$$F^* = \arg\max_{F} IG(S, F) \tag{2}$$

where IG is defined as follows,

$$IG(S, F) = E(S) - \sum_{v \in Value(F)} \frac{|S_v|}{S} E(S_v) \tag{3}$$

$E(S)$ is the entropy of S, defined as

$$E(S) = -p_{GI} log_2 p_{GI} - p_{NG} log_2 p_{NG} \tag{4}$$

where p_{GI} is the percentage of GIs in S, and conversely, p_{NG} is the percentage of non-GIs in S. $Value(F)$ is the set of all possible values for the feature F. S_v is the subset of S in which feature F has the value of v (*i.e.,* $S_v = \{s \in S | F(s) = v\}$).

Bagging. The Bagging model consists of N decision trees as base classifiers. The training set used for constructing each tree is sampled by bootstrap sampling, *i.e.,* randomly selecting a subset of given dataset with replacement. The final classification of bagging takes the majority votes of all decision tree classifications, *i.e.,*

$$H(x) = sgn(\sum_{i}^{N} H_i(x)) \tag{5}$$

where $H_i(x)$ is the ith classification result.

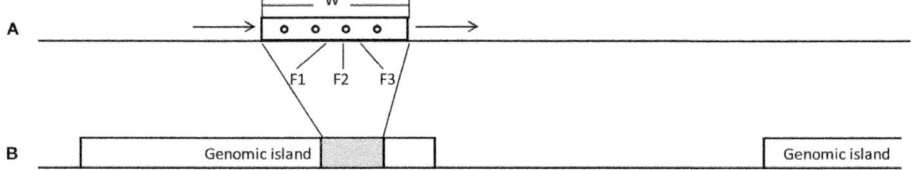

Fig. 2. A schematic view of the genome scale genomic island prediction. (A). A sliding-window based approach for predicting genomic island. The feature values (F1, F2, and F3) within the window are extracted and used for genomic island classification, based on our genomic island model trained. (B). The resulting genomic islands predicted. The contiguous segments classified as GI segment from (A) are considered to be one genomic island.

2.5   Genome-Scale Genomic Island Prediction

The whole genome scale genomic island prediction for any prokaryotic genome is based on our decision-tree based bagging model built upon the training sets of 113 genomes. The inputs for the GI prediction are: (a) the whole genome sequence; and (b) the gene annotation of the genome. The schematic view for the GI prediction is represented in Fig. 2, and the whole procedure can be described as follows:

1. Slide a window of appropriate size (8 kb in this study) along the query genome. We chose the window size of 8 kb since 8kb-long sequence segment was the smallest GI found so far. For those longer GIs, we concatenate them in Step 4.
2. Calculate the feature values (IVOM score, HEG, and AverageID) for each window.
3. Classify the segment for each window we scan using our decision-tree based bagging model. The resulting value is either a GI segment or non-GI segment;
4. Postprocess the classified ones from Step 3. If we see several contiguous GI segments, we treat them as one big GI.

2.6   Performance Evaluation

To evaluate the performance our model, we compared the predicted GIs with the benchmark dataset from the literature. True positives (TP) are the nucleotides in the positive dataset predicted to be genomic islands. True negatives (TN) are the nucleotides in the negative dataset predicted to be non-genomic islands. False positives (FP) are the nucleotides in the negative dataset predicted to genomic islands. False negatives (FN) are the nucleotides within the positive dataset not predicted to be genomic islands. We focus on three validation measures, recall, precision and accuracy, which are defined as follows,

$$Recall = \frac{TP}{TP + FN} \tag{6}$$

$$Precision = \frac{TP}{TP + FP} \tag{7}$$

$$Accuracy = \frac{TP + TN}{TP + TN + FP + FN} \tag{8}$$

3   Experimental Results

3.1   Feature Analysis

Analysis of IVOM scores. We used the software, AlienHunter, to generate the IVOM scores of GI segments and non-GI segments for each of 113 genomes. For the purpose of illustration, we chose four genomes, *Bradyrhizobium* sp. BTAi1, *Pseudomonas syringae* pv. phaseolicola 1448A, *Pseudomonas putida* GB-1, and *Burkholderia cenocepacia* MC0-3, to discuss feature value differences between GI and non-GI datasets. We found the IVOM scores for non-GIs tend to be small while those of GIs tend to large (Fig. 3). The average IVOM scores for non-GIs of the genomes of *B.* sp. BTAi1, *P. syringae* pv. phaseolicola 1448A, *P. putida* GB-1, and *B. cenocepacia* MC0-3 are 8.352, 6.918, 9.291 and 8.298, while the average IVOM scores for corresponding GIs are 22.066, 33.654, 42.866, and 24.466.

Analysis of gene information. For each GI segment and non-GI segment, we analyzed the distribution of number of highly expressed genes (HEG). For the genome of *B.* sp. BTAi1, GI segments either do not contain highly expressed genes or at most one (Fig. 3). The distribution of the numbers of highly expressed genes for non-GIs seems to be wide, even though most of them contain no or a

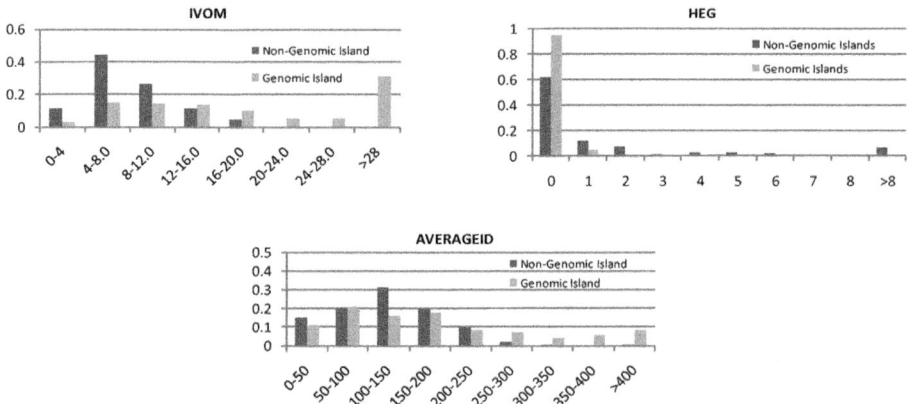

Fig. 3. Feature distribution comparison between GI segments and non-GI segments in *Bradyrhizobium* sp. BTAi1. The values in *x*-axis represent the IVOM scores, the number of HEGs, and the average intergenic distance among genes, and the values in *y*-axis represent the percentage for each range of values in *x*-axis.

few highly expressed genes. Similar distribution difference patterns between GIs and non-GIs are shown in other three genomes (http://www.esu.edu/cpsc/che_lab/software/GIHunter/SupplementaryData.pdf).

The gene information distribution difference between GIs and non-GIs can aid in classifying GIs. For instance, in the genome of *B*. sp. BTAi1, the IVOM score for the segment between position 5282656 and 5290656 is 19.76, dramatically different from that of the typical host genome. Thus, this region is predicted by AlienHunter to be a GI. However, the gene information shows that this region contains 14 ribosomal related genes, and this feature can be used to correct the classification by AlienHunter.

Analysis of intergenic distances. We also compared the average inter-genic distance for GI segments and non-GI segments. We found that the average intergenic distance for non-GIs, in general, tends to be short. For the genome of *B*. sp. BTAi1, there are only a few cases that the average intergenic distances for non-genomic islands are greater than 250 (Fig. 3). In contrast, we found that about 30% genomic island segments contain genes whose average intergenic distance are greater than 250 (Fig. 3). The intergenic distance distribution difference between GIs and non-GIs can also be seen in other three genomes shown in Supplementary Data.

While the mechanism of why the genes within GIs are widely spaced remains to be investigated, the average intergenic distance distribution difference between GIs and non-GIs seems to be able to discriminate GIs from non-GIs, when combining other features such as sequence information.

3.2   GI Structural Model

We have used the training dataset of 113 genomes, extracted three feature values (IVOM score, HEG and AverageID) for each training example, and generated the decision tree based bagging model described in Section 2 for genomic island prediction.

To evaluate the trained model, we measured the ROC (Receiver Operating Characteristic) curve, a graphic representation of tradeoff between sensitivity and specificity. The area under the curves (AUC) takes the value of between 0 and 1, and a random classifier has the AUC vlaue of 0.5. Theoretically, a well-performing classifier should have a high AUC value. The AUC value for our model is 0.8927, indicating that our model obtains high classification accuracy.

For the purpose of comparison, we also plotted ROC curves for the decision-tree based bagging models that used the sequence feature only (*i.e.*, IVOM), and two feature combinations, IVOM + AverageID and IVOM + HEG. The AUC value for the IVOM model is 0.783, and the AUC values for the IVOM + AverageID and IVOM + HEG models are 0.837 and 0.829 (Fig. 4). These AUC values strongly indicate that both intergenic distance (AverageID) and gene information (HEG) improve model accuracies for genomic island prediction, and the combination of all three features generates the most accurate model.

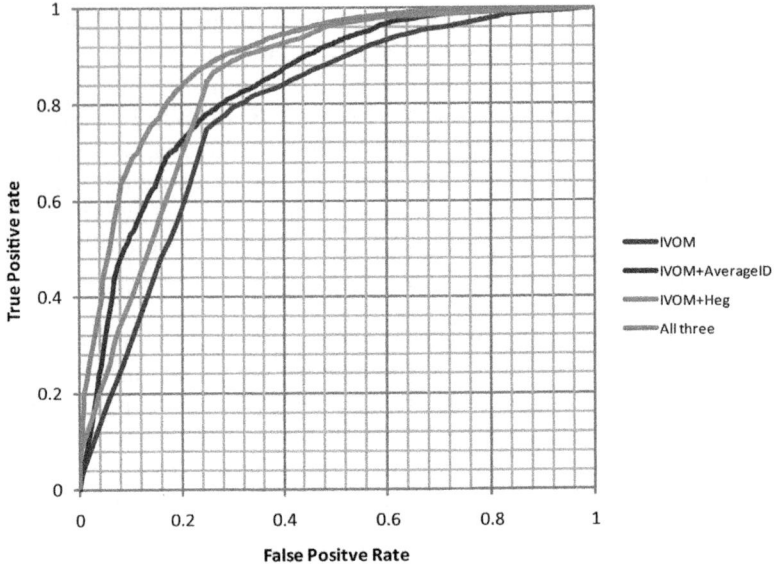

Fig. 4. The ROC curves of decision tree based bagging models. The legend of "All three" indicates all three features are combined to build the model.

3.3   Prediction Accuracy

To test the performance of our approach, we searched the literature, collected the genomic islands of three genomes, *Salmonella typhimurium* LT2 [24], *Streptococcus pyogenes* MGAS315 [23], and *Escherichia coli* O157:H7 str. Sakai [22], and compared these published ones with the ones predicted by our program GI-Hunter. It should be noted that these three genomes selected for performance testing were not included in 113 genomes trained in our model, so that the performance on these three genomes should be applicable to other genomes in general.

In the genomes of *S. typhimurium* LT2 and *E. coli* O157:H7 str. Sakai, our approach has the precisions of 100%. The corresponding recalls are 37.8% and 53.4%. In the genome of *S. pyogenes* MGAS315, our approach has the precision of 42.1%, and the recall of 75.1%. The overall accuracies for *S. typhimurium* LT2, *S. pyogenes* MGAS315 and *E. coli* O157:H7 str. Sakai are 91.5%, 93.3% and 80.4%.

For the comparison purpose, we also evaluated other sequence composition based tools, including SIGI-HMM [17], Centroid [14], PAI-IDA [16], IslandPath[1] [15], and AlienHunter [13]. As shown in Fig. 5, Centroid and PAI-IDA cannot detect any genomic island in the genomes of *S. typhimurium* LT2 and *S. pyogenes* MGAS315. IslandPath cannot detect any genomic island in the genome of

---

[1] We chose the dinucleotide and mobile gene information in IslandPath.

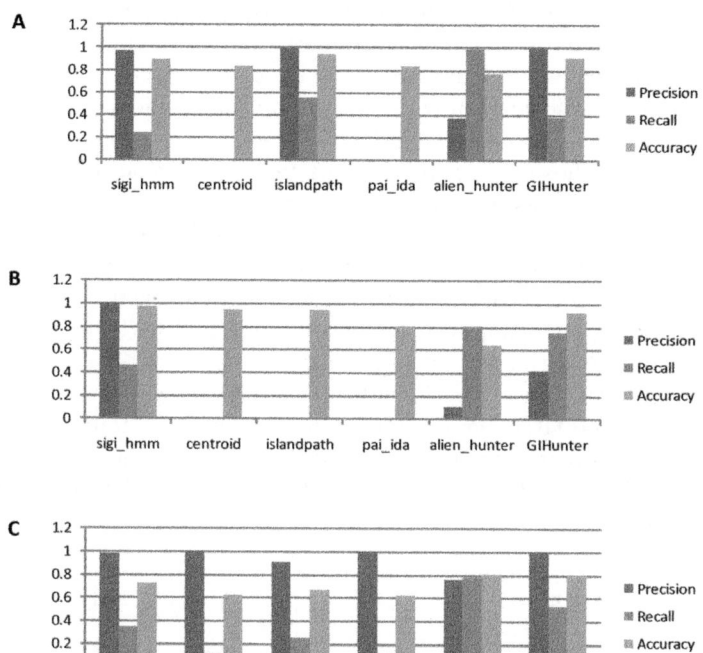

Fig. 5. Precision, recall and accuracy of GI computational tools on the genomes of (A) *Salmonella typhimurium* LT2; (B) *Streptococcus pyogenes* MGAS315; and (C) *Escherichia coli* O157:H7 str. Sakai

S. pyogenes MGAS315. This indicates that these three approaches are relative conservative in GI detection, which was also reported in previous studies [10].

When we compared our program GIHunter with SIGI-HMM, we found that GIHunter performs better than SIGI-HMM in the genomes of *S. typhimurium* LT2 and *E. coli* O157:H7 str. Sakai with all three metrics. In the genome of *S. pyogenes* MGAS315, our program GIHunter has lower precision but with higher recall when compared with SIGI-HMM. Finally, when compared with Alien-Hunter, GIHunter dramatically increase precision and accuracy, but with losing some recall. Overall, our program GIHunter performs better than other sequence composition methods on the tests of these three genomes, further indicating that multiple sources could improve prediction accuracy.

4 Conclusion and Discussion

In this paper, we have analyzed the gene information and intergenic distance for GIs and non-GIs, and reported the distribution differences of these features between GIs and non-GIs. By combining these two features into sequence composition, we have developed an ensemble of decision tree model for genomic island prediction. Our program, GIHunter, based on the trained model, has shown to be more accurate than other sequence composition based approaches in general.

The better performance of our approach over previous ones is due to the incorporation of gene information (*i.e.*, HEG) and intergenic distance into the sequence composition information, which can be seen clearly in their ROC curves. We must be aware, however, using a single feature of gene information for GI prediction may not be as effective as sequence composition information. The reason is most of 8kb long sequence segments, including both GI segments and non-GI segments, do not contain any HEG genes. In this perspective, the gene information feature is a complement to the feature of sequence composition, and could improve the prediction accuracy when combined with sequence composition. This is similar to the feature of inter-genic distance.

In our future study, we will include other GI-associated features, such as integrase, phages and tRNA genes to see whether we could improve our genomic island model. We hope the incorporation of these GI-associated features, as well as the ones we discovered in this study, will make the model more accurate for genomic island prediction.

Acknowledgment

This research was partially supported by President Research Fund (3012101113 and 3012101100) at East Stroudsburg University of Pennsylvania, and Shanghai White Magnolia Talent Fund (2010B127), China.

References

1. Hacker, J., Kaper, J.B.: Pathogenicity islands and the evolution of microbes. Annu. Rev. Microbiol. 54, 641–679 (2000)
2. Hacker, J., Bender, L., Ott, M., et al.: Deletions of chromosomal regions coding for fimbriae and hemolysins occur in vitro and in vivo in various extraintestinal Escherichia coli isolates. Microb. Pathog. 8(3), 213–225 (1990)
3. Hacker, J., Blum-Oehler, G., Muhldorfer, I., et al.: Pathogenicity islands of virulent bacteria: structure, function and impact on microbial evolution. Mol. Microbiol. 23(6), 1089–1097 (1997)
4. Lawrence, J.G., Ochman, H.: Amelioration of bacterial genomes: rates of change and exchange. J. Mol. Evol. 44(4), 383–397 (1997)
5. Karlin, S., Mrazek, J., Campbell, A.M.: Codon usages in different gene classes of the Escherichia coli genome. Mol. Microbiol. 29(6), 1341–1355 (1998)
6. Karlin, S.: Detecting anomalous gene clusters and pathogenicity islands in diverse bacterial genomes. Trends Microbiol. 9(7), 335–343 (2001)
7. Hensel, M.: Genome-based identification and molecular analyses of pathogenicity islands and genomic islands in *Salmonella* enterica. Methods Mol. Biol. 394, 77–88 (2007)
8. Cheetham, B.F., Katz, M.E.: A role for bacteriophages in the evolution and transfer of bacterial virulence determinants. Mol. Microbiol. 18(2), 201–208 (1995)
9. Langille, M.G., Hsiao, W.W., Brinkman, F.S.: Detection of genomic islands using bioinformatics approaches. Nature Reviews Microbiology 8(5), 373–382 (2010)

10. Langille, M.G., Hsiao, W.W., Brinkman, F.S.: Evaluation of genomic island predictors using a comparative genomics approach. BMC Bioinformatics 9, 329 (2008)
11. Ou, H.Y., He, X., Harrison, E.M., et al.: MobilomeFINDER: web-based tools for in silico and experimental discovery of bacterial genomic islands. Nucleic Acids Res. 35, W97–W104 (2007)
12. Vernikos, G.S., Parkhill, J.: Resolving the structural features of genomic islands: a machine learning approach. Genome Res. 18(2), 331–342 (2008)
13. Vernikos, G.S., Parkhill, J.: Interpolated variable order motifs for identification of horizontally acquired DNA: revisiting the *Salmonella* pathogenicity islands. Bioinformatics 22(18), 2196–2203 (2006)
14. Rajan, I., Aravamuthan, S., Mande, S.S.: Identification of compositionally distinct regions in genomes using the centroid method. Bioinformatics 23(20), 2672–2677 (2007)
15. Hsiao, W., Wan, I., Jones, S.J., et al.: IslandPath: aiding detection of genomic islands in prokaryotes. Bioinformatics 19(3), 418–420 (2003)
16. Tu, Q., Ding, D.: Detecting pathogenicity islands and anomalous gene clusters by iterative discriminant analysis. FEMS Microbiology Letters 221, 269–275 (2003)
17. Waack, S., Keller, O., Oliver, A., et al.: Score-based prediction of genomic islands in prokaryotic genomes using hidden Markov models. BMC Bioinformatics 7(1), 142 (2006)
18. Karlin, S., Mrazek, J.: Predicted highly expressed genes of diverse prokaryotic genomes. J. Bacteriology 182(18), 5238–5250 (2000)
19. Brieman, L.: Bagging Predictors. Machine Learning 24, 123–140 (1996)
20. Che, D., Hockenbury, C., Marmelstein, R., Rasheed, K.: Classification of genomic islands using decision trees and their ensemble algorithms. BMC Genomics 11(Suppl 2), S1 (2010)
21. Quinlan, J.R.: C4.5 Programs for Machine Learning. Morgan Kaufmann Publishers, San Mateo (1993)
22. Perna, N.T., Plunkett, G., Burland, V., et al.: Complete genome Sequence of Enterohaemorrhagic Escherichia coli O157:H7. Nature 409, 529–533 (2001)
23. Beres, S.B., Sylva, G.L., Barbian, K.D., et al.: Genome Sequence of a serotype M3 strain of group A Sreptococcus: Phage-encoded toxins, the high-virulence phenotype, and clone emergence. Proceedings of National Academy of Science 99, 10078–10083 (2002)
24. McClelland, M., Sanderson, K.E., Spieth, J., et al.: Complete genome Squence of Salmonella enterica serovar Typhimurium LT2. Nature 413, 852–856 (2001)

A Systematic Comparison of Genome Scale Clustering Algorithms
(Extended Abstract)

Jeremy J. Jay[1], John D. Eblen[2], Yun Zhang[2], Mikael Benson[3],
Andy D. Perkins[4], Arnold M. Saxton[2], Brynn H. Voy[2],
Elissa J. Chesler[1], and Michael A. Langston[2,*]

[1] The Jackson Laboratory, Bar Harbor ME 04609, USA
[2] University of Tennessee, Knoxville TN 37995, USA
[3] University of Göteborg, SE40530 Göteborg, Sweden
[4] Mississippi State University, Mississippi State MS 39762, USA
langston@cs.utk.edu

Abstract. A wealth of clustering algorithms has been applied to gene co-expression experiments. These algorithms cover a broad array of approaches, from conventional techniques such as k-means and hierarchical clustering, to graphical approaches such as k-clique communities, weighted gene co-expression networks (WGCNA) and paraclique. Comparison of these methods to evaluate their relative effectiveness provides guidance to algorithm selection, development and implementation. Most prior work on comparative clustering evaluation has focused on parametric methods. Graph theoretical methods are recent additions to the tool set for the global analysis and decomposition of microarray data that have not generally been included in earlier methodological comparisons. In the present study, a variety of parametric and graph theoretical clustering algorithms are compared using well-characterized transcriptomic data at a genome scale from *Saccharomyces cerevisiae*.Clusters are scored using Jaccard similarity coefficients for the analysis of the positive match of clusters to known pathways. This produces a readily interpretable ranking of the relative effectiveness of clustering on the genes. Validation of clusters against known gene classifications demonstrate that for this data, graph-based techniques outperform conventional clustering approaches, suggesting that further development and application of combinatorial strategies is warranted.

ISBRA Topics of Interest: gene expression analysis, software tools and applications.

1 Background

Effective algorithms for mining genome-scale biological data are in high demand. In the analysis of transcriptomic data, many approaches for identifying clusters of genes with similar expression patterns have been used, with new techniques frequently

---

* Corresponding author.

J. Chen, J. Wang, and A. Zelikovsky (Eds.): ISBRA 2011, LNBI 6674, pp. 416–427, 2011.
© Springer-Verlag Berlin Heidelberg 2011

being developed (for reviews see [1, 2]). Many bench biologists become mired in the challenge of applying multiple methods and synthesizing or selecting among the results. Such a practice can lead to biased selection of "best" results based on preconceptions of valid findings from known information, which begs the question of why the experiment was performed. Given the great diversity of clustering techniques available, a systematic comparison of algorithms can help identify the relative merits of different techniques [3, 4]. Previous reviews and comparisons of clustering methods have often concluded that the methods do differ, but offer no consensus as to which methods are best [3, 5-9]. In this paper, we compare a broad spectrum of conventional, machine-learning, and graph-theoretic clustering algorithms applied to a high quality, widely used reference data set from yeast.

A popular and diverse set of clustering approaches that have readily available implementations were employed in this analysis (Supplementary Table 1). These include five traditional approaches: k-means clustering [10], and the de facto standard hierarchical clustering, on which we tested four agglomeration strategies: average linkage, complete linkage, and the methods due to McQuitty [11] and Ward [12]. These approaches create clusters by grouping genes with high similarity measures together. Seven graph-based approaches are examined: k-clique communities [13], WGCNA [14], NNN [15], CAST [16], CLICK [17], maximal clique [18-20], and paraclique [21]. These methods use a graph approach, with genes as nodes and edges between genes defined based on a similarity measure. Finally, two other approaches are included: self-organizing maps [22], and QT Clust [23]. SOM is a machine learning approach that groups genes using neural networks. QT Clust is a method developed specifically for expression data. It builds a cluster for each gene in the input, outputs the largest, then removes its genes and repeats the process until none are left.

Many issues influence the selection and tuning of clustering algorithms for gene expression data. First, genes can either be allowed to belong to only one cluster or be included in many clusters. Non-disjoint clustering conforms more accurately to the nature of biological systems, but at a cost of creating hundreds to thousands of clusters. Second, because each method has its own set of parameters for controlling the clustering process, one must determine the ideal parameter settings in practice. There are many different metrics for this problem, which have been evaluated extensively [6, 24]. Because there is no way of measuring bias of one metric for a particular clustering method and data set, most clustering comparisons evenly sample the reasonable parameter space for each method [7, 25].

Metrics for comparing clusters can be categorized into two types: internal and external [26]. Internal metrics are based on properties of the input data or cluster output, and are useful in determining how or why a clustering method performs as it does. It provides a data-subjective interpretation that is typically only relevant to a single experimental context. Examples of internal metrics include average correlation value, Figure of Merit (FOM) [27], or diameter[23] which are difficult to compare. External metrics, on the other hand, provide an objective measure of the clusters based on data not used in the clustering process, such as biological annotation data. An external metric does not depend on the experimental context that produced it. Such metrics enable a comparison of the relative merits of these algorithms based on performance in a typical biological study, and can be compared regardless of the annotation source.

External metrics have been used in many previous studies of clustering performance. Some comparisons use receiver operating characteristic (ROC) or precision-recall curves [5]. These metrics are simple to calculate, but they provide too many dimensions (two per cluster) for a straightforward comparison of the overall performance of the methods. Many other studies [7, 25, 27, 28] have used the Rand Index [29], which generates a single value to measure similarity of clustering results to a known categorization scheme such as GO annotation. Rand, however, it is subject to many sources of bias, including a high number of expected negatives typically confirmed when comparing clustering results to categorized gene annotations [5, 26]. The Jaccard similarity coefficient ignores true negatives in its calculation, resulting in a measure less dominated by the size of the reference data, particularly in the large number of true negatives that are often confirmed. This idea has been raised in early work using the Rand Index and other partition similarity measures [30] and in the context of comparing clustering algorithms [1]. The Jaccard coefficient has not been widely adopted, due in part to the historical sparseness of annotations in reference sets for comparison. With deep ontological annotation now more widely available, however, external metrics such as the Jaccard coefficient provide a much more relevant, objective, and simplified basis for comparison.

A variety of tools are available to calculate functional enrichment of biological clusters [31-33]. Most are not suitable for high-throughput genome-scale analysis due to interface, speed or scalability limitations. (They often cannot easily handle the large number of clusters produced from whole genome clustering.) Several of the tools are meant exclusively for Gene Ontology terms, precluding the use of the large variety of publicly available annotation sources.

In the present study, we perform an evaluation of both combinatorial and conventional clustering analyses using an evenly distributed parameter set and biological validation performed using Jaccard similarity analysis of KEGG and GO functional gene annotations. For this analysis we developed a new enrichment analysis tool, specifically designed to handle genome-scale data (and larger) and any gene category annotation source provided. Results of clustering algorithms were compared across all parameters and also in a manner that simulates use in practice by selection of the optima generated from each method.

2   Results

For each clustering method and parameter, clusters of different sizes were obtained. Because there is an exponential distribution of annotation category sizes, matches among small categories are more readily detectable. We binned these results into three size categories (3-10, 11-100, and 101-1000 genes), and ranked the clusters based on Jaccard similarity scores. In practice, users generally are only interested in the few highest-scoring clusters. In many biological studies, only the top 5 to 10 clusters are scrutinized. Data-torturing is not uncommon in microarray studies due to the wealth of tools available, and in practice, some individuals may perform clustering until a satisfactory result is found. To simulate this practice, we therefore focused on the top five cluster scores for each size grouping (i.e., those with highest Jaccard similarity to annotations), whether derived from match to GO or KEGG annotations, and computed their average score (Average Top 5 or AT5). We chose AT5 as a comparison score because most of the methods produce at least five clusters of each size bin, but for some of the methods, cluster scores drop off quickly after these top five results, making a larger average meaningless. It is also significant that in practice users often adjust parameter settings to improve clusters. Accordingly, for each choice of method and cluster size category, we chose the highest AT5 values across all parameter settings (Best Average Top 5 or BAT5) for that method-size combination. These values are reported in Figure 1. It should be noted that for AT5 and BAT5, maximal clique, like any method that allows non-disjoint cluster membership, creates bias in this score by including results from similar clusters.

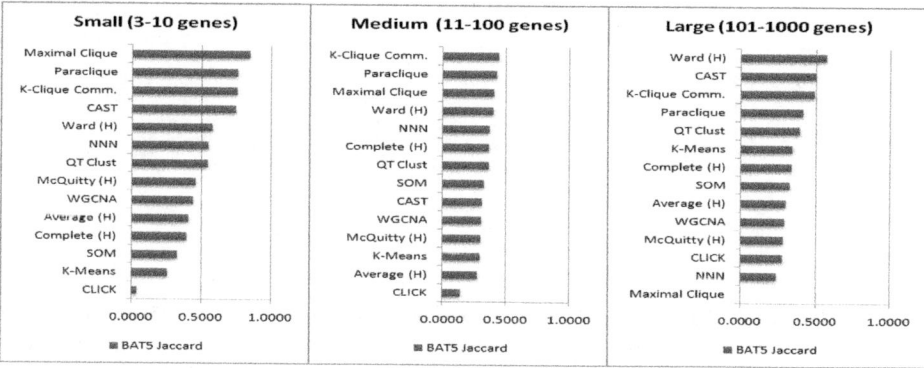

Fig. 1. Algorithms Ranked by Best Average Top 5 Clusters. BAT5 Jaccard values are shown for each clustering method and cluster size classification. (H) = Hierarchical clustering agglomeration method.

Another metric of clustering performance is whether a given method is able to find clusters that are readily identified by other methods. This is a direct comparison of the consistency of clustering algorithms. We identified any annotation category that received a Jaccard score greater than 0.25 in any of the hundreds of clustering runs we performed over all parameter settings. This produced a list of 112 annotation categories, 97 from Gene Ontology and 15 from KEGG. We then found the best category match score that each clustering method received on each of these selected annotations and averaged them. Graph based methods scored highest on this internal consistency metric. The best scoring among these were CAST, maximal clique and paraclique. For each clustering algorithm, we averaged the BAT5 scores from the three size bins, and ranked them by their average. High-averaging methods not only found good results, but they found them in all three size classifications, indicating robustness to variation in cluster size. These values provide a straightforward way to compare clustering methods irrespective of cluster size.

3   Discussion

In our comparison of clustering results by size, we found that maximal clique and paraclique perform best for small clusters; k-clique communities and paraclique perform best for medium clusters; Ward and CAST are the top performing methods for large clusters (Figure 1). Combined analysis of the clustering results across result sizes based upon the quartile of the results reveals that the performance is best for k-clique communities, maximal clique, and paraclique, shown in red in Supplementary Table 2.

For the analysis of consensus of clustering methods for cluster matches to annotation, we found that CAST, maximal clique and paraclique are best at identifying clustering results found by any other method (Figure 2).

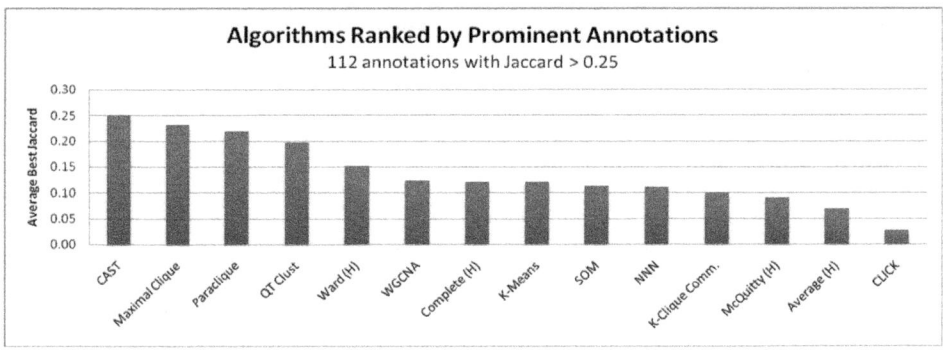

Fig. 2. Algorithms Ranked by Prominent Annotations. 112 annotations received a Jaccard score above 0.25. Each clustering method was ranked by the average of its highest Jaccard score for each of these annotations. (H) = Hierarchical agglomeration method.

This analysis of the performance of diverse clustering algorithms reveals a performance distinction between graph-based and conventional parametric methods. In our study, the best ranking methods are almost uniformly graph-based, building upon the rigorous cluster definition provided by cliques. Traditional methods suffer from relatively poor performance due to their less rigorous cluster definition or their heuristic nature, which often rely on growth of clusters around individual genes in a neighborhood joining or centroid proximity. These methods do not necessarily result in high correlation among all cluster members, whereas clique and other combinatorial algorithms typically require this by definition.

Conventional clustering algorithms frequently focus on details before looking at the bigger picture, by starting from a single gene instead of the full genome. Thus, these methods, such as hierarchical and k-means, lack a global consideration of the data set. Clusters are built incrementally at each step, beginning with a single gene's neighborhood, not a highly correlated geneset. Clusters will therefore tend to converge to a local optimum. This is why repeated randomized, as is frequently done with k-means, can typically improve results simply by selecting genes with larger neighborhoods. Neural network approaches like SOM suffer from a similar problem, as training takes place with incomplete views of the full data. Even QT Clust suffers from these issues, but overcomes to some extent them through additional computation. Clusters are built incrementally for each gene, but only the highest scoring cluster is partitioned from the rest of the data, at which time its individual genes are partitioned out. The process continues iteratively until no genes remain.

Thus, QT clust avoids bias introduced by arbitrarily selecting genes, but still has the same problem of incremental local growth of clusters.

Three of the combinatorial approaches, CAST, CLICK and WGCNA, represent data as graphs, but compute only heuristic solutions to the underlying graph-theoretic metrics. CAST approximates a solution to cluster editing. CLICK approximates a minimum weight cut. WGCNA represents data as a weighted graph, but applies hierarchical clustering to compute its final set of clusters.

NNN computes exact solutions to somewhat arbitrary problem parameters. The poor performance that we observe for it on this data may be because its objectives differ from those of other clustering algorithms. NNN connects genes based on relative correlation among gene pairs. Two genes G and G' are considered related if their correlation is high compared to all other correlations involving G and G', as opposed to all correlations for all genes. Thus, NNN may find clusters that have less pronounced relationships among all cluster members. These clusters may not be present in the high-level GO annotations, but may have biological relevance through more distant and indirect functional relationships.

Our work suggests that graph-based algorithms provide superior reliability and a highly promising approach to transcriptomic data clustering. Most of these methods attempt to find and exploit cliques), with the exception of CLICK which uses minimum cut. It has been suggested that clique-based approaches possess the best potential for identifying sets of interacting genes, due to the highly inter-correlated nature of the clusters produced [34]. The results reported here appear to corroborate that, given that four of the six best performing clustering methods in Supplementary Table 2 are based on clique. It should be noted that we applied the algorithms on a single data set for which both deep experimental data and strong biological ground truth are available, and that results may differ when a different data set is used. It is challenging, however, to conceive of a correlation matrix that would be fundamentally biased toward one type of algorithm, especially given that we provided the selection of parameters over a large range of values.

Graph-based problems relevant to clustering are often thought to be difficult to solve (that is, they are *NP*-hard) because globally optimal solutions are required. This can explain both the effectiveness of exact solutions and also why so few algorithms rely on exact solutions. Our results suggest that exact solutions are truly valuable in practice, and that continued research into computing exact solutions to *NP*-hard problems is probably worthwhile.

Though combinatorial approaches to clustering may perform better, implementation challenges have limited widespread adoption to date. Maximal clique's stand-alone utility is rather limited. Even with the best current implementations, it can take a staggering amount of time to run to completion. It tends to overwhelm the user by returning an exhaustive collection of vast numbers of overlapping clusters, even for a small genome such as yeast. Paraclique and k-clique communities are appealing alternatives due to the more manageable nature of their results. They employ a form of soft thresholding, which helps to ameliorate the effects of noise and generate nicely-enriched clusters without excessive overlap. From a sea of tangled correlations, they produce dense subgraphs that represent sets of genes with highly significant, but not necessarily perfect, pair-wise correlations. Paraclique relies on maximum clique, and thus operates in a top-down fashion. It generates impressive results through the use of its rigorous cluster definition followed by more lenient expansion, leading to very high average intra-cluster correlations. By avoiding the enumeration problems of maximal clique, it is also highly scalable. Moreover, through its complementary duality with vertex cover, it is amenable to advances in fixed-parameter tractability [35]. Paraclique's main drawback is its use of multiple parameters, making algorithm tuning more challenging. In contrast, k-clique communities relies on maximal clique and so operates in a bottom-up manner. It also generates impressive results, but its dependence on maximal clique severely restricts its scalability. Even for a small genome such as that of S. cerevisiae, and even for graphs in which there are no large maximal cliques, we could not run k-clique communities to completion without resorting to our own maximal clique implementation. A faster version of community's CFinder exists (I. Farkas, pers. comm.), but it achieves speed only by setting timeout values for maximal clique computations, thereby creating an approximation method rather than an optimization method. Thus, given the exponential growth rate of maximal cliques, exact algorithms that rely on such cliques are hobbled by memory limitations on larger genomes and denser correlation graphs. We are rather optimistic, however, that approaches exploiting high performance architectures [20] may have the potential to change this picture. The fourth clique-based approach, this time via cluster editing, is CAST. Although its execution is relatively fast, its heuristic nature ensures only mediocre results and difficult tuning. CAST is simply not an optimization technique. It seems able to detect pieces of important clusters (as evidenced by its prominence in Figure 2), but it is often not comprehensive. Given the extreme difficulty of finding exact solutions to cluster editing [36], we think an optimization analog of CAST is unlikely to be feasible in the foreseeable future.

4 Conclusions

Using Jaccard similarity for clustering results to gene annotation categories, we performed a comparative analysis of conventional and more recent graph-based methods for gene co-expression analyses using a well studied biological data set. Jaccard similarity provides a simple and objective metric for comparison that is able to distinguish between entire classes of clustering methods without the biases associated with the Rand Index. Our analysis revealed that the best performing algorithms were graph based. Methods such as paraclique provide an effective means for combining mathematical precision, biological fidelity and runtime efficiency. Further development of these sorts of algorithms and of user-friendly interfaces is probably needed to facilitate more wide-spread adoption.

5 Materials and Methods

5.1 Data

Saccharomyces cerevisiae was fully sequenced in 1996 and has been extensively studied and annotated since. It is therefore an ideal source for biological annotation. We compared the performance of the selected clustering techniques using the extensively studied gene expression data set from Gasch et al. [37]. This data was created to observe genomic expression in yeast responding to two DNA-damaging agents: the methylating agent methylmethane sulfonate (MMS) and ionizing radiation. The set includes 6167 genes from seven yeast strains, collected over 52 yeast genome microarrays.

The microarray data for yeast gene expression across the cell cycle was obtained from http://www-genome.stanford.edu/mec1. These data are normalized, background-corrected, log2 values of the Cy5/Cy3 fluorescence ratio measured on each microarray. We performed clustering either directly on this preprocessed data or on the correlation matrix computed from the data. In the latter case, correlations for gene pairs with five or fewer shared measurements were set to zero. The absolute values of Pearson's correlation coefficients were used, except when a particular clustering approach demands otherwise.

5.2 Clustering Methods

In order to evaluate a wide spectrum of approaches likely to be used in practice, and to avoid the difficult task of choosing the arbitrary "best" parameter setting, we selected roughly 20 evenly distributed combinations of reasonable parameter settings for each implementation. To facilitate comparison, we reduced the myriad of output formats to simple cluster/gene membership lists, grouped into three sizes (3-10, 11-100, and 101-1000 genes).For example, hierarchical clustering produces a tree of clusters, which we simply "slice" at a particular depth to determine a list of clusters.

5.3 Comparison Metrics

Given the prevalence of publicly available gene annotation information, we compared the computationally-derived clusters with manually curated annotations. Yeast annotation sources include Gene Ontology [38], KEGG Pathways [39], PDB [40], Prosite [41], InterPro [42] and PFAM [43]. For clarity and brevity, and to take advantage of their evenly distributed annotation sizes, the results presented here employ only the Gene Ontology and KEGG Pathways as sources.

We used Jaccard similarity as the basis for our analysis. It is easy to calculate, and concisely compares clusters with a single metric. Jaccard similarity is usually computed as the number of true positives divided by the sum of true positives, false positives, and false negatives.In the case of cluster comparisons, this equates to the number of genes that are both in the cluster and annotated, divided by the total number of genes that are either in the cluster or annotated.Thus Jaccard measures how well the clusters match sets of co-annotated genes, from 0 meaning no match to 1.0 meaning a perfect match.

We implemented a simple parallel algorithm to search all annotation sources for the genes in each cluster. For each annotation source, we found all annotations that match at least 2 genes in a given cluster. We then computed the number of genes in the cluster that match the annotation (true positives), the number of genes with the annotation but not in the given cluster (false negatives), and the number of genes in the cluster that did not match the annotation (false positives). We ignored genes in the cluster not found in the annotation source. Finally, the highest matching Jaccard score and annotation is assigned to the cluster.

We grouped Jaccard computations by method and parameter settings and then separated each grouping into three cluster size bins: 10 or fewer genes ("small"), 11-100 genes ("medium"), and 101-1000 genes ("large"). When running a clustering algorithm to validate a hypothesis, one generally has an idea of the desired cluster size, which we try to account for with these size classifications. A researcher looking for small clusters is often not interested in a method or tuning that produces large clusters, and vice versa. It is important to note that the use of average cluster size to determine cluster number (as is needed in k-means, hierarchical and SOM clustering) does not mean that all clusters will be of average size. Thus we find that these methods still generate clusters with small, medium and large sizes.

Each individual cluster was scored against the entire annotation set, and the highest Jaccard score match was returned for that cluster. This list of scores was then grouped by cluster size and sorted by Jaccard score. The highest 5 Jaccard scores per cluster size class were averaged to get the Average Top 5 (AT5). This process was then repeated for each parameter setting tested, amassing a list of around 20 AT5 scores per size class. From each list of AT5 scores, the largest value was selected and assigned to the Best Average Top 5 (BAT5) for that size class. This process is then applied to the next clustering algorithm's results. When all data has been collected, the BAT5 scores are output to a summary table, averaged, and sorted again (Supplementary Table 2).

Community Resources

Data and implementations are available upon request.Supplementary material may be found at: http://web.eecs.utk.edu/~aperkins/clustercomp/

Acknowledgments

This research was supported in part by the National Institutes of Health under grants R01-MH-074460, U01-AA-016662 and R01-AA-018776, by the Department of Energy under the EPSCoR Laboratory Partnership Program, and by the National Science Foundation under grant EPS-0903787.The research leading to these results has received funding from the European Community's Seventh Framework Programme ([FP7/2007-2013] under grant agreement number 223367. This research used resources of the National Energy Research Scientific Computing Center, which is supported by the Office of Science of the U.S. Department of Energy under Contract No. DE-AC02-05CH11231. Illes Farkas provided us with useful information about the CFinder software. Khairul Kabir and Rajib Nath helped generate sample results.

References

1. Jiang, D., Tang, C., Zhang, A.: Cluster analysis for gene expression data: a survey. IEEE Transactions on Knowledge and Data Engineering 16(11), 1370–1386 (2004)
2. Quackenbush, J.: Computational analysis of microarray data. Nature Reviews Genetics 2(6), 418–427 (2001)

3. Kerr, G., Ruskin, H.J., Crane, M., Doolan, P.: Techniques for clustering gene expression data. Computers in Biology and Medicine 38(3), 283–293 (2008)
4. Laderas, T., McWeeney, S.: Consensus framework for exploring microarray data using multiple clustering methods. Omics: A Journal of Integrative Biology 11(1), 116–128 (2007)
5. Myers, C., Barrett, D., Hibbs, M., Huttenhower, C., Troyanskaya, O.: Finding function: evaluation methods for functional genomics data. BMC Genomics 7(1), 187 (2006)
6. Giancarlo, R., Scaturro, D., Utro, F.: Computational clustering validation for microarray data analysis: experimental assessment of Clest, Consensus Clustering, Figure of Merit, Gap Statistics and Model Explorer. BMC Bioinformatics 9(1), 462 (2008)
7. de Souto, M., Costa, I., de Araujo, D., Ludermir, T., Schliep, A.: Clustering cancer gene expression data: a comparative study. BMC Bioinformatics 9(1), 497 (2008)
8. Mingoti, S.A., Lima, J.O.: Comparing SOM neural network with Fuzzy c-means, K-means and traditional hierarchical clustering algorithms. European Journal of Operational Research 174(3), 1742–1759 (2006)
9. Datta, S., Datta, S.: Methods for evaluating clustering algorithms for gene expression data using a reference set of functional classes. BMC Bioinformatics 7(1), 397 (2006)
10. Hartigan, J.A., Wong, M.A.: Algorithm AS 136: A K-Means Clustering Algorithm. Applied Statistics 28(1), 100–108 (1979)
11. McQuitty, L.L.: Similarity Analysis by Reciprocal Pairs for Discrete and Continuous Data. Educational and Psychological measurement 26(4), 825–831 (1966)
12. Ward, J.H.: Hierarchical Grouping to Optimize an Objective Function. Journal of the American Statistical Association 58(301), 236–244 (1963)
13. Palla, G., Derenyi, I., Farkas, I., Vicsek, T.: Uncovering the overlapping community structure of complex networks in nature and society. Nature 435(7043), 814–818 (2005)
14. Zhang, B., Horvath, S.: A General Framework for Weighted Gene Co-Expression Network Analysis. Statistical Applications in Genetics and Molecular Biology 4(1) (2005)
15. Huttenhower, C., Flamholz, A., Landis, J., Sahi, S., Myers, C., Olszewski, K., Hibbs, M., Siemers, N., Troyanskaya, O., Collier, H.: Nearest Neighbor Networks: clustering expression data based on gene neighborhoods. BMC Bioinformatics 8(1), 250 (2007)
16. Ben-Dor, A., Shamir, R., Yakhini, Z.: Clustering gene expression patterns. Journal of Computational Biology: A Journal of Computational Molecular Cell Biology 6(3-4), 291–297 (1999)
17. Sharan, R., Maron-Katz, A., Shamir, R.: CLICK and EXPANDER: a system for clustering and visualizing gene expression data. Bioinformatics 19(14), 1787–1799 (2003)
18. Abu-Khzam, F.N., Baldwin, N.E., Langston, M.A., Samatova, N.F.: On the Relative Efficiency of Maximal Clique Enumeration Algorithms, with Applications to High-Throughput Computational Biology. In: Proceedings of the International Conference on Research Trends in Science and Technology (2005)
19. Bron, C., Kerbosch, J.: Algorithm 457: finding all cliques of an undirected graph. Communications of the ACM 16(9), 575–577 (1973)
20. Zhang, Y., Abu-Khzam, F.N., Baldwin, N.E., Chesler, E.J., Langston, M.A., Samatova, N.F.: Genome-Scale Computational Approaches to Memory-Intensive Applications in Systems Biology. In: Gschwind, T., Aßmann, U., Wang, J. (eds.) SC 2005. LNCS, vol. 3628. Springer, Heidelberg (2005)
21. Chesler, E.J., Langston, M.A.: Combinatorial Genetic Regulatory Network Analysis Tools for High Throughput Transcriptomic Data. In: RECOMB Satellite Workshop on Systems Biology and Regulatory Genomics (2005)

22. Tamayo, P., Slonim, D., Mesirov, J., Zhu, Q., Kitareewan, S., Dmitrovsky, E., Lander, E.S., Golub, T.R.: Interpreting patterns of gene expression with self-organizing maps: Methods and application to hematopoietic differentiation. Proceedings of the National Academy of Sciences of the United States of America 96(6) (1999)
23. Heyer, L.J., Kruglyak, S., Yooseph, S.: Exploring Expression Data: Identification and Analysis of Coexpressed Genes. Genome Research 9(11), 1106–1115 (1999)
24. Milligan, G., Cooper, M.: An examination of procedures for determining the number of clusters in a data set. Psychometrika 50(2), 159–179 (1985)
25. Thalamuthu, A., Mukhopadhyay, I., Zheng, X., Tseng, G.C.: Evaluation and comparison of gene clustering methods in microarray analysis. Bioinformatics 22(19), 2405–2412 (2006)
26. Handl, J., Knowles, J., Kell, D.B.: Computational clustering validation in postgenomic data analysis. Bioinformatics 21(15), 3201–3212 (2005)
27. Yeung, K.Y., Haynor, D.R., Ruzzo, W.L.: Validating clustering for gene expression data. Bioinformatics 17(4), 209–318 (2001)
28. Yao, J., Chang, C., Salmi, M., Hung, Y.S., Loraine, A., Roux, S.: Genome-scale cluster analysis of replicated microarrays using shrinkage correlation coefficient. BMC Bioinformatics 9(1), 288 (2008)
29. Hubert, L., Arabie, P.: Comparing partitions. Journal of Classificiation 2(1), 193–218 (1985)
30. Wallace, D.L.: A Method for Comparing Two Hierarchical Clusterings: Comment. Journal of the American Statistical Association 78(383), 569–576 (1983)
31. Beissbarth, T., Speed, T.P.: GOstat: find statistically overrepresented Gene Ontologies within a group of genes. Bioinformatics 20(9), 1464–1465 (2004)
32. Dennis, G., Sherman, B.T., Hosack, D.A., Yang, J., Gao, W., Lane, H.C., Lempicki, R.A.: DAVID: Database for Annotation, Visualization, and Integrated Discovery. Genome Biology 4(9), R60 (2003)
33. Khatri, P., Draghici, S.: Ontological analysis of gene expression data: current tools, limitations, and open problems. Bioinformatics 21(18), 3587–3595 (2005)
34. Butte, A.J., Tamayo, P., Slonim, D., Golub, T.R., Kohane, I.S.: Discovering functional relationships between RNA expression and chemotherapeutic susceptibility using relevance networks. Proceedings of the National Academy of Sciences of the United States of America 97(22), 12182–12186 (2000)
35. Abu-Khzam, F.N., Langston, M.A., Shanbhag, P., Symons, C.T.: Scalable Parallel Algorithms for FPT problems. Algorithmica 45(3), 269–284 (2006)
36. Dehne, F., Langston, M., Luo, X., Pitre, S., Shaw, P., Zhang, Y.: The Cluster Editing Problem: Implementations and Experiments. In: Parameterized and ExactComputation (2006)
37. Gasch, A.P., Huang, M., Metzner, S., Botstein, D., Elledge, S.J., Brown, P.O.: Genomic Expression Responses to DNA-damaging Agents and the Regulatory Roleof the Yeast ATR Homolog Mec1p. Molecular Biology of the Cell 12(10), 2987–3003 (2001)
38. Ashburner, M., Ball, C.A., Blake, J.A., Botstein, D., Butler, H., Cherry, J.M., Davis, A.P., Dolinski, K., Dwight, S.S., Eppig, J.T., et al.: Gene Ontology: tool for the unification of biology. Nature Genetics 25(1), 25–29 (2000)
39. Kanehisa, M., Araki, M., Goto, S., Hattori, M., Hirakawa, M., Itoh, M., Katayama, T., Kawashima, S., Okuda, S., Tokimatsu, T., et al.: KEGG for linking genomes tolife and the environment. Nucleic Acids Research 36(Suppl 1), D480–D484 (2008)
40. Berman, H.M., Westbrook, J., Feng, Z., Gilliland, G., Bhat, T.N., Weissig, H., Shindyalov, I.N., Bourne, P.E.: The Protein Data Bank. Nucleic Acids Research 28(1), 235–242 (2000)

41. Hulo, N., Bairoch, A., Bulliard, V., Cerutti, L., Cuche, B.A., Castro, E., Lachaize, C., Langendijk-Genevaux, P.S., Sigrist, C.J.A.: The 20 years of PROSITE. Nucleic Acids Research 36(Suppl 1), D245–D249 (2008)
42. Mulder, N.J., Apweiler, R., Attwodd, T.K., Bairoch, A., Bateman, A., Binns, D., Bork, P., Bulliard, V., Cerutti, L., Copley, R., et al.: New developments in theInterPro database. Nucleic Acids Research 35(Suppl 1), D224–D228 (2007)
43. Finn, R.D., Tate, J., Mistry, J., Coggill, P.C., Sammut, S.J., Hotz, H.-R., Ceric, G., Forslung, K., Eddy, S.R., Sonnhammer, E.L.L., et al.: The Pfam protein familiesdatabase. Nucleic Acids Research 36(Suppl 1), D281–D288 (2008)

Mining Biological Interaction Networks Using Weighted Quasi-Bicliques

Wen-Chieh Chang[1], Sudheer Vakati[1], Roland Krause[2,3], and Oliver Eulenstein[1]

[1] Department of Computer Science, Iowa State University, Ames, IA, 50011, U.S.A.
{wcchang,svakati,oeulenst}@iastate.edu
[2] Max Planck Institute for Molecular Genetics, Berlin, Germany
[3] Free University Berlin, Berlin, Germany
roland.krause@molgen.mpg.de

Abstract. Biological network studies can provide fundamental insights into various biological tasks including the functional characterization of genes and their products, the characterization of DNA-protein interactions, and the identification of regulatory mechanisms. However, biological networks are confounded with unreliable interactions and are incomplete, and thus, their computational exploitation is fraught with algorithmic challenges. Here we introduce quasi-biclique problems to analyze biological networks when represented by bipartite graphs. In difference to previous quasi-biclique problems, we include biological interaction levels by using edge-weighted quasi-bicliques. While we prove that our problems are NP-hard, we also provide exact IP solutions that can compute moderately sized networks. We verify the effectiveness of our IP solutions using both simulation and empirical data. The simulation shows high quasi-biclique recall rates, and the empirical data corroborate the abilities of our weighted quasi-bicliques in extracting features and recovering missing interactions from the network.

1 Introduction

Proteins are the elementary building blocks of molecular machines that mediate cellular processes such as transcription, replication, metabolic catalyses, or the transport of substances. In these processes proteins interact with each other and can form sophisticated modules of protein-protein interaction (PPI) networks. Analyzing these networks is a thriving field in proteomics [14] and has extensive implications for a host of issues in biology, pharmacology [14], and medicine [6]. Computationally capturing the modularity of PPI networks accurately will gain insights into cellular processes and gene function. Yet, before such modularities can be reliably identified, challenging problems have to be overcome.

These problems are caused by incomplete and error-prone PPI networks, which can largely obfuscate the reliable identification of modules [11,13]. Protein-protein interaction can not be measured to the accuracy of the genome sequences, leaving some guesswork in identifying modularities correctly. Some interactions are highly transient and can only be measured indirectly, while others withstand

J. Chen, J. Wang, and A. Zelikovsky (Eds.): ISBRA 2011, LNBI 6674, pp. 428–439, 2011.

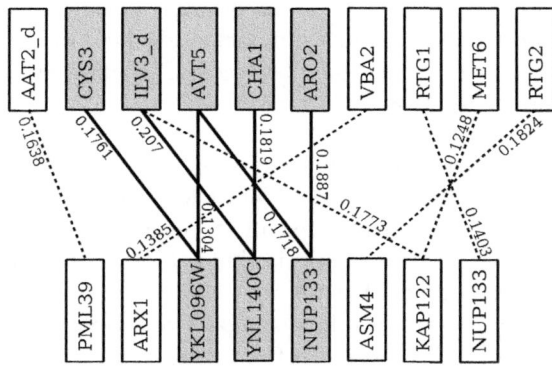

Fig. 1. A quasi-biclique (solid nodes/edges) identified from a gene interaction network in one of our experiment set where the edge weights are interaction scores. The bipartite graph is unweighted if only the existence of edges are considered.

denaturing agents. Functional interaction does not even have to be realized via physical interactions. Thus, computational methods for capturing modularity can not directly rely on presence or absence of interactions in PPI networks and need to be able to cope with substantial error rates.

Unweighted quasi-biclique approaches have been used in the past to identify modularity in PPI networks when presented as bipartite graphs that are spanned between different features of proteins, e.g. binding sites and domain content function [10, 11]. An example is depicted in Fig. 1. While these approaches aim to solve NP-hard problems using heuristics they were able to identify some of the highly interactive protein complexes in PPI networks [4, 9].

Unweighted quasi-biclique approaches are sensitive to the quantitative uncertainties intrinsic to PPI networks. In particular, protein-protein interactions are only represented by an unweighted edge in the bipartite graph if they are above some user-specified threshold. Therefore, unweighted quasi-biclique approaches are prone to disregard many of the invaluable protein-protein interactions that are below the threshold, and treat all protein-protein interactions above the threshold the same. Further, some interactions may or may not be represented due to some seemingly insignificant error in the measurement. Consequently, many of the crucial modules may be concealed and remain undetected by using unweighted quasi-biclique approaches.

Here we introduce novel weighted quasi-biclique problems that incorporate the level of protein-protein interactions by using bipartite graphs where edges are weighted by the level of the corresponding protein-protein interactions, e.g., Fig. 1. We show that these problems are, similar to their unweighted versions, NP-hard. However, in practice, exact Integer Programming (IP) formulations can tackle very efficiently many NP-hard real-world problems [3]. Therefore, we describe exact Integer Programming (IP) solutions for our weighted quasi-biclique problems. Furthermore our IP solutions exploit the typical sparseness of PPI networks when represented as bipartite graphs. This allows us to verify the ability of our IP solutions using a moderately sized PPI network that was published in the literature [2], and simulation studies. In addition our IP solutions can provide exact results for instances of the unweighted quasi-biclique approaches that were previously not possible.

Related Work. PPI networks can be represented as bipartite graphs and maximal bicliques in these graphs are self-contained elements characterizing highly interactive protein complexes that typically represent modules in PPI networks (e.g., [9,11]). *Bipartite graphs* are special kinds of graphs whose vertices can be bi-partitioned into sets X and Y such that each edge is incident to vertices in X and Y. A biclique is a subgraph of a bipartite graph where every vertex in one partition is connected to every vertex in the other partition by an edge. A biclique is *maximal* if it is not properly contained in any other biclique, and it is *maximum* if no other maximal bicliques have larger total edge weights. The problem of finding maximum bicliques is well studied in the literature of graph theory and is known to be NP-complete [12] and effective heuristics for this problem have been described and used in various applications [1]. However, bicliques are too stringent for identifying modules in PPI networks. For example a module is not identified through a biclique that is incomplete by one single edge. *Quasi-bicliques* are partially incomplete bicliques that overcome this limitation. They allow a specified maximum number of edges to be missed in order to form a biclique [16]. While quasi-bicliques are less stringent for the identification of modules, they might contain proteins that are interacting with only a few or none other of the proteins. Such situations occur when the missing edges are not homogeneously distributed throughout. The *δ-quasi-bicliques (δ-QB)* [15] allow to control the distribution of missing edges by setting lower bounds, parameterized by δ, on the minimum number of incident edges to vertices in each of the quasi-biclique's vertex sets.

Our Contributions. Here we define a "weighted" version of δ-QB, called α,β-weighted quasi-bicliques (α,β-WQB), to improve on the identification of modules in PPI networks by using the interaction levels between proteins. Thus, α,β-WQB's may be better applicable to handle noisy data sets as they distribute the overall missing information across the vertices of the quasi-biclique as shown in our simulations. Finding a maximum α,β-WQB in a given edge weighted bipartite graph is NP-hard, since it is a generalization of the NP-hard problem to find δ-QBs in unweighted bipartite graphs [11]. We also introduce a "query" version of the α,β-WQB problem that allows biologists to focus their analyzes on proteins of their particular interest. Given a PPI network and specific proteins from this network, called *query*, the query problem is to find a maximum weighted α,β-WQB that includes the query. We prove this problem is NP-hard. While the α,β-WQB problem and its query version are NP-hard, we provide exact IP solutions to solve both problems. By reducing the number of required variables and exploiting the sparseness of bipartite graphs representing PPI networks, our solutions solve moderate-sized instances. This allowed us to verify the applicability of α,β-WQB by analyzing the most complete data set of genetic interactions available for the Eukaryotic model organism *Saccharomyces cere-visiae*. Our results not only extract meaningful yet unexpected quasi-bicliques under functional classes, but also suggest higher possibilities of recovering missing interactions not presented in the input.

2   Basic Notation, Definitions and Preliminaries

A *bipartite graph*, denoted by $(U + V, E)$, is a graph whose vertex set can be partitioned into the sets U and V such that its edge set E consists only of edges $\{u, v\}$ where $u \in U$ and $v \in V$ (U and V are independent sets). Let $G := (U + V, E)$ be a bipartite graph. The graph G is called *complete* if for any two vertices $u \in U$ and $v \in V$ there is an edge $\{u, v\} \in E$. A *biclique* in G is a pair (U', V') that induces a complete bipartite subgraph in G, where $U' \subseteq U$ and $V' \subseteq V$. Since any subgraph induced by a biclique is a complete bipartite graph, we use the two terms interchangeably. A pair (U, V) *includes* another pair (U', V') if $U' \subseteq U$ and $V' \subseteq V$. In such case, we also say that the pair (U', V') is *included* in (U, V). A pair (U, V) is non-empty if both U and V are non-empty. A *weighted bipartite graph*, denoted by $(U + V, E, \omega)$, is a complete bipartite graph $(U + V, E)$ with a weight function $\omega \colon E \to [0, 1]$.

2.1   Maximum Weighted Quasi-Biclique (α,β-WQB) Problem

Definition 1 (α,β-WQB). *Let $G := (U + V, E, \omega)$ and $\alpha, \beta \in [0, 1]$. An α,β-weighted quasi-biclique, denoted as α,β-WQB , in G is a non-empty pair (U', V') that is included in (U, V) and satisfies the two properties:*
(1) $\forall u \in U'$: $\sum_{v \in V'} \omega(u, v) \geq \alpha|V'|$, and (2) $\forall v \in V'$: $\sum_{u \in U'} \omega(u, v) \geq \beta|U'|$.

The *weight* of an α,β-WQB is defined as the sum of all its edge weights.

Definition 2 (Maximum α,β-WQB). *A α,β-WQB, is a maximum weighted α,β-WQB of a weighted bipartite graph $G := (U + V, E, w)$, if its weight is at least as much as the weight of any other α,β-WQB in G.*

Problem 1 (α,β-WQB)
Instance: A weighted bipartite graph $G := (U + V, E, \omega)$, and values $\alpha, \beta \in [0, 1]$.
Find: A maximum weighted α,β-WQB in G.

Note that, we use the same notation (α,β-WQB) for a α, β-weighted quasi-biclique and maximum weighted α, β-weighted quasi-biclique problem. The context in which we use the notation will make the difference clear.

2.2   Query Problem

A common requirement in the analysis of networks is to provide the environment of a certain group of genes, which translates into finding the maximum weighted α,β-WQB which includes a specific set of vertices. We call this the query problem and is defined as follows.

Problem 2 (Query)
Instance: A weighted bipartite graph $G := (U + V, E, \omega)$, values $\alpha, \beta \in [0, 1]$, and a pair (P, Q) included in (U, V).
Find: The α,β-WQB which includes (P, Q) and has a weight greater than or equal to the weight of any α,β-WQB which includes (P, Q).

2.3   Time Complexity Results

Here we prove the NP-hardness of the α,β-WQB problem by a reduction from the *maximum edge biclique* problem, and of the query problem by reductions from the partition problem.

Lemma 1. *The α,β-WQB problem is NP-hard.*

Proof. Given a bipartite graph $G := (U + V, E)$ and an integer k, the *maximum edge biclique* problem asks if G contains a biclique with atleast k edges. The *maximum edge biclique* problem is NP-complete [12]. Let $G' := (U + V, E', \omega)$ be a weighted bipartite graph where $\omega(u, v)$ is set to 1 if $(u, v) \in E$ or is set to 0 otherwise. Note that, there is a biclique with k edges in G if and only if the maximum weighted α,β-WQB in G' has a weight of atleast k when α and β are set to 1. Therefore, the α,β-WQB problem is NP-hard.   □

To prove the hardness of query problem we need some auxiliary definitions. A *modified weighted bipartite graph*, denoted by $(U + V, E, \Omega)$, is a complete bipartite graph $(U + V, E)$ with a weight function $\Omega : E \to [0, 1]$ where, for any two edges e and e', $|\Omega(e) - \Omega(e')| \leq 1$.

Definition 3 (Modified α,β-WQB (MO-WQB))
Let $G := (U + V, E, \Omega)$ be a modified weighted bipartite graph. A non-empty pair (U', V') included in (U, V) is a MO-WQB of G, if it satisfies the three properties: (1) (U', V') includes $(\emptyset, V)$, (2) $\forall u \in U': \sum_{v \in V'} w(u, v) \geq 0$, and (3) $\forall v \in V': \sum_{u \in U'} w(u, v) \geq 0$.

Problem 3 (Existence)
Instance: A weighted bipartite graph $G := (U + V, E, \omega)$, values $\alpha, \beta \in [0, 1]$.
Find: If there exists a α,β-WQB (U', V') in G which includes the pair $(\emptyset, V)$.

Problem 4 (Modified Existence)
Instance: A modified weighted bipartite graph $G := (U + V, E, \Omega)$.
Find: If there exists a *MO-WQB* in G.

The series of reductions to prove the hardness of the query problem are as follows. We first reduce the *partition* problem, which is NP-complete [5], to the *modified existence* problem. The *modified existence* problem is then reduced to the existence problem. The existence problem reduces to the query problem.

Lemma 2. *The modified existence problem is NP-complete.*

Proof. The proof of *MO-WQB* $\in$ *NP* is omitted for brevity. We are left to show that *partition* $\leq_p$ *MO-WQB*. Given a finite set A, and a size $s(a) \in Z^+$ associated with every element a of A, the partition problem asks if A can be partitioned into two sets (A_1, A_2) such that $\Sigma_{a \in A_1} s(a) = \Sigma_{a \in A_2} s(a)$.

a. Construction: Let *SUM* be the sum of sizes of all elements in A. Build a modified weighted bipartite graph $G := (U + V, E, \Omega)$ as follows. For every element a in A there is a corresponding vertex u_a in U. The set V contains two vertices v_+ and v_-. For every vertex $u_a \in U$, $\Omega(u_a, v_+) = s(a)/(2 \times SUM)$ and $\Omega(u_a, v_-) = -s(a)/(2 \times SUM)$. Add an additional vertex u_{sum} to set U. Set $\Omega(u_{sum}, v_+)$ to -1/4 and $\Omega(u_{sum}, v_-)$ to 1/4. Note that, the weights assigned to edges of G satisfy the constraint on Ω for a modified weighted bipartite graph.

b. $\Rightarrow$: Let (A_1, A_2) be a partition of A such that the sum of the sizes of elements in A_1 is equal to the sum of the sizes of elements in A_2. Let $U_1 = \{u_a : a \in A_1\}$. The sum of weights of all edges from v_+ to the vertices in U_1 is equal to 1/4. Let $U' = U_1 \cup u_{sum}$. The sum of weights of all edges from v_+ to vertices of U' is 0. Similarly, the sum of weights of all edges from v_- to vertices of U' is 0. Thus, (U', V) is a *MO-WQB* of G.

$\Leftarrow$: Let (P, V) be a *MO-WQB* of G. The edge from v_- to u_{sum} is the only positive weighted edge from vertex v_-. So, P will contain vertex u_Σ. Since $\Omega(v_+, u_{sum})$ is negative, set P will also contain vertices from $U - u_{sum}$. The sum of the weights of edges from v_- to vertices in $P - u_{sum}$ cannot be smaller than -1/4. Similarly, the sum of the weights of edges from v_+ to vertices in $P - u_{sum}$ cannot be smaller than 1/4. So, the sum of all elements in A corresponding to the vertices in $P - u_{sum}$ should be equal to $SUM/2$. This proves that if G contains a *MO-WQB*, set A can be partitioned.   □

Lemma 3. *The* existence *problem is NP-complete.*

Proof. Similarly, the proof of *existence* $\in$ *NP* is omitted for brevity. Next we show *MO-WQB* $\leq_p$ *existence*. We prove this problem to be NP-complete by a reduction from the modified existence problem. The reduction is as follows.

a. Construction: Let $G := (U + V, E, \Omega)$ be the modified weighted bipartite graph in an instance of the modified existence problem. We build a graph $G' := (U + V, E, \omega)$ for an instance of existence problem from G. Notice that the partition and vertices remain the same. If the weight of every edge in the G is non negative, set $\alpha = \beta = 0$ and $\omega(u, v) = \Omega(u, v)$ for every edge $(u, v) \in E$. Otherwise, set α and β to $|x|$ and $\omega(u, v) = \Omega(u, v) - x$ for every edge $(u, v) \in E$, where x is the minimum edge weight in G.

b. $\Rightarrow$ and $\Leftarrow$: Let (U', V) be a *MO-WQB* of graph G. If weights of all edges in G are non negative, the constraints for both the problems are the same. If G has negative weighted edges, the constraints of both the problems will be the same when α, β and ω for the existence problem instance are set as mentioned in the construction. It can be seen that there is a *MO-WQB* in G if and only if there is a α, β-WQB in the graph G' which includes the pair $(\emptyset, B)$.   □

The *existence* problem can be reduced to the query problem by setting (P, Q) to $(\emptyset, V)$. The next lemma follows.

Lemma 4. Query *problem is NP-hard.*

3   IP Formulations for the α,β-WQB Problem

Although greedy approaches are often used in problems of a similar structure, e.g., multi-dimensional knapsack [8], δ-QB [11], our early greedy approach did not identify solutions close enough to the optimal. Here we present integer programming (IP) formulations solving the α,β-WQB problem in exact solutions. Our initial IP requires quadratic constraints, which are then replaced by linear constraints such that it can be solved by various optimization software packages. Our final formulation is further improved by adopting the implication rule to simplify variables involved. This improved formulation requires variables and constraints linear to the number of input edges, and thus, suits better for sparse graphs. Throughout the section, unless stated otherwise, $G := (U + V, E, \omega)$ represents a weighted bipartite graph, and $G' = (U', V')$ represents the maximum weighted α,β-WQB of G and E' represents the edges induced by G' in G.

Quadratic Programming. For each $u \in U$ $(v \in V)$, a binary variable x_u (x_v) is introduced. The variable x_u (x_v) is 1 if and only if vertex u (v) is in U' (V'). The integer program to find the solution G' can be formulated as follows.

$$\begin{array}{lll} \text{Binary variables:} & x_u, \text{ s.t. } x_u = 1 \text{ iff } u \in U' & \text{for each } u \in U \quad (1) \\ & x_v, \text{ s.t. } x_v = 1 \text{ iff } v \in V' & \text{for each } v \in V \quad (2) \\ \text{Subject to:} & \sum_{v \in V} \omega(u, v) \cdot x_v x_u \geq \alpha \sum_{v \in V} x_v x_u & \text{for all } u \in U \quad (3) \\ & \sum_{u \in U} \omega(u, v) \cdot x_u x_v \geq \beta \sum_{u \in U} x_u x_v & \text{for all } v \in V \quad (4) \\ \text{Maximize:} & \sum_{(u,v) \in U \times V} x_u x_v \cdot \omega(u, v) & \quad (5) \end{array}$$

The quadratic terms in the constraints are necessary because, α and β thresholds apply only to vertices in U' and V'. This formulation uses variables and constraints linear to the size of input vertices, i.e., $O(|U| + |V|)$. Since solving a quadratic program usually requires a proprietary solver, we reformulate the program so that all expressions are linear.

Converted Linear Programming. A standard approach to convert a quadratic program to a linear one is introducing auxiliary variables to replace the quadratic terms. Here we introduce a binary variable y_{uv} for every edge (u, v) in G, such that, $y_{uv} = 1$ if and only if $x_u = x_v = 1$, i.e., the edge (u, v) is in G'. The linear program to find the solution G' is formulated as follows.

Binary variables: Same as in (1) and (2)

$$y_{uv}, \text{ s.t., } y_{uv} = 1 \text{ iff } x_u = x_v = 1 \qquad \text{for all } (u,v) \in E \quad (6)$$

Subject to: $\quad y_{uv} \leq (x_u + x_v)/2 \qquad\qquad\quad\ \text{for all } (u,v) \in E \quad (7)$

$$y_{uv} \geq x_u + x_v - 1 \qquad\qquad\quad\ \text{for all } (u,v) \in E \quad (8)$$

$$\sum_{v \in V} \omega(u,v) \cdot y_{uv} \geq \alpha \sum_{v \in V} y_{uv} \quad \text{for all } u \in U \quad (9)$$

$$\sum_{u \in U} \omega(u,v) \cdot y_{uv} \geq \beta \sum_{u \in U} y_{uv} \quad \text{for all } v \in V \quad (10)$$

Maximize: $\quad \sum_{(u,v) \in U \times V} \omega(u,v) \cdot y_{uv} \qquad\qquad\qquad\qquad\qquad (11)$

Expressions (7) and (8) state the condition that $y_{uv} = 1$ if and only of $x_u = x_v = 1$. Expression (8) ensures that, for any edge whose end points (u,v) are chosen to be in G', y_{uv} is set to 1. Due to the use of y_{uv} variables, this formulation requires $O(|U||V|)$ variables and constraints.

Improved Linear Programming. Observe that constraint (7) becomes trivial if $y_{uv} = 0$. In other words, this constraint formulates implications, e.g., for binary variables p and q, the expression $p \leq q$ is equivalent to $p \rightarrow q$. Expanding on this idea, we eliminate the requirement of variables y_{uv} in constraints (9) and (10) in the next formulation while sharing the rest of the aforementioned linear program.

Subject to: $\quad \sum_{v \in V} (\omega(u,v) - \alpha) \, x_v \geq |V|(x_u - 1) \qquad \text{for all } u \in U \quad (12)$

$$\sum_{u \in U} (\omega(u,v) - \beta) \, x_u \geq |U|(x_v - 1) \qquad \text{for all } v \in V \quad (13)$$

There is a variable x_v for every vertex v in G. There is a variable y_{uv} for every edge (u,v) in G whose weight is not 0. The variable y_{uv} is set to 1 if and only if both x_u and x_v are set to 1. For any vertex $u \in U$ ($v \in V$), the variable x_u (x_v) is set to 1 if and only if vertex u (v) is in G'. Constraint (12) can also be explained as follows. If $x_u = 1$, the constraint transforms to the second constraint in the α,β-WQB definition. If $x_u = 0$, constraint (12) becomes trivial. Constraint (13) can be explained in a similar manner.

If there are n vertices in U and m vertices in V, there will be a total of $m + n + 2k$ constraints and $m + n + k$ variables where k is the number of edges whose weight is not equal to 0. The above formulations can be modified to solve the query problem by adding an additional constraint $x_v = 1$ to the formulation, for every vertex $v \in P \cup Q$.

4 Results and Discussion

Finding appropriate values for α and β is a critical part in an application. With no specific standard set to compare to, we simulated data sets to explore the problem. We then use our IP model to explore α,β-WQB's in a real world application, a recent data set of functional groups formed in genetic interactions. The filtered data set, compared to the raw data, served to investigate the role of non-existing edges in the input bipartite graph. While mathematically equivalent in the modeling step, a non-edge in a PPI graph can represent either a true non-interaction or a false-negative. Assuming the input consists of meaningful features, our preliminary results show that α,β-WQB's may recover missing edges with potentially higher weights better than δ-quasi-bicliques.

Simulations. We evaluated two different methods of choosing α and β values, by trying to retrieve a known α,β-WQB from a weighted bipartite graph. In each simulation experiment we do the following. The pair (U, V) represents the vertices of a weighted bipartite graph G. We randomly choose $U' \subseteq U$ and $V' \subseteq V$ as vertices of the known quasi biclique in G. The sizes of both U' and V' are set the same and is picked randomly. Random edges between the vertices of U' and V' in G are introduced according to a pre-determined edge density d. The edges between vertices of $U \backslash U'$ and $V \backslash V'$ of G are also generated randomly according to a pre-determined density d'. The edge weights of the known quasi-biclique (U', V') are determined by a Gaussian distribution with a mean mn and standard deviation dev. Weights of the edges of G not present in the quasi-biclique are also determined by a Gaussian distribution with a lower mean mn' and standard deviation dev'.

The values of α and β are then chosen in two different ways. As part of the simulation we evaluate the performance of both methods. The first method sets both α and β to the mean of the weights of the edges of the quasi biclique. In the second method, α and β are calculated as given below:

$$\alpha = \min \left\{ C_{u'} \mid C_{u'} = \left(\sum_{v' \in V'} w(u', v') \right) / |V'| \text{ for all } u' \in U' \right\}$$
$$\beta = \min \left\{ C_{v'} \mid C_{v'} = \left(\sum_{u' \in U'} w(u', v') \right) / |U'| \text{ for all } v' \in V' \right\}$$

The corresponding ILP model of the α,β-WQB problem is generated in Python, and solved in Gurobi 3.0 [7] on a PC with an Intel Core2 Quad 2.4 GHz CPU with 8 GB memory.

For the evaluation, let (U'', V'') represent the maximum weighted α,β-WQB returned by the ILP model. The number of vertices of U' in U'' and V' in V'' is our evaluation criterion. The percentage of vertices of U' (V') in U'' (V'') is called the recall of U' (V'). For a specific graph sizes experiments were run by varying the values mn and mn'. The values dev and dev' were set 0.1. The densities d and d' are set to 0.8 and 0.2. The experiments were run for graphs of size 16×16, 32×32 and 40×40. Each experiment is repeated thrice and the average number of recalled vertices is calculated. The recall of the experiments can be seen in Table 1. As the difference between the means increases, so does the average recall. The second method of choosing α and β yields a consistently higher recall.

4.1   Genetic Interaction Networks

A comprehensive set of genetic interaction and functional annotation published recently by Costanzo *et al.* [2] is amongst the best single data sources for weighted biological networks. The aim of our application is to identify the maximum weighted quasi-bicliques consisting of genes in different functional classes in the Costanzo dataset.

Table 1. Recall of vertices in the simulation. For every experiment, the value in the $A_U(A_V)$ column represents the average recall of $U'(V')$.

mn	mn'	16 × 16				32 × 32				40 × 40			
		Method 1		Method 2		Method 1		Method 2		Method 1		Method 2	
		A_U	A_V	A_U	A_V	A_U	A_V	A_U	A_V	A_U	A_V	A_U	A_V
0.5	0.5	60	50	60	50	16	11	33	33	58	40	41	20
0.55	0.45	55	33	55	33	42	43	53	23	53	31	66	37
0.6	0.4	50	27	66	55	53	50	66	66	50	50	58	60
0.65	0.35	100	49	100	49	60	93	73	100	51	48	60	61
0.7	0.3	72	88	100	88	100	73	100	86	69	75	100	100
0.75	0.25	66	72	100	83	76	93	100	100	75	71	100	94
0.8	0.2	55	100	72	100	100	83	100	100	91	49	100	73

Pairwise comparisons of the total 18 functional classes provide 153 sets. For every distinct pair (A, B) of such classes, we build a weighted bipartite graph $(U_A + V_B, E, \omega)$ where genes from functional class A are represented as vertices in U_A from functional class B are represented as vertices in V_B.

The absolute values of the interaction score ε, are used as the edge weights. Values greater than 1 are rounded off to 1. Any gene present in both the functional classes A and B is represented as different vertices in the partitions U_A and V_B and the edge between those vertices is given a weight of 1. We build LP models for the bipartite graphs to identify the maximum weighted quasi-bicliques.

Biological Interpretation and Examples. Genes with high degree and strong links dominate the results. In several instances, the quasi-bicliques are trivial in the sense that only one gene is present in U', and it is linked to more than 20 genes in V'. Such quasi-bicliques are maximal by definition but provide limited insight. A minimum of $m = 2$ genes per subset was included as an additional constraint to the LP model. It might be sensible to implement such restrictions in the application in general.

Given the low overall weight, the data set generated with the parameters $\alpha = \beta = 0.1$ and $m = 2$ provided the most revealing set of maximum weighted quasi-bicliques. A notable latent set that was obtained identified genes involved in amino acid biosynthesis (SER2, THR4, HOM6, URE2) and was found to form a 4 × 10 maximum weighted quasi-biclique with genes coding for proteins of the translation machinery, elongation factors in particular (ELP2, ELP3, ELP4, ELP6 , STP1, YPL102C, DEG1, RPL35A, IKI3, RPP1A). These connections, to our knowledge, are not described and one might speculate that this is a way how translation is coupled to the amino-acid biosynthesis.

In some cases the maximum weighted quasi-biclique is centered around the genes that are annotated in more than one functional class as they provide strong weights. These genes are involved in mitochondrial to nucleus-signaling and are examples where our approach recovers known facts. Using the query approach, it is possible to obtain quasi-bicliques around a gene set of interest quickly and extend the approach proteins of interest.

Table 2. A comparison of e under various QB parameters showing improvements of recovered edge weight expectation in α,β-WQB's

QB	D05/M1	D05/M2	AB/M2	A005/M2	A01/M2	A02/M2	A03/M2	A04/M2	A05/M2
avg(e)	0.0855	0.0844	0.0850	0.0806	0.0830	0.0867	0.0905	0.0934	0.1169

Recovering Missing Edges. The published data sets have edges under different thresholds removed. To sample such missing edges, we calculate the average weight of all the edges removed in the 153 bipartite graphs (generated above), and the calculated average weight is 0.0522. For each of the 153 maximum weighted quasi-bicliques, the missing edges induced by the quasi-bicliques are then identified, and the average missing edge weight e of each is calculated, and e is always greater than 0.0522. In other words, we observe that a missing edge in a maximum weighted quasi-biclique has a higher expected weight than the weight of a randomly selected missing edge.

We further compare e from our approach to e from the δ-quasi-bicliques (δ-QB) described by Liu *et al.* [11]. All quasi-bicliques (including exact δ-QB using our IP formulation) used to induce average missing edge weight e are:
(1) D05/M1: δ-QB with $\delta = 0.5$ and minimum node size is 1, i.e., $m = 1$.
(2) D05/M2: δ-QB with $\delta = 0.5$; $m = 2$.
(3) AB/M2: α,β-WQB using the minimum average edge weights found from D05/M2 as α and β; $m = 2$.
(4) AX/M2: α,β-WQB where $X = \alpha = \beta \in \{0.05, 0.1, 0.2, 0.3, 0.4, 0.5\}$; $m = 2$.

Comparing the averages of e from A005/M2 to A05/M2, we see a steady increase. Since α and β can be seen as expected edge weights of the resulting QB, the changes in e shows that QB identifies subgraphs of expected edge weights. In this particular case, the removed edge weights are at most 0.16, hence e can never approach closely to the parameter α.

5   Conclusions and Outlook

We address noise and incompleteness in biological networks by introducing a graph-theoretical optimization problem that identifies weighted quasi-bicliques. These quasi-bicliques incorporate biological interaction levels and can improve on the usage of un-weighted quasi-bicliques. To meet demands of biologists we also provide a query version of (weighted) quasi-biclique problems. We prove that our problems are NP-hard, and describe IP formulations that can tackle moderate-sized problem instances. Simulations solved by our IP formulation suggest that our weighted quasi-biclique problems are applicable to various other biological networks.

Future work will concentrate on the design of algorithms for solving large-scale instances of weighted quasi-biclique problems within guaranteed bounds. Greedy approaches may result in effective heuristics that can analyze ever-growing biological networks. A practical extension to the query problem is the development of an efficient enumeration of all maximal α,β-WQB's.

Acknowledgments. We thank Heiko Schmidt for discussions that initiated the concept of weighted quasi-bicliques. Further, we thank Nick Pappas and John Wiedenhoeft for valuable comments. WCC, SV, and OE were supported in part by NSF awards #0830012 and #1017189.

References

1. Alexe, G., Alexe, S., Crama, Y., Foldes, S., Hammer, P.L., Simeone, B.: Consensus algorithms for the generation of all maximal bicliques. Discrete Appl. Math. 145(1), 11–21 (2004)
2. Costanzo, M., Baryshnikova, A., Bellay, J., Kim, Y., Spear, E., Sevier, C., Ding, H., Koh, J., Toufighi, K., Mostafavi, S., et al.: The genetic landscape of a cell. Science 327(5964), 425 (2010)
3. Dietrich, B.: Some of my favorite integer programming applications at IBM. Annals of Operations Research 149(1), 75–80 (2007)
4. Ding, C., Zhang, Y., Li, T., Holbrook, S.: Biclustering Protein Complex Interactions with a Biclique Finding Algorithm. In: ICDM, pp. 178–187 (2006)
5. Garey, M.R., Johnson, D.S.: Computers and Intractability: A Guide to the Theory of NP-Completeness. W H Freeman, New York (1979)
6. Goh, K., Cusick, M., Valle, D., Childs, B., Vidal, M., Barabási, A.: The human disease network. PNAS 104(21), 8685 (2007)
7. Gurobi Optimization Inc.: Gurobi Optimizer 3.0 (2010)
8. Kellerer, H., Pferschy, U., Pisinger, D.: Knapsack Problems. Springer, Heidelberg (2004)
9. Li, H., Li, J., Wong, L.: Discovering motif pairs at interaction sites from protein sequences on a proteome-wide scale. Bioinformatics 22(8), 989 (2006)
10. Liu, H., Liu, J., Wang, L.: Searching maximum quasi-bicliques from protein-protein interaction network. JBSE 1, 200–203 (2008)
11. Liu, X., Li, J., Wang, L.: Modeling protein interacting groups by quasi-bicliques: Complexity, algorithm, and application. IEEE TCBB 7(2), 354–364 (2010)
12. Peeters, R.: The maximum edge biclique problem is NP-complete. Discrete Appl. Math. 131(3), 651–654 (2003)
13. Sim, K., Li, J., Gopalkrishnan, V.: Mining maximal quasi-bicliques: Novel algorithm and applications in the stock market and protein networks. Analysis and Data Mining 2(4), 255–273 (2009)
14. Waksman, G.: Proteomics and Protein-Protein Interactions Biology, Chemistry, Bioinformatics, and Drug Design. Springer, Heidelberg (2005)
15. Wang, L.: Near Optimal Solutions for Maximum Quasi-bicliques. In: Thai, M.T., Sahni, S. (eds.) COCOON 2010. LNCS, vol. 6196, pp. 409–418. Springer, Heidelberg (2010)
16. Yan, C., Burleigh, J.G., Eulenstein, O.: Identifying optimal incomplete phylogenetic data sets from sequence databases. Mol. Phylogenet. Evol. 35(3), 528–535 (2005)

Towards a Characterisation of the Generalised Cladistic Character Compatibility Problem for Non-branching Character Trees

Ján Maňuch[1,2], Murray Patterson[2], and Arvind Gupta[2]

[1] Department of Mathematics, Simon Fraser University, Burnaby, BC, Canada
[2] Department of Computer Science, UBC, Vancouver, BC, Canada

Abstract. In [3,2], the authors introduced the *Generalised Cladistic Character Compatibility* (GCCC) Problem which generalises a variant of the Perfect Phylogeny Problem in order to model better experiments in molecular biology showing that genes contain information for currently unexpressed traits, e.g., having teeth. In [3], the authors show that this problem is NP-complete and give some special cases which are polynomial. The authors also pose an open case of this problem where each character has only one generalised state, and each character tree is non-branching, a case that models these experiments particularly closely, which we call the *Benham-Kannan-Warnow* (BKW) Case.

In [18], the authors study the complexity of a set of cases of the GCCC Problem for non-branching character trees when the phylogeny tree that is a solution to this compatibility problem is restricted to be either a tree, path or single-branch tree. In particular, they show that if the phylogeny tree must have only one branch, the BKW Case is polynomial-time solvable, by giving a novel algorithm based on PQ-trees used for the consecutive-ones property of binary matrices.

In this work, we characterise the complexity of the remainder of the cases considered in [18] for the single-branch tree and the path. We show that some of the open cases are polynomial-time solvable, one by using an algorithm based on directed paths in the character trees similar to the algorithm in [2], and the second by showing that this case can be reduced to a polynomial-time solvable case of [18]. On the other hand, we will show that other open cases are NP-complete using an interesting variation of the ordering problems we study here. In particular, we show that the BKW Case for the path is NP-complete.

1 Introduction

Here we study the problem of constructing a phylogenetic tree for a set of species [7]. A *qualitative character* assigns to each species a *state* from a set of states, e.g., "is a vertebrate", or "number of legs". When the evolution of the states of the character is known, e.g., evolution from invertebrate to vertebrate is only forward, the character is called *cladistic*. This evolution of the states is usually represented by a rooted tree, called a *character tree*, on the set of

J. Chen, J. Wang, and A. Zelikovsky (Eds.): ISBRA 2011, LNBI 6674, pp. 440–451, 2011.

states. The *Qualitative Character Compatibility* Problem, or *Perfect Phylogeny* Problem, is NP-complete [4,23], while it is polynomial-time solvable when any of the associated parameters is fixed [1,15,16,19]. When characters are cladistic, the problem, called the *Cladistic Character Compatibility* Problem, is the problem of finding a perfect phylogeny tree on the set of species such that it can be contracted to a subtree of each character tree. This problem is polynomial-time solvable [6,12,25].

Experimental research in molecular biology [14,17,24,8] shows that traits can disappear and then reappear during the evolution of a species, suggesting that genes contain information about traits that are not always expressed. In [3,2], the authors argue that a new model for characters is needed in order for the resultant phylogenetic trees to capture this phenomenon. The authors thus devise the *generalised character*, which assigns to each species a *subset* of a set of states, where we only know that the expressed trait (state) is in this subset. The *Generalised Cladistic Character Compatibility* (GCCC) Problem is then the Cladistic Character Compatibility Problem on a set of species with generalised characters where we first have to pick one state from the subset for each character. Interestingly, generalised characters capture also the case of qualitative characters with missing data (the "Incomplete Perfect Phylogeny" Problem). Here, missing data can be replaced by a "wildcard" generalised state containing all possible states of the character. This problem was shown to be NP-complete even if the number of states is constant in [13].

The authors of [3,2] give a polynomial-time algorithm for the case of the GCCC Problem where for each character, the set of states of each species forms a directed path in its character tree. It thus follows that if the character trees are non-branching, then the Incomplete Cladistic Character Compatibility Problem can be solved in polynomial time. The complexity of this case when each character has at most two states was further improved in [22]. In [3,2], it was shown that the GCCC Problem is NP-complete using a construction involving character trees that are branching. However, the authors argued that in this setting the situation when a trait becomes hidden and then reappears does not happen, hence in [3] they posed an open case of the GCCC Problem where each character tree has one branch $0 \to 1 \to 2$ and the collection of sets of states for each species is $\{\{0\}, \{1\}, \{2\}, \{0, 2\}\}$. We call this the *Benham-Kannan-Warnow* (BKW) Case. They then showed in [2] that if a "wildcard" set $\{0, 1, 2\}$ is added to the collection, the problem is NP-complete.

In [18], the authors then study the complexity of cases of the GCCC Problem for non-branching character trees with 3 states and set of states chosen from the set $\{\{0\}, \{1\}, \{2\}, \{0, 2\}, \{0, 1, 2\}\}$ when the phylogeny tree that is a solution to this problem is restricted to be (a) any single-branch tree, (b) path or (c) tree, cf. Table 1. In [11], the authors state that searching for path phylogenies is strongly motivated by the characteristics of human genotype data: 70% of real instances that admit a tree phylogeny also admit a path phylogeny. In [18], the authors have the following results. For (5a–b) of Table 1 they show that the problem is equivalent to the *Consecutive-Ones Property* (C1P) Problem [9,20]. They then

Table 1. Complexity of all cases of the GCCC Problem for the character tree $0 \rightarrow 1 \rightarrow 2$ and set of states chosen from the set $\mathcal{Q} \subseteq \{\{0\}, \{1\}, \{2\}, \{0,2\}, \{0,1,2\}\}$. The BKW Case is marked with *.

	$\mathcal{Q}$\soln	(a) branch	(b) path	(c) tree
(1)	$\mathcal{Q} \subseteq \{\{0\}, \{1\}, \{2\}\}$	P [18]	P [18]	P [3,2]
(2)	$\{\{0,1,2\}\} \subseteq \mathcal{Q} \subseteq \{\{0\}, \{1\}, \{2\}, \{0,1,2\}\}; |\mathcal{Q}| \leq 2$	trivial	trivial	trivial
(3)	$\{\{0,1,2\}\} \subseteq \mathcal{Q} \subseteq \{\{0\}, \{1\}, \{2\}, \{0,1,2\}\}; |\mathcal{Q}| \geq 3$	P (Th. 1)	NP-c (Th. 3)	P [3,2]
(4)	$\mathcal{Q} \subseteq \{\{0\}, \{0,2\}, \{0,1,2\}\}$ or $\mathcal{Q} \subseteq \{\{2\}, \{0,2\}, \{0,1,2\}\}$	trivial	trivial	trivial
(5)	$\{\{1\}, \{0,2\}\}$	P [18]	P [18]	?
(6)	$\{\{0\}, \{1\}, \{0,2\}\}$	P [18]	NP-c (Th. 3)	?
(7)	$\{\{0\}, \{2\}, \{0,2\}\}(\cup\{\{0,1,2\}\})$	P (Th. 1)	NP-c (Th. 3)	P [3,2]
(8)	$\{\{1\}, \{2\}, \{0,2\}\}$	P [18]	P (Cor. 1)	?
(9)	$\{\{0\}, \{1\}, \{2\}, \{0,2\}\}$ *	P [18]	NP-c (Th. 3)	?
(10)	$\{\{1\}, \{0,2\}, \{0,1,2\}\} \subseteq \mathcal{Q}$	NP-c [18]	NP-c (Th. 2)	NP-c [2]

show that the BKW Case is polynomial-time solvable by giving an algorithm based on PQ-trees [5,20] used for the C1P Problem, giving also the entries (6a), (8a) and (9a) of Table 1. They show that (10a) is NP-complete by reduction from the *Path Triple Consistency* (PTC) Problem (cf. Section 3). Finally, they observe that (1a–b) are special cases of the Qualitative Character Compatibility Problem that are polynomial by [1].

In this work, we characterise the complexity of the remaining cases considered in [18] of the GCCC Problem for non-branching character trees for (a) and (b), completing these two columns in Table 1. This paper is structured as follows. In Section 2 we formally define the Generalised Cladistic Character Compatibility Problem. In Section 3 we study several types of ordering problems, some being polynomial, while others are NP-complete; one of them is then used to determine the complexity of several cases in Table 1. Section 4 contains the tractability results of this work. Section 4.1 gives a polynomial-time algorithm based on that of [3,2] for the case of the GCCC Problem for (a) where for each character, the set of states of each species forms a directed path in its character tree, giving entries (3a) and (7a) of Table 1. In Section 4.2 we show that case (8b) is polynomial by showing that any instance of this case can be reduced to solving an instance of polynomial case (8a) of [18]. In Section 5, we show that case (10b) is NP-complete by reduction from the PTC Problem of Section 3, and that cases (3b), (6b), (7b) and (9b) are NP-complete by reduction from the LEF-PTC Problem of Section 3. Note that this last result includes the fact that case (9b), the BKW Case of the GCCC Problem for (b) is NP-complete. Finally, Section 6 concludes the paper with some open problems and future work.

2   Generalised Cladistic Character Compatibility Problem

Let S be a set of species. A *generalised (cladistic) character* [3,2] on S is a pair $\hat{\alpha} = (\alpha, T_\alpha)$, such that:

(a) α is a function $\alpha : S \rightarrow 2^{Q_\alpha}$, where Q_α denotes the set of states of $\hat{\alpha}$.
(b) $T_\alpha = (V(T_\alpha), E)$ is a rooted character tree with nodes bijectively labelled by the elements of Q_α.

The *Generalised Cladistic Character Compatibility* (GCCC) Problem is to find a perfect phylogeny [4] of a set of species with generalised characters:

Generalised Cladistic Character Compatibility (GCCC) Problem
Input: A set S of species and a set C of generalised characters on S.
Question: Is there a rooted tree $T = (V_T, E_T)$ and a "state-choosing" function $c : V_T \times C \to \bigcup_{\hat{\alpha} \in C} Q_\alpha$ such that the following holds:

(1) For each species $s \in S$ there is a vertex v_s in T such that for each $\hat{\alpha} \in C$, $c(v_s, \hat{\alpha}) \in \alpha(s)$.

(2) For every $\hat{\alpha} \in C$ and $i \in Q_\alpha$, the set $\{v \in V_T \mid c(v, \hat{\alpha}) = i\}$ is a connected component of T.

(3) For every $\hat{\alpha} \in C$, the tree $T(\alpha)$ is an induced subtree of T_α, where $T(\alpha)$ is the tree obtained from T by labelling the nodes of T only with their α-states (as chosen by c), and then contracting edges having the same α-state at their endpoints.

Essentially, the first condition is that each species is represented somewhere in the tree T, and the second condition is that the set of nodes labelled by a given state of a given character form a connected subtree of T, just as with the Character Compatibility Problem. Finally, condition three is that the state transitions for each character $\hat{\alpha}$ must respect its character tree T_α.

The GCCC Problem is NP-complete [3,2], however it is polynomial for many special cases of the problem [3,2,18]. We will consider the following variants of the GCCC Problem. The GCCC Problem with non-branching character trees (GCCC-NB Problem) is a special case of the GCCC Problem in which character trees have a single branch, i.e., each character tree T_α is $0 \to 1 \to \cdots \to |T_\alpha| - 1$. If we restrict the solution of the GCCC-NB Problem (a phylogeny tree) to have only one, or two branches starting at the root, we will call this problem the Single-Branch GCCC-NB (SB-GCCC-NB) Problem, and the Path GCCC-NB (P-GCCC-NB) Problem, respectively. In addition, if in any of these problems, say in problem X, we restrict the set of states to be from the set Q, we will call this problem the Q-X Problem. Table 1 summarises the cases studied here.

3   Ordering Problems

In this section, we discuss several different types of ordering problems. These problems are related to the Single-Branch and Path GCCC-NB Problems. We will use one of these variants to obtain a hardness result in Section 5.

The *Path Triple Consistency* (PTC) Problem is a simplified version of the extensively studied *Quartet Consistency* (QC) Problem [23]. In the QC Problem, given a set S and the collection of quartets $(a_i, b_i : c_i, d_i)$, where $a_i, b_i, c_i, d_i \in S$, the task is to construct a tree T containing vertices S such that for each quartet there is an edge of T whose removal separates vertices $\{a_i, b_i\}$ from vertices $\{c_i, d_i\}$. This problem was shown to be NP-complete in [23]. In [18], we have observed that the problem remains NP-complete when we restrict the tree to be a path. In this case it is easy to see that (i) we can assume the path contains only vertices in S and (ii) each quartet $(a_i, b_i : c_i, d_i)$ can be replaced with the three triples $(a_i, b_i : c_i)$, $(a_i, b_i : d_i)$ and $(c_i, d_i : a_i)$. The PTC Problem can be viewed as a total ordering problem with negative constraints $c_i \notin [a_i, b_i]$, where $[a_i, b_i]$ is the set of all elements between a_i and b_i in the total ordering. The *Total Ordering* (TO) problem with positive constraints $c_i \in [a_i, b_i]$ was shown to be NP-complete in [21].

Here, we study two subclasses of the PTC Problem and one subclass of the TO Problem in which one element of each constraint is fixed:

Left Element Fixed Path Triple Consistency (LEF-PTC) Problem
Input: A set $S = \{1, \ldots, n\}$, an element $r \notin S$, and a set of triples $\{(a_i, r : c_i)\}_{i=1}^k$ where $a_i, c_i \in S$ for every $i \in \{1, \ldots, k\}$.
Question: Is there a path (an ordering) P on vertices $S \cup \{r\}$ such that for each $i \in \{1, \ldots, k\}$, there is an edge of P whose removal separates $\{r, a_i\}$ from c_i.

Right Element Fixed Path Triple Consistency (REF-PTC) Problem
Input: A set $S = \{1, \ldots, n\}$, an element $r \notin S$, and a set of triples $\{(a_i, b_i : r)\}_{i=1}^k$ where $a_i, b_i \in S$ for every $i \in \{1, \ldots, k\}$.
Question: Is there a path (an ordering) P on vertices $S \cup \{r\}$ such that for each $i \in \{1, \ldots, k\}$, there is an edge of P whose removal separates $\{a_i, b_i\}$ from r.

One Element Fixed Total Ordering (OEF-TO) Problem
Input: A set $S = \{1, \ldots, n\}$, an element $r \notin S$, and a set of triples $\{(a_i, b_i, c_i)\}_{i=1}^k$ where for every $i \in \{1, \ldots, k\}$, either $a_i, c_i \in S$ and $b_i = r$, or $a_i, b_i \in S$ and $c_i = r$.
Question: Is there a path (an ordering) P on vertices $S \cup \{r\}$ such that for each $i \in \{1, \ldots, k\}$, b_i appears between a_i and c_i on P.

In what follows, we will show that the first problem (LEF-PTC) is NP-complete, while the other two problems (REF-PTC and OEF-TO) are solvable in polynomial time. Thus, the LEF-PTC Problem seems to be the simplest version of the problem which is still intractable.

Lemma 1. *The LEF-Path Triple Consistency Problem is NP-complete.*

Proof. Here, we give a reduction from *Not-All-Equal-3SAT* (NAE-3SAT) [10]. The NAE-3SAT Problem is: given a set of Boolean variables $X = \{x_1, \ldots, x_n\}$ and a set of clauses $\{C_1, \ldots, C_m\}$, where each clause contains three literals, is there a truth assignment to the set of variables such that in no clause, its three literals are all true or all false. Given an instance of NAE-3SAT, let S be the union of variable symbols $\{v_1, \bar{v}_1, \ldots, v_n, \bar{v}_n\}$ and literal symbols $\{\ell_1^1, \ell_1^2, \ell_1^3, \ldots, \ell_m^1, \ell_m^2, \ell_m^3\}$.

The basic principle of the reduction is the following observation. The triple $(a_i, r : c_i)$ is equivalent to the following condition on the elements in $S \cup \{r\}$:

$$r < c_i \Leftrightarrow a_i < c_i. \tag{1}$$

The Boolean value of predicate $r < v_i$ will represent the value of variable x_i, for $i \in \{1, \ldots, n\}$. First, we introduce the triples $(v_i, r : \bar{v}_i)$ and $(\bar{v}_i, r : v_i)$, for

$i \in \{1, \ldots, n\}$. These triples are equivalent to the following logical statement: $r < \bar{v}_i \Leftrightarrow v_i < \bar{v}_i \Leftrightarrow v_i < r$. Hence, they enforce $\bar{v}_i < r$ iff $r < v_i$, and hence the Boolean value of predicate $r < \bar{v}_i$ represents the value of $\neg x_i$.

Now, let clause C_j contain variables x_{k_1}, x_{k_2} and x_{k_3}. We will use symbols $\ell_j^1, \ell_j^2, \ell_j^3$ to represent the values of the three literals of C_j: the Boolean value of the i-th literal of C_j will be equal to the value of predicate $r < \ell_j^i$. To achieve this, we will reuse the above constraints. For each variable x_{k_i} with positive occurrence in C_j, we introduce the triples $(\ell_j^i, r : \bar{v}_{k_i})$ and $(\bar{v}_{k_i}, r : \ell_j^i)$, and for each variable x_{k_i} with a negated occurrence in C_j, triples $(\ell_j^i, r : v_{k_i})$ and $(v_{k_i}, r : \ell_j^i)$. These triples will guarantee that predicate $r < \ell_j^i$ represents the Boolean value of the i-th literal of C_j. The reason why we have a symbol for each literal is that the position of the literal symbol ℓ_j^i and the position of the variable symbol v_{k_i} (or $\bar{v}_{k_i}$) are only very weakly dependent: one is smaller than r if and only if the other is, but otherwise they are independent. This is important, since the clause gadgets introduced in the next paragraph might put some ordering restrictions on its literal symbols, and hence if we would use the variable symbols v_{k_i} ($\bar{v}_{k_i}$) in several clause gadgets, the ordering restrictions from different clause gadgets might not be compatible.

The clause gadget for clause C_j will contain the three triples $(\ell_j^1, r : \ell_j^2)$, $(\ell_j^2, r : \ell_j^3)$ and $(\ell_j^3, r : \ell_j^1)$. The purpose of these constraints is to guarantee that in any ordering at least one and not all literals in the clause C_j are true. For instance, assume that all literals are true, i.e., $r < \ell_j^i$ for $i \in \{1, 2, 3\}$. By (1), this is equivalent to $\ell_j^1 < \ell_j^2$, $\ell_j^2 < \ell_j^3$ and $\ell_j^3 < \ell_j^1$, which leads to a contradiction. Similarly, if literals are false in the ordering, all three inequalities will reverse their direction, and we get a contradiction again. Hence, each clause is satisfied and predicates $r < v_i$ define a solution to the instance of NAE-3SAT.

Now, assume that the instance of NAE-3SAT has a solution $\psi : X \to \{\text{false}, \text{true}\}$. Consider the ordering of elements of $S \cup \{r\}$ satisfying the following conditions:

(a) for each $v_i \in \{v_1, \ldots, v_n\}$, v_i appears to the right of r, i.e., $r < v_i$ in the ordering, if and only if $\psi(x_i) = \text{true}$ for the x_i corresponding to v_i;

(b) for each clause C_j, the relative ordering of the literal symbols $\ell_j^1, \ell_j^2, \ell_j^3$ and r is one of the following: $(\ell_j^1, r, \ell_j^2, \ell_j^3)$, $(\ell_j^3, \ell_j^2, r, \ell_j^1)$, $(\ell_j^2, r, \ell_j^3, \ell_j^1)$, $(\ell_j^1, \ell_j^3, r, \ell_j^2)$, $(\ell_j^3, r, \ell_j^1, \ell_j^2)$ and $(\ell_j^2, \ell_j^1, r, \ell_j^3)$.

Note that for any valid combination of truth assignments to the literals of C_j, there is one ordering in the list above. This ordering imposes a restriction on the relative ordering of the two literal symbols appearing on the same side of r, the reason why we created the literal symbols. It is easy to see that for each $s \in S$, other than on which side of r the s appears, there is at most one constraint specifying its relative ordering to another element. Hence, it is always possible to find an ordering satisfying the above conditions.

Let us verify that this ordering satisfies all triple constraints. The constraints $(v_i, r : \bar{v}_i)$ and $(\bar{v}_i, r : v_i)$ (respectively, $(\ell_j^i, r : v_{k_i})$ and $(v_{k_i}, r : \ell_j^i)$; $(\ell_j^i, r : \bar{v}_{k_i})$ and $(\bar{v}_{k_i}, r : \ell_j^i)$) are satisfied just by the placement of symbols to the correct sides of r. For instance, if $r < v_i$ then the relative ordering of $v_i, \bar{v}_i, r$ is $\bar{v}_i, r, v_i$ and this ordering satisfies both triples. For the constraints for clause C_j, only the relative ordering of elements $\ell_j^1, \ell_j^2, \ell_j^3$ and r is important. It is easy to check that any of the six orderings of these elements listed above satisfies all three triples for C_j. Hence, the constructed ordering is a solution to the corresponding instance of the LEF-PTC Problem. $\square$

Lemma 2. *Any instance of the REF-Path Triple Consistency Problem always has a solution, and thus the problem is solvable in constant time.*

Proof. Consider any ordering of $S \cup \{r\}$ with r as the first (last) element. Then the first (last) edge separates r from any pair of elements in S. Thus, such an ordering is a solution to any instance of the REF-PTC Problem. $\square$

Lemma 3. *The OEF-Total Ordering Problem can be solved in linear time.*

Proof. The algorithm will work in two stages. In the first stage the elements will be clustered into parts each appearing on different sides of r. In the second stage, we will determine the ordering of the elements in each part.

Constraint (a_i, r, c_i) is satisfied if and only if a_i and c_i appear on opposite sides of r. Constraint (a_i, b_i, r) is satisfied if and only if (i) a_i and b_i appear on the same side of r, and (ii) b_i is closer to r than a_i, which we write as $b_i \prec a_i$. Consider the graph with vertex set S and edges between any two vertices $u, v \in$ such that u and v appear together in some triple (a_i, b_i, c_i). Let C be a connected component of this graph. It is easy to see that once we fix the side of one element in the component, the side of all elements in the component will be determined. Hence, we can uniquely partition C into two (paired) clusters such that all edges from constraints of type (a_i, r, c_i) are between two clusters and all edges from constraints of type (a_i, b_i, r) are inside one of the two clusters. Now, pick one cluster from each pair and place all its elements on one side of r and all other clusters to the other side. Note that there can many ways how to do this, the number of ways is exponential in the number of pairs of clusters.

It remains to satisfy the precedence conditions. These conditions $(b_i \prec a_i)$ define a partial ordering on each side of r. Any total ordering compatible with these partial orderings will form a solution to the problem. Such an ordering can be found in time $O(n + k)$. $\square$

4   Tractability Results

4.1   An Algorithm for Cases of the SB-GCCC Problem

Here we show that when each $\alpha(s)$ induces a directed path in T_α, for each $\hat{\alpha} \in C$, $s \in S$, the SB-GCCC Problem is polynomial-time solvable. The algorithm we use, while much simpler, is based on the algorithm given in [2].

Theorem 1. *The SB-GCCC Problem is solvable in time $O(|S| \sum_{\hat{\alpha} \in C} |Q_\alpha|)$, if each $\alpha(s)$ induces a directed path in T_α, for each $\hat{\alpha} \in C$, $s \in S$.*

Proof. Consider an instance of the SB-GCCC Problem (S, C) with the required property. Let $\text{start}_\alpha(s)$ and $\text{end}_\alpha(s)$ be the first and the last node on the directed path induced by $\alpha(s)$. We define the partial ordering on the nodes of T_α by saying $v \preccurlyeq_\alpha w$ if the directed path from the root r_α of T_α to w passes through v. Similarly, for each solution (T, c) we define the partial ordering $\preccurlyeq_T$ on S based on T. Since T has a single branch, $\preccurlyeq_T$ is a total ordering, i.e., for every $s_1, s_2 \in S$, s_1 and s_2 are comparable by $\preccurlyeq_T$. Hence, for every $\hat\alpha \in C$, $c(s_1, \hat\alpha)$ and $c(s_2, \hat\alpha)$ are comparable by $\preccurlyeq_\alpha$. Therefore, for all $s \in S$, $c(s, \hat\alpha)$ lie on a single branch (directed path starting in the root) P_α of T_α. Since $\text{start}_\alpha(s_1) \preccurlyeq_\alpha c(s_1, \hat\alpha)$ and $\text{start}_\alpha(s_2) \preccurlyeq_\alpha c(s_2, \hat\alpha)$, we can assume that for all $s \in S$, $\text{start}_\alpha(s)$ lie on a subpath P'_α of P_α starting in the root r_α of T_α and ending in $\text{start}_\alpha(\ell_\alpha)$, where $\ell_\alpha \in S$ and $\text{start}_\alpha(s) \preccurlyeq_\alpha \text{start}_\alpha(\ell_\alpha)$ for every $s \in S$. If that is not the case, there is no solution. This can be checked in time $O(|S||Q_\alpha|)$ for each $\hat\alpha \in C$.

Next, we will argue that it is enough to consider only solutions in which c maps all elements in S to P'_α. Consider a solution (T, c). Any $c(s, \hat\alpha) \notin P'_\alpha$ must lie on the subpath of P_α ending at vertex $\text{start}_\alpha(\ell_\alpha)$. Since $\text{start}_\alpha(s) \preccurlyeq_\alpha \text{start}_\alpha(\ell_\alpha)$, we can remap $c(s, \hat\alpha)$ to $\text{start}_\alpha(\ell_\alpha)$. It is easy to check that conditions (1)–(3) of the GCCC Problem remain satisfied after mapping all such $c(s, \hat\alpha)$ to $\text{start}_\alpha(\ell_\alpha)$. Hence, we can assume that $c(s, \hat\alpha) \in \alpha'(s) = \alpha(s) \cap P'_\alpha$, for each $\hat\alpha \in C$ and $s \in S$. Note that for all $s \in S$, $\alpha'(s)$ induce directed subpaths of P'_α.

Now, we are ready to present the algorithm for solving the SB-GCCC problem with the required property. First, we will build a set $\mathcal{C}$ of constraints on the ordering of the nodes of T which have to be satisfied in any solution (T, c). If for $s_1, s_2 \in S$ and $\hat\alpha \in C$, the paths induced by $\alpha'(s_1)$ and $\alpha'(s_2)$ are disjoint, and the path induced by $\alpha'(s_1)$ is closer to the root r_α, then we must have $s_1 \prec_T s_2$. Therefore, we add this constraint to the set $\mathcal{C}$. Let T be a single branch tree that satisfies all these constraints in $\mathcal{C}$ and let $s_1 \prec_T s_2 \prec_T \cdots \prec_T s_{|S|}$ be the elements of S ordered according to this tree. (If such a tree does not exist, there is no solution.) For each character $\hat\alpha \in C$, we will map $c(s_i, \hat\alpha)$ to $\alpha'(s_i)$ using Algorithm 1, where $\max(a, b)$ is the element (a or b) further from the root if a and b are comparable, and undefined otherwise.

Algorithm 1. Iterative algorithm that assigns to each species a state

---
1: $c(s_1, \hat\alpha) \leftarrow \text{start}_\alpha(s_1)$
2: **for** $i = 2$ up to $|S|$ **do**
3:     $c(s_i, \hat\alpha) \leftarrow \max(\text{start}_\alpha(s_i), c(s_{i-1}, \hat\alpha))$
4: **end for**

Let us verify that (T, c) is indeed a solution. First, note that since all $\text{start}_\alpha(s_i)$ lie on the path P'_α, the arguments of the max function are always comparable. Furthermore, it is easy to see that all $c(s_i, \hat\alpha)$ are assigned to the set $\{\text{start}_\alpha(s); s \in S\}$, and that $c(s_1, \hat\alpha) \preccurlyeq_\alpha c(s_2, \hat\alpha) \preccurlyeq_\alpha \cdots \preccurlyeq_\alpha c(s_{|S|}, \hat\alpha)$. It remains to show that for each i, $c(s_i, \hat\alpha) \in \alpha'(s_i)$. Let i be the smallest index for which $c(s_i, \hat\alpha) \notin \alpha'(s_i)$. We must have that $\text{end}_\alpha(s_i) \prec_\alpha c(s_i, \hat\alpha)$. Since $c(s_i, \hat\alpha) = \text{start}_\alpha(s_j)$ for some $j < i$, the subpath of P'_α induced by $\alpha'(s_i)$ is closer to the root than the subpath induced by $\alpha'(s_j)$. Hence, $\mathcal{C}$ must contain the constraint $s_i \prec_T s_j$, which contradicts the fact that T satisfies all these constraints. It follows that (T, c) is a solution.

Finally, let us analyse the running time of the algorithm. We can verify whether this set $\mathcal{C}$ of constraints defines a partial ordering and find a total ordering T compatible with this partial ordering in time $O(|S| + m)$, where m is the number of constraints. For each $\hat{\alpha} \in C$, we can have at most $|Q_\alpha|$ disjoint induced paths, and it is enough to consider the constraint between the neighbouring disjoint induced paths only. Hence, $m = O(\sum_{\hat{\alpha} \in C} |Q_\alpha|)$.   □

We remark that this type of theorem does not hold for the case of path phylogeny, cf. Table 1.

4.2 The $\{\{1\}, \{2\}, \{0, 2\}\}$-P-GCCC-NB Problem

We will show that if there is a solution to an instance (S, C) of the $\mathcal{Q}^*$-P-GCCC-NB Problem then there is a solution to the instance (S, C) of the $\mathcal{Q}^*$-SB-GCCC-NB Problem, and vice versa, where $\mathcal{Q}^* = \{\{1\}, \{2\}, \{0, 2\}\}$. Since the single branch version of this problem can be solved in polynomial time [18], it follows that also the path version is polynomial-time solvable.

Lemma 4. *An instance (S, C) of the $\{\{1\}, \{2\}, \{0, 2\}\}$-P-GCCC-NB Problem has a solution if and only if the instance (S, C) of the $\{\{1\}, \{2\}, \{0, 2\}\}$-SB-GCCC-NB Problem has a solution.*

Proof. Let $\mathcal{Q}^* = \{\{1\}, \{2\}, \{0, 2\}\}$. Obviously, a solution to the instance (S, C) of the $\mathcal{Q}^*$-SB-GCCC-NB Problem is also a solution to the instance (S, C) of the $\mathcal{Q}^*$-P-GCCC-NB Problem. Now, assume that (T, c) is a solution to the instance (S, C) of the $\mathcal{Q}^*$-P-GCCC-NB Problem. Let P_1 and P_2 be two branches of T starting at the root r. Let T' be the tree obtained by attaching P_2 to the last vertex of P_1. To define the state-choosing function c' we only need to determine the values of $c'(s, \hat{\alpha})$ when $\alpha(s) = \{0, 2\}$. Consider $s \in S$ and $\hat{\alpha} \in C$ such that $\alpha(s) = \{0, 2\}$. If there is a species $s' \prec_T s$ such that $\alpha(s') = \{1\}$ then we set $c'(s, \hat{\alpha}) = 2$, otherwise we set $c'(s, \hat{\alpha}) = 0$. We will show that (T', c') is a solution to the instance (S, C) of the $\mathcal{Q}^*$-SB-GCCC-NB Problem.

For each $\hat{\alpha} \in C$, the set of species $S_{\hat{\alpha}, \{1\}} = \{s \in S | \alpha(s) = \{1\}\}$ must induce a connected component in T. Since $\alpha(r) = 0$, this component lies entirely in P_1 or in P_2. Hence, the set $S_{\hat{\alpha}, \{1\}}$ induces a connected component K in T' as well. By the definition of c', all species that lie below K in T' are assigned value 2 and all species s such that $\alpha(s) = \{0, 2\}$ that lie above K in T' are assigned value 0. Hence, the only possible violation is if there is a species s such that $\alpha(s) = \{2\}$ that lies above K in T'. This species s either lies above K in T or lies in the branch that does not contain K in T. In either case, (T, c) cannot be a solution to the instance (S, C) of the $\mathcal{Q}^*$-P-GCCC-NB Problem, a contradiction.   □

Corollary 1. *The $\{\{1\}, \{2\}, \{0, 2\}\}$-P-GCCC-NB Problem is polynomial-time solvable.*

5 Hardness Results

In this section, we show that the $\{\{1\}, \{0, 2\}, \{0, 1, 2\}\}$- and $\{\{0\}, \{1\}, \{0, 1\}\}$-P-GCCC-NB Problems are NP-complete. We will show the first result by reduction from the *Path Triple Consistency* (PTC) Problem [18] and the second from the *Left Element Fixed Path Triple Consistency* (LEF-PTC) Problem (Lemma 1).

In [18], it was shown using a reduction from the PTC Problem that the $\{\{1\}, \{0, 2\}, \{0, 1, 2\}\}$-GCCC-NB and -SB-GCCC-NB Problems are NP-complete. We will use the same reduction to show that the $\{\{1\}, \{0, 2\}, \{0, 1, 2\}\}$-P-GCCC-NB Problem is NP-complete as well.

Theorem 2. *The $\{\{1\}, \{0, 2\}, \{0, 1, 2\}\}$-P-GCCC-NB Problem is NP-complete.*

Proof. Let $\mathcal{Q}^\triangle = \{\{1\}, \{0, 2\}, \{0, 1, 2\}\}$. Let S and $\{(a_i, b_i : c_i)\}_{i=1}^k$ be an instance of the PTC Problem. We will construct an instance of the $\mathcal{Q}^\triangle$-P-GCCC-NB Problem as follows. Let S be the set of species and $C = \{\hat{\alpha}_1, \ldots, \hat{\alpha}_k\}$ the set of characters. For every $\hat{\alpha} \in C$, we let $\alpha_i(a_i) = \alpha_i(b_i) = \{1\}$, $\alpha_i(c_i) = \{0, 2\}$ and for all $s \in S \setminus \{a_i, b_i, c_i\}$, $\alpha_i(s) = \{0, 1, 2\}$.

We will show that the instance of the PTC Problem has a solution if and only if the constructed instance of the $\mathcal{Q}^\triangle$-P-GCCC-NB Problem has a solution. First, consider a path P containing vertices S which is a solution to the constructed instance. Consider the ordering of elements in S as they occur on P starting from the leaf on one branch of P and ending with the leaf on the other branch. For every $i \in \{1, \ldots, k\}$, all elements in $[a_i, b_i]$ must have state 1 for character $\hat{\alpha}_i$, hence, $c_i \notin [a_i, b_i]$, i.e., this ordering is a solution to the PTC Problem.

On the other hand, let ordering O be a solution to the PTC Problem. Consider a tree T with a single branch consisting of the all-zero root followed by vertices in S ordered by O. Note that, for every $i \in \{1, \ldots, k\}$, c_i appears either above both a_i and b_i, or below them. The state-choosing function is defined as follows. For every node in S, we choose for character $\hat{\alpha}_i$ state 0 if they are above both a_i and b_i, state 1 if they are between a_i and b_i, and state 2 otherwise. Clearly, this tree is compatible with all character trees and it is easy to see that each $c(s, \hat{\alpha}) \in \alpha(s)$, i.e., T is a solution to the $\mathcal{Q}^\triangle$-P-GCCC-NB Problem. □

We now show that the $\{\{0\}, \{1\}, \{0, 1\}\}$-P-GCCC-NB Problem is NP-complete, by giving a reduction from the LEF-PTC Problem (Lemma 1).

Theorem 3. *The $\{\{0\}, \{1\}, \{0, 1\}\}$-P-GCCC-NB Problem is NP-complete.*

Proof. Given an instance of the LEF-PTC Problem $S = \{1, \ldots, n\}$, r, and the set of k triples $(a_i, r : c_i)$, let S be the set of species, and $C = \{\hat{\alpha}_1, \ldots, \hat{\alpha}_k\}$ be the set of characters. For each $\hat{\alpha}_i \in C$, we let $\alpha_i(a_i) = \{0\}$ and $\alpha_i(c_i) = \{1\}$, while for all other $s \in S \setminus \{a_i, c_i\}$ we let $\alpha_i(s) = \{0, 1\}$.

Let path phylogeny T be a solution to this instance of the $\{\{0\}, \{1\}, \{0, 1\}\}$-P-GCCC-NB Problem. Let r be the root of T, i.e., r is the all-zero vertex. Consider the ordering of elements in $S \cup \{r\}$ based on the ordering of vertices on path T starting in the leaf of one branch and ending in the leaf of the other branch.

Assume the triple $(a_i, r : c_i)$ is not valid, i.e., c_i appears between a_i and r. However, this is not possible since vertex a_i is then below c_i in T and we have a transition from 1 to 0 somewhere on the path from c_i to a_i for character $\hat{\alpha}_i$. Hence, the ordering is a solution to the LEF-PTC Problem.

Conversely, let path/ordering P be a solution to the LEF-PTC Problem. Consider the path phylogeny obtained from P by rooting it at r and the state-choosing function assigning 1 to c_i and all nodes below c_i and 0 to all other nodes for character $\hat{\alpha}_i$. Clearly, this tree is compatible with all character trees. The state choosing function could only fail, if a_i is below c_i, in which case $c(a_i, \hat{\alpha}_i) = 1$, but $\alpha_i(a_i) = \{0\}$. However, this is not possible as then c_i would be between r and a_i on P which violates the constraint $(a_i, r : c_i)$. The claim follows by Lemma 1.   □

Note that Theorem 3 implies NP-completeness of several cases of the P-GCCC-NB Problem. In fact, any case of the problem in which set $\mathcal{Q}$ contains two distinct state singletons $\{a\}$ and $\{b\}$, and a set containing states 0, c and d such that $a \preccurlyeq_\alpha c \preccurlyeq_\alpha b$ and $b \preccurlyeq_\alpha d$ in T_α is NP-complete. For instance, for $a = c = 0$, $b = 1$ and $d = 2$, we have that the $\{\{0\}, \{1\}, \{0, 2\}\}$-P-GCCC-NB Problem is NP-complete (((6b) in Table 1).

6   Conclusions and Open Problems

We have characterised the complexity of the remainder of the cases of the $\mathcal{Q}$-SB-GCCC-NB and $\mathcal{Q}$-P-GCCC-NB Problems for $\mathcal{Q} \subseteq \{\{0\}, \{1\}, \{2\}, \{0, 2\}, \{0, 1, 2\}\}$. This leaves open, however, some interesting cases of the GCCC-NB Problem. In [18], the authors show that when $\mathcal{Q}' = \{\{1\}, \{0, 2\}\}$, the input corresponds to a binary matrix M, hence the $\mathcal{Q}'$-SB-GCCC-NB Problem is equivalent to the C1P Problem. That is, the $\mathcal{Q}'$-SB-GCCC-NB (resp., $\mathcal{Q}'$-GCCC-NB) Problem is to find a single-branch path (resp., tree) with vertex set containing the columns of M such that for each row of M, the set of vertices labelled 1 by this row forms a connected subpath (resp., subtree), i.e., M has the C1P (resp., a "connected-ones property" of trees). If we can determine in poly-time that the connected-ones property holds (like we can for the C1P), it might provide an answer to the BKW Case. Finally, it would be interesting to study these problems for all subsets of $2^{\{0,1,2\}}$, as it would complete the study for all possible inputs to the GCCC-NB Problem when character trees are $0 \to 1 \to 2$.

References

1. Agarwala, R., Fernandez-Baca, D.: A polynomial-time algorithm for the perfect phylogeny problem when the number of character states is fixed. SIAM J. Computing 26(6), 1216–1224 (1994)
2. Benham, C., Kannan, S., Paterson, M., Warnow, T.: Hen's teeth and whale's feet: Generalized characters and their compatibility. J. Computational Biology 2(4), 515–525 (1995)

3. Benham, C., Kannan, S., Warnow, T.: Of chicken teeth and mouse eyes, or generalized character compatibility. In: Galil, Z., Ukkonen, E. (eds.) CPM 1995. LNCS, vol. 937, pp. 17–26. Springer, Heidelberg (1995)
4. Bodlaender, H., Fellows, M., Warnow, T.: Two strikes against perfect phylogeny. In: Kuich, W. (ed.) ICALP 1992. LNCS, vol. 623, pp. 273–283. Springer, Heidelberg (1992)
5. Booth, K.S., Lueker, G.S.: Testing for the consecutive ones property, interval graphs, and graph planarity using PQ-tree algorithms. J. Computer and System Sciences 13(3), 335–379 (1976)
6. Estabrook, G., McMorris, F.: When is one estimate of evolutionary relationships a refinement of the another? J. Mathematical Biology 10, 327–373 (1980)
7. Felsenstein, J.: Numerical methods for inferring evolutionary trees. The Quarterly Review of Biology 57(4), 379–404 (1982)
8. Figuera, L., Pandolfo, M., Dunne, P., Cantu, J., Patel, P.: Mapping the congenital generalized hypertrichosis locus to chromosome Xq24-q27.1. Nature 10, 202–207 (1995)
9. Fulkerson, D., Gross, O.: Incidence matrices and interval graphs. Pacific J. Mathematics 15, 835–855 (1965)
10. Garey, M.R., Johnson, D.S.: Computers and Intractability: A Guide to the Theory of NP-Completeness. W.H. Freeman, New York (1979)
11. Gramm, J., Nierhoff, T., Sharan, R., Tantau, T.: Haplotyping with missing data via perfect path phylogenies. Discrete Applied Mathematics 155, 788–805 (2007)
12. Gusfield, D.: Efficient algorithms for inferring evolutionary trees. Networks 21, 19–28 (1991)
13. Gusfield, D.: The multi-state perfect phylogeny problem with missing and removable data: Solutions via integer-programming and chordal graph theory. In: Batzoglou, S. (ed.) RECOMB 2009. LNCS, vol. 5541, pp. 236–252. Springer, Heidelberg (2009)
14. Janis, C.: The sabertooth's repeat performances. Natural History 103, 78–82 (1994)
15. Kannan, S., Warnow, T.: Inferring evolutionary history from DNA sequences. SIAM J. Computing 23(4), 713–737 (1994)
16. Kannan, S., Warnow, T.: A fast algorithm for the computation and enumeration of perfect phylogenies. In: Proc. of SODA 1995, pp. 595–603 (1995)
17. Kollar, E., Fisher, C.: Tooth induction in chick epithelium: Expression of quiescent genes for enamel synthesis. Science 207, 993–995 (1980)
18. Maňuch, J., Patterson, M., Gupta, A.: On the Generalised Character Compatibility Problem for Non-branching Character Trees. In: Ngo, H.Q. (ed.) COCOON 2009. LNCS, vol. 5609, pp. 268–276. Springer, Heidelberg (2009)
19. McMorris, F., Warnow, T., Wimer, T.: Triangulating vertex colored graphs. SIAM J. Discrete Mathematics 7(2), 296–306 (1994)
20. Meidanis, J., Porto, O., Telles, G.P.: On the consecutive ones property. Discrete Applied Mathematics 155, 788–805 (2007)
21. Opatrny, J.: Total ordering problem. SIAM J. Computing 8(1), 111–114 (1979)
22. Pe'er, I., Pupko, T., Shamir, R., Sharan, R.: Incomplete directed perfect phylogeny. SIAM J. Computing 33, 590–607 (2004)
23. Steel, M.: The complexity of reconstructing trees from qualitative characters and subtrees. J. Classification 9, 91–116 (1992)
24. Trowsdale, J.: Genomic structure and function in the MHC. Trends in Genetics 9, 117–122 (1993)
25. Warnow, T.: Tree compatibility and inferring evolutionary history. J. Algorithms 16, 388–407 (1994)

Author Index